Purchased with funds provided by the

*Jerome S. and Meta K. Howard
Endowment Fund*

The
Gull Guide
North America

The Gull Guide
North America

Amar Ayyash

Princeton University Press
Princeton and Oxford

Copyright © 2024 by Amar Ayyash

Princeton University Press is committed to the protection of copyright and the intellectual property our authors entrust to us. Copyright promotes the progress and integrity of knowledge. Thank you for supporting free speech and the global exchange of ideas by purchasing an authorized edition of this book. If you wish to reproduce or distribute any part of it in any form, please obtain permission.

Requests for permission to reproduce material from this work should be sent to permissions@press.princeton.edu

Published by Princeton University Press
41 William Street, Princeton, New Jersey 08540
99 Banbury Road, Oxford OX2 6JX

press.princeton.edu

All Rights Reserved
ISBN (pbk.) 978-0-691-19589-6
ISBN (e-book) 978-0-691-26345-8
Library of Congress Control Number: 2024909930
British Library Cataloging-in-Publication Data is available

Editorial: Robert Kirk and Megan Mendonça
Production Editorial: Karen Carter
Text Design: D & N Publishing, Wiltshire, UK
Jacket/Cover Design: Wanda España
Production: Steven Sears
Publicity: Caitlyn Robson and Matthew Taylor
Copyeditor: Penny Mansley and Patricia Fogarty

Cover Images: (front) © Ryan Sanderson; (back) Adult Black Kittiwakes © Simon Colenutt

This book has been composed in Brandon Grotesque

Printed in China

10 9 8 7 6 5 4 3 2 1

CONTENTS

Dedication	viii
Acknowledgments	viii

Introduction — 1

Taxonomy — 4

Gull Topography — 7
Feather Tracts — 7
Bare Parts — 12
Kodak Gray Scale — 14

Aging and Molt — 17
Aging — 17
Molt — 18

Identification — 30
Distribution and Expectations — 30
Related Ages — 30
Categorizing Field Marks — 31
Nuances, Caveats, and Pitfalls — 34
Variation — 38
Relative Size Comparisons — 49
Aberrations — 54

Species Accounts — 57

1. Small Tern-like and Hooded Gulls — 59
1. Sabine's Gull (*Xema sabini*) — 60
2. Swallow-tailed Gull (*Creagrus furcatus*) — 68
3. Black-legged Kittiwake (*Rissa tridactyla*) — 74
4. Red-legged Kittiwake (*Rissa brevirostris*) — 82
5. Ivory Gull (*Pagophila eburnea*) — 87
6. Ross's Gull (*Rhodostethia rosea*) — 92
7. Little Gull (*Hydrocoloeus minutus*) — 98

8. Bonaparte's Gull (*Chroicocephalus philadelphia*) 106
9. Black-headed Gull (*Chroicocephalus ridibundus*) 114
10. Gray-hooded Gull (*Chroicocephalus cirrocephalus*) 121
11. Laughing Gull (*Leucophaeus atricilla*) 127
12. Franklin's Gull (*Leucophaeus pipixcan*) 135

2. *Larus* Gulls 143

13. Heermann's Gull (*Larus heermanni*) 144
14. Belcher's Gull (*Larus belcheri*) 152
15. Black-tailed Gull (*Larus crassirostris*) 158
16. Short-billed Gull (*Larus brachyrhynchus*) 165
17a. Common Gull (*Larus canus canus*) 174
17b. Kamchatka Gull (*Larus canus kamtschatschensis*) 183
18. Ring-billed Gull (*Larus delawarensis*) 192
19. California Gull (*Larus californicus*) 201

Herring Gull Complex 213

20. American Herring Gull (*Larus smithsonianus*) 215
21. European Herring Gull (*Larus argentatus*) 236
22. Vega Gull (*Larus vegae*) 247

23a. Yellow-legged Gull (*Larus michahellis michahellis*) 261
23b. Azores Gull (*Larus michahellis atlantis*) 276
24. Lesser Black-backed Gull (*Larus fuscus*) 289
25. Kelp Gull (*Larus dominicanus*) 305
26. Western Gull (*Larus occidentalis*) 317
27. Yellow-footed Gull (*Larus livens*) 328
28. Great Black-backed Gull (*Larus marinus*) 338
29. Slaty-backed Gull (*Larus schistisagus*) 349
30. Glaucous-winged Gull (*Larus glaucescens*) 364
31. Glaucous Gull (*Larus hyperboreus*) 376
32. Iceland Gull (*Larus glaucoides*) Complex 387
32a. Thayer's Gull (*Larus glaucoides thayeri*) 389
32b. Kumlien's Gull (*Larus glaucoides kumlieni*) 404
32c. Iceland Gull (*Larus glaucoides glaucoides*) 418
32d. Commentary 427

3. Hybrids	435
H1. Glaucous-winged × Western (Olympic) Gull	439
H2. Glaucous-winged × American Herring (Cook Inlet) Gull	446
H3. Glaucous × American Herring (Nelson's) Gull	454
H4. Glaucous × Glaucous-winged (Seward) Gull	460
H5. Glaucous × Great Black-backed Gull	464
H6. American Herring × Great Black-backed (Great Lakes) Gull	467
H7. American Herring × Lesser Black-backed (Appledore) Gull	471
H8. Other Hybrids	478
Appendix	484
A1. Heuglin's Gull (*Larus fuscus heuglini*)	485
A2. Taimyr Gull (*taimyrensis*)	489
A3. Pallas's (Great Black-headed) Gull (*Ichthyaetus ichthyaetus*)	494
A4. Gray Gull (*Leucophaeus modestus*)	497
Glossary	502
References	505
Index	517

Dedicated to Inaam, Ibrahim, Asmaa, and Ismail.

May your curiosity of the world always be unyielding.

ACKNOWLEDGMENTS

When I officially began writing *The Gull Guide* in 2018, I reached out to several bird authors who shared invaluable information and advice. They wished me luck in my endeavor but did not hesitate to remind me that gulls were the toughest group of birds to take on. Knowing their words were without exaggeration, I fully embraced the challenge ahead and spent the next five years researching, writing, and amassing thousands of photographs from all around the world. And although I've spent the better part of my 40s on this great undertaking, there's a peculiar feeling of the time having run off all too fast.

Many people are to thank for their help and support along the way.

I would like to express my deepest gratitude to Peter Adriaens for sharing his wealth of knowledge over the years. Peter's command of gull identification is unparalleled, and his field studies are of foremost authority. I'm also indebted to Mars Muusse and Peter Pyle for helping to crystalize my thoughts on molt over the last fifteen years. A special thank you to Ted Floyd, who was instrumental in my first attempts at writing for a bird audience. To my friend Greg Neise, your birding acumen and insights on fieldcraft have left a deep impression on me since the days of the Illinois Birders' Forum—thank you. I'd also like to thank Steve N. G. Howell, whose work, including *Gulls of the Americas*, has influenced me and many others so greatly. Martin Reid, Alvaro Jaramillo, and Bruce Mactavish have also been constant sources of inspiration, and it is these three whom I secretly blame for instigating my obsession with gulls. Thank you to Liam Singh, a talented young mind whose zeal for gull study is a breath of fresh air. Our work on winter adult Thayer's Gulls, and our countless conversations on gulls of the Pacific Northwest, was a most welcome distraction during the last couple of years of the project.

I could not have conducted much of my research without the universities and museums which graciously allowed access to their collections. I want to especially thank the following curators and collections managers for their hospitality and assistance: Robert C. Faucett, Kimball L. Garret, David Willard, Ben Marks, Gary Shugart, Jason Weckstein, Christopher Milensky, Anna Sveinsdottir, Dawn Roberts, Brett Benz, Ben Winger, Jack Withrow, Michel Gosselin, and Gregory Rand. Special thanks to Donna L. Dittmann and Steven W. Cardiff, who invited me to review their novel collection of Kelp x American Herring hybrids from the Chandeleur Islands at LSUMNS.

I'm very grateful for the biologists and ornithologists who allowed visits to observe and study breeding birds in their colonies. I thank Dr. Julie Ellis and William Clark for generously welcoming me to witness their model work on Appledore Island. A very special thank you to Bruce Buckingham, who has involved me in his research on American Herring Gulls on Lake Erie. I eagerly await my annual visits with Bruce, which are always warm, enlightening, and humbling.

Many, many people generously provided use of their photographs. Those deserving special mention include Chris Gibbins, Alex A. Abela, Steven G. Mlodinow, Liam Singh, Mars Muusse, Bruce Mactavish,

ACKNOWLEDGMENTS

Osao & Michiaki Ujihara, Richard Bonser, Steve Heinl, Hans Larsson, Phil Pickering, Steve Arena, Lancy Cheng, Kirk Zufelt, Alex Lamoreaux, and the late Tom Johnson, who is sorely missed.

To the members of the North American Gulls Group, serving as your Administrator for the last 12 years has been the greatest of educations—you are the pulse of everything gull-related across the continent.

I wish to thank and acknowledge the following individuals: Robert D. Hughes, Dan Kassebaum, Matthew Winks, Doug Stotz, the late Wes Serafin, Michael L. P. Retter, Adam Sell, Tom Kelly, Ethan Gyllenhall, Chris West, Bruce Heimer, Steve Ambrose, Geoff Williamson, Paul Sweet, Andy Sigler, Joel Greenberg, Josh Engel, Walter Marcisz, David Johnson, Joe Lill, Kenneth J. Brock, John Kendall, John Cassady, Jeff McCoy, Michael Topp, Ryan Sanderson, Mike Bourdon, Jen Brumfield, Chuck Slusarczyk Jr., Tim Jasinski, Kenn Kaufmann, Ted Keyel, James Pawlicki, Kevin McLaughlin, Tyler L Hoar, Jean Iron, the late Ron Pittaway, Dave Brown, Jared Clarke, Chris Corben, Michael Brothers, Noah Arthur, Michael Donahue, Paul Hurtado, Guy McCaskie, Marteen van Kleinwee, Charles Sontag, Rebecca Sher, George Armistead, Nikolas Haass, Steve Hampton, Laura Burke, Skye Christopher G. Haas, Nathan Swick, Klaus Malling Olsen, Jan Jörgensen, James Kennerley, Peter Kennerley, Hannu Koskinen, Daniel López Velasco, Gabriel Martín, Will Chatfield-Taylor, Byron Chin, Joanna Chin, Caleb Putnam, Brandon Holden, Karl Bardon, Peder Svingen, Chris Brown, Robbye Johnson, Nicholas Komar, Jeffrey Gordon, Andrew Haffenden, Declan Troy, Peter Post, Robert Lewis, Nial Moores, Yann Kolbeinsson, Harry Hussey, Dominic Mitchell, Kevin Karlson, Glenn Coady, Jeremy Gatten, Timothy B Roadcurlew, Justin Peter, Wendy Tatar, Nathan Dubrow, Joe Bourget, Carl Baggott, Lou Bertalan, Alex Boldrini, Carlos Pacheco, Dirk Van Gansberghe, Chris van Rijswijk, Thibaut Chansac, Emily Weiser, Blair Nikula, Joachim Bertrands, Nicholas Komar, Yann Muzika, Akimichi Ariga, Silas Olofson, Lars Witting, Ian Paulsen, Malia Kai De Felice, Asier Aldalur, Chris Hill, Alix d'Entremont, Amir Ben Dov, Delfín González, Antonio Gutierrez, Suzanne Sullivan, Shonn Morris, Jamie Spence, Frank Lin, Blake Matheson, Brian Sullivan, and Woody Goss.

My deepest thanks to Steve N. G. Howell and Alvaro Jaramillo for reviewing the manuscript, Peter Adriaens for reviewing the accounts on European taxa, and to Paul Lehman for his assistance with the sections on Distribution. For their remarkable expertise and commitment to the book, I thank Robert Kirk and Karen Carter, and the rest of the staff at Princeton University Press who helped see me through the layers of this project.

Writing is a long and sometimes lonely process, and periods of seclusion are inevitable. This was especially so in my case, having written a good deal of the book during a global pandemic. I'm eternally grateful for the love and support of my family. To my little "big" brother, Emad, your endless encouragement has meant the world to me. To my parents, Ibrahim and Inaam, you are my backbone and my greatest inspiration. The older I get, the more I understand your wisdoms. Finally, to my wife, Olivia, to whom I am beholden, words cannot express my gratitude. You have taught me more than you can ever know. Thank you!

INTRODUCTION

Gulls are a cosmopolitan group of birds found in a variety of landscapes. Among them are some of the most coveted bird species on the planet. Others are commonplace and have gained admission into our immediate surroundings. It is known that gulls are much more widespread in the northern hemisphere than in the southern hemisphere. While other families of birds have seen precipitous declines in recent decades, many gull populations have become increasingly resilient, as they find ways to capitalize on anthropogenic practices; few bird groups are as suited for such conditions.

For good reason, gulls are noticeably absent from jungles, many mountainous regions, and throughout the Indonesian archipelago. Some are highly localized with restricted ranges, such as the bug-eyed Red-legged Kittiwake. Some are widespread with circumpolar distributions, such as Glaucous Gull. There are populations that are mostly sedentary with little to no migration, as we find in Yellow-footed Gull. And others are true long-distance migrants traveling well north and south of the equator every year, such as Franklin's Gull. Many are colonial breeders; others nest in small, loose groups; and some breed solitarily.

They're equally at home on land, in the air, and on water, and we don't have to go very far to find them. Some are all-purpose seabirds. Others are much more landbound and seldom take to the sea. Their bills range from small and pointed like daggers to large and bulbous with hooked tips. They're equipped with webbed feet for swimming and with legs mostly centered under the body, which allow them to walk with ease. Gulls are master opportunists, and on the whole, I would assert that they're more versatile than any other group of birds.

Whether it be a pair of Bonaparte's Gulls nesting in a spruce tree in the muskeg wilderness, a flock of California Gulls gorging on brine flies on the shores of the Great Salt Lake, a flotilla of kittiwakes wheeling through a fierce storm tens of miles out at sea, a club of Ivory Gulls scavenging polar bear kills on the pack ice of the Arctic, or Ring-billed Gulls stalking tourists from a rooftop in Chicago, gulls are arguably the greatest evolutionary success story in the avian world.

1 Black-legged Kittiwake is the most abundant gull in the world and is strictly pelagic. It is now increasingly found nesting farther south on human-made structures. AMAR AYYASH. HOMER, ALASKA. AUG.

2 Laughing Gulls waiting for a trawler to bring up its nets. Congregating here often means a guaranteed meal. CHERI PHILLIPS. DUVAL COUNTY, FLORIDA. MAY.

3 Young Kelp Gull on an ice floe in Antarctica. No other gull breeds this far south, and populations here are thriving. ALAN KWOK. FOURNIER BAY, ANTARCTICA. DEC.

4 Birders take in point-blank views of winter gulls at the annual IOS Gull Frolic on Lake Michigan. AMAR AYYASH, ILLINOIS. FEB.

No gull species is known to have gone extinct for as long as modern taxonomy has kept records (Dee 2018). In many ways, they are similar to *Homo sapiens*: omnivores exploiting and consuming whatever they cross paths with.

Inhabiting all of the Earth's continents, they are a remarkable group of some 50 species, yet they've seemingly spawned more identification debates than the rest of the planet's bird taxa combined. This is one of the powerful attractions when it comes to gulls. They are everywhere we turn. Some are quite distinctive and almost never present identification concerns, while others are truly Gordian knots of the avian world. Anyone and everyone can easily have a go at identifying them. They're accessible to the average layperson and to field ornithologists alike. Once we delve into their plumages and variability, a somewhat addictive element takes over, which can only be tamed with increased observation.

Fortunately, gulls are relatively large and accommodating, are readily on display, and often allow close study for extended periods of time. They are photogenic, and as digital photography has become common among legions of birders and naturalists, individual birds are now regularly recognized and tracked for several consecutive years as they travel hundreds of miles, whether it be through a plumage aberration, a distinct wingtip pattern, or a leg band. The widespread sharing of photographs on eBird and social media has provided us with invaluable information.

Through digital photography, we have concrete evidence that gulls show extraordinary site fidelity. Like clockwork, many individuals return to the same breeding and wintering quarters year after year. A fine example of this is an adult Great Black-backed Gull believed to be the same individual that by the winter of 2023 had returned to a small lake in Pueblo County, Colorado, for a 30th year (discovered as a first cycle in March 1993)! Much information on molt, distribution, and age-specific variation has been collected in this impressive wave of citizen science.

It is accepted that no general field guide can present the layers of variation found in gulls—hence the need for specialty guides. The material found in this guide's introductory chapters will greatly bolster

INTRODUCTION 3

5 This adult Lesser Black-backed (green F05) became a celebrity for several years as it migrated north and south along the Atlantic Coast from Maine to Florida. It was found breeding with Herring Gulls for several seasons on Appledore Island. Some of its hybrid offspring were assigned similar green, field-readable bands and were also found along the Atlantic for several years. The closest known nesting colonies for this species are in southwest Greenland. AMAR AYYASH. FLORIDA. JAN.

6 At the time, this was a potential second state record Glaucous-winged Gull for Texas, but field marks could not be exacted. Although this individual would go unscrutinized in its expected range, great caution must be taken with vagrant white-winged gulls, especially when worn and bleached. CAMERON JOHNSON. TEXAS. JUNE.

your understanding of the information in the species accounts. And although readers are often inclined to move straight to the photos, synthesizing the ideas in the introductory chapters will considerably strengthen your approach to identification.

Finally, every gull flock encountered is an opportunity to learn something new, and the opportunities are endless. Most times, we're able to confidently label a bird in question, but some will defy identification. It is important to accept that identifying every gull 100 percent of the time is an impractical undertaking. The sooner we come to terms with this, the sooner we're able to enjoy gulls for what they are. Struggling with an identification should be looked at as an opportunity to grow and cultivate our craft. By no means should leaving a bird unidentified ever be written off as a loss. Of more importance is the information we gather from wrestling with identification problems, building a data bank of sorts, and honing our skills for a future identification. This is the very essence of practice, and with time, it translates into experience.

TAXONOMY

Our innate desire to name and sort organisms is the impetus behind taxonomy. There is a need to separate birds based on their differences, but equally important is to relate them based on derived similarities. The study of evolutionary relationships known as "cladistics" is a rigid science, although not without some arbitrary decision-making. Taxonomic relations in this book are presented almost entirely based on the recommendations of the American Ornithological Society, with a few exceptions (e.g., the Herring Gull complex). When relevant, the species accounts provide insight on alternative taxonomic treatments from around the world.

Gulls belong to the order Charadriiformes. They, along with the terns, noddies, and skimmers, make up the family Laridae. Although this guide loosely speaks of "family" when collectively referring to gulls, bear in mind that the gulls are, strictly speaking, a subfamily. The two most divergent gull species are, however, more closely related to one another than either is to a tern, noddy, or skimmer. Thus, gulls make up the subfamily Larinae, which is divided into 11 genera, 6 of which are monotypic, and 10 of which have been recorded in North America (see table 1).

For decades, the majority of gull species were placed in the genus *Larus*. Utilizing what they had at their disposal, ornithologists piecing together gull relationships throughout much of the 19th and early to mid-20th centuries relied primarily on morphology and behavior (as opposed to voice, molt, and genetics). Later workers began to point out the marked differences found in *Larus*, hinting it was too broad a genus (Voous 1975; Chu 1998). Gull lineages have only recently become clearer, with the use of genetic data. We know well that selection pressures quite often lead to convergence of species that aren't that closely related but sometimes lead to rapid divergence of species that are very closely related. Consequently, some taxa resemble each other outwardly but have genetic makeups that suggest they're not each other's closest relatives, while others share fewer characteristics but have surprisingly similar DNA (Sonsthagen et al. 2016).

In 2005, Pons et al. provided an in-depth molecular phylogeny for all known larids, in which 53 species of gulls were identified. That study supported the long-standing belief that gulls are indeed a monophyletic clade (all descendants sharing a common ancestor) but proposed that the traditional genus *Larus* was best partitioned into multiple genera. The overhaul was profound and has since moved gull classification in a new direction. Some of the relationships that had been suspected previously were confirmed through this phylogeny, while others remained unresolved.

Gull species of midlatitude and tropical regions are, overall, more genetically distinct, with lineages that have diverged significantly when compared to those in the far north near the arctic circle. From a taxonomic standpoint, *Larus* remains the most unsettled genus, with multiple questions awaiting answers. Modern genetic analysis on this genus indicates that recent speciation events (radiation) have unfolded in a relatively short period of time. Recent studies suggest that these gulls, in general, have diversified faster than any other group of birds (Jetz et al. 2014). In North America, some large white-headed gull species exhibit remarkable genetic relatedness—similar to that expected of populations within a single species. Several studies confirm that at higher latitudes in North America, individuals of different species are clustered by locality rather than by species (Sternkopf et al. 2010; Sonsthagen et al. 2016). For example, Glaucous Gulls found in North America appear to be more closely related to American Herring Gull than they are to Glaucous Gulls in Europe. Recent and ongoing hybridization via secondary contact is often cited as the primary source for this limited genetic drift (Sonsthagen et al. 2016).

The genetic relatedness of our northern gulls presents a dilemma in defining species, and reconstructing their relationships has not been met with great success. Certainly, most—if not all—species concepts are currently incompatible when applied to a number of large white-headed gulls. Further studies at the subgenomic level may reveal that many of the phenetic differences we find in some species are simply managed by just a few genes. And although some may argue that as field observers we need not concern ourselves with the findings of molecular systematics and that our focus should be on what we can diagnose morphologically, I contend that both efforts are inseparable and intertwined. To adopt one without the other is to construct a system on "shifting sands." Instead, we should enthusiastically welcome the findings of modern genetics. The problems lie in how the results

GULL TOPOGRAPHY

Having some knowledge of the various parts and feather tracts of a gull is the first way to orient ourselves with this group of birds. Many of the terms found here are rooted in ornithology, while others are informal and unique to the gull-watcher's lexicon. There are some who would like a gull book with no jargon and claim that this terminology only confuses readers. This is demonstrably false. Gull identification has a good deal of nuance, and as with any other specialized subject, it is impossible to communicate effectively without introducing a set of terms. These terms are not meant to complicate identification but instead to facilitate understanding. Implementing a common language fosters clarity. There is no need to worry about knowing all of this terminology at once. The images in this guide crystallize the ideas, and with practice and exposure the terms become very useful tools.

With gulls, the mosaic of plumages at hand invites detailed examination that often goes beyond determining species. If we are to begin pointing out and discussing field marks, then surely we must be able to refer to them with names. Instead of saying "white spots on the wingtip," I would want to specify whether I'm referring to mirrors, tongue tips, or apicals. Or when referencing the tail, I may want to make a distinction between actual rectrices and tail coverts. Feather minutiae and the opportunities they provide for fervent discussion and pinpoint accuracy are among the primary reasons many people are so strongly drawn to this group of birds. Fortunately for us, gulls are relatively large and out in the open, allowing for convenient observation at length. Most of the elements described here are quite easy to make out in the field, once we know what to look for.

FEATHER TRACTS

Two categories of feathers are of foremost importance: flight feathers and body feathers. Flight feathers consist of remiges and rectrices. The remiges are the longest wing feathers, made up of primaries and secondaries. All gulls have 10 primaries on each hand, numbered from innermost to outermost, away from the body, from p1 to p10. Depending on the size of the species, gulls can have 16–23 secondaries, numbered inward, toward the body, from s1 to s23. The innermost "secondaries," closest to the body, make up a distinct group of feathers known as "tertials," which are not true flight feathers. Gulls have 12 rectrices (tail feathers), numbered outward from the center of the tail, from r1 to r6, along each half.

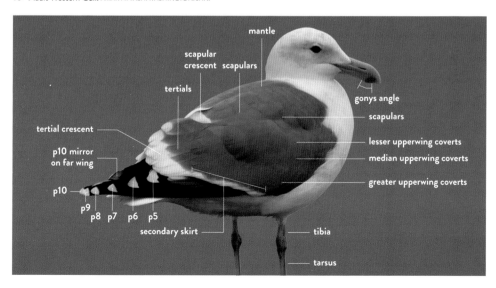

10 Adult Western Gull. AMAR AYYASH. WASHINGTON. JAN.

GULL TOPOGRAPHY

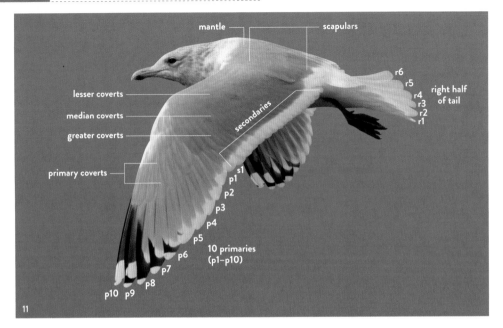

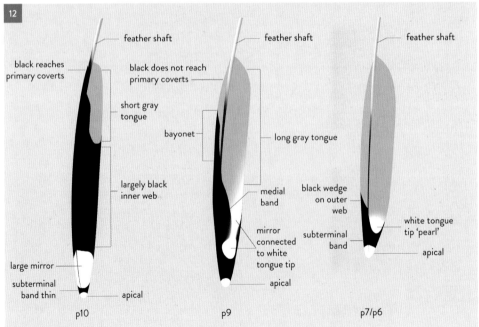

11 Adult Thayer's Gull. The lesser, median, and greater coverts labeled here are secondary coverts. The primaries are numbered distally, and the secondaries are numbered proximally. AMAR AYYASH. WISCONSIN. MARCH.

12 Individual primary feathers of no particular species. Very helpful with some taxa is describing if black reaches the primary coverts or if a pale tongue is long or short.

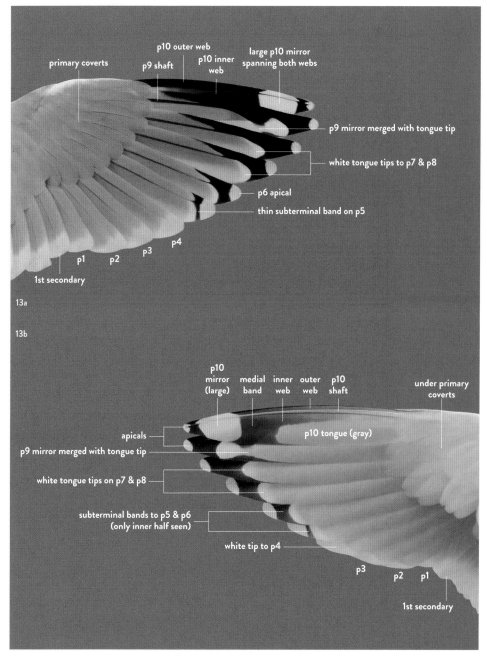

13A American Herring Gull. Upperside of right wingtip. AMAR AYYASH, ILLINOIS. FEB.
13B Underside of right wingtip. The extent of a pale tongue is generally easier to see from the underside.

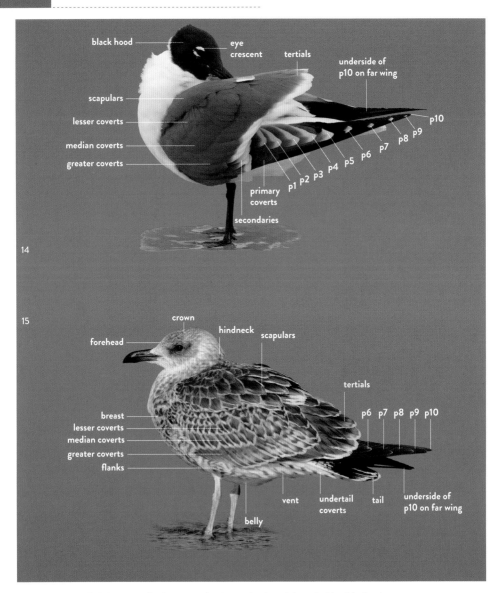

14 Adult Laughing Gull. Preening gulls often expose their inner primaries and the underside of the far wing. AMAR AYYASH. FLORIDA. JAN.
15 Juvenile Lesser Black-backed Gull. Note the pointy primary tips and pale-edged upperparts, which are expected at this age. AMAR AYYASH. MICHIGAN. NOV.

GULL TOPOGRAPHY

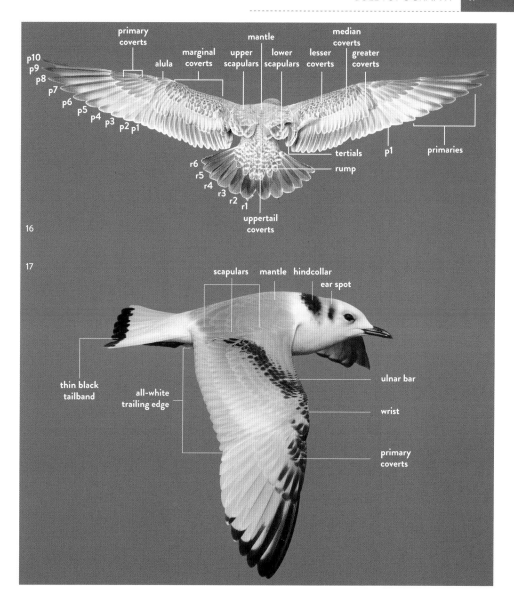

16 1st-cycle Short-billed Gull. The outer primaries on this bird form a contrasting venetian-blind pattern of dark outer webs and pale inner webs. The marginal coverts envelope the leading edge of the wing. The upper scapulars are shorter feathers compared to the lower, "longer" scapulars. AMAR AYYASH. WASHINGTON. DEC.

17 Juvenile Black-legged Kittiwake. A number of smaller gulls show an ulnar bar running diagonally across the wing coverts. AMAR AYYASH. ALASKA. AUG.

The remaining feathers—essentially anything not a flight feather—are referred to as "body feathers" in this guide. These include the feathers on the head, neck, and underparts; the scapulars; and the coverts. The scapulars, mantle, and upperwing coverts are often referred to as "upperparts."

GULL TOPOGRAPHY

18 Adult Short-billed Gull has medium gray upperparts. The underside of the remiges shows a contrasting gray shadow effect. **(1)** The wing is fully spread, and important features on the wingtip can be examined. **(2)** As the wing begins to fold, note how the inner webs of the primaries become concealed on the upperwing, while the outer webs are concealed on the underwing. **(3)** As the wing folds, p1 sits over p2, p2 sits over p3, and so on, with the longest primary, p10, situated on the bottom of the stack. **(4)** Note here the delineation between the scapulars and the tertials, each with their respective white crescents (tips). **(5)** At rest, p6 is typically the first primary seen beyond the tertial tips. The longest visible primary is p9, with p10 typically not seen and tucked underneath. On the far wing, the underside of p10 is visible.

BARE PARTS

The unfeathered regions of a bird are known as the "bare parts" or "soft parts." With gulls, these include the bill, gape, legs, and feet, as well as the eye and surrounding orbital ring. Bare parts go through a gradual progression of color changes as a bird matures and becomes an adult. As a rule, gulls begin life with dark bills and dark eyes. With two-year gulls, the changes in bare-part colors are generally orderly and predictable. With three- and four-year gulls, the progression is more variable, and sometimes remarkably so, depending on the individual, species, and even time of year.

In addition to age-related changes from one year to the next, adult bare parts can undergo striking seasonal transformations. Generally speaking, the legs, bill, and orbital ring often go from having a lackluster appearance to being bright and colorful as the breeding season approaches. These colors are ornamental in nature and beautifully complement what are otherwise plain-colored plumages. It is believed the intensity in coloration of these parts may have evolved as an important signal to

19 Adult Herring Gull. AMAR AYYASH. OHIO. APRIL.

GULL TOPOGRAPHY

20 California Gulls. The center bird and left bird are in basic plumage, but with varying bill colors. The individual on the right is in alternate plumage and in breeding condition. LEFT: ALEX A. ABELA. CALIFORNIA. SEPT. CENTER: AMAR AYYASH. WASHINGTON. JAN. RIGHT: GAVIN MCKINNON. ALBERTA. APRIL.

indicate the potential quality of a mate. Studies have shown a positive correlation between bare-part color condition and body condition, as well as larger egg and clutch sizes (Kristiansen et al. 2008). Unlike plumage coloration, which is more static and retained for months on end, bare parts have been shown to change color within weeks, days, hours, or even seconds across various bird species (Iverson & Karubian 2017). With gulls, the changes are reasonably gradual and typically take place within weeks. When adult bare parts are most vivid, we say the bird is in "high breeding condition." Overall, nonbreeding adults average duller bare parts. These factors should be taken into consideration when reading the species accounts. Although there are always exceptions, it should be assumed that bare-part colors are annually transformed to some degree, and that they are generally in flux between the breeding and nonbreeding seasons.

BILL COLOR

Bill colors in gulls are mostly limited to shades of yellow, red, and black. Ivory Gull is an exception with a distinctive greenish-gray bill base. In nonbreeding condition, it is common for many species to take on black markings near the bill tip. This may recall patterns commonly seen on subadult birds. The orange-red gonys spot in many *Larus* species typically becomes duller and diminished in size at this time. Other species, particularly those with bill colors consisting of browns and reds, develop an entirely black tip in nonbreeding condition—for example, Heermann's and Black-headed Gulls. The bill of Franklin's Gull often goes from a bright maroon red in breeding condition to entirely black with red tip in nonbreeding condition. In the most extreme seasonal bill transformation, the bill changes entirely from one color to another, as found in Laughing Gull (maroon red to all black). Exceptionally, some species have bill colors that remain unchanged throughout the year, such as Ross's Gull, which has an all-black bill year-round.

GAPE

The region where the bill base and mouth lining come together is known as the "gape." This unfeathered line of skin extends onto the face and can appear to flare up in size and color in breeding condition. In some species, such as California Gull, the gape is almost always exposed and noticeable, while in others, it is insignificant and not very noticeable.

EYE COLOR

When considering eye color in gulls, and in birds in general, we're usually referring to the iris: the region surrounding the pupil. Recall that gulls hatch with dark eyes. Some species retain these dark eyes as adults and almost never deviate from this—for example, the kittiwakes. Others develop pale eyes (usually a clear yellow), such as Glaucous Gull. Most interesting are species with variable eye color as adults, ranging from clear yellow to honey brown to unmistakably dark. Common examples of this are Iceland Gulls from the Canadian Arctic, Short-billed Gull, and Western Gull.

21 The bill color on adult Laughing Gulls is seasonally transformed from all black in low breeding condition to a reddish maroon in high breeding condition. LEFT: JOHN CHARDINE. FLORIDA. FEB. RIGHT: STEVE ARENA. MASSACHUSETTS. MAY.

22 Orbital-ring color can be useful when apparent. Besides having a shorter bill and more refined appearance, the Kumlien's Gull in the front has a raspberry-pink orbital, while the American Herring shows an obvious orange-yellow orbital. AMAR AYYASH. ILLINOIS. MARCH.
23 A mixed flock of Western and California Gulls. Western Gull has pink legs in all ages, but California Gull transitions from flesh-colored to grayish-green to yellow as an adult. JERRY TING. CALIFORNIA. APRIL.

ORBITAL RINGS

The bare skin immediately surrounding each eye is known as the "orbital ring." Orbital rings (skin) should not be confused with eye-rings (feathers). With gulls, especially large white-headed gulls, the orbital ring and gape typically match in color and become well saturated in prime breeding condition. These colors are often helpful from an identification standpoint but should be used as supporting field marks and not as diagnostic. Some hybrids show a blend of two colors, and this could be very helpful in supporting a hybrid identification. As a rule, orbital-ring color wanes outside the breeding season, often becoming decidedly dull. Sometimes, an opaque, brownish-black color develops, rendering this feature useless as an identification field mark. Still, even in the nonbreeding season, orbital-ring color is often discernible under satisfactory viewing conditions.

LEG COLOR

Leg color is very helpful in making identifications and can mean the difference between having a handful of species to select from versus just one or two. Most species acquire pink or yellow legs as adults. Regardless of whether the species is yellow legged or pink legged as an adult, both types begin life with pinkish legs. In species which acquire yellow legs as they mature, three-year gulls typically begin to do so at about one year of age, and four-year gulls at about two years of age. However, allow for variation in this regard, as there are no rules or exact timings for changes in bare-part colors. Some species retain dark legs throughout every life stage, but this is much more an exception.

KODAK GRAY SCALE

With the many shades of gray found in gulls, how does one begin to categorize such a neutral color? In the field, we would agree that an adult Franklin's Gull and an adult Laughing Gull are the same shade of gray when seen side by side, and that a Lesser Black-backed Gull is noticeably darker than a California Gull. But are these differences always obvious? And what about the difference in gray coloration between a California Gull and an American Herring Gull? We would be correct in saying the former is darker; but

24 The Kodak Gray Scale has tonal values ranging from 0 to 19, with 0 equivalent to white and 19 being jet black. These 20 steps allow us to catalog relative shades of gray and black without accounting for the silvery, bluish, or brown hues found in some gulls. In the species accounts, a range of values is given for each taxon.

again, is this always the case? Some populations of California Gull approach and even match the paler upperparts of American Herring Gull, and this can have implications for identification. What metric, if any, is used to quantify these different shades?

There have been several attempts by ornithologists to standardize color usage and nomenclature in birds, including American ornithologist Robert Ridgway's (1912) noteworthy contributions in which he provided over 1,100 color names to assist the naturalist. The most widely used instrument to compare the grays in gulls is the Kodak Gray Scale (KGS). This simple tool assigns numbers to relative values of darkness on a scale from 0 to 19. Zero is equivalent to white, and 19 is black. Ivory Gull is the only species assigned a 0. No taxon is quite the jet black worthy of a 19, but some come quite close.

Measuring grays in gulls was normalized by Steve N.G. Howell (2003), who scored the upperparts of adult specimens in controlled museum conditions using a KGS card. In 2013, I began a similar undertaking, recording KGS values of specimens throughout various collections in North America. Comparing my results with those available in the literature, I consistently found myself within published range values, with the exception of a few taxa, which are noted in the species accounts (i.e., Slaty-backed, Vega, American Herring, Glaucous-winged, and Kumlien's Gulls). In instances where I was in doubt, I often called on museum staff and volunteers to independently score specimens and found that we derived similar values, with a margin of error of 0.5–1.0 points. A shade difference of 0.5 is negligible in the field and would take some effort to discern with the naked eye. A 1.5–2.0-point difference is readily detectable by most people, especially given favorable lighting conditions. Therefore, the values that are currently available in the literature are, to a great extent, reproducible.

Only adults have been measured, with little to no consideration given to their silver, blue, or brown hues. Subadults are not included in these measurements. Subadults sometimes score a higher KGS value than adult conspecifics, due to the blended brown tones in their feathers contributing to a darker

25 A pair of adult California Gulls (front) with variable gray upperparts and size. Nominate *californicus* types (left) have a KGS value of 6.0–7.5, while paler and larger *albertaensis* has KGS value of 5–6. PHIL PICKERING. OREGON. OCT.

26 Three shades of gray of three ubiquitous species across North America: Adult Glaucous-winged Gull (KGS: 5–7), American Herring Gull (KGS: 4–5), California Gull: (5.0–7.5). STEVE HEINL. KETCHIKAN, ALASKA. OCT.

appearance. Other subadults sometimes show lighter gray upperparts with an inherently paler aspect to their feathers, generating a lower score than adults (e.g., in Slaty-backed Gull). Therefore, it is not practical to suggest consistent range values for subadults, but we still can loosely categorize their grays with the naked eye, and many come close to those of adult conspecifics.

Equally usefully, the KGS highlights variation in upperpart coloration found in polytypic species. For instance, the palest subspecies of Lesser Black-backed Gull (*L. fuscus graellsii*) has a KGS range of 9–11, while the darkest race (nominate *L.f. fuscus*) is typically at 13–17, with some individuals purportedly scoring a perfect 19 in the field (Gibbins & Baxter n.d.). This is perhaps the most extreme example of variability found in the upperpart coloration of any gull species.

Most North American species score somewhere between 3 and 12 on the KGS. Great Black-backed Gull is an exception and is our darkest regularly occurring species, with a range of 13–15. It's important to stress that few people actually use these numbers with any sort of rigidity while out gull-watching. They are most useful for unifying tonal shades in the literature, when assessing museum specimens, and when studying digital photos. In the field, we employ plain and simple descriptions, such as "pale," "medium gray," "slate black," and "black." These terms are also inherently subjective, however. In any case, the reader should have a few benchmarks in mind to help place KGS differences, and in table 2, the shades are grouped in four broad categories.

TABLE 2	FOUR BROAD RANGES FROM THE KODAK GRAY SCALE TO CATEGORIZE UPPERPART TONES OF ADULT GULLS	
KGS RANGE	DESCRIPTION	EXAMPLE
3–5	Very pale gray to pale gray	American Herring Gull
6–8	Medium gray	Short-billed Gull
9–11	Dark gray	Yellow-footed Gull
12–19	Slate black to jet black	Great Black-backed Gull

AGING AND MOLT

AGING

Few families of birds provide observers with the opportunity to study plumage in quite the way gulls do. A standard practice in gull-watching is to assign an age to your subject. A gull can be aged based on the appearance of its plumage and molt patterns. Bare parts can also assist in aging but less reliably than plumage. Note that the age we assign to a gull is no more than a crude estimate, based on previously established data from known-age birds. Without a leg band or other contraption to reveal actual life history, there is no way to be positively certain which year a gull hatched, save for juveniles. Nonetheless, using common knowledge, we can reasonably agree on the correct "age group" of most gulls.

Gulls are generally categorized as two-, three-, or four-cycle species; this refers to the number of years it takes to attain an adult-like plumage (see table 3). Once an individual takes on adult plumage, it's assumed that plumage aspect is permanent, without any radical changes likely in the future. In general, most gulls do not begin breeding until they're adults, but there are individuals that depart from the norm and breed earlier than expected. Furthermore, there are individuals that reach "adult age" with regard to years but still show some remnants of subadult plumage. For the most part, adults are rather straightforward to categorize. The nitty-gritty of aging often revolves around non-adult birds.

Recording the age of a gull—or of any bird, for that matter—offers insight on population dynamics, breeding success, occurrence, longevity, and migration patterns. Such information is critical to conservation management. In years when birders report an absence of juvenile Heermann's Gulls on the Pacific Coast, we can be sure that large-scale breeding failure has occurred on Isla Rasa, where an estimated 90 percent of the species breeds. A local population of Short-billed Gulls with an overabundance of second-cycle individuals is an indication that last year's crop had a good survival rate. And although a few random reports of this may be insignificant, tens of observers sharing such information over a wide area provide compelling data.

Consider the fact that the overwhelming majority of Slaty-backed Gulls found in North America are adult types, and that the same can be said for Kelp Gull. This presents an interesting question: Are we missing the criteria needed to identify young birds, or are they indeed extremely rare here? The long-

TABLE 3	THE NUMBER OF MOLT CYCLES TYPICALLY TAKEN TO ACQUIRE ADULT PLUMAGE		
TWO	THREE	FOUR	
Black-headed Gull	Belcher's Gull	American Herring Gull	Iceland Gull
Bonaparte's Gull	Black-legged Kittiwake	Black-tailed Gull	Kelp Gull
Gray-hooded Gull	Common Gull	California Gull	Lesser Black-backed Gull
Ivory Gull	Franklin's Gull	European Herring Gull	Pallas's Gull
Ross's Gull	Gray Gull	Glaucous Gull	Slaty-backed Gull
Sabine's Gull	Laughing Gull	Glaucous-winged Gull	Taimyr Gull
Swallow-tailed Gull (more study needed)	Little Gull	Great Black-backed Gull	Vega Gull
	Red-legged Kittiwake	Heermann's Gull	Western Gull
	Ring-billed Gull	Heuglin's Gull	Yellow-legged Gull
	Short-billed Gull		
	Yellow-footed Gull		

27 This individual is a "subadult" Black-legged Kittiwake, which is easily overlooked as an adult. Note the black markings on the bill base and black streaks on the primary coverts and along the outer edges of p8–p10. AMAR AYYASH. ILLINOIS. JAN.

standing enigma of Lesser Black-backed Gull's presence in North America also raises some age-related questions. It's easy to find many one-year-old individuals summering throughout the continent during the breeding season, but adults virtually escape discovery by late April. It's believed adults carry out long-distance migrations to breed in remote regions such as Greenland. But where do most of our nonbreeding second- and third-cycle Lesser Black-backeds go in the boreal summer? Taking note of these age-related nuances can make an ordinary observation much more meaningful.

In some instances, there are age groups whose numbers have been inadvertently inflated due to observer oversights. A good example of this is adult Black-legged Kittiwakes reported in the interior. Careful scrutiny of these reports often reveals a subadult plumage. Here, "subadult" refers to a bird with an overall adult-like appearance but with some obvious features that clearly point away from a "perfect" adult. Many of these kittiwakes would likely be in their second cycle. Adults, on the other hand, are truly pelagic and less often stray from the open seas. This is somewhat mirrored by another species, Sabine's Gull. Juvenile Sabine's traverse the interior for their first southbound migration, whereas adults (similar to jaegers) have learned more suitable migration routes and are overwhelmingly pelagic.

Bear in mind that there are terms employed with other families of birds that have limited usage when aging gulls. One such term is "immature." This label is often avoided, unless we are collectively referring to non-adult gulls in a general sense. Gulls take on successive plumages as they mature, and labeling a gull as an immature conceals some of the useful information discussed above. If I report an "immature California Gull," my audience is left with an uninformed image. This ambiguity is unnecessary, especially in this day and age when bird identification has developed into a form of artistry. If I specify "third-cycle California Gull," the mental image suddenly shifts, and clarity is gained. Noting the age or plumage of your subject is most welcome, and for this we need not make any apologies.

I have been loosely using the term "cycle" without any indication of its meaning. This is intentional, in order to prime the reader. To understand the notion of cycle when it comes to aging, we have to delve into molt.

MOLT

Feathers are remarkably sophisticated structures, but their greatest limitation is that they break down and weaken with time. Just as birds fit breeding and migration into their annual cycle, they must also grow new feathers (i.e., molt). Feather molt is a recurring event integral to a bird's annual cycle. Birds cannot survive without molting, and we can often detect whether individuals are healthy or not by inspecting their plumage and general state of molt. Here, I introduce a brilliant system that is used to catalog molt, known as the modified Humphrey-Parkes, or the Humphrey-Parkes-Howell (H-P-H) system (Howell 2010). This system aims to unify molt in birds by establishing homologies. Here, special attention is given to molt in gulls.

Gulls are semiprecocial when they hatch. At this stage, they are dressed in downy feathers and are flightless. In the following weeks, they begin growing their first "true" feathers, resulting in juvenile plumage, also known as the *first basic plumage*; the two terms are used interchangeably. The molt that gives way to juvenile plumage is the prejuvenile molt, also known as the *first prebasic molt*. The prefix "pre" implies active molt that is leading into a plumage. Molts produce plumages and have a one-to-one relation; that is, the first prebasic molt results in first basic plumage, the second prebasic molt results in second basic plumage, and so on. Prebasic molts are *complete* molts, in which all body and flight feathers are replaced.

AGING AND MOLT

31 American Herring Gull with much of its juvenile feathers in place, but note its renewed scapulars (1st alternate). The molt gap where p1–p2 have been dropped signals the onset of the 2nd prebasic molt. Thus this individual would now be labeled a 2nd cycle. AMAR AYYASH. ILLINOIS. MAY.
32 American Herring Gull in high molt with its 2nd prebasic molt well underway. Note that p1–p5 are new (2nd basic), although most of the secondaries and tail feathers are retained (1st basic). AMAR AYYASH. WISCONSIN. JULY.
33 2nd-cycle Lesser Black-backed Gull with only p10 juvenile. Most of the secondaries, the upperwing coverts, and the entire tail have been renewed in the 2nd prebasic molt. AMAR AYYASH. NEW JERSEY. AUG.
34 2nd-cycle California Gull in the last stages of its 2nd prebasic molt. Note that p9–p10 are new but not fully grown; otherwise, this bird is essentially in 2nd basic plumage. AMAR AYYASH. CALIFORNIA. SEPT.

Figure 1. The Molts and Plumages of Bonaparte's Gull in 1st and 2nd Cycles
In CAS, there is a partial preformative and partial prealternate molt inserted in 1st cycle. In every subsequent cycle, there is a partial prealternate molt and a complete prebasic molt. Bonaparte's typically takes on an adult-like appearance in 2nd basic plumage, which is generally referred to as "adult" plumage, provided there are no features of immaturity.

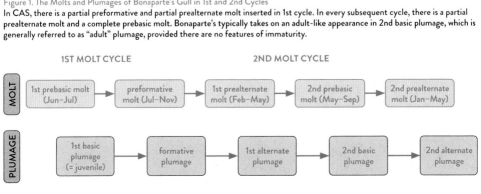

35 1st-cycle Bonaparte's Gull. **Left:** Juvenile plumage in mid-Aug. **Center:** Undergoing preformative molt in Sept. Note the gray scapulars and reduced neck and head markings. **Right:** Formative plumage in Nov. The scapulars are entirely gray, and the lower hindneck has taken on a gray wash, with white crown. No flight feathers are replaced in 1st cycle in this species.

36 Bonaparte's Gull. **Left:** Now with new, crisp scapulars as well as neck and head feathers that have come in via another inserted molt—the 1st prealternate molt (May). **Center:** 1st alternate plumage (May). All juvenile flight feathers still in place. **Right:** The 2nd prebasic molt has commenced, as signaled by the molt gap at the inner primaries (June).

37 Bonaparte's Gull. **Left:** The 2nd prebasic molt is well underway with mostly new, 2nd-generation primaries, upperwing coverts, and tail feathers. The lesser coverts, inner secondaries, and p9–p10 are still juvenile (Aug.). **Center:** Adult type in basic plumage (Dec.). **Right:** Adult in alternate plumage (April). Once adult plumages are acquired, stasis is reached, and we can't be sure which molt cycle an adult is undergoing thereafter. See Figure 1, previous page.

If the species exhibits an alternate strategy (i.e., SAS or CAS), the second prealternate molt will produce second alternate plumage, which is often most noticeable toward the conclusion of the second cycle and into the third molt cycle.

In three-cycle species, such as Ring-billed Gull, the third prebasic molt produces third basic plumage, which is generally indistinguishable from an adult plumage. In four-cycle species, such as American Herring Gull, the fourth prebasic molt produces fourth basic plumage, which corresponds to an adult plumage. When adult plumage is acquired, stasis is more or less reached, and we don't expect any marked changes in appearance thereafter.

TO BREED OR NOT TO BREED

As a gull acquires adult plumage, we should notice other features that are taking adult form, such as the patterns and colors of bare parts. Plumage and bare-part color patterns are strongly correlated, but there is no causal relation. That is, bare-part color and molt are controlled by different mechanisms.

Importantly, H-P-H does not presuppose any knowledge of breeding phenology. Therefore, this system is not hindered by connecting plumages to presumed breeding status. Not all birds breed every

AGING AND MOLT   23

38 This Black-headed Gull was photographed several months apart in the same parking lot. In alternate plumage (left), an associated breeding aspect is often found. In basic plumage (right), a nonbreeding aspect comes together. AMAR AYYASH. MARYLAND. MARCH, DEC.

year, and furthermore, all breeding birds don't necessarily adhere to a typical 4-season or 12-month cycle (e.g., Swallow-tailed Gull). Although the triggers for molt may be influenced by breeding and migration, our primary interest lies in which set of feathers is produced by which molt event. This is one of the reasons why H-P-H is a beautifully simple system, and it remains the instrument of choice to teach and learn about molt. H-P-H does not deny, however, that birds often exhibit a breeding or nonbreeding condition when acquiring certain plumages. For this, we use the term "aspect" to refer to overall appearance (Humphrey & Parkes 1959; Howell & Dunn 2007; Howell 2010; Howell & Pyle 2015). The combination of alternate plumage and vivid bare parts produces a "breeding aspect" (which many refer to as "breeding plumage"). Further, basic plumage in gulls accompanied by dull bare parts is referred to as "nonbreeding aspect."

DISTINGUISHING MOLTS

Both prebasic and prealternate molts in gulls commonly overlap, so that as one is wrapping up the other may already be in full motion. We can sometimes delineate generational feathers based on their pattern and condition. Contrasts seen between different generations of feathers are known as *molt limits*. In general, these limits allow us to describe which molt produced which feathers. However, keep in mind

39 1st-cycle Yellow-footed Gull that has replaced all of its scapulars, many innerwing coverts, a number of tail feathers, and perhaps 1–3 inner secondaries in its 1st prealternate molt. Note the obvious contrast between these feathers (old=juvenile, new=1st alternate). Many large gull species have less-extensive wing covert molt and rarely replace tail feathers in the 1st prealternate molt. ALVARO JARAMILLO. MEXICO. JAN.

40 1st-cycle Slaty-backed Gull. Note the color and pattern difference in the scapulars. Although they differ in appearance, these feathers are all presumably products of the 1st prealternate molt, with the darker feathers having grown in most recently. MIKIYA OIKAWA. JAPAN. MARCH.

that feathers with different colors or patterns are not always products of different molts. A good example of this can be seen on the scapulars and wing coverts of immature gulls (see plate 19.8). Feathers grown early in a given molt may appear delayed in pattern, while those grown late in that same molt may appear more advanced (Howell & Dunn 2007). Conversely, feathers grown early in a particular molt may appear advanced in color and pattern, and those grown later may regress in pattern. An example of this can be found on the primaries of some second- and third-cycle-type large white-headed gulls. In these individuals, several inner primaries have an advanced, adult-like appearance, while the outer primaries show a delayed aspect. Such differences in the appearance of feathers produced by the same molt have been associated with hormones (Howell 2010). Consequently, it may prove difficult at times (if not impossible) to determine whether some feathers are from the same molt or from different molts (see plate 20.54).

EMBRACING H-P-H

Anyone with a curiosity about feather replacement can follow these progressions by simply learning a few precepts of molt and aging. The symmetries are profound and easily observed in gulls, and utilizing H-P-H is remarkably handy for this. Important to many readers will be connecting molt nuances to field-identification questions. To that end, much knowledge can be applied. We know very well that molting birds can have altered wing and body shapes, along with unfamiliar plumage patterns that don't quite match up with the neat illustrations in our field guides. Noticing a Lesser Black-backed Gull has dropped its outermost primaries prompts us to not put much emphasis on its primary projection, whereas this structural field mark is quite useful when the species has fully grown primaries. We can greatly narrow down our search for a first-cycle Thayer's Gull in the northern winter if we understand it should be sporting a plumage that's largely juvenile. Even an identification as routine as separating Franklin's Gull and Laughing Gull may involve an age-related question tied to plumage. Many first alternate Franklin's Gulls have an upperwing pattern that resembles that of an adult Laughing Gull far more than that of an adult Franklin's.

The superiority of H-P-H lies in its universality. H-P-H harmonizes differences in austral and boreal seasons when describing molts and plumage while remaining biologically meaningful.

41 1st-cycle Thayer's Gull (far left) with three similar-aged American Herrings. In addition to the paler primaries and tertials, the scapulars on early 1st-cycle Thayer's are often largely juvenile, which makes for a more uniform look. The Herrings have already replaced their scapulars via a partial 1st prealternate molt. AMAR AYYASH. ILLINOIS. NOV.

42 2nd-cycle Glaucous Gull. Now starting its 2nd prebasic molt (p1–p2 dropped). The entire plumage seen here is essentially juvenile; thus this bird is going into its second complete molt without undergoing any apparent inserted molts in 1st cycle. ROBERT GUNDY. FLORIDA. MAY.

Thus, H-P-H avoids the use of seasons to label a plumage (e.g., first summer, third winter). For instance, a *second-winter* Laughing Gull can technically mean two different things, creating unnecessary ambiguity. But a *second basic* Laughing Gull is a second basic Laughing Gull, regardless of its being observed in New Jersey or Brazil, in September or February.

Using seasons to describe plumages can also be inexact. For example, an advanced two-month-old Ring-billed Gull sporting its first alternate scapulars in late August would be labeled "first winter" by this logic. But this isn't very intuitive, given that the time of year is still "summer." A first-cycle Glaucous Gull retaining first basic plumage in January is still a "juvenile." Yet those who adhere to using seasons call such birds "first winter" because of the time of year, despite there being no new plumage. This outdated system is painfully inconsistent, and although teaching and learning an imprecise system may seem easier at first, it hinders an observer's ability to genuinely understand how a bird's molt and plumage are related.

The "Identification" chapter discusses the extents of some of the molts mentioned above, and the species accounts provide more details, including photographic examples to accompany the text. If you are just beginning to learn about molt and molt terminology, you can always come back to this section and digest it in stages. It's important to know which species are two-cycle, three-cycle, or four-cycle gulls (see table 3). Consider which inserted molts are found in the first molt cycle, and make a distinction between complete and partial molts. Finally, keep in mind the importance of associating a plumage with the molt that produced it.

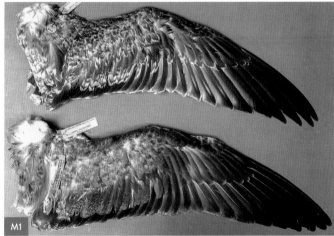

M1 1st-cycle (top) and 2nd-cycle California Gull wings. 1st-generation primaries generally have pointy tips. Subsequent ages show more rounded and blunt-tipped primaries. LSUMNS. AMAR AYYASH. CALIFORNIA. OCT.

M2 1st- and 2nd-cycle (right) American Herring Gulls. In addition to the rounded primary tips and pale eye, note the more muted and marbled pattern to the upperparts on the 2nd cycle. AMAR AYYASH. MICHIGAN. SEPT.

M3 1st-cycle Lesser Black-backed Gull with most scapulars and several wing coverts replaced in a partial, 1st prealternate molt. These new feathers are 1st alternate feathers. The dark brown wing coverts and tertials are 1st basic (i.e., juvenile). AMAR AYYASH. YUCATAN, MEXICO. DEC.

M4 2nd-cycle Lesser Black-backed Gull. The adult-like gray feathers were presumably grown in via a protracted (and likely ongoing) 2nd prealternate molt, which occurs somewhat regularly from late fall through early spring in some large gulls. AMAR AYYASH. FLORIDA. JAN.

M5 2nd-cycle Heermann's Gull undergoing its 2nd prebasic molt. Note the new 2nd basic primaries (black) below the tertials. The outer primaries, as well as a number of lesser coverts (pale and faded) are old juvenile feathers. The upperparts are presumably a mixture of 1st alternate and 2nd basic feathers. ALEX A. ABELA. CALIFORNIA. JULY.

AGING AND MOLT

M6 Some immature Glaucous Gulls can be tricky to age, especially without an open wing. The middle individual has a pale eye, a fairly advanced bill, and more adult-like gray upperparts growing in. In mid-June, we might assume it is undergoing its 3rd prebasic molt (i.e., early 3rd cycle), and the other two are undergoing a 2nd prebasic molt (i.e., early 2nd cycle). DAVID J. RINGER. ALASKA. JUNE.

M7 2nd-cycle Slaty-backed Gull. The 2nd prebasic molt has begun, as indicated by the molting primaries. The scapulars are 1st alternate. The median coverts are largely missing, exposing dark bases to the greater coverts. The marginal coverts are juvenile feathers, which still appear relatively fresh after almost one full year. MIKIYA OIKAWA. JAPAN. MAY.

M8 2nd-cycle Iceland Gull (subspecies unknown) undergoing 2nd prebasic molt. Note that p9–p10 and most secondaries are 1st basic (juvenile). Despite being a year old, these retained juvenile feathers are in fairly good condition. AMAR AYYASH. WISCONSIN. JULY.

M9 3rd-cycle Olympic Gull (Glaucous-winged × Western) undergoing its 3rd prebasic molt. Here, p1–p4 are adult-like gray with broad white tips (3rd basic). The outer primaries, secondaries, and tail feathers are 2nd generation (i.e., 2nd basic). The gray scapulars and median coverts may be 2nd alternate and/or 3rd basic feathers. LIAM SINGH. BRITISH COLUMBIA. JULY.

M10 Presumed 3rd-cycle Glaucous-winged × American Herring Gull. The innermost primaries and innermost secondaries have an ordinary, 3rd basic appearance (adult-like). The outer primaries and most secondaries are delayed in appearance (2nd basic–like). Despite this difference, it's believed these feathers were all produced by the 3rd prebasic molt, but perhaps at different time intervals, under varying conditions. AMAR AYYASH. BRITISH COLUMBIA. MARCH.

M11 Adult California Gull undergoing prebasic molt. The outermost secondary, s1, and p1–p6 are fully grown; p7 is almost fully grown, and p8 has not emerged beyond the primary coverts. Note the faded black appearance of p9–p10 (older), as well as the increased wear suffered by the white mirrors (less durable than dark pigments). AMAR AYYASH. WASHINGTON. AUG.

AGING AND MOLT

M12 1st-cycle Kumlien's Gull (Orange 3Z). Banded as a 1st cycle in St. John's, Newfoundland (Jan. 2015). A pale bill base at this age is not rare. LANCY CHENG, NEWFOUNDLAND; JAN. 2015.

M13 Orange 3Z, now as a 2nd cycle with faint, adult-like gray on some scapulars, marbled pattern on tertials and greater coverts, demarcated black tip on the bill, and paling iris. LANCY CHENG, NEWFOUNDLAND, DEC. 2015.

M14 Orange 3Z, now as a 3rd cycle. Retarded outer primaries and greater coverts may invite confusion when aging. Note that the entire back is solid, adult-like gray, including several upper tertials that show broad white tips. If one looks closely, a bold and well-defined p9 mirror is transparent, supporting identification as an older bird. LANCY CHENG, NEWFOUNDLAND, DEC. 2016.

M15 Orange 3Z, now as a 4th cycle. Suggests an adult, and without life history, we would age it as such. An open wing may reveal subadult features, such as diffuse pigment on the wingtip or streaking on the primary coverts. Perched, the only suspicious feature is the dull bill base, with a dirty pattern on the tip. LANCY CHENG, NEWFOUNDLAND, DEC. 2017.

M16 Orange 3Z, now as a 5th cycle. The outer primaries are 5th-generation but look remarkably similar to the previous generation. Without life history, there would be no way to estimate this adult's age. The bill pattern is more advanced now, but note that even older adults may show dull bill patterns similar to M15. LANCY CHENG, NEWFOUNDLAND, DEC. 2018.

IDENTIFICATION

Gull identification is best achieved using the two-pronged approach of, first, attending to size and structure and, second, evaluating plumage and bare-part patterns. We rely on size and structure to quickly narrow down possibilities while simultaneously using plumage and bare-part details to resolve an identification. Size and structure are sometimes confused with one another. Although related, they are different ideas. A Great Black-backed Gull can be characterized as being massively large (size) with a flat, blocky head (structure). Ivory Gull is a smaller bird (size) with compact wings and pigeon-like body (structure). A Glaucous-winged Gull may be described as having beady eyes (size) with a bulbous-tipped bill (structure). Two species may have identical body lengths but appreciably different structures. As with plumages, there are some patterns that are highly distinctive and unique to certain taxa, while other features show considerable overlap from one species to the next. From an identification standpoint, it is unsatisfactory to obsess over plumage without attending to size and structure. But similarly, it is not prudent to commit to size and structure without learning plumages. They are individually valuable but should be employed together.

DISTRIBUTION AND EXPECTATIONS

There are a few steps we can take to increase our potential in the field. First, a mixed-species flock will be daunting without a proper set of expectations. Range maps heavily shape our expectations, and knowing which species to expect goes a long way. Studying seasonal distributions and peak migration timings prepares us for arriving and departing birds. Most regions throughout the continent have no more than four common gull species in a given season, often differing in size and favoring slightly different niches. Gull flocks are seldom segregated by species in the nonbreeding season, although it's not uncommon for smaller species to cluster together and larger species to cluster together.

Recognize the great propensity of gulls to stray, and develop an inkling for those which are uncommon but regularly occurring versus those which are truly rare. Assume a rare gull can be found just about anywhere you set foot, as they can. From parking lots to fish markets to agricultural fields, rarities appear without notice when you least expect them. But as tempting as it is to concentrate on rarities, making this our sole focus becomes limiting, especially if the objective is to improve identification skills. Hoping to turn every other gull into a rarity is distracting and, frankly, is a backward approach to identification, especially for beginners.

It is our common gulls, often taken for granted, that provide the reference points needed to recognize and identify more-unusual species. In fact, when we use descriptions such as "paler," "longer winged," or "daintier," we are often making an indirect comparison with common species. And so, it almost goes without saying that the first steps in any problematic identification should be eliminating the expected. Once expected taxa are confidently eliminated, the possibility of something more unusual can be pursued. Naturally, probability wins on most days. In the East, Great Black-backed, American Herring, Ring-billed, and Bonaparte's Gulls are great reference species. In the West, Glaucous-winged, Western, California, and Short-billed Gulls fill that role.

RELATED AGES

There is a great lesson to be learned in demography when we look at age ratios in almost any gull flock. Having the highest survival rates, adults are most ubiquitous, and first cycles are the next most abundant age group encountered. Intermediate ages are the least numerous—an inherent consequence of mortality rates. A large percentage of birds perish within their first year of life, whether it be from disease, predation, or inadequate food sources (Ryder 1980). From the surviving cohort, birds in a smaller subgroup perish during the next year, as they still suffer from inexperience and are prone to natural defects. A gull that makes it through the first few years of life has favorable odds of becoming an adult that will breed and contribute to the gene pool. Generally speaking, an adult replaces itself in its lifetime if the population is stable. This ensures fit and thriving populations, and it is quite observable in gulls.

With gulls, comparisons must be made at an age-related level. For instance, in their first cycle, Lesser Black-backed and American Herring Gulls can present an identification challenge, but as adults, they're

43 An exciting component to gull-watching is mixed-species flocks that often allow extended study in the open. The majority of this flock is made up of Heermann's Gulls, with at least four other species for interest. PETER HAWRYLYSHYN. CALIFORNIA. DEC.

44 A recently fledged Ring-billed Gull in juvenile plumage (ghost type). Note the short wing projection and petite bill. The outer primaries and bill will undoubtedly develop and grow in the upcoming weeks. AMAR AYYASH. INDIANA. JULY.

unlikely to be confused. The age group we typically learn to identify first is adults, not only because they're most abundant, but because their plumages are less variable than those of other age groups. Studying adults gives us a good sense of size and overall structure for a species. It should be noted, though, that size isn't age dependent. That is, a free-flying juvenile has a fully grown body that might exceed the size and weight of its adult parent. Do keep in mind, however, that the outermost primaries on a recently fledged gull may still be in their final stages of growth and that the bill depth may continue to develop for several months in some individuals.

CATEGORIZING FIELD MARKS

The species accounts provide a plethora of field marks to consider that can generally be divided into three types: diagnostic, extremely indicative, and supporting. It is necessary to reiterate a basic caveat here: field marks used in gull identification are almost always applied in an age-related or seasonal context. The black underwing of an adult Little Gull is diagnostic, with no other gull showing such dark underwings. However, this is an example of an *age-related* field mark, as it doesn't apply to first-cycle Little Gulls, which have pale underwings. Adult California Gulls acquire a distinctive black-to-red bill pattern around the bill tip in nonbreeding condition. This trait is supporting at best, as other four-year gulls may show a similar bill pattern. Furthermore, the pattern is lost in breeding condition and as such is an example of a *seasonal* field mark. Some features are of little to no value. For example, all juvenile gulls found in North America can be expected to have dark eyes. The ability to weigh the significance of a single field mark and correctly make use of it increases with experience.

45 Two Little Gulls—an adult and a 1st cycle—in a flock of Bonaparte's Gulls. The adult's black underwing is diagnostic, but the 1st cycle can easily go overlooked with its paler underwing (roughly four birds to the left of the adult).
EVAN SPECK. INDIANA. MARCH.

Other field marks are variable to the extent that they're found in a subset of a population and are absent in others. For example, any large black-backed gull showing a string-of-pearls and a broad white trailing edge is extremely indicative of Slaty-backed Gull. But some adult Slaty-backeds have a lackluster pattern on the outer primaries, showing little trace of the ornate string-of-pearls associated with the species. Learning the nuances of variation is by and large the cornerstone to large-gull identification. This topic is discussed in greater detail near the end of this chapter.

Another crafty approach to identification is knowing which traits are never expected. Constants are far less common than variables in gull identification, but they do exist. For example, Ring-billed Gulls never have red on the bill, adult Glaucous Gulls never develop mirrors or show pigment on the wingtip, and Red-legged Kittiwakes never acquire a tailband in any plumage. Although we take these features for granted, noting the absence of a characteristic can be surprisingly helpful.

Finally, recognizing *confusing pairs* is a familiar strategy not just in gull identification but in bird identification in general, and this is one thing that separates those with experience from beginners. When an experienced observer is faced with an identification challenge, an automated list of possible birds is

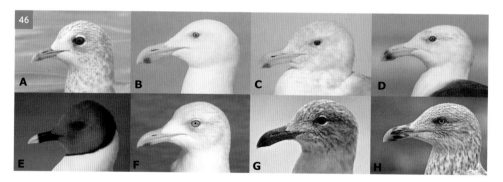

46 Describing What You See:
A Large, darkish eye on a small head. Petite, greenish-yellow bill with tapered tip and light smudging. No gonys spot. The nape is densely smudged. Adult Short-billed. Jan.
B Honey-colored eye with bold yellow orbital. The eye is disproportionately small and sits high on the face. Large, bright yellow bill with reddish gonys spot. Entirely clean white head. Adult Western Gull. Oct.
C Dark beady eye on a large face. Strong yellow bill with small gonys spot. The entire head shows a smudgy gray wash. Adult Glaucous-winged Gull. Sept.
D Darkish eye with orange-red orbital. A very stout bill with blobbed tip. Fleshy bill base with black marks near the tip. Blocky head shows faint streaks on crown. Adult Great Black-backed Gull. Sept.
E Dark eye with red orbital. No white eye crescents. Petite bill with black base and yellow tip. Slate-black hood bordered by a black necklace. Adult Sabine's Gull. June.
F Pale eye with light flecking. Hints of a pinkish orbital. The bill is thin and straight with minimal gonys expansion. The face and hindneck show light gray streaking. Adult Iceland Gull. Feb.
G Dark eye with red orbital and thin, white eye crescents. Long, droopy red bill with black tip. The head has gray ground color with moderate black markings. Adult Heermann's Gull. Nov.
H Pale eye. Dull yellow bill with much black along the cutting edge. The head and neck are heavily streaked. 3rd-cycle Lesser Black-backed. Dec.

IDENTIFICATION

47 Variation found in large gulls includes differences in bill and body size, upperpart coloration, and, here, wingtip patterns. The adult Slaty-backed Gull on the left shows extensive white on the wingtip, whereas the individual on the right has a relatively "dark" wingtip. LEFT: MARTEN MULLER, KOREA, FEB. RIGHT: GRAHAM GERDEMAN, JAPAN, JAN.

48 2nd-cycle American Herring and Thayer's Gull (right). This "confusion pair" shows some overlap in every plumage, but note here the paler tertials and primaries on Thayer's, as well as the pale underside to the far wing (p10). AMAR AYYASH. ILLINOIS. FEB.

generated, and a handful of similar species are then compared to one another. At the very least, knowing which species are in the running greatly focuses our efforts.

WHAT TO LOOK FOR

A gull in question should promptly be sized as small, medium, or large. Nearby birds and stationary objects can be used for comparison to help place your subject. Consider the size and shape of the bill with respect to the body's proportions. Is the bill relatively long or short? Thin or stout? Is the tip pointed or blunt? Parallel edged or bulbous? What color are the eyes? Is there any distinctive pattern to the head markings? Consider the shade of gray to the upperparts: Are they pale, medium gray, slate colored, or black? Take note of leg color and other bare-part details. Structurally, does the body have a front-heavy or shallow-

49 Another "confusion pair," 1st-cycle Ring-billed and Short-billed Gull (front), both with gray postjuvenile scapulars. Ring-billed averages a large body and bill, but also note it has replaced upper tertials and inner coverts. Short-billed typically retains these juvenile feathers until its 2nd molt cycle. ROBERT RAKER. COLORADO. APRIL.

50 Adult California Gull (left) and American Herrings. Although often described as being smaller than Herring, some California Gulls are just as large. AMAR AYYASH. MICHIGAN. NOV.

51 The adult Slaty-backed Gull (far right) immediately "pops," but more subtle are the "gull-gray" species, including two adult-type Herrings (center), adult Thayer's (far left with black wingtips), and three Glaucous-winged Gulls. These paler species are separated by wingtip coloration, eye color, and head and bill proportions. LIAM SINGH. BRITISH COLUMBIA. MARCH.

chested appearance? Is the gull potbellied or sleek in the rear? Are the legs noticeably long or short? Do the tips to the secondaries droop far below the wing coverts, or are they completely hidden? Do the primaries look truncated, or does the bird appear to have a long wing projection? What patterns are visible on the folded wingtips, including the underside? Evaluating these features are all essential first steps to making an identification. And although all of these questions do not need answers for every bird encountered, they are nonetheless questions that provide the framework for your identification.

NUANCES, CAVEATS, AND PITFALLS

LIGHTING

Lighting can have a dramatic effect on colors and tones, and it plays a significant role in how we see and interpret field marks. In general, the best viewing conditions for evaluating the gray upperparts of a gull are days with a thin cloud cover. This allows for sufficient—but not too strong or too sparse—sunlight to filter through. As a rule, direct sunlight makes for harsh and unpleasant gull-watching conditions. Too much light results in a halo effect across the upperparts. Grays become washed out, and tones can't be interpreted correctly. The opposite extreme—low light brought on by heavy overcast—causes grays to look muted and darker than they are. When considering pigments in the field, you want to be aware of these effects and mentally correct for any differences brought on by lighting conditions. These phenomena are easily observed on days when clouds are rapidly shifting, yielding moments of bright sunlight and moments of thick cloud cover. A medium gray mantle can change from appearing silvery to dark gray in a matter of seconds. Photographers know this well and will admit gulls can be some of the trickiest bird subjects to accurately compensate for. This is in part due to their high-contrast plumages, which combine whites, grays, and blacks all in individual birds.

With this in mind, consider your position in the field and the time of day. Ideally, you'd like the sun to your back and the gulls you're watching in profile. A gull viewed from behind will appear darker, while a gull viewed head-on may look paler. We can make reliable comparisons of gray values with gulls standing side by side at the same angle. Take any individual that starts to slightly turn away and notice how its gray tone is suddenly altered.

The potential for such pitfalls is not to be underestimated. The difference between the KGS values 6 and 9, for instance, is substantial in gull identification. It is fairly straightforward to manage these effects in a gull flock, where comparisons can be made from one individual to the next. If a known species appears a shade or two darker than it should be, then I would expect to make a corresponding adjustment to other species present. Needless to say, it's more challenging with birds viewed alone. A bird of interest that captures your attention should be observed from various angles to ensure an accurate assessment. To get a sense of how pale or dark its upperparts are, compare them to any black pigments on the wing. Also,

52 Lighting and camera settings hugely impact what we "see." This adult Glaucous-winged Gull was photographed at different times of the day by the same observer. The photo on the left shows dark wingtips, as found in some Glaucous-winged hybrids. The image on the right shows a perfectly fine Glaucous-winged. FRANK LIN. BRITISH COLUMBIA. NOV.

use the white feathers of the head and neck to estimate whether lighting conditions are favorable. If the white head and neck are blown out, or too much shadowing is observed, then the grays you're seeing are likely unreliable. If these feathers are evenly white throughout, then you can assume the shade of gray being observed is consistent.

Beware heavy shadow effects that are produced on various surfaces, such as snow, ice, and white sands. Shadows cast back onto your subject result in what looks like a much darker bird. Upperparts are altered in a similar respect on water, but the effect may be multiplied due to unavoidable glare. I recall an aggravating morning spent searching for an adult Black-tailed Gull (KGS: 8.0–9.5) in a large congregation of Ring-billed Gulls (KGS: 4–5) on Lake Erie. The intense sunlight made picking out this individual a real challenge. The upperparts of all the gulls present appeared to be glowing lightbulbs, while the sides of their bodies were covered in shadow. Once spotted, the Black-tailed Gull stuck out like a sore thumb, but only with high-magnification optics. Trying to find it with the naked eye was futile. As new birders arrived and scanned for the bird, not a single observer could find it without being directed by others. We returned to see this same individual the following weekend on a partly cloudy day: the Black-tailed Gull was effortlessly spotted in seconds!

Related circumstances arise with gulls in flight. In addition to colors being muted and appearing darker in low-light conditions, shape and structure may be impacted. Body size can appear bulkier, and wings can look noticeably broader. In brighter lighting conditions, wings tend to look thinner, as their outlines coalesce with their background. From below, various plumage features are highlighted, such as a tailband or a window. A negative result of this is noticing "more" of a window on species where it typically isn't expected (i.e., first-cycle California or Lesser Black-backed Gull). Some gulls, such as adult Glaucous-winged, exhibit a translucent quality to their flight feathers, and bright light shining down on the wing exaggerates this look. The never-ending conundrum of labeling pale Thayer's / dark Kumlien's is quite relevant to this discussion too. The folded wingtip on a perched individual may give the impression of a standard Thayer's type, but when the bird is in flight, the impression can shift to a standard dark gray Kumlien's. The reason for this is that when the wings are folded, there are layers of pigment overlapping one another with little to no light permeating the feathers, creating a cumulative dark effect, whereas the spread wing—specifically in bright conditions—assumes a surge of light through each individual primary, resulting in an apparently lighter shade of pigment that might look slate gray rather than black. How this is reconciled in the field is left to the observer.

In neutral lighting, the underside of the remiges can often be used to help pick out darker gray-backed and black-backed species. The secondaries and primaries usually reveal a darker contrasting row of feathers, or a shadow bar, when compared to those of paler gray-backed species. But beware the natural reflective properties of feathers. The black on the underside of the primaries, for example, can be seemingly lost as a bird banks or turns away.

Finally, under clear, blue skies, adult gulls flying at high altitudes can seem to "lose" their black wingtips, inviting thoughts of white wingers. As can happen when we watch hawks at a distance, the colors on the ventral side of the body often reveal themselves as our eyes adjust to the lighting.

53 Adult Kelp Gull in photographs taken seconds apart. The black underside to the flight feathers (left) is the correct color. A slight change in angle (right) causes the light to reflect a silvery underwing. AMAR AYYASH. PERU. NOV.

BEHAVIOR

Behavior can assist in recognizing a species. But it can also alter our perception of what a species should or shouldn't look like. It's important to ask if the size and structure you're seeing is in part due to behavior. Preening birds and birds attempting to cool down in warm temperatures commonly fluff out their feathers, making the body appear plump and stocky. An agitated or alarmed gull will often outstretch its neck, slightly shift its wings down, and produce a hunched-forward posture. This will make for a bird that looks larger and more powerful. Conversely, gulls confronting stiff winds will deliberately compress their contour feathers tightly against the body, giving them a streamlined figure. This can create a sleeker and longer-winged appearance. Feathers held down against the crown make a flatter looking head that is associated with males, while raised feathers portray a rounder head shape, giving the gull a more delicate, female-type appearance. Feathers held tightly in the loral region often give the bill a longer, exaggerated feel. A similar effect is produced when feathers around the bill base are matted down or missing. Gulls walking or resting along the shoreline with waves waxing and waning beneath them will often compress their belly feathers tightly against the body. This behavior exposes much of the tibia and results in what looks like a long-legged bird.

As for species that have forked tails or modest divots to the central tail feathers, note that the fork shape is lost when the tail feathers are completely fanned out. Also, beware molting rectrices on such species, which may show outer tail feathers equal in length to or even shorter than the central tail feathers.

For various reasons, gulls sometimes hold their mandible and maxilla ever so slightly apart, giving the impression of a larger bill. This is revealed by closely examining the cutting edge. A Thayer's Gull on the West Coast with mandible and maxilla slightly separated may be accused of being a Glaucous-winged × Herring hybrid simply because the bill looks too deep. It's also quite common to find gulls that have full crops—or, more accurately, full gullets—which can give the breast and upper neck a much chunkier appearance. We often see this on porked-out individuals arriving from feeding sites. Upon critical examination, the upper neck will look bulgy and sometimes misshapen. Be cognizant of such circumstances, and understand how they may dramatically change the apparent size or shape of an individual. Sufficient observation time usually dispels skewed impressions.

Like other birds, gulls engage in habitual behaviors. These can be specific to a certain species or a handful of species, particularly when it comes to feeding methods. Gulls feeding in crab apple trees in the East will likely be Ring-billed Gulls. A dark, medium-sized gull pirating a pelican on the West Coast is very likely a Heermann's. A flock of small gulls following a plow on the Great Plains immediately suggests Franklin's. A small gull with boomerang-shaped wings wheeling over the open sea will almost certainly be a kittiwake. These are but a few behavioral traits that we've learned to associate with particular species.

Although not necessarily helpful for identification but relevant to this topic, some behaviors are associated with specific age groups. For instance, handling inanimate objects with the bill is a common behavior of first-cycle gulls. It is not unusual for gulls less than a year old to be found picking up and experimenting with rocks, straws, bottle caps, cigarette butts, and a whole host of other items. In my experience, it seems these objects are utilized for play but may actually be serving as practice for manipulating prey.

54 3rd-cycle American Herring Gull in photographs taken seconds apart. Head and bill shape are useful in picking out some taxa, but note that this feature changes drastically depending on behavior and how the feathers are being held against the body. AMAR AYYASH. ILLINOIS. DEC.

IDENTIFICATION

DISTANCE

Can a gull half a mile away be identified with confidence? The answer is a qualified yes! Observers routinely make such calls in the field, given there are sufficient field marks to support an identification. Some species are so distinct that they are effortlessly identified at long distances. However, this is more the exception than the rule with gull identification. Many distant identifications are educated guesses based on probabilities of what is expected at a certain date and location. The skill set for identifying distant birds is cultivated by working backward. Learning how to first identify a species at close range, and then gradually increasing the distance, works best. With experience, we develop an imprint of size and proportions, wing-beat rhythm, and overall color patterns for a species, and we check off these elements when identifying distant birds with little effort or analysis.

BLEACHING AND WEAR

Bleaching (i.e., fading) and wear negatively alter the integrity of feathers. Bleaching is caused by prolonged exposure to the sun's ultraviolet light. While some birds can shield themselves from the adverse effects of ultraviolet radiation, most gulls are subject to constant sunlight exposure. As a result, colors lose their richness over time, leaving faded feather patterns. Black feathers initially fade to brown, browns fade to tan, and in the most extreme cases dark feathers can bleach to a sullied white. Within the primaries, the outer primaries are usually the first to fade, because they are most exposed to the elements.

In general, gulls in regions with abundant sunlight throughout the year are more prone to bleaching. The combination of salt water and sandblasting in coastal areas might intensify this problem. This is evident, for example, in midwinter when comparing the plumages of Lesser Black-backed populations on the Gulf Coast with those in the Northeast. Those farther north appear darker and less worn, while those farther south along coastal areas average paler and exhibit more wear. It is also evident after February when considering the plumages of Glaucous-winged and Thayer's Gulls, whose condition will be worsening on the central California coast but fairly intact in SE Alaska and British Columbia. Nevertheless, bleached gulls

55 By late winter and early spring, 1st-cycle large gulls can become extensively worn and bleached. Glaucous Gull (left) commonly bleaches to a sullied white. The smaller gull on the right is an Iceland Gull, presumably Thayer's, based on probability. Identifying it to subspecies out of range would prove problematic. PHIL PICKERING. OREGON. APRIL.

56 Known-origin 1st-cycle American Herring Gull. Pictured here as a juvenile in early July (left) and then in mid-Feb. Note how the pale tips to the upperwing coverts and tertials have worn down and become frayed. The juvenile scapulars have been replaced in a partial 1st prealternate molt. Also note how the primaries and tertials have faded in six months' time. KURT WRAY. OHIO. JULY. JOE BAILEY. INDIANA. FEB.

are found at all latitudes, including, but to a lesser extent, in the interior, away from coastal habitats. There is evidence indicating that feather bleaching in gulls is gradual throughout much of the fall and winter, with a suddenly accelerated rate in early to mid-spring (Howell 2001b). This helps explain why there's an abrupt surge in bleached first-cycle gulls once April approaches. By April and May, the effects of bleaching are vastly on display throughout the entire continent, and it is further compounded by excessive wear.

Wear is the other chief agent of feather degradation and can be accelerated if feathers have been weakened through fading. By "wear," we mean mechanical abrasion resulting in feather material being broken and lost. Like a broom's bristles or a tire's treads, when feathers rub against or otherwise come into contact with various elements and objects, they sustain some degree of reduction and breakage. Studies suggest darker feathers resist wear much better than pale feathers (Bonser 1995). Darker feathers, which are enriched with melanin, have harder surfaces, and this increases their durability. Juvenile gulls' brown upperparts often have pale edges, providing a crisp look to their plumage when fresh. These pale fringes are usually the first to fade, break down, and wear away. Rarely is wear evenly distributed throughout the upperparts, and this can make for an untidy appearance. In adult-type gulls, black regions to the primaries may remain more intact, while the white apicals and mirrors suffer more from abrasion. This can be due partly to their physical location (exposure) and partly to their weaker structure (lack of melanin). And although black feathers are better at resisting damage than adjacent white regions, they too have limited strength and are susceptible to wear (Ayyash 2016).

Common daily maintenance of feathers involves rifts being repaired by gentle preening or "brushing" with the bill. But over time, feather barbs are fractured and removed, making it impossible for them to lock together. At this stage, feathers are beyond repair. At their worst, feathers may eventually look like coarse strands of loose hair, as they completely wear down to their shafts.

Many bleached or worn gulls are identifiable, but in some instances, distressed plumages and their lack of color patterns only compound what might be an already-problematic identification. Attempting to competently identify hybrids and vagrants or to assign subspecies under such circumstances is best avoided. Also, be aware that the combination of wear and bleaching, as well as missing feathers and newer feathers growing in, can create uncharacteristic patterns that are seldom portrayed in field guides.

VARIATION

Years ago, I attended a morning walk in Cape May led by a prominent birding figure. It was late summer, and as participants began to congregate, someone noticed our first black-backed gull of the day. The bird was alone, preening, and was clearly a subadult. It was far enough down the beach to present some ambiguity for our group. With our binoculars, a few of us began to pick out field marks, when our trip leader immediately dismissed it as a Great Black-backed Gull. I suggested we consider a Lesser Black-backed Gull, based on the heavily marked head and the overall plumage aspect. At the time, Lesser was somewhat rare here in the summer, so expectations were heavily tilted one way. Our trip leader proceeded to explain it could only be a Great Black-backed, based on its rotund body, shorter wings, and hefty bill. Indeed, the impression, based on structure, was of a stocky gull, but there were no other birds around for comparison.

The gull then got up and flew closer, putting down near enough for everyone to confidently agree on its identification: a second-cycle Lesser Black-backed. The bill was mostly black and appeared deceptively large from a distance. This often happens against paler, contrasting backgrounds. The short-winged look was due to a couple of missing outer primaries, as it was undergoing wing molt. It now appeared much sleeker in the rear, and we attributed the "rotund" body shape to its raised feathers while it preened. Surprising to the entire group was a distinct straw-yellow quality to the legs, which had seemed convincingly pink from down the beach. It proved to be a life bird for several participants and served as a great learning moment for everyone present, especially our trip leader, who had overlooked some important identification points.

Fortunately, bird identification is a "low-stakes game." Much of the reward lies in the journey of watching ourselves move along a learning curve. With gulls, the inflection points on this curve consist of learning feather topography as well as aging and molt. Over the years, I have watched friends attempt to learn gulls without a basic understanding of these topics, only to land themselves in a minefield. In fact, I would submit that a fair number of birders who "can't do gulls" have either adamantly avoided these topics or in fact don't realize these topics are requisites to grasping this group of birds. The problem, at root, lies in the misconception that gulls should conform to the plates in a general field guide, and this simply isn't the case with a fair number of individuals we encounter.

We often read in bird-identification literature that gulls are notoriously variable. Is this true? Yes and no. It's a statement that has become clichéd and needs some clarification. All species exhibit degrees of variation, even your local American Crows. But rarely is this variation striking enough to present identification problems. Extensive individual variation—the type that is immediately evident in even the smallest sample—is the variation we're interested in here. This sort of variation often gives us pause and almost always has implications for identification, especially when a particular feature (or features) can be found in several similar species. With gulls, more than 90 percent of identification problems come from less than 10 percent of birds.

Variation in two-year gulls is fairly straightforward and doesn't necessarily fall in the "gulls are notoriously variable" category. Extensive variation is often found in species that have a large geographic distribution, in which various "types" or subspecies show noteworthy plumage and size differences. This is generally found in some three- and four-year species, especially the latter. First and second cycles of several *Larus* species present variation that borders on downright Feral Pigeon madness. A prime example of this is the Herring Gull complex. I could carry on for pages describing the individual variation found in their early plumages and still come up short (e.g., see the number of first-cycle American Herring plates included in that account). This scope of variation generally can't be taught, but it can be reasonably learned in a contextual setting. The photographic examples in this guide are predominantly of "typical" individuals, with some examples of extremes that point out identification pitfalls. But it is typical birds that we're after. This is the soundest and sanest approach to identification. Once we're comfortable identifying run-of-the-mill individuals of a certain species, we then proceed to slightly widen our goal posts and allow for more variation, as we learn what features may be encountered. Your local species and the variability associated with them should be a priority. For the rest of this chapter, I discuss some of the variation one can expect in a same-species flock.

SIZE

Close scrutiny of most gull flocks will reveal size differences among conspecifics. At the species level, these are generally attributed to sex or geography. Gulls, particularly three- and four-year gulls, show appreciable size variation, with males averaging larger bodies than females (the opposite of raptors). This is also noticeable in bill size, neck and body girth, and, to a lesser extent, leg length and wing dimensions. It is not unusual, for example, for a female-type Western Gull to appear up to 25 percent smaller than male conspecifics. One can often observe this *sex-related variation* during the breeding season, when gulls are courting or paired up on or near the nest. Outside of these circumstances, identifying the sex

57 An example of size variation: This adult Thayer's (left) and Ring-billed Gull appear to be the same size. Thayer's is generally larger than most Ring-billeds, so we might assume this one is a small female type with a large male Ring-billed. Allow for size variation in large, white-headed gulls. AMAR AYYASH. ILLINOIS. MARCH.

of an individual is speculation, and for that reason we qualify our statements with "type" (i.e., "female type" or "male type"). Being cognizant of size differences also helps us understand differences between species. A male-type Iceland Gull may very well approach the size of a female-type Glaucous Gull. Such identifications, although supposed to be simple to work out in theory, require careful assessment of structure, as well as plumage and bare-part details.

Another explanation for size differences within a species is *geographic variation*. For example, the subspecies of California Gull in south-central Canada (*Larus californicus albertaensis*) averages larger than the nominate race found in the Great Basin (*Larus californicus californicus*). These races come together between the Colorado Front Range and the Pacific Coast in the nonbreeding season, and differences in body size, among other features, are readily observed at this time. And although there is great overlap, the difference in size between a typical female-type *L.c. californicus* and a male-type *L.c. albertaensis* can be doubly impressive. A similar observation can be made with American Herring Gull. When visiting the Northeast in the breeding season, I'm always struck by the size and structural differences between Atlantic Coast Herrings and those I'm accustomed to seeing on the Great Lakes. In addition to averaging smaller bodies, Great Lakes birds show smaller heads and bills and appear more compact. Those in the East, especially in places like Maine and New Hampshire, appear to me as having been raised on growth hormones and are rather intimidating.

There are times when the size of an individual appears to be beyond the extreme end of "normal" variation, and this inevitably challenges our notions of how small or how large a particular species can be. It helps to see these birds among conspecifics, and if they check all the other boxes, we write them off as being atypical in size. Repeatedly encountering individuals that appear way too small or way too big for a particular species may just mean you need to widen your bell curve.

PLUMAGE

No gull species exhibits sexual dichromatism, so separating males and females by plumage isn't possible. Instead, *age-related*, *seasonal*, and *geographic* plumage differences are central. In many respects, these distinctions are the crux of gull identification. It goes without saying that age-related plumage differences are to be expected. If I'm scanning a flock of Great Black-backed Gulls, I should be prepared to sort them into four age groups, while if I'm watching Short-billed Gulls, I anticipate three age groups, and if I'm looking at Bonaparte's Gulls, just two age groups.

58 An example of intraspecies variation, the adult Glaucous Gull on the right is presumably male, with a female type on the left. In addition to size, males average shorter wing projection. BRIAN SULLIVAN. ALASKA. AUG.

59 Trio of Ring-billed Gulls. The center and right birds are 1st cycles displaying remarkable variation. The center bird has juvenile scapulars with noticeably dark aspect. The bird on the right has paler juvenile wing coverts, replaced scapulars, and a bicolored bill. ROBBYE JOHNSON. WISCONSIN. SEPT.

First Cycles: The simplest reason for differences in plumage appearances in first cycles is hatch date. Consider, for example, that by mid-June there are fully fledged juvenile Ring-billed Gulls that have already taken wing in some parts of the continent, while in other regions adult Ring-billeds are still building nests. Striking differences in hatch dates can commonly occur in the same colony too. By early September, a mishmash of plumages can be seen as these same juvenile Ring-billeds begin to migrate and stage at various sites. Some individuals will have entirely crisp juvenile plumes with well-marked heads and bodies, while others are already becoming white headed, showing signs of wear and beginning to molt.

An extra layer of variation is added when we consider the nature of juvenile plumages. Some gulls fledge with inherently weak feathers that very soon begin to show wear and bleaching, while others have plumages that naturally hold up well through the winter and into early spring. Some birds go through a rapid preformative or first prealternate molt early on, in order to replace weaker feathers. With Ring-billeds, for example, it's not unusual to see all of the scapulars replaced, along with some wing coverts and even tertials, in just a couple of months. The differences in these "postjuvenile" molts (i.e., preformative or first prealternate molts) are so pronounced at times that birders mistake what they're seeing for different age groups, when in fact it is variation within the same age group.

Interestingly, northern gull populations generally come equipped with more durable juvenile plumages, while populations at more-southern latitudes have weaker juvenile plumages. There is much individual variation involved, however, and more study is needed to figure out the physiological causes behind this, but the phenomenon can be observed in the waves of newly arriving gulls in late fall and early winter. This is one of the reasons why a cohort of first-cycle Herrings may be all over the board in appearances in late winter.

But surely hatch dates and the integrity of juvenile plumages are not entirely satisfactory reasons for why we find so much variation in some first cycles? Numerous bird groups have rather dramatic differences in their hatch dates and originate from varying latitudes, yet in time, many come together to look more or less the same. To add to what is already remarkable plumage variation in some first-cycle gulls, also consider "types" and individual variation. Some juveniles naturally have dark or pale plumage aspects. A "ghost-type" juvenile Ring-billed thrown into a group of "brown types" is sure to trip up the casual observer who isn't aware of these things. Add size variation to the equation, and it results in some interesting variety. In addition, some individuals may have upperparts that are more boldly marked, while others may be less patterned. Some may show a wide tailband in a cohort of individuals showing thinner tailbands. These differences in plumage are plain to see, especially as we increase our sample sizes. Each bird comes with a suite of variables that contribute to its appearance, and being aware of this answers many of the questions arising in gull identification.

Adults: Variation in adult plumages is more clear-cut or, at least, more orderly than that found in first cycles. Questions related to adult plumages often revolve around upperpart coloration and outer-primary patterns. Upperpart color differences at the species level are often a result of regional variation. Some taxa show no appreciable variation in their KGS values, such as Ring-billed Gull, while others exhibit a wide range of variation, such as Western and Lesser Black-backed Gulls. There are instances when the slightest deviation in upperpart coloration is problematic, while at other times it's entirely expected.

60 One-year-old Lesser Black-backed Gulls undergoing their 2nd prebasic molt. The individual on the right has a fairly typical appearance for that species, but the bird on the left is unusually pallid. AMAR AYYASH. WISCONSIN. JULY.

61 From these five Lesser Black-backeds, the individual on the far right has upperpart coloration matching Laughing Gull; this is considered the palest gray within range for that species. AMAR AYYASH. FLORIDA. JAN.

62 A duo of male Glaucous Gulls nicely displaying size and color variation found in this species. On the right is a member of the smallest and darkest subspecies, *barrovianus* (North Slope, Alaska. May). On the left is a *leuceretes* type (Labrador, Canada. May). In time, Glaucous Gull may prove to be multiple species. AMAR AYYASH. BURKE MUSEUM, WASHINGTON.

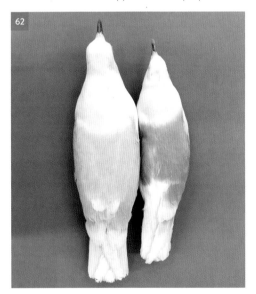

An American Herring Gull that appears a couple of shades paler or darker than usual is promptly flagged as a potential hybrid, whereas a Lesser Black-backed Gull that's paler or darker than surrounding Lesser Black-backeds is not a problem in itself. Differences observed in upperpart coloration should have a reasonable explanation. Whether it be lighting, bleaching, possible hybridization, or regional variation, we're charged with deciphering it.

Outer-primary patterns are in many ways the defining feature of large white-headed gulls. Although adults of two-year species develop some rather stunning wingtips, they don't show any remarkable variation. Variation in primary patterns is mostly relevant to three- and four-year species. Some of this variation is geographic, some is standard individual variation, and some may even be related to age. Differences related to the degree of pigment on the wingtip and primary patterns are of most importance. By "degree of pigment," we mean the extent of paleness or darkness. Iceland Gull shows a complete gradation from 0-pigmented, white wingtips to black wingtips. Much of this variation is believed to be geographic. Those wintering on the Pacific are generally darkest and assigned to *Larus glaucoides thayeri*, while those wintering off Greenland are palest and assigned to *Larus glaucoides glaucoides*. The most variable winter populations are Atlantic and Great Lakes birds, which display every shade in between the two extremes. Many of these birds are relegated to *Larus glaucoides kumlieni*. The amount of variation here is disturbing, because it is unmatched by any other gull species, and more so because it is not completely understood. Glaucous-winged Gulls also exhibit variably pigmented wingtips. The pigment found on some adults is concolorous with their gray upperparts, but many commonly show wingtips that are slightly darker than their upperparts, and a small percentage show wingtips that are paler. These differences are often chalked up to individual variation, at least by this author, but they're inconveniently found in several hybrid zones, creating potential for much confusion. Therefore, Glaucous-winged Gulls are routinely scrutinized for signs of hybridization, probably more than any other gull species. So long as other field marks are within range, we cautiously label them Glaucous-wingeds, or Glaucous-winged "types."

In addition to the degree of pigment found on wingtips, there is an array of appreciable primary-pattern differences at the species level. Comparing wingtip patterns from one adult to the next is rather amusing. But it is also fascinating to see how the wingtip on a particular individual can advance from one year to the next. Multiple case studies have demonstrated that adults regularly show less black on their wingtips as they age, taking on increasing regions of white (e.g., see Sauvage & Muusse 2013). This plasticity has also been recorded in mirrors, which generally increase in size with time.

Size reduction in mirrors is seldom recorded in any species, but it has been documented (Maarten van Kleinwee, pers. comm.). A study that measured mirror size on a group of adult Common Gulls from 1997 to 2007 found mirrors averaged largest when birds were in the middle of their reproductive age and actually decreased in size on older females (Sepp et al. 2017). Burgeoning data from known-age birds continues to broaden our understanding of just how variable adult wingtip patterns could be. Despite this variability, there are unifying patterns unique to most species.

Beyond wingtip patterns, there are other plumage features worth considering, such as head streaking in adult types. White-headed adults typically acquire maximum head and neck markings in basic plumage, which often coincides with the nonbreeding season. Some species display distinct patterns made of small, dark spots. Others have streaking that may be coarse or fine. Some appear blotted or densely smudged. Some show markings that run horizontally, others vertically. And some species show a combination of these patterns concentrated on various parts of the head and neck. California Gull, for example, maintains a rather unmarked foreneck and breast, with more of its head markings restricted to the sides and rear of the neck. A few species, such as Great Black-backed and Western Gulls, seldom show heavily marked heads, with the latter remaining largely white headed throughout the year. Incidentally, except in fall, when it can be quite heavily streaked dusky for a short while, Glaucous Gull is an interesting case, with adults in North America showing minimal if any head markings, while populations in Europe and Asia often acquire well-marked heads in basic plumage.

All in all, the patterns to these markings are quite consistent across whole populations of a species, but the extent is extremely variable from one individual to the next. At any given time of the year, especially in basic plumage, you can survey a handful of conspecifics that range from some being completely white headed to some having a pseudo-hooded appearance. An interesting question to ponder is whether these markings are consistent on the individual level from one year to the next. I recall an adult Kumlien's Gull that took up a small marina as its winter quarters in Hammond, Indiana, for seven consecutive years. It was a reliable bird that one was virtually guaranteed to see almost any day throughout the winter. The extent of its head markings was remarkably similar from one year to the next, and I personally know of several other cases that show head markings are fairly consistent on individual adults from one year to the next.

There is a phenomenon of hooded gulls taking on completely dark heads at the "wrong" time of year. Dark hoods are trademarks of alternate plumage, which often coincides with the breeding season. But occasionally, we encounter anomalous individuals with hoods at times of the year when they're not expected. To see a Bonaparte's Gull with a complete hood in December is always a brainteaser. There are some working theories for what may be happening with these birds. One idea is that the triggers for a hooded-head pattern may swap at a certain point in time. This implies these individuals are without a hood when the rest of the population normally has one. There is little evidence supporting this, however. Another theory is that these are birds in basic plumage that have produced head patterns identical to

63 Adult Laughing Gull with a complete hood in the boreal winter, likely the result of an early prealternate molt, as suggested by the whitish neck and breast. AMAR AYYASH. FLORIDA. JAN.

those found in alternate plumage. Finally, it is suggested that some individual birds may simply be running ahead of conventional molt schedules. This is almost certainly the case with Franklin's Gull. Franklin's undergoes a complete prealternate molt on the nonbreeding grounds in South America. A handful of adults with complete hoods begin to inch northward as early as January, surprising North American observers at first glance. Accelerated prealternate molts put these birds slightly ahead of schedule and at the top of their class. A similar phenomenon is observed with Laughing Gulls every winter at Daytona Beach Shores, Florida. Here, we find the largest wintering population of Laughing Gulls in North America: an estimated 75,000–100,000 individuals along a single stretch of beach. I have found that it's not unusual to see 1 in 2,000 adults with a complete hood as early as January. As is often the case, peculiarities become less peculiar with larger sample sizes. The best explanation for such birds is an early prealternate molt.

BARE PARTS

Bare parts are just as labile as plumage. I've noted that all North American gulls begin with dark eyes, dark bills, and dark to pinkish legs. Eye color is a prominent feature, and it's usually one of the first things that strikes us when examining a bird. All of our regularly occurring two-year gulls are dark eyed, and, incidentally, none has yellow legs. At the risk of overstating this truism, two-year species do not show much variation in this arena either. Knowing which three- and four-year species are dark eyed and which are pale eyed as adults will help in identification. Further, being familiar with those taxa that have variable eye color is of paramount importance. The species accounts highlight these differences with modifiers such as "somewhat variable," "variable," and "highly variable." Figure 2 is provided for quick reference.

Of those species that become unmistakably pale eyed as adults, allow for some variation with timing. If a third-cycle-type gull has all the characteristics of an American Herring Gull but shows darker eyes, then it's very likely just an American Herring Gull with delayed eye-color maturation. Conversely, you might encounter a one-year-old American Herring Gull that has already developed surprisingly pale eyes—all expected variation. Usually, the greater concern is finding with pale eyes a gull that is "supposed" to be dark eyed, such as Glaucous-winged or California Gull. California Gull, for instance, is a consistently dark-eyed gull, but on rare occasions a pale-eyed adult is reported. Whether this falls under variation or should be rightfully considered an aberration is up for debate.

I've mentioned that orbital rings are naturally altered throughout the seasons in a fairly predictable manner. Breeding condition results in vibrant colors, and nonbreeding condition often renders them dull. Be mindful of vague colors as the orbital ring transitions during nonbreeding condition to breeding condition and the reverse. This challenge is a theme every year, with observers hoping to turn Thayer's Gulls into Vega Gulls, for instance. The former has a pinkish orbital, while Vega's is a crimson red.

IDENTIFICATION 45

Figure 2. North American Gull Identification Chart.

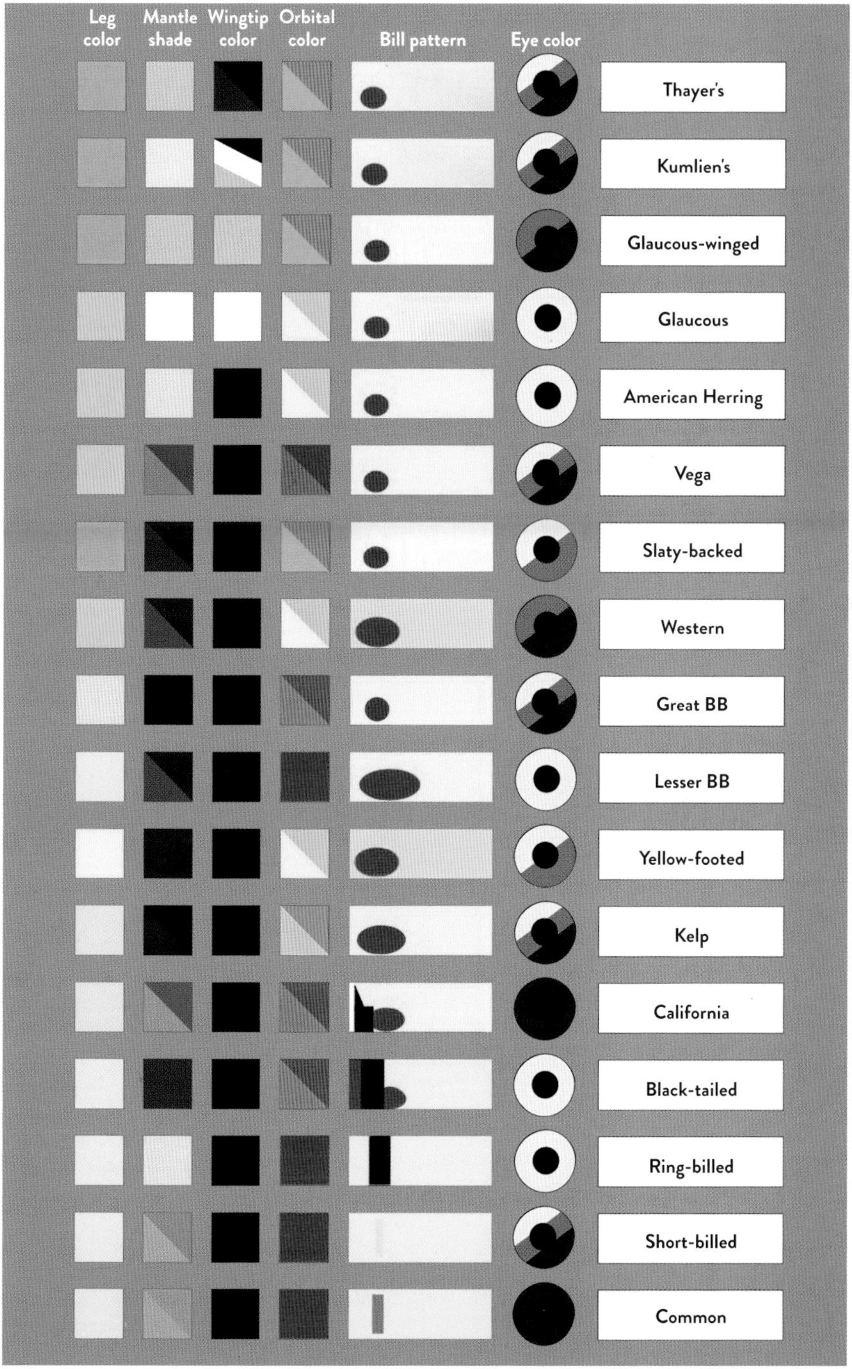

As these colors begin to wane or take form, they sometimes develop a deceiving hue, especially in less-than-ideal lighting. In such cases, it's better to not use orbital rings as support for an identification. Finally, some species can show more than one "standard" orbital color across a population. For example, American Herring Gulls are commonly seen with either yellow or fiery-orange orbital rings. These differences aren't necessarily regional or seasonal, as I've sometimes found them in neighboring adults in the same colony. But overall, there is good consistency in orbital-ring color at the species level, and it has proven to be a useful feature when used properly.

Bill patterns and bill color can be appreciably consistent. A good many Iceland, Lesser Black-backed, and Glaucous-winged Gulls retain mostly dark bills for much of their first cycle. To see one with a pale bill in early first cycle is relatively rare. Others, such as California and Glaucous Gulls, take on a pale bill with a sharply demarcated black tip early in their first cycle, and when they don't show this by mid-fall, we go to the drawing board. And then, of course, there are those species which defy consistency, such as American Herring Gull: a few first cycles in this species acquire bicolored bills at just a few months old, while others remain dark billed into their third cycle—and everything in between is also possible. More often than not, bill-color transformations in four-year species are gradual, with the most drastic changes occurring between the second and third cycles.

Leg color is another bare-part feature that requires careful inspection. The idea is to distinguish yellow-legged and pink-legged species. Yellow-legged species generally begin to show yellowing legs late in their second cycle, with considerable individual variation. California Gulls take on a sickly blue-gray color as their legs transition to yellow. Others acquire a dull straw color before becoming yellow. Beware older "yellow-legged" species with retarded leg color. A classic example of this is third-cycle-type Lesser Black-backed Gulls. Some have adult-like plumages but may retain pinkish legs. The expectation is for these birds to have mustard-yellow legs, but this isn't always the case, so approach them with caution before evoking a rarer species or a hybrid.

Historically, bird-identification literature put much emphasis on leg-color brightness in several species, inflating the use of this feature. It was widely believed by field observers that adult Iceland and Slaty-backed Gulls always showed bright, bubblegum-pink legs and that they were invariably brighter than other accompanying pink-legged species. But although these two do often show brighter legs in the nonbreeding season, it is not a rule by any means. Another pink-legged species, Glaucous-winged Gull, is known for having a dark purplish hue to its legs, but to see one without this coloration shouldn't be a deal-breaker. The takeaway is to not overvalue variable color features.

Uncommon but regularly occurring in gulls are individuals displaying unusually bright bare parts—so bright that they are borderline neon. The causes for hypervivid bare parts are not altogether clear, but the two suspected factors are diet and hormones, which may not be mutually exclusive. Carotenoids are responsible for many of the red, orange, and yellow pigments found in birds, and they're only obtained through feeding (Goodwin 1984). In a study on Lesser Black-backed Gulls, carotenoid supplementation in female diets resulted in a significant increase in bare-part color indices: legs, bills, and orbital rings became significantly brighter and more saturated on birds that were fed carotenoids (Blount et al. 2002). Interestingly, poultry farmers in parts of Asia commonly add carotenoids to chicken feed to influence these desired colors (Wang et al. 2023). At any rate, gulls with ultrabright bare parts often show perfectly normal plumages, further reinforcing the fact that bare-part color and plumage are not always in harmony.

An interesting phenomenon which we see fairly regularly is some pink-legged species showing yellowish legs as adults. Generally, these individuals have the correct field marks to identify them, except for their confusing leg color. Popular thought has it that these individuals may have some hormonal circumstances causing the legs to take on the "wrong" color. Western Gull, which has pink legs as an adult, is a good example of this. Every year, a few adults along the Pacific Coast are reported with the wrong leg color, particularly in early spring, when in high breeding condition. Often, the bills on these birds are also super vibrant and extra bright. These individuals are rarely, if ever, reported in the nonbreeding season, so they may indeed be experiencing a hormonal episode associated with breeding.

64 It's not rare to find adult Western Gulls with yellow legs like this, particularly at the onset of the breeding season. Whether this color change is due to diet, hormones, or some other factor is not fully understood. VIVEK KHANZODE. CALIFORNIA. APRIL.
65 3rd-cycle-type Lesser Black-backed Gull with pinkish legs at an age when yellow legs are generally expected. The best explanation for this is a delay in bare-part maturation. Adults with this leg color are suspected hybrids with American Herring. AMAR AYYASH. FLORIDA. JAN.

It is also quite possible that this phenomenon of wrong leg color is due to an individual's inability to absorb carotenoids or enzymatic degradation of carotenoids (Wang et al. 2023). Another theory is that this aberration is caused by a rare allele. Quite a few Herring and Great Black-backed Gulls along the Eastern Seaboard display dull yellow legs year-round. I have knowledge of, and have personally recorded, several American Herring Gulls from Massachusetts and Florida which have shown yellowish legs for several consecutive years, giving credence to the allele theory. Still, I am unaware of any published studies on wrong leg color specifically in gulls.

A miscellaneous note related to feet: they're generally concolorous with the legs, but there are instances in which yellow-legged species show obvious pinkish tones concentrated around the feet, especially on the underside. You may ask, Who looks at the underside of a gull's feet? My first experience with Yellow-footed Gulls is a memorable one, standing beside the godfather of California birding, Guy McCaskie, at the Salton Sea. Near the end of a wretchedly long and hot August day, we settled in on a massive gull flock and were determined to find a Western Gull for our day list. As gulls would fly in and fly off, we mused over and over at the number of adult Yellow-footeds with pink-bottomed feet. I found it slightly ironic, as this is the only gull whose English name commits to foot color.

MOLT

Variation in the timing of molt is quite evident across different age groups. Younger ages begin their prebasic molts ahead of adults. A one-year-old Bonaparte's Gull, for instance, could begin replacing primaries at the end of May, whereas an adult might not start until July, once a portion of its breeding duties have been fulfilled. There are direct and indirect costs associated with molt, which is quite taxing to carry out while breeding. Some adults will begin flight feather molt and then entirely pause it until the end of the breeding season. As a result, it's not unusual to see adults finishing their flight feather molt very late in the fall in the northern hemisphere. A gull growing flight feathers beyond this period, however, can be an indication of something more unusual. Vega Gulls come to mind, as they commonly molt later than most North American gull taxa, including American Herring Gulls. In fact, one of the supporting field marks for identifying adult Vegas in winter is late primary molt.

In other instances, prebasic molts may be drawn out or protracted, due to some unusual circumstance. There are years when numerous adult-type Black-legged Kittiwakes along the Pacific Coast are documented with retarded flight feather molt. These molts extend through the winter and into the following spring season, presumably taking almost an entire year to complete! It's suspected that a combination of climate- or weather-related variables may reduce their energy intake, and this inevitably slows down flight feather replacement (Howell & Corben 2000b). This seems, however, to be an annual

66 A mixed flock of Great Black-backed, Lesser Black-backed, and American Herrings. The northern summer provides an excellent opportunity to learn molt as our gulls are at the height of their prebasic molts. AMAR AYYASH. MASSACHUSETTS. JULY.

67 Adult Black-legged Kittiwake with two retained outermost primaries and outer secondaries in Jan. Although found regularly on the Pacific, suspended (or protracted) prebasic molts are seldom seen in the Atlantic subspecies. ALEX LAMOREAUX. MAINE. JAN.

occurrence, and it's not rare to find beached kittiwakes along the Pacific that are still molting outer primaries in midwinter. Many large gulls have protracted prealternate molts that carry over through winter, albeit at a slow and subtle pace. Some large gulls systematically pause their prealternate molts over much of the winter and then resume later in the season as migration and breeding approach. Observation of these differences is illuminating, and it increases our awareness of why a gull looks the way it does at a certain time of year.

Recall that the onset of a prebasic molt signals a new plumage cycle, which in effect is indicated by the start of primary molt. This approach of using the shedding of p1 to signal the start of the next plumage cycle is without question binary. From an observer's perspective, a gull is in either its first cycle or its second cycle, its second cycle or its third cycle, and so on. Consequently, a pressing question arises: If the difference between being in one plumage cycle and being in the next is determined simply by the presence or absence of p1, then how does one distinguish between two same-aged birds that are in different plumage cycles and two birds that are a year apart in age in different plumage cycles? The simple answer is: examine the open wing. Of course, this may not always be possible, but more times than not it is, particularly with gulls. Suppose we have a report of two Short-billed Gulls in June, a first cycle and a second cycle. Without photographic support or accompanying notes, there isn't a sure way to know whether both birds hatched in the same year, with the "second-cycle" individual having just very recently graduated to its second plumage cycle (i.e., the second prebasic molt has been initiated), or whether they're legitimately one year apart in age, and neither has begun its prebasic molt. Another remote possibility is the first cycle being a recently hatched juvenile and the second cycle being days away from beginning its third prebasic molt. That would put them at two years apart in age! It is this reason you'll find some people who are versed in plumage cycles specifying any active molt that they observe.

This can be as simple as noting "new primaries growing in" or "some secondaries dropped." Or a more detailed note, such as "active second prebasic molt," could be given. If no visible molt is observed, we can specify the apparent plumage seen at that point in time—for example, "first alternate" or "third basic."

There is a novel system in use to deal with these nuances, known as the "cycle-based aging system" or Wolfe-Ryder-Pyle, shortened to WRP (Wolfe et al. 2010). It's based on a three-letter code, with the first letter describing the current cycle, the second indicating active molt or not molting, and the third representing observed plumage. The system is popular with students of molt and bird-banders, and it has practicable application with gulls.

Finally, it is not uncommon for extensive and complete prealternate molts that include rectrices and remiges to begin with the tail feathers and secondaries, which are then followed by the primaries (see plates 12.23 and 25.23). This is quite different from the typical prebasic molt sequence, which commences with the shedding of p1. We find this in some populations of various species, such as Franklin's and Yellow-footed, and in some Lesser Black-backed and Kelp Gulls (Adriaens et al. 2023).

RELATIVE SIZE COMPARISONS

I had always known Sabine's Gull was a small gull, but for a long time my experience with the species was limited to peering at distant singletons as they zipped by on lake watches. I hadn't fully realized their diminutive size until observing tens of them alongside other species from the stern of a boat. Given that size is one of the first things we process (or at least should process) when looking at a bird, having an accurate sense of "true" size is critical. No matter how many measurements we commit to memory or read in a field guide, until we've seen a species in multiple scenarios, our perception of its size is mostly an abstraction. Optics have a big hand in this, and the result of stepping back from time to time and judging a bird's size with the naked eye is often surprising.

An effective way to absorb size is by direct comparison. Seeing a gull with a petrel or tern—even a raptor—not only teaches size but illustrates flight style and reinforces shape. Plates S.1–S.30 attempt to showcase sizes of select species. It helps, of course, to be familiar with at least one of the species in each image. These side-by-side comparisons should help you place the taxa presented in this guide into three broad categories: small, medium, and large.

S1 1st-cycle Bonaparte's with adult Short-billed and Ring-billed Gulls. GAIL WEST. CALIFORNIA. NOV.
S2 Adult Ring-billed and Bonaparte's Gull. CHRISSY MCCLARREN. MISSOURI. NOV.
S3 Adult California, American Herring, and Ring-billed Gull. KEITH CARLSON. WASHINGTON. DEC.

S4 Adult Bonaparte's Gull and Parasitic Jaeger. JEFF DYCK. MANITOBA. JUNE.
S5 Adult Little Gull and Black-legged Kittiwake. LENART JEPPSSON. SWEDEN. DEC.
S6 1st-cycle Heermann's and Common Raven. AMAR AYYASH. CALIFORNIA. JAN.
S7 Adult Common Gull (right of center) with Black-headed Gulls and Kumlien's Gulls. BRUCE MACTAVISH. NEWFOUNDLAND. JAN.
S8 Adult Glaucous Gull with Black-legged Kittiwakes. CAROLINE LAMBERT. ALASKA. JUNE.
S9 Adult Ivory Gull and Glaucous Gull. BRUCE MACTAVISH. NEWFOUNDLAND. FEB.
S10 Adult Glaucous-winged Gull and Thayer's Gull (right). LIAM SINGH. BRITISH COLUMBIA. MARCH.

S11 Adult Thayer's Gull (left) and Lesser Black-backed Gull with American Herring (background). AMAR AYYASH. ILLINOIS. DEC.
S12 Adult American Herring and Laughing Gull. AMAR AYYASH. FLORIDA. JAN.
S13 3rd-cycle Black-tailed Gull (underexposed) with adult Bonaparte's and 1st-cycle Ring-billed. CHRISSY MCCLARREN. ILLINOIS. JAN.
S14 Adult Slaty-backed Gull, 1st-cycle Glaucous Gull and adult Glaucous-winged Gull. TYLER HOAR. JAPAN. FEB.
S15 Adult Great Black-backed Gull with Canada Geese. CHARMAINE ANDERSON. ONTARIO. FEB.
S16 Adult Great Black-backed Gull with Bald Eagle. JIM TAROLLI. NEW YORK. FEB.
S17 Adult Yellow-footed Gull (center) with four California Gulls and White Pelicans. MARK CHAPPELL. CALIFORNIA. OCT.

S18 Adult Laughing Gulls in pursuit of Sandwich Tern. PETER BRANNON. FLORIDA. NOV.
S19 Adult Heermann's and Western Gull. BYRON CHIN. CALIFORNIA. JULY.
S20 Adult Swallow-tailed Gull with California and Heermann's Gulls. DAVE BEEKE. WASHINGTON. SEPT.
S21 Adult Franklin's Gull (center) with California Gulls and Ring-billed (lower left). ALAN KNOWLES. ALBERTA. MARCH.
S22 1st-cycle *sibiricus* Black-headed Gull (center) with 1st-cycle Franklin's Gull (right) and Bonaparte's Gulls. AARON MALIZLISH. CALIFORNIA. DEC.
S23 Juvenile Sabine's Gulls with Red-necked Phalaropes. JOEL BEYER. UTAH. SEPT.
S24 Adult Bonaparte's Gulls with 1st-cycle Great Black-backed Gull. AMAR AYYASH. MASSACHUSETTS. APRIL.

S25 1st-cycle Glaucous and Lesser Black-backed Gull. RICHARD SMITH. NEWFOUNDLAND. FEB.
S26 1st-cycle Western and California Gull (right). ALEX A. ABELA. CALIFORNIA. AUG.
S27 1st-cycle Laughing Gull (left of center) with three Parasitic Jaegers and Common Tern. STEVE ARENA. MASSACHUSETTS. OCT.
S28 2nd-cycle Great Black-backed Gull with Long-tailed Jaeger. BRANDON HOLDEN. ONTARIO. SEPT.
S29 1st-cycle Kumlien's Gull with Northern Fulmar. RONNIE D'ENTREMONT. NOVA SCOTIA. JAN.
S30 Juvenile American Herring Gull with Bald Eagle. JASON DAIN. NOVA SCOTIA. OCT.

ABERRATIONS

As ubiquitous as gulls are, we often notice anomalies in them so slight that they wouldn't be spotted on smaller birds or those with more elusive lifestyles. An interesting point to ponder is that the more variable a species' diet is, the more likely aberrations will be found in that species. For instance, there is data that suggests urban bird populations more commonly exhibit leucism than their rural counterparts, but it's not clear whether this disparity is due to an observer bias (Rollins 1953). An example of this may be the relatively high frequency with which white patches are recorded on the primary coverts of adult Heermann's Gulls. Is this a result of more people watching them intently combined with their highly contrasting plumages? Or is it a deep-seated genetic mutation within the species? Plates AB.1–AB.15 show common aberrations and abnormalities that are regularly reported in gulls. Some have the potential to present confusion, such as an unexpected dark eye on a pale-eyed species, while others are simply intriguing oddities. Much rarer oddities, such as a gull having an extra primary on one wing, or an extra tail feather, have been excluded from this section, but the reader should be aware of such quirks.

AB1 Melanistic Laughing Gull of unknown age. Melanism is rare in gulls and birds in general. More common is to see various tracts with extra melanin (see 7.13 and 8.11). Recalls Lava Gull, which averages a thicker and more blunt-tipped bill, without an obvious white throat. TROY HIBBITTS. TEXAS. JUNE.

AB2 Leucistic Yellow-footed Gull, identified by leg color, bill and body size, and location. Leucism is a generic term that's more or less useless when trying to understand pigment deficiencies, but it is a term widely understood by the birding and ornithological community. Leucism is by far the most common color aberration we find in gulls. NEIL CLARK. MEXICO. MAY.

AB3 Leucistic Short-billed Gull. A small white gull that immediately evokes thoughts of Ivory Gull, but bill size and color are important here. Leucistic individuals often have normal bare-part colors. MARIO BALITBIT. CALIFORNIA. NOV.

AB4 Dilute Laughing Gull with a lesser form of albinism throughout the upperparts. AMAR AYYASH. FLORIDA. JAN.

AB5 Calico Ring-billed Gull. This aberration is recorded annually in one or two Ring-billed Gulls. Interestingly, many have grayish underparts and a black bill with a yellow tip. TIM REEVES. NEW MEXICO. FEB.

AB6 Alternate Franklin's Gull. The intense pink suffusion on the body is not an aberration and is regularly found in smaller gulls, especially Franklin's and Ross's. This coloration is caused by a carotenoid known as astaxanthin. Carotenoids must be eaten; the effects are brought on only by diet. JULIAN HOUGH. TEXAS. APRIL.

AB7 1st-cycle Western Gull with broad stress bar across the primaries, greater coverts, and tail feathers. It has been suggested that this aberration, loosely known as "Willet Wing," is more common in years of poor food availability. STEVE HAMPTON. CALIFORNIA. OCT.

AB8 Adult American Herring Gull with heterochromia. Two dark eyes are very rare in American Herring Gulls. It's much more common to find one dark and one pale eye. Some reasons for an unexpected dark eye include injury, a genetic mutation, or possibly avian flu infection. AMAR AYYASH. MICHIGAN. OCT.

AB9 1st-cycle Ivory Gull with a handful of postocular lice spots. Seen somewhat regularly, especially on young gulls; the obligate ectoparasites feed on the feathers, skin, and blood of their host. A large lice load can be an indication that a bird's health is in peril. DARREN CLARK. MONTANA. FEB.

AB10 Adult Lesser Black-backed Gull with a sublingual fistula, brought on when a bird's tongue protrudes through the floor cavity of the mouth. Recorded in at least 17 gull species. Does not appear to be detrimental to survival in most cases. AMAR AYYASH. ILLINOIS. OCT.

AB11 1st-cycle Glaucous-winged type with bill color suggesting avian keratin disorder (AKD). The elongated hooked tip is a separate aberration found mostly in 1st cycles, sometimes with "crossbill" pattern. DONALD PENDLETON. CALIFORNIA. MARCH.

AB12 Adult Laughing Gull with bare-part depigmentation. Particularly frequent in this species, resulting in variable orange on the bill and legs, with otherwise ordinary plumage. AMAR AYYASH. FLORIDA. JAN.

AB13 1st-cycle American Herring Gull with depigmentation restricted to the bill. Normal leg color. AMAR AYYASH. ILLINOIS. DEC.

AB14 Oiled 3rd-cycle-type American Herring Gull. Caused by external chemicals and oils in the water, oiled feathers appear matted, glazed, and untidy. Oiling often impedes flight and the ability to keep the body insulated. AMAR AYYASH. WISCONSIN. JULY.

AB15 Soiled, adult-type Glaucous Gull found in a Russian coal-mining settlement. Likely exposed to coal dust and oiled. Soiling is easily mistaken for melanism. This bird would likely never be confidently identified out of context, but note the short wing projection with white tips, orange orbital ring, and evenly proportioned bill. NOAH STRYCKER. SVALBARD. JUNE.

SPECIES ACCOUNTS

The species accounts are systematically arranged as follows:

Header: Each species account is numbered, with the common and scientific names, followed by body (L) and wingspan (W) lengths. I have found some of the measurements given here in collections across North America, but the overwhelming majority were recorded by Olsen and Larsson (2004). Also included is the number of molt cycles typically required to acquire adult plumage, the accepted molt strategy— Simple Basic Strategy (SBS), Simple Alternate Strategy (SAS), or Complex Alternate Strategy (CAS)— and a range of KGS values.

Overview: A general overview is given, with information ranging from natural history, feeding habits, behavior and nesting preferences to population estimates and conservation concerns for the species.

Taxonomy: Historic and current notes on taxonomy are provided, especially when relevant to modern classifications. Some simply state, "Monotypic," while others go into more detail. If more than one subspecies is recognized, pertinent comments are made on distinguishing features and geographic distributions.

Range: A summary of breeding and nonbreeding ranges, with notes on known migration timings and peak movements, is given for each species. Seasons are sometimes used to generalize the time of year, and these should be understood as seasons of the northern hemisphere unless otherwise noted. General range maps are provided for most taxa but typically don't reflect isolated sites or occurrences where a species isn't regularly expected. Red indicates breeding range; blue, nonbreeding range; yellow, migration when this is widespread or noteworthy; and green, regions where a species can be found year-round, which encompasses breeding.

Identification: These sections begin by describing adults and then discuss first cycles. These are the only ages given for two-cycle species. Second cycle is described for three-cycle species, and third cycle is described for four-cycle species. Presumptions on subadult characteristics leading to definitive plumages are sometimes provided, through photographic examples when noteworthy. These can be generalized by light markings on the primary coverts or tail, more extensive pigment on the wingtip, brown wash on the coverts, dusky blemishes on the wing linings, and/or suspiciously delayed bare parts. Ultimately, it is impossible to know the age of such birds without life-history data.

Similar Species: Here, the reader is given a brief overview of differences to consider when those field marks are noteworthy. These sections are not meant to be exhaustive; they aim to highlight overarching differences that can be supported by averages. It is often difficult to describe some features on paper, especially in four-year gulls, given the limitations of print publications and the variability in those features. In some sections, I refer the reader to a different species, implying that more details are given in that taxon's account. These sections should be read with reference to the provided photographs.

Molt: The molt section provides a general description of any recognized molt strategy. This information is mostly from birds of unknown origin (i.e., those without leg bands) and hence should be viewed as tentative. Much of this data is in agreement with Pyle (2008), Howell and Dunn (2007), and Howell (2010) and with data compiled on the Gull Research Organisation (n.d.) website and is from personal observations from both the field and museum collections. The summaries focus on adults and first cycles, as these ages are requisites to understanding the bigger picture. Second- and third-cycle individuals in this guide are assumed to be those ages based on overall patterns that match known-age birds.

Hybrids: Information is given on known and / or suspected hybridization in the wild, along with relevant notes on hybridization in captivity. The overwhelming majority of suspected hybrid gulls are of unknown origin and as such should be considered putative. To say that our knowledge of hybrid gulls is in its "developing" stages is an understatement. Some hybrid populations, such as the Glaucous-winged × Western hybrids (so-called Olympic Gulls) of the Pacific Northwest, are much better known than others.

Photographic Plates: Each species account provides a series of photographs, beginning with first-cycle and progressing to adult plumages. Plates are referenced beginning with the species account number. For example, plate 29.7 is plate 7 in species account 29 (i.e., Slaty-backed Gull). Where subspecies have their own accounts, a letter is also found. For instance, 32b.5 is plate 5 in species account 32b (i.e., Kumlien's Iceland Gull). Care has been taken to ensure images give an accurate portrayal of upperpart coloration, but despite this, it is impossible to reproduce true-to-life colors and shades in every instance. The images are captioned beginning with molt cycle, followed by relevant notes on plumage and identification. The photographer, general location, and month are also listed.

SECTION 1

SMALL TERN-LIKE & HOODED GULLS

1 SABINE'S GULL *Xema sabini*

L: 12.5"–14.0" (32–36 cm) | W: 33.5"–35.5" (85–90 cm) | Two-cycle | CAS | KGS: 7–9

OVERVIEW Arguably the most elegant gull species in the world, with adults having a dazzling geometry of black, gray, and white on the upperwing. This Holarctic breeder is highly pelagic in the nonbreeding season, with the longest migration of all gulls. Flight is powerful but graceful, with frequent dives. Feeding habits are much more similar to those of terns and phalaropes than gull-like. Often found feeding in locations where phalaropes concentrate, forming concentric circles on the water's surface while foraging. Walks with a slight waddle along the water's edge and on mudflats. Associates with Arctic Terns in migration and prefers to nest near that species in the tundra zone. Like terns, males carry prey items to females in bill (not regurgitated, as in many other gulls). Sabine's is often kleptoparasited by jaegers. The genus name, *Xema*, is apparently devoid of meaning, with no etymology (Jobling 2010). The future English name for this species must be held to the highest standard.

Trans-equatorial migration, mainly oceanic to the Pacific Coast of South America and Atlantic Coast of South Africa. Large numbers are attracted to the cool, rich waters of the Humboldt Current in W South America and of the Benguela Current off W South Africa. Presumably a first in avian research, members of a nesting pair of Sabine's Gulls fitted with geotrackers in the Canadian Arctic showed "divergent migratory pathways," migrating to different continents, wintering in different oceans, and then reuniting to nest together again in the northern summer (Davis et al. 2016). Apparent longevity record for the species is an individual banded as a chick in 1999 on South Hampton Island which was found breeding at the same site as a 24-year-old adult in July 2023 (Lain Stenhouse, pers. comm.).

TAXONOMY Four races have been considered (Portenko 1939; Burger & Gochfeld 1996). Smallest and palest birds from N Alaska to Greenland said to be nominate *Xema sabini sabini*. Larger and darkest birds in NE Siberia and W Alaska are *X.s. woznesenskii*. Birds from Spitsbergen and east to the Taimyr Peninsula are *X.s. palaearctica* (but sometimes placed with nominate). More study of fresh specimens and larger samples from purported *X.s. tschuktschorum* populations on Chukotskiy Peninsula needed. Some authorities regard differences clinal and maintain Sabine's is monotypic (Cramp & Simmons 1983; Harrison et al. 2021). Interestingly, Swallow-tailed Gull is not closely related to Sabine's, despite their superficial resemblances. Instead, Ivory Gull is believed to be sister taxon to Sabine's (Pons et al. 2005).

RANGE
Breeding: Holarctic breeder, from NE Russia east to Svalbard. In North America, nests in small colonies in marshy, low-lying tundra with pools and ponds, and rarely on barrier islands, coastally from W Alaska eastward to the high Arctic of Nunavut, as well as locally in central and N Greenland. Grass nest often placed in moss or wet ground close to water's edge.
Nonbreeding: Mostly pelagic during nonbreeding season. Seen rarely from or on shore along Pacific and Atlantic Coast. Southbound migrants well south of the breeding

---

1 Juvenile. Scaly upperparts with pale edging, dark hindneck, and fleshy legs. BRIAN C. JOHNSON. ARIZONA. SEPT.
2 Juvenile. Pale edging to upperparts showing slight wear, not as boldly marked as 1.1. DARREN CLARK. IDAHO. OCT.
3 Juvenile with similar-aged Bonaparte's Gull (left). Sabine's averages a smidgen smaller. Note thin white secondary skirt. DAVID TURGEON. QUEBEC. SEPT.
4 Juvenile. Pale edging to upperparts largely worn away, although retaining dark hindneck. RYAN O'DONNELL. UTAH. NOV.
5 Some gray formative feathers on scapulars now, with adult Bonaparte's. JAMES PAWLICKI. NEW YORK. NOV.
6 1st cycle showing moderate wear on the upperparts. More gray has grown in on the mantle and upper scapulars. Note petite bill, thin white secondary skirt, and white edges to inner webs of outer primaries. REINHARD GEISLER. FLORIDA. DEC.
7 1st cycle. Now showing much adult-like gray on scapulars and some upperwing coverts (replaced via preformative molt). Visible primaries are still juvenile. ROGER AHLMAN. ECUADOR. FEB.
8 1st cycle. Black bill pattern and retained juvenile outermost primaries. Much of the upperparts are adult-like now (formative), with visible signs of primary molt. White head with black hindneck and gray wash (see also 1.15–1.16). RICHARD BONSER. CHILE. APRIL.

range are occasionally common off the Pacific Coast, peaking between late August and early October, but small numbers of birds appear as early as late July, and a few may linger well into November, exceptionally through early December. Recent tracking of Pacific Coast migrants found southbound birds staged in the Northern California Current off British Columbia to Oregon for approximately one month before continuing to migrate offshore to wintering areas in Peru. These Alaskan birds were found to migrate coastally or traversing the state. Northbound migrants were found to stage in Baja California and then again off Washington to British Columbia. Some cross mainland Alaska to reach arctic Canada, while others cross the Rockies toward Hudson Bay and then proceed north (Gutowsky et al. 2021). In the NW Atlantic, very small numbers occur fairly regularly, during late August and September, south to the waters off Labrador and along the lower St. Lawrence estuary in Quebec (and then southeastward across the Atlantic to the waters off SW Africa). In the W Atlantic south of these areas, as well as onshore, Sabine's Gulls are very rare to casual. A small number of southbound migrants are also widespread and annual throughout much of the North American interior, with most birds found around the Great Lakes, Upper Midwest, across the Great Plains, and throughout the interior West. Peak occurrence is between early or mid-September and early October, but a handful of birds are seen regularly well into November, including at such sites as the lower Great Lakes and Niagara River.

In spring, most migrants off the North American Pacific Coast are seen between late April and late May, primarily on pelagic trips. Smaller numbers of early transients can be seen as early as mid- to late March as far north as Washington, at least during abnormally warm water conditions. A few laggards, mostly presumed nonbreeders, may still be passing north in early and mid-June. Away from the West Coast, this species is strictly very rare in spring (late April to early June) through the interior and along the East Coast. Found somewhat regularly in small numbers at larger lakes in the N Canadian Prairies, the Yukon, and locally in interior Alaska, where flocks have been recorded on many occasions, mostly in late May. Presumed nonbreeders are seen very rarely or casually off the Pacific Coast, well south of the breeding range, during the early summer.

Winter range is mainly in the Humboldt Current off W South America (mostly Peru, also Ecuador and northernmost Chile) and in the Benguela Current off SW Africa. However, small to moderate numbers of birds are found in that season in the Pacific North to off Central America and Mexico, as far north as off the southern tip of Baja California. Also, there is a midwinter record in the Gulf Stream off central Florida, as well as one of a bird that wintered on Lake Erie in Ohio.

Casual in W Pacific. Vagrants have been recorded at scattered sites throughout much of the world!

IDENTIFICATION

Adult: In alternate plumage, adults show a dark charcoal-gray head with a black necklace bordering the hood. Lacks white eye crescents seen in some other hooded gulls. Small black bill with distinctive yellow tip. Dark eyes. Red orbital and red gape. Blackish legs, becoming dull brown with grayish-red tones in nonbreeding condition. Upperwing pattern is tricolored with triangle pattern in black, gray, and white. In early spring (March–June), larger white apicals are usually found on the primaries, soon giving way to wear

---

9   Same individual as 1.2. Brownish-gray upperparts with bold white triangle pattern on wing. Thin black tailband. DARREN CLARK. IDAHO. OCT.
10 Tricolored pattern of black, brown, and white on upperwing. All-white secondaries and inner primaries. Note forked tail with thin black tailband. STEVE YOUNG. ENGLAND. SEPT.
11 Juvenile. Underwing shows white triangle pattern with gray imprint of dark upperwing. Grayish-brown on face and side of neck. Forked tail less apparent here. CHRISSY MCCLARREN. MISSOURI. OCT.
12 Juvenile with similar-aged Laughing Gull. STEVE ARENA. MASSACHUSETTS. SEPT.
13 A late southbound migrant undergoing preformative molt. Note gray scapulars and several inner-wing coverts that have grown in. JAMES RIEMAN. TEXAS. DEC.
14 1st cycle. Whiter head and black hindcollar. All juvenile scapulars and some wing coverts replaced (formative). Active inner-primary molt evident. KEN BEHRENS. SOUTH AFRICA. JAN.
15 Same individual as 1.8. The preformative molt in Sabine's Gull is a complete molt. Most tail feathers, some secondaries, and p1–p6 are new. RICHARD BONSER. CHILE. APRIL.
16 Same individual as 1.8 and 1.15. Note basic adult-like head pattern. Most secondaries juvenile, as well as the three outermost primaries (p8–p10). RICHARD BONSER. CHILE. APRIL.

throughout the breeding season. All-white tail with forked shape, appreciably more obvious when tail is not fully spread. All-white underparts. Underwing shows imprint of the upperwing's triangles (seen best when backlit or in overcast conditions). Adults begin head molt on southbound migration, showing blotchy white around bill base and forehead, becoming largely white with irregular black hindcollar on the wintering grounds (Oct.–Feb.). Light gray wash may develop on lower neck in basic plumage. Legs at this time more likely to become duller black. *Similar Species:* Most likely to be confused with young Black-legged Kittiwake in flight, but the upperwing pattern is strikingly different on Sabine's and should be unmistakable upon close inspection. Black-legged Kittiwake is a larger bird and has an M-pattern with black carpal bar extending onto the inner wing coverts. At rest, compare to other hooded species. Sabine's is roughly the size of Bonaparte's Gull but without eye crescents and shows yellow on bill tip. Outer primaries are all black and lack the white wedge seen on the open wing of Bonaparte's. If you're lucky enough to be confronted with the question of whether you're looking at a Swallow-tailed or a Sabine's Gull, note that Swallow-tailed is considerably larger with a longer and more pronounced fork to the central tail. The outer wing on Swallow-tailed shows more white and less black at the wrist, primary coverts, and mid-primaries. *1st Cycle:* Juveniles have a chalky gray-brown aspect to the upperparts with pale, scaly feather edges when fresh. This dusky pigment carries on to the hindneck, sides of the head, and crown, while the face

17 Age uncertain. Brownish outer primaries and primary coverts, head pattern, blackish bill tip, and remnants of what appear to be brown juvenile wing coverts suggest 1st cycle that replaced all flight feathers on the nonbreeding grounds in complete preformative molt. MARTIN D. PARR. ENGLAND. JULY.
18 Age uncertain. Likely 2nd cycle with suspended primary molt (p5/p6). Reduced black on head, leg color, brownish outermost primaries, and absence of yellow bill tip point away from adult. STEVE ROTTENBORN. CALIFORNIA. AUG.
19 Dull yellow bill tip, frayed white tips to primaries, and incomplete hood suggest 2nd cycle, but ruling out 1st cycle not possible. Note fleshy legs, unlike black legs of most adults. THOMAS A. BENSON. CALIFORNIA. OCT.
20 2nd-cycle type. White tips on primaries not as prominent as in adult, although in part due to wear. Subterminal band on p5 and black on base of p4 suggest subadult. CATHY SHEETER. COLORADO. SEPT.

is largely white. Folded primaries usually show thin pale crescents on their tips. In flight, the pattern to the upperwing is similar to that in adults, but the back and upperwing coverts are without the adult's dark gray (this may be difficult to discern at long distances). Juveniles possess a narrow black tailband, which also may go unnoticed at long distances. Black on outer web of outer tail feathers usually doesn't reach outer edges. Bill is all black and legs dull pink. Transition to formative plumage occurs mainly on wintering grounds, but some upperpart, head, and neck feathers may be replaced north of the equator during migration. On rare occasions, some individuals found in North America in winter may show gray formative backs, white heads, and thick black hindcollars. At this time, upperwing coverts are highly susceptible to wear. One-year-old birds in the spring typically acquire a mottled, patchy hood, but some may be all black and adult-like (more study needed). The bill tip may be dark or a duller yellow on such individuals. Both p4 and p5, especially the latter, can show irregular black pattern. The legs become darker at this age, but some still rather pinkish gray. *Similar Species:* Compare with 1st-cycle Bonparte's, Little and Franklin's Gull. The entire inner wing on Sabine's is uniformly brown, unlike Little Gull's defined carpal bar forming a bold, contrasting M-pattern. Also note the entirely white midpanel to the trailing edge on Sabine's Gull. See also juvenile Ross's Gull, especially when swimming.

21 Subadult type. Dark, slate-colored hood. Dull yellow bill tip, diffuse black, dull orbital, diffuse black necklace, and white head spotting suggest subadult, or an adult in low breeding condition commencing prebasic molt. MIKE DANZENBAKER. ALASKA. JULY.
22 Adult in alternate plumage. Bright yellow bill tip, vivid red orbital ring, complete hood, and well-defined black necklace. IAN DAVIES. ALASKA. JULY.
23 Alternate adult. Unmistakable. Note bold white tips to primaries, slate-colored hood, red orbital, and bright yellow bill tip. IAN DAVIES. ALASKA. JUNE.
24 Adult. Banded as a chick in 1999 on Southampton Island and found again at the same site 24 years later as a breeding adult, establishing an apparent new longevity record for the species. BRENDAN KELLY. NUNAVUT. JULY.

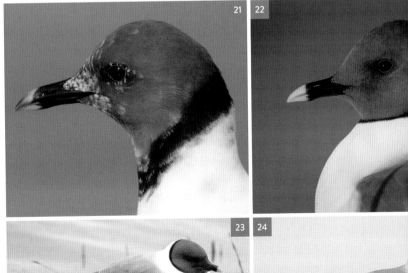

1 SABINE'S GULL

MOLT Many Sabine's seen in the ABA Area are either largely in juvenile plumage or adults mostly in alternate plumage. Molt strategy is unique among gulls, paralleling that of Arctic Tern and Long-tailed Jaeger. Preformative molt is complete in most birds. First prealternate molt is limited (or absent; more study needed). Flight feathers are replaced on the wintering grounds, typically from December onward (as early as late Sept. in a few early-molting individuals). End of adult prebasic molt overlaps with prealternate molt on the wintering grounds, to the extent that outermost primaries and innermost secondaries may still be in last stages of growth while head and neck feathers are becoming alternate. There appears to be no consistent method for aging many 2nd-cycle-type individuals, although an irregular subterminal band on p5 with some pigment on the base of p4 may indicate 2nd cycle (more study needed). A black bill tip, or a dull yellow tip with diffuse black, is found in some late 1st cycles, which may support age. These same individuals may show legs that are dull brown and pinkish gray (as in some adults in nonbreeding condition). It is believed individuals with prominent white tips to the primary coverts are adults, although this has not been confirmed (see plates 1.20 and 1.32). Furthermore, wear and variation obscure the usefulness of this feature, so it should be used cautiously when aging.

HYBRIDS None.

25 Alternate adults. Unmistakable. Distinctive red gape (right), thin white secondary skirt (left), black legs, and thin black necklace. SHAILESH PINTO. ALASKA. JUNE.

26 Adult. First obvious indication of prebasic molt is patchy head pattern. Worn tips to primaries common on southbound migrants. Flight feather molt has not begun, however, and typically commences on the "wintering" grounds in the southern hemisphere. RYAN SANDERSON. INDIANA. OCT.

27 Adults. Still with bright yellow bill tip, soon to become duller. Basic head pattern often with patchy blotches and dark hindcollar. Scapulars and some wing coverts already replaced (basic). XABIER PUMARINO. SPAIN. SEPT.

28 Immaculate adult displaying bold black, gray, and white upperwing pattern. Forked tail. Note unmarked p5 tip and white slivers on inner webs of outer primaries. PETER FLOOD. MADEIRA. MAY.

29 The prominent white tips on the primary coverts are thought to be an adult feature; however, see 1.17. KIRSTEN SNYDER. ALASKA. JUNE.

30 Adult underwing. Note gray shadow bar above secondaries, longer outer tail feathers, black legs, and bright yellow bill tip. CAMERON RUTT. ALASKA. JUNE.

31 Adult beginning head molt on southbound migration (prebasic molt). White primary tips are completely worn down. Obvious triangle pattern seen on underwing in favorable lighting. JAY MCGOWAN. NEW YORK. NOV.

32 Age uncertain. Either 2nd cycle or adult type with late prealternate molt (thus the minimally marked head). Adult-like features include white tips to primaries and unmarked p5 tip. Subadult features include black bill tip and missing hood at this date. SEAN CROCKETT. CALIFORNIA. MAY.

33 Adult type. Flight feather molt is carried out on the wintering grounds, typically from Dec. forward. The prebasic molt here shows p1–p3 new, p4–p5 dropped, and p6–p10 retained. REGARD VAN DYK. SOUTH AFRICA. DEC.

34 Adults in alternate plumage with Western Sandpipers. The upperwing pattern is an eye-catching geometry of black, gray, and white. LIAM SINGH. ALASKA. JUNE.

35 Three adult Sabine's, four adult Bonaparte's Gulls, and a bonus gull (far right). Leg color is helpful here, although more useful is the underside to the outer primaries. JAY MCGOWAN. MANITOBA. JUNE.

2 SWALLOW-TAILED GULL *Creagrus furcatus*

L: 22"–24" (56–61 cm) | W: 49"–55" (124–139 cm) | Two-cycle (?) | molt strategy (?) | KGS: 7.0–8.5

OVERVIEW Rightfully placed in its own genus, Swallow-tailed Gull is a unique gull with an unusual biology. Primarily endemic to the Galapagos Islands, where it favors the warmer waters of its eastern shores. Nests built on cliff ledges and in small caverns. A necessary requirement is that the nest be close to a boulder or crevice where young can take cover when needed. The species breeds asynchronously, any month of the year, with a breeding cycle of 9–10 months (Hailman 1964a; Snow & Snow 1967; Harris 1970). Lays a one-egg clutch. Mostly feeds at night; its extraordinarily large eyes are well adapted to low light. Much of the day is spent resting and on feather upkeep, with members of the colony leaving in unison at dusk with a loud, raucous departure (Hailman 1964b). Lumbered flight pattern. The species has a peculiar voice, which includes a variety of peeps, whistles, and rattles. Main prey items are squid and clupeid fish. *Creagrus*, meaning "meat hook," refers to its drooping bill with pointed tip, and *furcatus* is derived from *furca*, meaning "two-tined fork," a reference to its characteristic forked tail. Estimated population of 35,000 individuals (BirdLife International n.d.-a).

TAXONOMY Monotypic. Previously placed in *Larus*. Also formerly in *Xema*, with Sabine's Gull; it was assumed the two shared a close relationship based on morphology, but Pons et al. (2005) clarified this in their extensive molecular phylogeny.

RANGE
Breeding: Mostly restricted to the Galapagos Islands, where it may be found breeding year-round. Small breeding colony on Isla Malpelo, Colombia (250–300 individuals).
Nonbreeding: Away from the breeding grounds, almost always observed offshore, away from land. Disperses to the eastern tropical Pacific, to the waters of S Colombia, Ecuador, Peru, and south to central Chile. Occurs rarely and irregularly, especially during warm-water events (El Niño), north to Panama. As of this writing, Mexico has three records, all from late March to early April (Howell, forthcoming), and recently, found regularly off Cocos Island, Costa Rica (Arias 2023). Records from the United States are increasing, but it is still very rare, with six accepted records in California—all adults. One Washington record of a roaming adult: King County and Snohomish County (31 Aug.–10 Sept. 2017). The sighting in Snohomish County (Everett, Washington) is the northernmost occurrence for the species worldwide. Thus far, vagrants to the north have all been adult types in alternate/near-alternate plumage.

IDENTIFICATION
Adult: Medium gray upperparts with a charcoal-colored hood in alternate plumage. All-white tail with a deep fork, unmatched by any other gull. Vermillion orbital ring. Large, dark eyes with a distinctly bug-eyed quality. Legs coral red to pink. Relatively long black bill with a pearl-gray tip. Hooked, pointed tip

---

1. Juvenile. Scaly upperparts, black "panda-bear" eye patch, and brownish earspot. ANDREW TAYLOR. GALAPAGOS. SEPT.
2. Juvenile. Long black bill, white underparts, and short, fleshy legs. DARRYL RYAN. GALAPAGOS. MARCH.
3. A darker juvenile showing some wear. Black eye-patch distinctive. ANDREW W. JONES. GALAPAGOS. APRIL.
4. The genus name *Creagrus*, meaning "meat-hook," refers to the droopy, sharp bill tip. Some gray postjuvenile scapulars are beginning to emerge. RICHARD BONSER. GALAPAGOS. APRIL.
5. 1st cycles. Postjuvenile molt has replaced most scapulars and some upperwing coverts. White head and black eye-patch distinctive. GARTH V. RILEY. PERU. NOV.
6. 1st cycle with 2nd-cycle-type Sabine's Gull. Note the significantly larger body and bill of Swallow-tailed Gull. CHARLY MORENO. CHILE. AUG.
7. Juvenile. Black outer primaries with large white triangle pattern on mid-wing. Distinctive forked tail with thin tailband. DAVID MARQUES. GALAPAGOS. JULY.
8. A superficial resemblance to the upperwing pattern found on Sabine's Gull, but note the large white divide on the inner-primary coverts and alula. JAY MCGOWAN. GALAPAGOS. AUG.

gives the bill a droopy appearance. A white lore spot and a smaller pale spot adjacent to the base of the mandible are always present on hooded adults (plate 2.15). Often shows a velvety gray wash that fades between the dark hood and mantle. When eyes are closed, a pale patch of feathers simulates a "fake" eyespot. The long tail, long wings, and short tarsus give the species a tern-like stance in which the posterior half of the body often appears raised. The outer scapulars are fringed in white, seen on resting birds and in flight. Broad wing. The open wing has a spectacular pattern, with black outer primaries and a large gray triangle formed by the wing coverts and scapulars. No mirrors, but instead shows a piano-like pattern of black outer webs and white inner webs to the outermost primaries. A contrasting gray wash is found across the inner primaries and outer secondaries of some individuals. The outer-primary coverts are sometimes with black streaks. Basic adults are white headed with a dusky earspot and large black circle around the already-large eye. Newer primaries typically show larger white apicals. *Similar Species:* Sabine's Gull shares a superficial resemblance in upperwing pattern and dark hood in alternate plumage, but the two are, in reality, quite dissimilar. Swallow-tailed is significantly larger, with a pale lore spot and a pale bluish-gray (not yellow) tip to the bill. Black on the outer primaries on Swallow-tailed does not meet the gray upperwing coverts at the wrist. In flight, Swallow-tailed Gull's tail is longer and much more deeply forked. Also note the distinctive white fringes to the outer scapulars, which Sabine's never

9   1st cycle. White inner webs to the outer primaries, white triangle pattern on upperwing, and thin tailband. DEBORAH FORD. GALAPAGOS. JULY.
10 1st cycle. Note white region around alula (far wing) and, on this individual, dark outer webs on underside of secondaries. PETER ADRIAENS. PERU. NOV.
11 1st cycle. Obvious divot in center of tail and imprint of triangle pattern of upperwing. MICHAEL J. ANDERSEN. PERU. DEC.
12 1st cycle. Most scapulars replaced, postjuvenile. This individual shows dark marks on all 10 primaries. ALVARO JARAMILLO. CHILE. NOV.

shows. At a distance at sea, its slow, labored flight in combination with white appearance overall recalls a Great Egret.

1st Cycle: Juveniles are largely white bodied with uniform scaly brown upperparts with pale edging. Youngest birds show brownish smudging around lower neck and breast. White head has a disproportionately large black eye, accentuated by a heavy-mascara look. Faint earspot. All-black bill is long and droopy with a slight gonydeal expansion. Legs a dusky gray pink. More extensive black in outer wing than adults, lacking white apicals on the outer primaries. Primary coverts with variable black streaking. The brown upperwing coverts form a weak half-length carpal bar. Underparts white. Uppertail is largely white but may show brown-tipped coverts in heavily marked individuals. Thin black tailband. Postjuvenile molt soon gives way to gray, adult-like upperwing coverts and scapulars, with little remnants of brown juvenile feathers. *Similar Species:* None.

MOLT Remarkable variation in molt timing among various colonies—another practical example of why employing molt cycles is more useful than seasons or "years" when aging. Molt cycle in adults approximately nine months, mirroring breeding period (Howell & Dunn 2007). More study needed on intraspecies variation from known-age birds. Asynchronous breeding and pelagic lifestyle make

13 2nd cycle. Retained juvenile outer primaries with new, incoming, 2nd-generation primaries. Brown juvenile wing coverts evident. FERNANDO DIAZ. CHILE. FEB.
14 2nd-cycle type (right) and adult. Brown wash throughout upperparts suggests subadult, although age uncertain. Differences in head patterns may simply be due to asynchronous molt timings. ROGER L. HORN. GALAPAGOS. APRIL.
15 Adult in alternate plumage. Large eye with reddish orbital, gray bill tip, and two white spots at bill base. BYRON CHIN. GALAPAGOS. JULY.
16 Adult in alternate plumage. Smoky gray neck and upper breast. Note thin white border to scapulars. AMAR AYYASH. WASHINGTON. SEPT.

2 SWALLOW-TAILED GULL

systematizing this species' molt strategy an arduous task. Likely SAS, with the 1st inserted molt being prealternate. Prebasic and prealternate molts can have substantial overlap, to the extent that as outer primaries are growing, a full alternate hood may be found. Primary molt commonly found in postbreeding birds out at sea. Second-cycle individuals may show patchy alternate hood, brown wash across upperparts, and an all-black bill, or a duller gray bill tip than adult's.

HYBRIDS None.

17 Alternate adults. Large bug-eyes befitting a nocturnal feeder. Medium gray upperparts and pink legs. SUSAN FLECK. GALAPAGOS. AUG.
18 White tail and gray bill tip suggest an adult type undergoing prebasic molt. Active primary molt. VICTOR FAZIO. GALAPAGOS. NOV.
19 Adult. White partition between gray outer wing coverts and outer-primary coverts, unlike Sabine's. RICHARD BONSER. GALAPAGOS. AUG.
20 Adult. Deeply forked tail, most pronounced when tail is held square with the body. FREDERIC PELSY. GALAPAGOS. SEPT.
21 Adult. Gray ghosting of upperwing pattern evident. Forked tail shape lost when feathers are fanned out. DAVID KIRSCHKE. GALAPAGOS. JULY.
22 Many adult types show black markings on the outer-primary coverts. Also, some individuals show a contrasting gray wash on the inner primaries and outer secondaries, seen best in neutral lighting. RUSS MORGAN. GALAPAGOS. JUNE.
23 Adult (on water) and 2nd-cycle type, both completing flight-feather molt (prebasic molt). The individual in flight is aged based on brown lesser covert (right wing) and black bill tip. BRIAN SULLIVAN. PERU. JULY.
24 Adult with alternate hood (left) and three basic adults. Swallow-tailed Gulls molt asynchronously, as breeding occurs year-round. The complete molt takes place mostly at sea after breeding. THIBAUD ARONSON. PERU. APRIL.

3 BLACK-LEGGED KITTIWAKE *Rissa tridactyla*

L: 16.5"–18.0" (42–46 cm) | W: 37.0"–41.5" (94–105 cm) | Three-cycle | SAS | KGS: 6.5–8.0

OVERVIEW A true seabird, nesting colonially on narrow cliffs of seashores. Highly pelagic outside the breeding season, wintering almost entirely at sea. Black-legged Kittiwake has a disjunct circumpolar distribution throughout the arctic and subarctic zones. Simply known as "Kittiwake" in Eurasia. Thought to be the most abundant gull species in the world, with an estimated 9 million adults and perhaps that many subadults, for a total world population that approaches 20 million birds (Coulson 2011). Numbers in flux, and now regularly uses human-made structures to nest (from window ledges, piers, and shipwrecks to bridge girders). Highly variable diet and able to adjust according to availability of food (Baird 1994). Prefers capelin and pollock, and less so mollusks but readily consumes them. Some birds wintering off the California coast have been recorded feeding on squid, anchovy, and euphausiids. Regularly seen and reported at sea watches along both coasts, and more commonly grounded on West Coast beaches and breakwalls in small numbers, especially before and after storms. A prize bird seen in substantially lower numbers at lake watches throughout the Great Lakes region and other inland regions near large bodies of water.

The specific scientific name, *tridactyla*, meaning "three-toed," highlights the often reduced or completely missing hind toe (hallux). An adept flier with extreme aerial agility when compared to most other gulls. Wingbeats rapid and stiff when strong winds are encountered. Wheeling flight patterns on par with petrels and shearwaters. Does much feeding in flight, stopping to pick prey from the water, as well as plunge-diving. Surface-seizes while treading water. Rather noisy at nesting sites, uttering a repeated series of agitated and nasal *kitti-yakhs*, from which we get its common name.

Although kittiwakes exhibit much interannual variation in breeding success, it has been suggested that E Atlantic populations have two to three times the rate of success in fledging young compared to those in the N Pacific (Hatch et al. 1993).

TAXONOMY Two subspecies recognized, and both present in North America. *Rissa tridactyla tridactyla* ranges in the Canadian Arctic, N Atlantic, and east to Russia. *Rissa tridactyla pollicaris* found in Gulf of Alaska through the N Pacific and likely NE Siberia. More study needed on populations in NE Russia, where both races are thought to converge. Out-of-range birds likely impossible to reliably identify to subspecies.

Banded chicks hatching in W France are resighted in W Greenland with some regularity (Jean-Yves Monnat, pers. comm.). A notable sighting is a second-calendar-year bird which hatched in Brittany, France, in June 2017 and was then found in Les Escoumins, Quebec, in August 2018, on the St. Lawrence River, associating with local kittiwakes. The sighting is the first known Black-legged Kittiwake to be found on mainland America since this particular banding study began, in 1979. Subspecies assumed to be nominate *tridactyla*.

R.t. tridactyla (KGS: 6.5–7.5): Averages paler upperparts with a shorter bill. Body measurements also indicate this race is smaller overall. The scapulars and upperwing coverts are often slightly darker than the outer wing, showing a noticeable contrast in flight. This pattern is also found in *pollicaris*, but not as frequently, and with less of a contrast.

R.t. pollicaris (KGS: 7–8): Adults average darker upperparts and a slightly longer bill and are more likely to have black markings on p5 (Chardine 2002). The upperwing and back are more uniform gray with less contrast from pale gray to medium gray found in nominate. Adults in basic plumage and 1st-cycle birds may show broader and thicker hindcollar, based on some preliminary field observations, but more data needed. Further, 1st-cycle *pollicaris* is more likely to have broken ulnar bar at the wrist (Ayyash, personal observation (pers. obs.)). From a sample of 104 juveniles on the Kenai Peninsula, 22 individuals had a discontinuous ulnar bar, showing "headlights" that are often associated with Bonaparte's Gull.

For many years, it was said that *pollicaris* more often showed a rudimentary hind claw, but this feature needs critical examination. The size of the hind claw is thought to vary geographically among various populations of the nominate subspecies. Furthermore, the hind claw cannot be used in the field to identify individuals to subspecies away from the breeding grounds (Coulson 2011).

RANGE

Breeding: Circumpolar. In North America, nests mostly on sea cliffs and sea stacks, rarely on buildings, shipwrecks, and bridges. In the Pacific region, breeds from NW Alaska (Chukchi Sea) south to the Aleutians and Gulf of Alaska, and very locally in inner SE Alaska waters (i.e., Glacier Bay). In the Atlantic region, nests in two largely disjunct areas, in the Canadian high Arctic in E Nunavut and Greenland, and farther south in the Gulf of St. Lawrence and around Newfoundland and N Nova Scotia. Also a few in Bay of Fundy, New Brunswick. Summering nonbreeders are regular farther south off S Alaska and Bay of Fundy, more rarely even farther south.

Nonbreeding: Found primarily in offshore waters. Seen from shore primarily during strong onshore winds, or during "invasion" years, in either case more often along Pacific Coast than along Atlantic. In the Pacific region, regularly wanders in during late summer and fall eastward into the Beaufort Sea (as far as the Yukon), westward to W Nunavut, and southward very rarely to S Hudson Bay and James Bay. Migration farther to the south begins off both northern coasts largely in September, rarely in late August, but birds do not become numerous until late October, November, or December. Numbers off and along the Pacific Coast south to California and NW Mexico (Baja Peninsula), and off and along the Atlantic Coast south through the mid-Atlantic region (e.g., North Carolina and Bermuda) are highly variable from year to year in both the timing and the abundance of birds—particularly in the more southerly parts of the winter range. In some years, the species may be absent from these waters, whereas in others, moderate numbers may arrive already by late fall. Most individuals have departed the southern winter range by late March. In a few years, there is a major incursion only late in the season, beginning in February or March. During late winters with high abundance, good numbers may be found onshore along the Pacific Coast, such as at piers and beaches (and casually up to several miles inland), and many birds may remain well into the spring and sometimes early summer. Along the Atlantic Coast, such lingering birds are more typically found only offshore and occur as far south as S New England, only casually farther south. Along both coasts, numbers drop substantially as the summer progresses. Casual in winter north to the N Bering Sea. Very rare migrant and winter visitor to the Great Lakes and off and along the Atlantic Coast south to NE Florida. Casual but widespread records in migration (mostly late fall) and winter elsewhere in the interior (mostly at dams and at large lakes and reservoirs, and where most involve 1st-cycle birds) and south to the Gulf Coast to NE Mexico and to the N West Indies, also to Hawaii. Accidental south as far as central Panama, Trinidad, and even Peru.

IDENTIFICATION

Adult: Upperparts are a neutral gray with a bright white head in alternate plumage. Bill is relatively long, all yellow, with a slightly pointed tip. Dark eyes, with reddish orbital and gape. Noticeably short, black legs, although a brown hue is not uncommon. Long winged, with a narrow, pointy hand. At rest, the back is held in an upward position, close to a 45-degree angle (unlike the near-horizontal posture of non-*Rissa* species). In flight, the outer primaries show a well-demarcated black tip often described as a "dipped-in-ink" pattern. Never shows mirrors on outer primaries. The white trailing edge to the wing is thin, most so on the inner to outer primaries. Long-tailed appearance in flight, with central rectrices gently forked, especially when tail feathers are held straight. Neck appears long and cormorant-like when outstretched during flight calls. In basic plumage, the hindneck becomes gray, with variably sized earspot and head markings. The postocular markings may form a blackish band wrapping over the crown, as seen in 1st-cycle birds. In nonbreeding condition, the bill often has a strong green tinge. *Similar Species:* Superficial resemblances to Short-billed Gull in body size, rounded head, and long-winged appearance, but Short-billed has longer legs, which are yellow. Wingtip on Short-billed has white mirrors, which kittiwakes never show. Also note the thick white trailing edge of Short-billed, as well as the wider tertial and scapular crescents. The bill on adult-type kittiwakes is always unmarked, unlike the faint black subterminal markings shown by Short-billed in nonbreeding condition. Compared to Red-legged Kittiwake, Black-legged is slightly paler and larger, with a substantially longer bill. The head on Red-legged averages rounder. Red-legged shows a broader white trailing edge, and overall, the underside to the flight feathers of Red-legged is a smoky gray, unlike the mostly white underwing of Black-legged. Leg color is a dead giveaway. Consideration should be given to aberrant Black-legged Kittiwakes which sometimes show orange to orange-red legs. One to two are reported annually, more frequently so in western populations. Hybridization with Red-legged Kittiwake is an unavoidable question, but there is no evidence to support it, nor do these individuals show any other feature inconsistent with Black-legged Kittiwake.

1st Cycle: Juveniles have an adult-like gray back; all-black bill; short, black legs; and dark eyes. Distinctive black hindcollar of variable thickness, complemented by a black earspot. At times, the black earspots are extensive, wrapping over the head to form a double-hindcollar appearance. White crown with a blocky head. Black M-pattern across the upperwing coverts and outer primaries. Bright white trailing edge. Underparts and tail white. Narrow black tailband, which doesn't typically reach the outer edges of the tail. The gentle fork to the central tail is even more conspicuous in this age group, as the effect is deepened by the black tailband. However, the fork is immediately lost when the tail feathers are spread. The black hindcollar is generally lessened in late winter, and the head develops a smaller earspot with grayer hindneck. The bill, especially around the base, begins to lose its pigment, giving way to a sickly green-

---

1   Two Black-legged Kittiwake nests. The young are acquiring their first set of true feathers (juvenile plumage) via a prejuvenile molt (i.e., 1st prebasic). Interestingly, kittiwakes have adult-like gray scapulars in juvenile plumage, unlike the mottled brown found in most gulls. PAUL JONES. NEWFOUNDLAND. JULY.
2   Juvenile. Black hindneck, black earspot, and black bill. A small percentage of 1st cycles have dull-colored legs, certainly not black. AMAR AYYASH. ALASKA. AUG.
3   Juveniles with variable head pattern. The individual above has a full black bar connecting earspots, a less common pattern. AMAR AYYASH. ALASKA. AUG.
4   1st cycle with faded hindcollar and reduced black on lesser coverts. Brownish tones to legs. PHIL PICKERING. OREGON. FEB.

3 BLACK-LEGGED KITTIWAKE

yellow color with some black speckling. *Similar Species:* Compared to other species with M-patterns across the upperwing, Black-legged is larger, but impression of size may not always be apparent, especially on single birds in flight. Note that the neck and head pigment on kittiwakes is always black and never the brown colors shown by other species. Little Gull typically shows a much bolder and thicker M-pattern on the upperwing, with dark pigment extending to much of the primary coverts and mid- to inner primaries. Little Gull has traces of a variable dark trailing edge to the flight feathers, although some may have an all-white trailing edge to the secondaries. Black-legged has a distinctive, long, boomerang-shaped wing, whereas Little has a short, broad wing with a rounded outer-primary look. The legs on Little are a dull pink color, and the crown often has a brownish cap. Kittiwakes don't show a "Wilson's Warbler" cap. Bill-size differences are unmistakable. Red-legged Kittiwake has an all-white tail without a black tailband, and the carpal bar, when present, typically has a weak pattern. At rest, note leg color and noticeably stubbier bill of Red-legged, which also has slightly darker gray upperparts. Sabine's Gull lacks a defined carpal bar and instead shows a solidly filled gray-brown triangle pattern on the wing coverts that is concolorous with the scapulars and mantle. The primary coverts and mid-primaries are dark, as opposed to white in Black-legged. The bill on Sabine's is noticeably more petite, and the tail generally shows a deeper fork.

2nd Cycle: Similar to adult with gray upperwing and white body, but perhaps separable by black pigment on alula and primary coverts, as well as random dark spots on upper secondary coverts. More black on

5   1st cycle with bill now largely yellow and noticeable wear on wing coverts. Short black legs and bold hindcollar are diagnostic. ANDREW THEUS. GEORGIA. MARCH.
6   Juvenile. White trailing edge to secondaries and inner primaries with black M-pattern across upperwing. Thin black tailband. DAVID FERNANDEZ. ICELAND. AUG.
7   Juvenile. A weaker ulnar bar than 3.6, with a break in black at the wrist. AMAR AYYASH. ALASKA. AUG.
8   Juvenile. Black tips to underside of outer primaries. AMAR AYYASH. ALASKA. AUG.

9   Juvenile (left) with similar-aged Bonaparte's (formative plumage). Compare trailing edge. Optical zoom makes it difficult to appreciate larger size of Kittiwake here. JESSE AMESBURY. DELAWARE. DEC.
10 2nd cycle lacking hindcollar now. Many wing coverts replaced with some brown lesser coverts still apparent. Yellow bill with black on edge of base and tip. STEVE ARENA. MASSACHUSETTS. JULY.
11 Same individual as 3.10, with dark juvenile lesser coverts more exposed. 2nd-generation primaries (juvenile p10 retained). Mostly white tail.
12 Both 2nd cycles with the individual on the right slightly more advanced. Note black on bill tip, dark spots on wing coverts (left), and black on tail (right). AMAR AYYASH. MASSACHUSETTS. JULY.
13 Note that the innermost primaries (p1–p2) have been dropped, signaling the start of the 2nd prebasic molt (thus this bird is now a 2nd cycle). Faded brown ulnar bar and primaries still juvenile. BRUCE MACTAVISH. NEWFOUNDLAND. JULY.
14 2nd prebasic molt well underway with new p1–p6. Tail and secondaries also mottling, and many new, adult-like gray wing coverts. Same age as 3.13, but more advanced molt. BRUCE MACTAVISH. NEWFOUNDLAND. JULY.

15 2nd cycle. Black on primary coverts and alula, and long black streaks on p8–p10. Nominate more often shows darker gray scapulars and wing coverts, with paler outer wing. RICHARD SMITH. ENGLAND. APRIL.
16 Alternate adult. Yellow bill, short black legs, and medium gray upperparts. Small white apicals typical. RUSS NAMITZ. ALASKA. MAY.
17 Alternate adult. Dark eye, red orbital and gape, and tapered bill tip. JOHN CHARDINE. GREENLAND. JUNE.
18 Adult showing considerable wear to upperparts and primary tips. Gray base to outer primaries (below tertials). A noticeably longer bill on this individual. AMAR AYYASH. ALASKA. AUG.
19 Basic adult. Greenish-yellow bill, smoky gray hindneck, and black on head. Upperparts are rather fresh. PHIL PICKERING. OREGON. JAN.
20 Adult with juvenile. AMAR AYYASH. ALASKA. AUG.

21 Adult type (right) with adult Short-billed Gulls, which are similar in size. Compare wingtip and leg color. AMAR AYYASH. ALASKA. AUG.
22 Adult (right of center) showing thick yellow bill, larger size, and slightly darker upperparts compared to Bonaparte's. JAMES PAWLICKI. NEW YORK. DEC.
23 Alternate adult. Wingtip lacks mirrors with sharp "dipped-in-ink" triangle pattern. DAN VICKERS. ALASKA. JUNE.
24 Adult. Demarcated black wingtip pattern and thin white trailing edge. Undergoing primary molt (p1–p3 new). AMAR AYYASH. ALASKA. AUG.

outer wingtip than in adult, especially inner web of p10 and outer webs of p9–p8. Black wingtip pattern generally untidy when compared to definitive adult demarcated dipped-in-ink appearance. Bill may show random black speckling, especially near the base and very tip.

MOLT Interannual variation in wing molt observed on West Coast. Presumed *pollicaris* adults along the Pacific Coast may show suspended or protracted wing molt, which is thought to be related to El Niño conditions. Howell and Corben (2000b) noted that in some years, primary molt doesn't conclude until May/June. Nominate adults average earlier completion of prebasic molt, likely due to earlier breeding season (Coulson 2011), and are less likely to show suspended/protracted prebasic molts late into the nonbreeding season (Ayyash, pers. obs.).

Juvenile kittiwakes are unique in that their juvenile mantle, and scapular feathers are an adult-like gray, unlike the variable browns found in other gulls. These gray feathers are 1st basic and should not be mistaken for 1st alternate or formative feathers. The first molt thereafter is considered a prealternate molt and is partial.

More study needed from known-age 2nd cycles to determine extent of molts, and age-related differences in plumage.

HYBRIDS None.

25 Adult type with basic head pattern already at this date. Wing molt delayed. AMAR AYYASH. ALASKA. AUG.
26 Adult. Suspended flight feather molt (p9–p10 retained, as well as several secondaries), a pattern undocumented on Atlantic Coast, but fairly common on the Pacific. ALEX LAMOREAUX. MAINE. JAN.
27 Winter adults on the Pacific average bolder black on back of head compared to Atlantic populations. AARON MAIZLISH. CALIFORNIA. MAR.
28 Adult underwing. Black "dipped-in-ink" wingtip, black legs, and boomerang-shaped wings are distinctive. STEVE ARENA. MASSACHUSETTS. NOV.

4 RED-LEGGED KITTIWAKE *Rissa brevirostris*

L: 15.5"–17.0" (39–43 cm) | W: 33"–35" (84–89 cm) | Three-cycle | SAS | KGS: 8.5–9.0

OVERVIEW An obligate seabird. This Bering Sea endemic breeds colonially on four main island groups. Largest colony found on St. George in the Pribilofs, with an estimated population of 220,000 individuals. World population thought to be under 500,000 (BirdLife International n.d.-c). Lays a one-egg clutch, unlike the two- to three-egg clutch of its larger relative, Black-legged Kittiwake, with which it is congeneric. Overall, Red-legged more skittish and wary than Black-legged Kittiwake. Although both species nest in close proximity in the same colonies, recent data has shed much light on Red-legged's disparate diet, foraging techniques, and winter range (Drummond et al. 2021). *Brevirostris* (meaning "short-billed") has much more specialized prey requirements than Black-legged, feeding mostly on northern lanternfish in the cold waters of N Pacific. A study aimed at understanding the activities and winter ranges of both *Rissa* species, using geolocators, found that Black-legged generally moves south to the subarctic N Pacific, where it is more of a generalist, while Red-legged remains primarily in the Bering Sea (Kokubun et al. 2015). The large eye of Red-legged is thought to be an adaptation for nocturnal feeding, but interestingly, most foraging activity is between an hour before sunrise and sunset, with birds remaining largely stationary at night during the boreal winter. The species has a strong association with sea ice in the nonbreeding season. With warming waters in the north in recent decades, the predictability of available prey has made it much more likely for breeding colonies to all but completely fail in some years. Currently listed as vulnerable by the International Union for Conservation of Nature. Climate change has only exacerbated this status. Shifting food sources throughout the deep ocean basin and Bering Sea shelf region may prove to be catastrophic to this species.

TAXONOMY Monotypic.

RANGE
Breeding: Local breeder on islands in the southern and central Bering Sea as well as on the Commander (Komandorski) Islands, Russia. Nests on sea cliffs on the Pribilof Islands (St. Paul, St. George, and Otter Islands, with largest numbers on St. George), locally in the Aleutians (Buldir Island, in W Aleutians; Bogoslof and Fire Islands, in E Aleutians; and a few birds on several other small islands). Most birds nest on the Pribilofs, but numbers are likely in decline. Recently found breeding on St. Matthew Island; first confirmed nesting was observed in 2018 (Robinson et al. 2020). The colony there is said to be between 100 and 150 pairs. Before that, the species was casual to accidental on St. Matthew in the breeding season.
Nonbreeding: Distribution away from breeding grounds is poorly known (Byrd & Williams 1993). Formerly presumed to winter in deep water well offshore in N Pacific Ocean and near ice edge in SE Bering Sea, but data very sparse. More recently, studies using geolocators discovered that at least some birds moved north after the breeding season and were found in the NW Bering Sea during October–December; they then moved southwest to waters off E and S Kamchatka and N Kuril Islands between December and February before returning to the breeding grounds in early spring (Drummond et al. 2021). Rare visitor to Aleutians in early summer—mostly adults, in small numbers, presumably failed breeders. Casual visitor both offshore and onshore south to Washington and Oregon, accidental to California. Casual to Japan. Also accidental well inland in Alaska and Yukon, and exceptionally to S Nevada).

IDENTIFICATION
Adult: Slaty-gray upperparts about as dark as Laughing Gull's. White head and neck in alternate plumage. Dark eye is disproportionately large on a small, rounded head with a noticeably steep forehead. Vermilion legs are exceedingly short, giving the bird a squat appearance. Unmarked yellow bill is stubby with a noticeable hook at the tip. Orbital ring and gape red. Wingtip has the signature dipped-in-ink effect with no mirrors, but proximal edges of black subterminal bands more jaggedly shaped than those of

4 RED-LEGGED KITTIWAKE

Black-legged. A relatively wide white trailing edge to the secondaries and inner primaries is visible from a distance. The underside to the flight feathers and the underside to the primary coverts are a dusky gray. Secondary underwing coverts and axillaries are white. All-white tail. A long wing projection at rest. In basic plumage, the eyes typically become encircled with dark markings; dusky earspot and variably sized hindcollar with a slight gray wash to the hindneck. Nonbreeding condition renders the legs a dull orange red, and the bill takes on a conspicuous green cast. *Similar Species:* Black-legged Kittiwake is larger, with a longer bill and paler gray upperparts. Leg color is diagnostic, but caution should be taken, as some Black-legged Kittiwakes may show dull orange legs at times. The white trailing edge on Black-legged is significantly thin, and the underside to the flight feathers is white—not the dusky gray seen on Red-legged. Therefore, the black wingtip on the underside of the wing on Black-legged contrasts much more with the rest of the wing.

1st Cycle: Juveniles are unique in that they have a white, adult-like tail with no tailband (shared by no other gull species at this age). Dark eyes encircled with dusky flecking. Decidedly short and petite black bill. Black earspot, complemented by a dark hindcollar and variable gray wash on hindneck. Leg color varies from dull pink colored to brown or blackish. The gray upperparts are fringed in white. Similar to adults in tone, but the upperwing has a white obtuse triangle formed by the outer secondaries and inner

1   Juvenile with Black-legged Kittiwake. Stubby black bill, black hindcollar, short dingy legs, and minimal black on wing coverts. STEVE HEINL. ST. PAUL ISLAND. SEPT.
2   Juvenile. Short black bill with large blocky head. Commonly shows adult-like wing coverts. Black shaft streaks on tertials. STUART PRICE. JAPAN. SEPT.
3   Juveniles. Medium to dark gray upperparts, black hindcollar, and largely unmarked wing coverts. STUART PRICE. JAPAN. SEPT.
4   Juvenile. Our only 1st-cycle gull with a white tail. Black on wing coverts doesn't typically form M-pattern found in other small gulls. RYAN O'DONNELL. ST. PAUL ISLAND. AUG.

primaries, recalling Sabine's Gull. The upperwing coverts have variable dark brown and black markings, ranging from lightly marked lesser and marginal coverts to, in some well-marked birds, a quasi carpal bar. Outer primaries and their coverts are black, with the mid-primaries becoming a tricolored pattern of black to gray to white. By early spring, the earspot and hindcollar are reduced, and overall the head is more white, similar to basic adult. The bill shows a yellow base with black. Legs also transition to dull pink or red, but less vivid than those of adults. *Similar Species:* Compared to similar-aged Black-legged Kittiwake, Red-legged is darker, with a more compact body. Leg color is usually enough to separate the two at rest. There are diagnostic differences in plumage, particularly Red-legged's all-white tail, lack of a defined carpal bar, and more extensive gray on the outer-wing coverts. The upperwing on Red-legged may cause confusion with young and/or adult Sabine's Gull, but Red-legged typically has fewer head markings, as

---

5 Juvenile. Large white triangle formed on mid-wing recalls pattern found on Sabine's, but note reduced black on outer wingtip. STEVE HEINL. ST. PAUL ISLAND. AUG.
6 1st cycle. Black on secondary coverts often limited to carpal edge. Smoky gray underside to outer primaries. Also note subtle white tongue tips to p5–p6, at times to p7/p8. JOHN CHARDINE. ST. MATTHEW ISLAND. SEPT.
7 1st cycles. Black bills, marked primary coverts, and smoky gray underside to primaries. STUART PRICE. JAPAN. SEPT.
8 2nd cycle. Bill now greenish-yellow, orange-red legs and 2nd-generation primaries emerging (brown outermost primary juvenile). MARK CHAPPELL. ST. PAUL ISLAND. JULY.
9 2nd cycle. Diffuse dark wash on bill, orange-red legs, and small brown streaks on outer lesser coverts. Active primary molt. RYAN O'DONNELL. ST. PAUL ISLAND. JULY.
10 Undergoing 2nd prebasic molt, now with some upperparts replaced, as well as inner primaries (p1–p4). Still showing dusky collar and earspot. LARS PETERSSON. ST. PAUL ISLAND. JUNE.
11 2nd-cycle type. Black on primary coverts and irregular black pattern on p8–p9. Many show complete p5 band (adults often have limited black mark only on outer web). Dirty bill. RYAN O'DONNELL. ST. PAUL ISLAND. MAY.
12 Alternate adult. White head, dark eye, short stubby bill, and vivid red legs. SCOTT SCHUETTE. ST. PAUL ISLAND. JUNE.
13 Adult. Medium to dark gray upperparts. Short red legs diagnostic. RICHARD BONSER. ST. PAUL ISLAND. JULY.
14 Adults with Black-legged Kittiwake in background. SHAILESH PINTO. ST. PAUL ISLAND. JULY.

4 RED-LEGGED KITTIWAKE

well as a tricolored pattern to the primaries—most notably, gray on the bases of the mid-primaries.
2nd Cycle: Plumage is adult-like save for some dark markings on the marginal coverts and outer-primary coverts. The wingtip also has more pigment, especially on p8–p10 and their coverts. The underside to the flight feathers now grayish, but perhaps not as contrasting as the adult's. May show remnants of an earspot and weak hindcollar, but an all-white head is typical in 2nd alternate plumage. The bill is largely greenish yellow, and the legs are dull red to bright red.

MOLT A moderate amount of molt takes place away from breeding colonies and is therefore poorly studied. White tail feathers of hatch-year birds are juvenile (1st basic) and not produced by an inserted

molt. First prealternate molt appears to be limited to head and body feathers, with few or no wing coverts replaced. Second prebasic molt is complete, giving way to an adult-like plumage.

HYBRIDS None.

15 Adult with Black-legged Kittiwake. Although difficult to appreciate here, Red-legged averages smaller with a shorter bill and shorter legs. WOLFE REPASS. ST. PAUL ISLAND. AUG.
16 Adult. Vivid red legs and smoky gray underside to remiges. RYAN O'DONNELL. ST. PAUL ISLAND. JUNE.
17 Alternate adult. No mirrors. "Dipped-in-ink" wingtip resembles Black-legged Kittiwake. Outer web of p10 commonly shows black to primary coverts, and black mark on outer web of p5. RYAN O'DONNELL. ST. PAUL ISLAND. MAY.
18 Adult (left) with Black-legged Kittiwake. Red-legged has darker upperparts and averages shorter wings. LIAM SINGH. ST. PAUL ISLAND. MAY.
19 Adult. Short, stubby bill. Broader trailing edge compared to Black-legged Kittiwake, showing greater contrast with darker upperparts. ERIC VANDERWERF. ST. PAUL ISLAND. JUNE.
20 Adult type beginning prebasic molt (note new p1–p2). Many adults complete primary molt away from the breeding grounds. NICK HAJDUKOVICH. SOUTHERN WATERS OF BERING SEA. SEPT.

5 IVORY GULL *Pagophila eburnea*

L: 17"–19" (43–48 cm) | W: 36"–38" (91–97 cm) | Two-cycle | SBS | KGS: 0

OVERVIEW A hardy, ivory-white gull of the high Arctic. This showstopper is the only gull with a KGS value of 0. Ivory Gull is the only species in its genus, *Pagophila*, which translates to "ice-loving." Movements highly regulated by drift ice, and often found near ice floes. Commonly scavenges carrion; also feeds on excrement of seals and walruses, and on small fish and crustaceans. Although most often found near water, Ivory Gull feeds terrestrially; feeds much less while swimming. The black-and-white spotting on juvenile birds is a feature shared with other species of the Arctic (e.g., some Gyrfalcons and Snowy Owls).

A paper highlighting the distribution of Ivory Gull in the high Arctic of Canada suggests a drastic decrease in North American populations (Gilg et al. 2016). The decrease is so extensive that Ivory Gull is no longer considered near threatened but instead endangered. The population trajectory in North America is not very promising. Canadian populations are believed to have suffered a loss of over 80% since the 1980s (Richards & Gaston 2018). However, an extensive aerial survey recently conducted in the north of Greenland suggests numbers there nearly double those from a decade earlier (Boertmann et al. 2019). Age distribution in vagrants to S Canada and the lower 48 states appears equal, with 1st cycles and adults alike straying. Vagrants turn up in a variety of settings, from dams and parking lots to being fed by fishermen and, the most bizarre sighting of all, a 1st-cycle "yard bird" in N Wisconsin found perched on a trampoline. Banding efforts indicate strong site fidelity in this species, with at least two individuals resighted at their natal colonies after 28 years (Mallory et al. 2012).

TAXONOMY Monotypic.

RANGE
Breeding: Holarctic. Breeds (or formerly bred) mostly in colonies, on rocky islands and inland from coasts (sometimes well inland) on barren uplands, steep ridges, and nunataks in the high Arctic, above the arctic circle, very locally from Nunavut (Ellesmere, Cornwallis, Seymour, Devon, Baffin (Brodeur Peninsula), and Perley Islands) eastward to NW and E Greenland, Spitsbergen, and several Russian archipelagos east to at least Severnaya Zemlya, in the central Russian Arctic. Some of the better-known sites in Canada, at least until recently, included Pond Inlet, Arctic Bay, Grise Fjord, and Resolute Bay. Major population decreases from some nesting sites have resulted from warming temperatures and disappearance of sea ice, from contaminants, and from hunting. The Canadian breeding population in the early 2000s was estimated at only about 1,000 birds, less than half the number estimated during the 1980s.

Nonbreeding: Occurrence closely tied to pack ice, drift ice, polynyas, and the ice edge—often deep within consolidated pack ice. Therefore, Ivory Gulls are often irregular in occurrence and have been shown undergoing bidirectional migration off E Greenland (Gilg et al. 2010). Found most often at food sources such as remains of dead marine mammals or other dead animals, or sources of fat. In Nunavut, seen somewhat regularly in Lancaster Sound, Barrow Strait, and Jones Sound, but only very rarely in Hudson Bay and the Foxe Basin. Autumn migration commences in September or early October, when numbers may pass Wrangel Island, in E Russian Arctic (into Nov.), or move east from NE Canadian Arctic. This first major movement in fall may be mostly west–east, in a circumpolar fashion, following the ice edge; then finally move south into the Bering Sea or Davis Strait beginning in November or later. Formerly occurred as early as late September, but now mostly in October or later, at Utqiagvik (formerly Barrow), Alaska, in the Chukchi and Beaufort Seas. Not regularly seen in N Bering Sea in late fall or early winter, as formerly, due to lack of sea ice. Surmised wintering grounds in the Sea of Okhotsk and Bering Sea are a long distance from known nesting areas. In the N Atlantic region, mostly seen in late fall and winter from Baffin Island and the Greenland Sea south to Davis Strait and Labrador Sea, but current arrival time in these waters somewhat uncertain. Winters mostly between N 50° and N 64°. Rare but perhaps regular in winter south to N Newfoundland when sea ice is present.

Spring movements in North America not well known. Most data from St. Lawrence Island, Alaska, in N Bering Sea, where influx of birds is often detected between late February and April or early May. Formerly regular in small numbers there (i.e., at Gambell) into late May or even early June in years when substantial sea ice was still present. Also rare in late spring farther east to the Seward Peninsula (e.g., Nome area). Very rare farther south in late winter and spring at Pribilofs and Aleutians.

Away from its normal range, Ivory Gull occurs very rarely in winter in W North America along the coast and inland from central Alaska and S Yukon south to Washington, and exceptionally to central and S California, SW Arizona, and Colorado. Recorded more regularly, though still very rare, and mostly in midwinter, east of the Rockies, from the Canadian Prairies east through the Great Lakes region (Ontario has over 30 accepted records). Also to Quebec, the Atlantic Provinces, and New England; exceptionally farther south to New Jersey, Tennessee, and Georgia. A vagrant to mainland Europe, Japan, Korea, and China.

IDENTIFICATION

Adult: Head-to-neck proportions somewhat pigeon-like, with compact body, plump breast, and short legs. The all-white plumage with no pigment is complemented by black legs and dark eyes. Grayish-green

---

1 Juvenile. Variable dark spotting throughout, short black legs, and dark grayish bill. DAVID DISHER. NEW JERSEY. DEC.
2 Juvenile with extensive mottling on lores and chin compared to 5.1. Grayish bill with yellowish tip. Compact, pigeon-like shape. RYAN BRADY. WISCONSIN. FEB.
3 Juvenile showing fresher black spots throughout, with adult in background. HENRIK HAANING NIELSEN. SVALBARD. SEPT.
4 For relative size comparison, from left to right: 1st-cycle Ross's, Black-legged Kittiwake, Ivory, and Sabine's Gull. CHICAGO ACADEMY OF SCIENCES. AMAR AYYASH.

bill with orange-yellow tip is unique among larids. In flight, the wing appears broad, with a long, pointy hand. Wing beats swift and powerful, almost jaeger-like. *Similar Species:* Caution should be taken with albino or leucistic individuals of other gull species. Leg and bill color and behavior are often sufficient for ruling out one or the other. All-white pigeons have been known to excite observers, especially when found in the correct habitat. Bare parts, and especially the short, stubby bill, help rule out Ivory.

1st Cycle: Ivory-white upperparts, flight feathers, and rectrices with variable black spots to their tips. The face, especially around the base of the bill and eyes, is peppered with black and appears grimy. Black legs, dark eyes, and a bill pattern similar to adult's but typically darker and duller. *Similar Species:* None.

MOLT Our only gull species known to exhibit SBS, with no preformative or prealternate molt inserted in 1st cycle. Subsequent cycles apparently lack prealternate molt. The 2nd prebasic molt is complete, resulting in adult-like plumage. Thus, Ivory Gull typically molts from juvenile plumage directly to adult basic plumage in its second calendar year. Adults undergo rapid primary molt before breeding duties (March–May). Primaries typically replaced out to p6/p7 and then suspend over the breeding season. Primary molt resumes in late August–September (Howell 2001a).

HYBRIDS None.

5   Juvenile. White underwing with the exception of black spotting on outer edge of hand and black tips to primaries. SAM KOENEN. MONTANA. FEB.
6   A consistently marked juvenile with dark tips to all flight feathers and greater coverts. Unmistakable. STEVE KOLBE. MINNESOTA. JAN.
7   Less marked than 5.6 with thinner black tailband. TOM JOHNSON. NEW JERSEY. NOV.
8   2nd cycle in primary molt (p1–p7 new, p8–p10 retained juvenile). Ivory Gull molts from one basic plumage to the next, with no inserted molts (Simple Basic Strategy). SAYAM CHOWDHURY. RUSSIA. JUNE.

5 IVORY GULL   91

9 Adult. Immaculate white plumage, short black legs, dark eyes, and grayish-green bill with orange-yellow tip. BRUCE MACTAVISH. NEWFOUNDLAND. FEB.

10 Adult. Bill pattern diagnostic. Fairly straight bill of medium length, showing no bulge at the gonys. Short wing projection and pigeon-like profile. BLAKE MATHESON. CALIFORNIA. NOV.

11 Adult in breeding condition. Scarlet orbital, grayish-yellow bill, and orange tip. Plumage is "basic," as Ivory Gull lacks a prealternate molt. SIMON COLENUTT. NORWAY. MARCH.

12 Adults. Medium-sized, grayish bill with orange-yellow tip. BRUCE MACTAVISH. NEWFOUNDLAND. FEB.

13 Adult. Entirely white plumage. Broad arm with pointed hand. Bill pattern unique. BRUCE MACTAVISH. NEWFOUNDLAND. FEB.

14 Adult. Broad base to wing with pointed hand. Black legs and orange bill tip. BRUCE MACTAVISH. NEWFOUNDLAND. FEB.

15 Adults, all with molt gap in mid- and outer primaries. Ivory Gull undergoes rapid primary molt before breeding and often suspends molt during nesting duties. Outer-primary molt resumes in Aug.–Sept. NICK HAJDUKOVICH. ALASKA. MAY.

16 Adult. Heart-stopper. Immaculate white larid with compact proportions. AMAR AYYASH. ILLINOIS. JAN.

17 Adult. Few references show this species in molt. Here p1–p6 are fully grown, p7 growing, and p8 dropped; p9–p10 are old. The tail feathers are molting from the center out. KIRK ZUFELT. ALASKA. MAY.

6 ROSS'S GULL *Rhodostethia rosea*

L: 12.5"–14.0" (32–36 cm) | W: 32"–34" (81–86 cm) | Two-/three-cycle | CAS | KGS: 4–5

OVERVIEW Among the most coveted bird species in the world and certainly the least known of all gulls. The Latin name *rosea* means "rosy" or "pink," and therefore, the future English name should highlight this. Ross's Gull was traditionally considered a breeder of the Siberian Arctic, but recent surveys and concerted efforts to track birds with geolocators and satellite transmitters have turned up several small breeding colonies in N Canada and Greenland. Still, it's believed that less than 1% of the world's breeding population is accounted for, making this arctic seabird a wonder among wonders (Maftei et al. 2012). Of importance is the species' preference for nesting on the outer periphery of Arctic Tern colonies. Oddly, Ross's Gull habitually engages in heterospecific courtship behaviors and has been observed displaying to at least six different species, primarily gulls. In the nonbreeding season, the species is strongly connected to the edges of pack ice in and around the Arctic Ocean. Exactly how Ross's Gull will be impacted by the decreasing arctic pack ice remains to be seen. Vagrants found well south of the arctic circle in Canada and the lower 48 states are primarily adults, and why so few 1st-cycle individuals are recorded is unknown.

TAXONOMY Monotypic.

RANGE
Breeding: Breeds very locally in the Arctic. Most birds nest along the arctic coast of Russia, in marshy tundra and deltas, between the Taimyr Peninsula and Kolyma River Delta. Nests placed on hummocks. In North America, a small number breed very locally in N Canada, with nesting first confirmed in late 1970s and during 1980s along coast of Hudson Bay at Churchill area, Manitoba, where perhaps as many as five pairs bred during the peak. Since then, found nesting in high Arctic of Nunavut: near Resolute Bay, along Queens Channel and Penny Strait (northwest of Cornwallis Island), in vicinity of Cheyne Islands (as early as 1976), at Nasaruvaalik Island (where up to 5 pairs and 12 adults have been seen), and at several smaller islets (Maftei et al, 2015). Also possibly at Prince Charles Island in Foxe Basin. Also breeds very locally and irregularly in west-central and NE Greenland. Has perhaps bred at Svalbard.
Nonbreeding: Casual visitor south of (poorly) known wintering areas. Fall migration pattern unique. Postbreeding dispersal from Russian breeding sites commences in late summer. Birds initially move east into the Chukchi Sea and continue into the W Beaufort Sea (late Sept.–Oct.), then presumably return west to the Chukchi Sea, then south into the NW Bering Sea (mostly late Oct.–Dec.). Well known from Utqiagvik, where eastbound birds have been noted in moderate to large numbers. Peak numbers formerly occurred between late September and mid-October, but in some recent years, warming ocean temperatures and little to no sea ice have resulted in a slight shift, to peaking in mid-October, with few or no birds seen earlier in the season. Formerly recorded there in numbers as high as around 20,000–40,000 birds annually, but recent maximums no more than half that. Marked interannual variation in numbers is the status quo, however. A return flight to the west, later in the season, past Point Barrow, has not been documented. Small numbers (up to a dozen) have been noted on multiple occasions in N Bering Sea at W St. Lawrence Island (e.g., Gambell), Alaska, in late fall (mostly in Nov. and early Dec., with the earliest arrivals at the end of Sept.). Main winter grounds poorly known but thought to be associated with pack ice from the W Bering Sea south to the Sea of Okhotsk and waters off extreme N Japan. Northbound migration in Russia appears to be, at least in part, overland from the Sea of Okhotsk and Gulf of Anadyr to the breeding grounds along the arctic coast. A few birds wander east during May–June in the Bering Sea, where the species is very rare, but there are many records from St. Lawrence Island, and it is casual farther south at the Pribilofs and Aleutians.
   In the N Atlantic Ocean region, probably regular in numbers in early fall in the Greenland Sea off NE Greenland. As in the Pacific, winter distribution largely correlated with pack ice. Any regular winter range

6 ROSS'S GULL

in the Atlantic poorly known and mostly the source of conjecture. Recent studies have shown, however, that at least a small number of birds may winter in the Davis Strait / Labrador Sea area, between Eastern Canada and Greenland (Gilg et al. 2010).

Wandering individuals have occurred casually at a wide range of sites both coastally and inland, mostly between October and early April, from Alaska to Newfoundland and south as far as the Salton Sea in SE California, and CO, MO, IL, MD, and DE. Some of these birds were associated with flocks of Bonaparte's Gulls. Also in NW Europe, N China, and central Japan.

IDENTIFICATION

Adult: An unmistakable, cartoonish-looking gull. Ross's Gull has the shortest and smallest bill of all gulls, even smaller than that of its closest relative, Little Gull. In alternate plumage, adults sport a diagnostic black necklace and can almost always be found with some degree of pink suffusion throughout the plumage. Black bill; short reddish legs; oversized dark eyes with red orbital. The white, wedge-shaped tail is another diagnostic feature. The pale gray upperparts contrast with a white trailing edge, which is broadest at the outer secondaries and inner primaries and then tapers off at the mid-primaries. The outermost primary (p10) shows a narrow black outer edge in flight. In basic plumage, adults forfeit their black necklace but may show remnants of a dark collar or a small, dusky earspot. Dark coloration may

1 1st cycle. Noticeably short bill, white crown with some dark around the eye. JOACHIM BERTRANDS. BELGIUM. JAN.
2 1st cycle. Earspot, dark eye-patch with faint border around white head, long wings with noticeable white on visible inner primaries. BENJAMIN DOUGLAS. MINNESOTA. NOV.
3 1st cycle with adult. Variable pink blush is common; white crown, earspot, and some dark around the eye. JOHN PUSCHOCK. ALASKA. OCT.
4 1st cycle. Alternate plumage with white hindneck and black necklace. Obvious pink blush on body. BOB FRIEDRICHS. ALASKA. JUNE.

6 ROSS'S GULL

surround the eye, especially below and in front. *Similar Species:* The underwing, especially in low light or when subjected to strong shadow, can look quite dark, inviting confusion with adult-type Little Gulls, which show dark, blackish underprimaries. Little Gull lacks the wedge-shaped, jaeger-like, central tail pattern of Ross's, although beware birds with molting tail feathers. The white trailing edge is sufficiently different and quite helpful when the open wing is observed. The outer primaries on Ross's have few if any white terminal tips, while Little shows a consistent white trailing edge throughout to the outer primaries. The hand on Ross's is pointier, while on Little, the hand appears broad and rounded.

1st Cycle: Complete juvenile plumage rarely recorded anywhere in the world. Recently fledged birds have a grayish-brown head, hindneck, and breast, with golden-buff edges to their brown upperparts (recalling a shorebird instead of a gull). Birds straying from their natal colonies will by then have adult-like gray

---

5   1st cycle and adult in alternate plumage, observed associating for a couple of weeks. TOM JOHNSON. ALASKA. JUNE.
6   Faint necklace and short bill distinctive. Visible inner primary (p5) dropped, thus 2nd cycle. TOMMY PEDERSEN. NORWAY. JUNE.
7   1st cycle with a mixture of juvenile (brown) and formative (gray) scapulars. Long, black central tail feathers and white secondaries are key features. Small black bill. AURELIEN AUDEVARD. FRANCE. JAN.
8   1st cycle. Short bill. No dark cap in formative plumage. Black limited to central tail feathers on this individual. White secondaries with deep white bar cutting into subterminal region of primaries. Averages more black-tipped primaries (p2/p3) than Black-legged Kittiwake (p5/p6). RUNE BISP CHRISTENSEN. DENMARK. FEB.
9   1st cycle. Tail pattern diagnostic. Brown on upper rump retained on many individuals. Deep white triangles formed on mid-wing with clean secondaries. No black cap. IAN DAVIES. NEW YORK. JAN.
10 1st cycle. Grayish wing linings (paler on Little Gull). Clean white secondaries, with white extending to inner primaries. Noticeably long central tail feathers. AMAR AYYASH. ILLINOIS. MARCH.
11 1st cycle (same as 6.4 and 6.5). Legs brighter in breeding condition. Black necklace variable in 1st alternate. CORY GREGORY. ALASKA. JUNE.
12 Adult in alternate plumage. Complete black necklace, red legs, and long gray wings. DUBI SHAPIRO. ALASKA. JUNE.
13 Alternate adult. Nests in marshy tundra as well as on small islets throughout the Arctic. ILYA UKOLOV. RUSSIA. JUNE.

scapulars with "basic-like" head. The face will have dusky patches around the eye, with a small, faint postocular spot. The ulnar bar is bold and dark throughout, and this is evident on resting birds showing much blackish brown on the upperwing coverts. The petite bill, short legs, and compact head make a Ross's Gull difficult to confuse with anything else. First alternate birds acquire partial to complete black necklace similar to adult's and replace a variable number of tail feathers, often taking on a completely white tail. *Similar Species:* The freshest juveniles recall juvenile Little or Sabine's Gull when perched, but Ross's has a stubbier bill, shorter neck, and more domed head. The upperwing on Sabine's lacks M-pattern, and the central tail feathers are inverted and fork shaped. Compared to similar-aged Little, Ross's has black pigment restricted to the central tail feathers, which are longer and form a wedge. The mid-outer primaries on Little show more consistent pigment all along the length of the outer webs. Ross's shows white secondaries and generally lacks any hints of a dark cap once in formative plumage.
2nd Cycle: Similar to adult, although may include dark streaks or spotting on the outer primaries.

MOLT Believed to have two partial molts in 1st cycle (CAS). Preformative molt produces gray postjuvenile scapulars and basic-like head and neck pattern. Prealternate molt brings on alternate head and neck pattern with thin black necklace in adults. The black necklace may be complete or partially formed in 1st alternate. Some replace tail feathers in 1st prealternate molt (see plate 6.19).

HYBRIDS None.

14 Adult. Basic plumage can still show deep pink suffusion. Short black bill, faint earspot, and dark around the eye. A noticeably pink individual. JOHN PUSCHOCK. ALASKA. OCT.

15 Adult. Remarkably short bill, round head, and long wings. Note black outer edge to outer primary (p10). STEVE YOUNG. ENGLAND. APRIL.

16 Adult. Unmistakable, long central tail feathers and black necklace. JAN VAN HOLTEN. RUSSIA. JUNE.

17 Adult. The pink ghost of the Arctic. White trailing edge tapers off at inner primaries. NICK HAJDUKOVICH. ALASKA. OCT.

18 Adult. Gray underwing can appear rather dark in various lighting conditions. Tail shape distinctive. JOHN PUSCHOCK. ALASKA. OCT.

19 Adult and 1st cycle. TOM JOHNSON. ALASKA. JUNE.

20 Adult. Grayish underwing. Black on outer edge of p10 (far wing). Central tail feathers with noticeable wear. JERRY TING. CALIFORNIA. JAN.

21 Adults and 1st cycle (right). Long, pointy hand. Deep, forward sagging breast on leftmost bird is a typical Ross's silhouette. Grayish underwing on adult can appear blackish due to lighting. Pink blush on underparts evident. TOM JOHNSON. ALASKA. JUNE.

22 2nd-cycle type with adult Black-legged Kittiwake. Aged by faint black spotting on primary coverts and outer-primary tips. SVEINN JONSSON. ICELAND. MAY.

7 LITTLE GULL *Hydrocoloeus minutus*

L: 10.5"–11.5" (27–29 cm) | W: 27"–29" (69–74 cm) | Three-cycle | CAS | KGS: 4.5–5.5

OVERVIEW The smallest member of the gull tribe. A relatively recent addition to the gulls of North America, Little Gull is an old-world species, first documented here during the Franklin expedition (1819–1820). However, first evidence of breeding was not found until 1962, in Oshawa, Ontario (Weseloh 1994). Records of banded individuals in North America support the belief that Little Gulls may still be undergoing a transatlantic colonization from Europe or perhaps even Asia. Quite remarkable is a banded adult recorded at Point Pelee, Ontario, in July 2001, which had come from Finland (Wormington 2015), while the remains of a 1st-cycle individual from Sweden were discovered in Westmoreland County, Pennsylvania, in June 1996 (McWilliams & Brauning 2018). North American population estimated at around 400 pairs (Tyler L. Hoar, pers. comm.). Whether this is a self-sustaining population is unclear (Ewins & Weseloh 2020). Its preference for nesting in vegetation near shallow water makes breeding sites unpredictable, as fluctuating water levels readily drive this species away. A specialist at hawking insects and gleaning from the water's surface while in flight. The fluttery flight pattern is distinctive, unlike the deeper and more forceful beats of Bonaparte's, which it often associates with. Also associates to a lesser degree with *Sterna* terns and Black Tern.

TAXONOMY Monotypic.

RANGE
Breeding: Eurasia and E North America, where it is patchily distributed (breeds mostly NE Europe, W Siberia, and E Siberia). In North America, single pairs or very small groups breed or have bred at small lakes and marshes in S Great Lakes region, peaking during 1980s, including along the upper St. Lawrence River basin, as well as in the S Hudson Bay and James Bay region. In Great Lakes region, it has bred primarily in Ontario, where first nest was found, near Lake Ontario at Oshawa, in 1962, but where breeding sites are mostly ephemeral and breeding success generally low. Most confirmed nesting records were made between 1962 and 1989. Sites included those on Lake Ontario (e.g., east of Toronto, Ontario), Lake Erie (e.g., Rondeau Provincial Park, Ontario), Lake Saint Clair, Lake Huron (e.g., Georgian Bay, Ontario), and Lake Michigan (e.g., Manitowoc, Wisconsin, and Escanaba, Michigan). Also found nesting on several occasions in S Quebec (e.g., Lachine Rapids, early 1980s). One nesting record in S Minnesota, at N Heron Lake. Farther north, presumed breeding around western shore of James Bay and in S Hudson Bay lowlands, with a few confirmed nest records (e.g., in area around Churchill, Manitoba, on at least three occasions, first in 1981). Also possibly nests to west of there, in NW Canada, where adults seen during nesting season. But access to almost all of these northern areas is very limited, and ornithological coverage is poor. Fresh juveniles found as far west as California in August and September, suggesting the species may nest well west of its "traditional" breeding range in North America. Less likely is some transpacific movement involving Russian breeders.
Nonbreeding: Almost always associated with Bonaparte's Gulls. Numbers have declined somewhat throughout North America since peak between the late 1960s and late 1980s. Stronghold has always been Great Lakes region, especially north shores of Lake Erie and Lake Ontario (e.g., Long Point, Oshawa, and Niagara River area, Ontario, and, less so, Rochester, New York; numbers at the first two Ontario sites have reached 200 or more individuals in fall, with a record 266 birds—virtually all adults—at Long Point on 7 Nov. 1988 (Alan Wormington, pers. comm.). Also recorded at many other sites on a somewhat regular basis in small numbers, from Milwaukee region east to Cornwall Dam and the Montreal area along the St. Lawrence River, as well as farther to the northeast in the lower St. Lawrence River lowlands and Lac Saint-Jean, Quebec. Small to moderate numbers appear first, as early as late July into September, but then larger numbers are typically found in late fall, as late as December. Spring migrants normally return beginning in April.

Along the Atlantic Coast, formerly more numerous. Now rare to very rare but still regular in migration and winter, mostly from S New England through mid-Atlantic (Massachusetts to North Carolina). Very rare to casual north (to Newfoundland) and south (to central Florida) of core area. Most often seen at coastal bays, inlets, sewage-treatment ponds, and offshore out to a distance of about 10 miles: areas where large numbers of Bonaparte's Gulls are found. Numbers peaked during 1970s and 1980s, when up to around 100 individuals might be found, and there has been a notable decline since that time, especially in winter, although the species still occurs annually in very small numbers in most states between Massachusetts and North Carolina. Probably most likely to be found during November–December and, especially, between late February and late April, when current peak numbers are noted. This early-spring "influx" also noted at some reservoirs and rivers inland. During the period of peak abundance, a few spring counts reached 15–20 individuals in both New Jersey and Delaware. Some favored former sites since the 1970s (only a few of which are still current) include Newburyport, Massachusetts; Oyster River/Old Saybrook, Connecticut; several inlets along south shore of Long Island, New York; South Amboy, Jersey City, and Cape May area, New Jersey; lower Susquehanna River, Pennsylvania; Little Creek to Indian River Inlet, Delaware; mouth of Chesapeake Bay region, Virginia (where former maximum count was 6 birds); and Cape Hatteras, North Carolina. A few individuals may very rarely linger to early June, but only casually into summer, until first of returning birds appear, as early as end of July.

Casual visitor farther inland through Great Plains (mostly at large reservoirs) from Prairie Provinces south to Gulf Coast (as far south as central Mexico) and Texas, mostly in migration, and again, mostly in spring. Rare to very rare in interior West with sporadic sightings slightly increasing in the last decade. Along West Coast, a very rare winter visitor, with most records from Washington (e.g., Point No Point, Kitsap County) and California (most from Central Valley, along and up to 10 miles off the coast, and at Salton Sea). Also very rare north to S Alaska, Yukon, Northwest Territories, Nunavut, and Greenland, and at Bermuda. One bird in the N Bering Sea at Gambell, Alaska, in September 2010, was equidistant between known breeding sites in Russia and Canada.

IDENTIFICATION

Adult: A graceful gull with noticeably delicate proportions; very short, reddish legs; and a short, thin, black bill. Some show a dark maroon color to the bill, particularly in high breeding condition. Adults in alternate plumage are striking. The black hood lacks white crescents, making it difficult to see the dark eye from a distance. The underwing on adults is uniformly dark—from a dusky dark gray to jet black in many individuals. The upperwing is a silvery gray, with white tips bordering all of the flight feathers. Adults show no black on the upperwing. The wings are broad for such a small species, and the wingtips

1 Juvenile. Short petite bill, dark crown, bold earspot, and dark brown upperparts. ROXANE FILION. ONTARIO. AUG.
2 Typical juvenile, similar to 7.1. Upperparts dark chocolate-brown. MARK CHAVEZ. COLORADO. SEPT.

are decidedly rounded (in adults). The tail and underbody are white. In basic plumage, a black earspot is visible, and very often a dark cap is retained on the crown. Gray wash on the hindneck regularly extends to the sides of the lower neck and upper breast. *Similar Species:* Bonaparte's is larger, lacks a dark underwing, and shows a large white wedge on the outer primaries. Sometimes placed in the same genus, Little may be confused with Ross's Gull from a distance. Ross's has a disproportionately shorter bill and will much more often show a deep pink hue to its white feathers. The tail on Ross's Gull is wedge shaped, with longer central feathers; also, there is no black hood in alternate plumage or dusky cap in basic plumage. Ross's shows a much broader white trailing edge along the outer secondaries and inner primaries; the outer primaries gradually lose the broader white tips seen on Little. Ross's may have a dusky underwing, but not black. Also, Ross's regularly has a black sliver along the outer web of p10, which is also sometimes found in Little Gull, but not consistently. Structurally, Ross's has a narrower wing with a pointy hand, whereas Little has a broad wing and a blunt-tipped hand.

1st Cycle: Juveniles have a dark chocolate-brown cap and earspot, with dusky pigment surrounding the eye. The hindcollar—wrapping around most of the neck and enveloping the upper mantle—has a similar dark brown-black coloration, complementing the scaly brown juvenile scapulars with pale fringes. Juvenile scapulars soon give way to an admixture of adult-like gray-and-brown pattern (Sept.–Oct.). By November, many individuals show all-gray scapulars. The tertials often remain dark brown, as do the lesser and median coverts. The greater coverts are a grayish white. First cycles have an all-black bill and dull pink legs. The open wing on 1st cycles is remarkably bolder than that of similar species, in that Little Gull has a fuller, darker, and complete carpal bar continuing through to the primary coverts. The primary coverts are mostly black with dark outer webs and white inner webs to the outer primaries. The primary tips are dark with insignificant white apicals. The secondaries more often show a weak, washed-out appearance to the dark trailing edge, but this is variable from nearly all white to completely dark brown. The uppertail is white, sometimes with dark barring of juvenile feathers on upper rump, with a narrow tailband widest at its center and tapering off along the outer rectrices; the innermost tail feathers sometimes appear shorter in flight (when tail is not widely spread), creating a slight divot to the middle of the tail (not as substantial as in Sabine's). The underwing coverts are mostly white, but sometimes with a dusky cast throughout the wing linings. On the folded primaries, the wing typically shows a slight sawtooth pattern, with black outer webs against white inner webs. In flight, 1st cycles have a more pointed hand (unlike adults) but still display a relatively broad wing for such a diminutive gull. First alternate birds rarely take on a full hood in spring; rather, a basic-like head with dusky cap and earspot is more typical, or a partially mottled hood. At this time, a variable number of tail feathers may be replaced to all white, and hence the dark tailband becomes fragmented. *Similar Species:* Black-headed and Bonaparte's Gulls are larger, with discontinuous carpal bars at the primary coverts and alula. These two also show a uniform and complete dark trailing edge to the secondaries. Black-legged Kittiwake juveniles have all-gray backs with a relatively thick black semiring on the hindcollar, and entirely white secondaries. Although some 1st-cycle Littles may retain a hindcollar, the pattern is weaker than that of Black-legged and averages

---

3 1st cycle, now with several upper scapulars replaced. Black-and-white pattern on underside of p10 (far wing). SEBASTIAN JONES. MASSACHUSETTS. OCT.
4 1st cycle. Preformative molt well underway with most scapulars replaced. Solid dark brown median and lower lesser coverts. JAMES PAWLICKI. OHIO. SEPT.
5 1st cycle. 1st prealternate molt produces black on head, white hindneck, and new, adult-like gray upperparts. AMAR AYYASH. WISCONSIN. JUNE.
6 1st cycle. Some 1st alternate individuals take on a complete black hood. Note black-and-white sawtooth pattern on juvenile primaries. AMAR AYYASH. WISCONSIN. MAY.
7 1st cycle with similar-aged Bonaparte's. Smaller in every respect with solid brown tertials. DAVID TURGEON. QUEBEC. JUNE.
8 1st cycle largely in juvenile plumage. Bold M-pattern across upperwing with some black on secondaries (variable). RYAN SANDERSON. INDIANA. SEPT.
9 1st cycle showing considerably dark inner primaries. This individual has replaced some outer wing coverts (gray). DELLA LACK. ENGLAND. AUG.
10 1st cycle. Replaced scapulars and lower neck (formative). Particularly bold black carpal bar and primaries. Typical tailband pattern. MERJIN LOEVE. THE NETHERLANDS. DEC.

more brown than black. The ulnar bar on juvenile Black-legged Kittiwake, although continuing on to form a complete M-pattern, averages thinner and weaker than Little's. Bill and body sizes of Little are unmistakable and noticeably smaller than those of Black-legged Kittiwake. Ross's Gull at any age is much rarer, except perhaps on the North Slope, other parts of Alaska, and the Arctic. Head pattern is generally

---

11 1st cycle. Largely white underwing. Dark shadow bar on underside of secondaries. Two-toned black-and-white outer primaries. JIM TAROLLI. NEW YORK. NOV.
12 1st cycle. Fairly pale inner primaries and secondaries. Trailing edge can appear all-white from a distance, inviting confusion with Black-legged Kittiwake or Ross's, but note piano pattern on outer primaries, paired with black cap. GLYN SELLORS. ENGLAND. APRIL.
13 1st cycle with melanistic juvenile wings. This aberration is uncommonly found in a number of small gull species. AMAR AYYASH. WISCONSIN. MAY.
14 1st alternate Little Gull (left) with similar-aged Bonaparte's. Compare extent of carpal bar at wrist and trailing edge. DAN DUSO. MICHIGAN. JUNE.
15 1st alternate with complete black hood. All tail feathers replaced (as in 7.14). Consistent black on secondaries and throughout inner primaries. STEVE YOUNG. ENGLAND. MAY.
16 2nd cycle now molting inner primaries (visible primaries juvenile). Note petite bill, short legs, and solid dark tertials. AMAR AYYASH. WISCONSIN. JUNE.
17 2nd cycle in basic plumage. Faint cap and earspot, and gray wash on hindneck. Black on outer primaries typical in this age class. SEBASTIAN JONES. MASSACHUSETTS. OCT.
18 2nd cycle in alternate plumage. Complete hood not uncommon at this age. Black on tertials and outer primaries rather extensive. HENREY DEESE. MASSACHUSETTS. MAY.
19 2nd cycle undergoing 2nd prebasic molt (p1–p5 2nd generation). Blackish wing linings and underside to new primaries. CHRISTOPH MONING. GERMANY. AUG.
20 2nd cycle in basic plumage. Aged by black on tips to outer primaries. Blackish underwing, typically not as bold as adult. STEVE YOUNG. ENGLAND. SEPT.
21 2nd cycle with adult-type Bonaparte's (right). Compare size and wing pattern. CHUCK SLUSARCZYK JR. OHIO. NOV.
22 2nd cycle. Safely aged by considerable black on the wingtip, with streaks on the alula and primary coverts. 2nd prealternate molt likely underway, as suggested by solid black cap merging with earspots. DELFIN GONZALEZ. SPAIN. DEC.

a small earspot with mostly white crown, and it has longer wings with more white on the mid-primaries. The wedge-shaped central tail feathers on Ross's are distinctive.

2nd Cycle: Similar to adult, in that the underwing is dark, but it averages less pigment than adults on the underwing coverts. Regularly seen with full hood, but also may have a mottled black-and-white head. Upperwing all gray with some dark streaking on the primary coverts and / or some pigment on the outer primaries. *Similar Species:* Confusion may arise in the midst of 2nd prebasic molt in midsummer. The inner primaries on Little have broad white tips, as opposed to the entirely gray inner primaries of similar-aged Bonaparte's.

MOLT A distinctive subadult, 2nd basic plumage is regularly recorded. Therefore, Little Gull is sometimes regarded as a three-cycle gull. Some individuals may become adult-like in 2nd basic plumage, however (more study needed). First cycles appear to have partial preformative molt, resulting in adult-like gray scapulars and whitish head and neck feathers, and a subsequent partial prealternate molt, in which a weak to full hood may be acquired. A variable number of tertials and tail feathers may be replaced in 1st prealternate molt. Some individuals (rare) may replace all tail feathers at this time (Ayyash, pers. obs.).

HYBRIDS None.

23 2nd cycle in alternate plumage. White freckling on head is not uncommon. GLYN SELLORS. ENGLAND. APRIL.
24 2nd cycle. Complete alternate hood, shown with adult Bonaparte's (left). Note smaller body, smaller bill, and lack of white eye crescents. DANIEL JAUVIN. QUEBEC. MAY.
25 Adult in alternate plumage. Aged by clean white primary tips. Bold black underwing. Short bill, bright red legs. ANDREJ CHUDY. SLOVAKIA. APRIL.
26 Adult in alternate plumage. Lacks white eye crescents. Some show a subtle dark maroon bill in breeding condition. ANDREJ CHUDY. HUNGARY. MAY.
27 Adult in basic plumage. Grayish wash to neck, darkish cap and earspot. Black on underwing noticeable. UKU PAAL. ESTONIA. DEC.
28 Adult with 2nd-cycle-type Bonaparte's. White primary tips, small head with blackish cap and short bill. JAMES PAWLICKI. OHIO. OCT.
29 Adult in alternate plumage. Complete white trailing edge extends to outermost primaries. DALIA GAMTA KINTU. LITHUANIA. AUG.
30 Adult. Diagnostic black underwing typical. ILYA UKOLOV. RUSSIA. APRIL.
31 Alternate adult. Some otherwise perfect-looking adults have duller underwing, perhaps age-related. Pink blush on underparts. STEVE YOUNG. ENGLAND. APRIL.
32 Adult still in basic plumage at this date. Unmistakable. Darkish cap and earspot, and gray wash on hindneck. DALIA GAMTA. LITHUANIA. APRIL.
33 Adult type in basic plumage. Dark streak on outer edge of p10 likely age-related. JOSHUA VANDERMEULEN. ONTARIO. NOV.

8 BONAPARTE'S GULL *Choroicocephalus philadelphia*

L: 12.5"–13.5" (32–34 cm) | W: 31.5"–34.0" (80–86 cm) | Two-cycle | CAS | KGS: 5–6

OVERVIEW The most widespread small gull in North America, found on both coasts and in the interior in migration and in winter. Bonaparte's Gull is a benchmark species that should be learned thoroughly. Very much pelagic, but equally at home on small lakes and on rivers, and regularly at sewage lagoons and dams. Breeds throughout wetlands and boreal forest edges in Alaska and interior Canada. The only gull that nests predominantly in trees, with a strong preference for black spruces. Thus, a future English name for this species may reference its unique arboreal-nesting habit. Understudied on its breeding grounds, which are somewhat remote from human activity. Whether plunge-diving or surface-seizing, its buoyant flight patterns and foraging habits are a skillful combination of gull, tern, and shorebird behaviors. Diet quite varied: feeds mainly on insects in the breeding season; in migration and in winter, feeds mostly on small fish and invertebrates in a wide range of microhabitats.

TAXONOMY Monotypic.

RANGE
Breeding: Typically nests singly or in small colonies in trees (more rarely on the ground) bordering ponds and marshes in open taiga forest and muskeg from the Alaska Peninsula and interior W Alaska eastward across central and N Canada to W Quebec, and locally to E Quebec (e.g., Lac Saint-Jean; casually to Magdalen Islands). Single pair confirmed breeding in Aroostook County, Maine, in 2016, 2020 (Bill Sheehan, pers. comm.).
Nonbreeding: Uncommon to locally abundant migrant in E North America (especially on E Great Lakes and Niagara River (until freeze-up), where as many as around 100,000 birds might occur, and along the St. Lawrence River), from the central Great Plains east to the Atlantic Coast, and along the Pacific Coast. Typically found in flocks. Largest numbers in interior are at large lakes, reservoirs, shallow rapids, and dams. On and near the coast, prefers bays, river mouths, inlets with strong currents, and sewage-treatment ponds. Uncommon to rare in much of interior West eastward through W Great Plains, except locally uncommon to fairly common at several large lakes and Salton Sea. The first fall migrants, including both adults and juveniles, arrive early to SE Alaska and to the Great Lakes, Canadian Maritimes, and Maine—by late July or early August. Large numbers, formerly up to 30,000, congregate in late summer in Bay of Fundy region (e.g., Passamaquoddy Bay, Maine / New Brunswick). Further substantial movements south and east do not take place, however, until October through December. Rare to very rare visitor north to Newfoundland and Labrador. Along the Pacific Coast, a few migrants may also appear in late summer, south as far as N California, but migrants are most numerous there from October to December.

In late fall and winter, large numbers (up to thousands) may be found through December, and in some mild years well into January, around S Great Lakes, especially Lakes Erie and Ontario (particularly near the Niagara River) and along the St. Lawrence River. Thereafter, all but a few birds are forced out by freeze-up of most open water, at which time large numbers move to the Atlantic Seaboard and to large lakes, reservoirs, and dams on the S Great Plains and in the interior Southeast. Along the Atlantic Coast, the largest numbers winter between Massachusetts and North Carolina, with small numbers north to S Nova Scotia, and locally moderate numbers (uncommon to fairly common) south to Florida (both along the coast and inland). Most numerous along and near the coast at inlets and bays with substantial tidal currents and rips, as well as at sewage-treatment ponds. Also found regularly in moderate numbers offshore, out to a distance of around 10 miles, sometimes even farther (e.g., at edge of Gulf Stream, about 20–60 miles offshore). Small to moderate numbers winter in the Gulf of Mexico south to NE Mexico, and small numbers to the Bahamas and Cuba.

"True" juveniles are rarely encountered together in large numbers during migration. Noteworthy is the record of 3,450 1st-cycle individuals in migration on 1 September 2010 at Point Pelee—an extraordinary account for the Great Lakes, where adults are usually the predominant age group at that time of year (Wormington 2013).

Along and near the Pacific Coast, uncommon to fairly common, only very locally common, in winter (Nov. to April) from N California to Baja California and NW mainland Mexico (e.g., Gulf of California). Habitats as in East Coast birds, preferring areas with tide rips, sewage-treatment ponds, dams, and offshore waters out to a distance of around 20 miles. May be numerous inland in Central Valley and Salton Sea, California. From NW California to S British Columbia, common at this season only locally in Puget Sound; mostly rare elsewhere. Uncommon to rare locally in winter in interior West, such as at large reservoirs in S New Mexico and W Texas. Casual in winter north to S Alaska.

Spring migration commences during late February or early March, with the first wave of migrants arriving on Great Lakes typically in late March, in central Canada in early or mid-May, and on the breeding grounds by late May. Peak movements mostly mid-March to mid-May, earliest in mid-Atlantic and S New England, on southern and central Great Plains, and along US West Coast. Rare in much of the winter range after mid-May, but a few birds often linger into June. Summering nonbreeders—mostly young birds, but also a few adults—occur regularly in small numbers through much of June, with the largest and most consistent numbers on the Great Lakes, locally along the West Coast, and along the East Coast in New England and the mid-Atlantic. Abundance thereafter declines substantially in most areas as the summer progresses, until the first early-fall migrants arrive. Rare to casual in summer at other inland sites south of the breeding range.

Rare to casual visitor west to Bering Sea islands in Alaska, Hawaii, Russia, Japan, and Taiwan; east to Bermuda, Greenland, much of Europe, and N and W Africa; north to Nunavut; and south to S Mexico and to most countries in Central America as far as central Panama, as well as to the Lesser Antilles. Casual or accidental to Azores, W Europe, Hawaii, NE Russia, Japan, and Taiwan.

Substantial declines in numbers have been noted in recent years at a number of wintering and migration stopover sites along the East Coast and the West Coast; they were formerly much more numerous on both coasts. In much of S California, for example, Bonaparte's Gulls are no longer found in good numbers on the mainland coast, and the most consistent flocks are present from 2 to 15 miles offshore. On the Great Lakes, numbers have fallen substantially during migration. In the mid-Atlantic region, numbers at several favored coastal inlets have noticeably declined.

IDENTIFICATION

Adult: In alternate plumage, adults sport a black hood with thin white eye crescents. In basic plumage, the hood is reduced to a small, dark earspot on a white head, which will often show a noticeable smoky-

1 Juvenile. Ginger-brown scapular centers and hindneck. Variable gray on wing coverts expected. AMAR AYYASH. ILLINOIS. AUG.
2 Juvenile. Some 1st cycles show a paler bill base, inviting confusion with Black-headed, which has a longer bill that's consistently paler. KEITH CARLSON. IDAHO. AUG.

gray wash to the nape and lower hindneck. Small, thin, tern-like bill has a pointy tip and is predominantly all black. In high breeding condition, the legs are an orange red that transitions to a duller pink in nonbreeding condition. Some birds show a pink blush, mainly on the white underparts. The upperparts are pale gray. The wing has a distinctive pattern, with white outer primaries and primary coverts forming a flashy wedge from above when the wing is open. The primaries (p3 / p4–p10) have black tips, forming a narrow black trailing edge to the outer wing. At rest, relatively small white apicals are seen against fresh black feathers, becoming indistinct when the outer primaries are worn. On the folded wing, a large white sliver may be seen on the primaries, adjacent to the tertials and inner greater coverts. The underside to the folded primaries is predominantly white. *Similar Species:* Black-headed and Little Gulls have some plumage and structural similarities. Both are rarer anywhere in North America but are often found

3 Juvenile with extensive gray on upperparts (1st basic), with juvenile Laughing Gull. GINA BEEBE NICHOL. CONNECTICUT. AUG.
4 1st cycle with many scapulars now replaced with plain, adult-like gray feathers (formative). Lower scapulars, upperwing coverts, and tertials juvenile. AMAR AYYASH. MICHIGAN. SEPT.
5 1st cycle with adult. White crown, gray wash on hindneck, reduced earspot, and entirely gray scapulars (formative plumage). Lesser coverts on Little Gull bolder brown, lacking pale edging. ALLAN CLAYBORN. OHIO. NOV.
6 1st cycle. Plain gray median coverts on this individual (juvenile). The lesser coverts are obscured by the long outer scapulars. AMAR AYYASH. ILLINOIS. NOV.
7 1st cycle in alternate plumage. Partial hood common at this age. Fresh scapulars (1st alternate) have grown in, along with white neck and body feathers. Dull fleshy-orange legs. JIM HULLY. WISCONSIN. MAY.
8 1st cycles. Many new coverts and scapulars. Note white flash on upperside of stretched primaries. AMAR AYYASH. WISCONSIN. MAY.
9 1st cycle. Brown wash on hindneck and mantle. Scapulars a mix of juvenile and formative feathers. AMAR AYYASH. INDIANA. SEPT.
10 1st cycle in formative plumage. Complete dark trailing edge, thin tailband, and M-pattern that breaks at the wrist and outer wing. AMAR AYYASH. INDIANA. NOV.
11 1st cycle with extensive melanin on upperwing, forging a pattern on the inner primaries and secondaries similar to Black-headed Gull, but note short black bill. AMAR AYYASH. WISCONSIN. MAY.

associating with Bonaparte's Gulls, giving observers one more reason to carefully scrutinize every bird in the flock. Black-headed Gull is slightly paler, larger, and longer legged, with a dark brown (not black) hood. The bill on Black-headed is appreciably longer and is a deep red that may be tinged with black or have a graduated black tip. The underprimaries on Black-headed show extensive black throughout, readily eliminating Bonaparte's. Little Gull is proportionately smaller, with shorter legs and bill. There are no contrasting white eye crescents against the black hood on Little Gull. Also, Little Gull lacks a white wedge on the upperwing along the outer primaries. The underwing (both flight feathers and coverts) on adult Little Gulls is largely black, unique among all gulls, and promptly grabs the attention of observers.

1st Cycle: Juveniles (July / Aug.–Sept.) regularly show a ginger-brown wash on the hindneck, breast, and crown. The bill is black but may show a little pink, paling near the base, mostly on the lower mandible (often more than in adults). Legs dull pink. The gray greater and median coverts show white and brown subterminal edges, soon becoming uniformly gray due to wear and fading. The lesser coverts are brownish black of varying intensity. Young Bonaparte's Gulls soon replace their ginger-brown scapulars with gray, adult-like formative feathers, often by mid-September through October. At this time, the head and neck commonly become all white with a black earspot. Folded primaries similar to adult's, although the white sliver near the tertials is less prominent, showing more black. The open wing on young Bonaparte's Gulls reveals an M-pattern across the upperwing, mostly through the lesser coverts, but also through portions of the inner median and greater coverts. This M-pattern is relatively weak in Bonaparte's, breaking at the wrist; hence, the primary coverts and alula show white "headlights" with variable black streaks. Quite distinctive on the open wing is the complete

12 1st cycle. Dark trailing edge across entire wing and thin tailband. Pale gray underside to primaries. HARTMUT WALTER. CALIFORNIA. DEC.
13 1st cycle in alternate plumage. Complete hood sometimes found at this age. Pale underprimaries (far wing). Inner primaries and inner-primary coverts gray (compare to Black-headed). AMAR AYYASH. WISCONSIN. MAY.
14 Both 2nd cycles now molting primaries (new primaries emerging on individual on right). Note variation in head pattern at this age. AMAR AYYASH. WISCONSIN. JULY.
15 2nd cycle with similar-aged Black-legged Kittiwake (right). Thin black bill and noticeably smaller. STEVE ARENA. MASSACHUSETTS. JULY.
16 2nd cycle in heavy wing molt (2nd prebasic molt). Two outer primaries are juvenile, as well as inner secondaries. AMAR AYYASH. WISCONSIN. JULY.
17 2nd cycle. Black markings on all primary tips (adults usually have p1–p2 with no black on the tips). Also aged by black streaks on primary coverts, secondaries, and tail. SCOTT JUDD. ILLINOIS. MARCH.
18 2nd cycle in basic plumage. Aged by black on p1–p2, spots on tail and secondary centers. AMAR AYYASH. OHIO. SEPT.
19 Adult in alternate plumage. Full hood with white eye crescents, thin black bill, and orange-red legs. AMAR AYYASH. MASSACHUSETTS. APRIL.

black trailing edge from the body to the outermost wingtip seen from above and below. The inner primaries are gray, becoming two toned with white at p5/p6, with the outer primaries having white inner webs and black outer webs. The underwing is largely white, except for the narrow black trailing edge. The white uppertail is complemented by a narrow black tailband. By May, 1st alternate plumage may result in a hooded appearance, although rarely full as in adults. *Similar Species:* In addition to Black-headed and Little Gulls, consideration should be given to Black-legged Kittiwake or other species showing the so-called M-pattern across the upperwing. Black-headed displays larger proportions, with a pale bill and dark tip. Overall, the M-pattern is weaker and often paler across the upperwing. Note the outer greater primary coverts are largely white on Black-headed but streaked with black on Bonaparte's. On Black-headed, the inner primaries also show more pigment along the shaft streaks, whereas they are usually all gray on Bonaparte's. The dark underprimaries on Black-headed rule out Bonaparte's, but some Black-headed Gulls in late spring to early summer can show paler underprimaries due to fading. Compare size and structure. Little Gull is noticeably smaller, with shorter legs and a head pattern often showing a more distinct black cap. The tertials and upperwing coverts are typically bolder and more solidly filled than Bonaparte's. In flight, Little Gull's carpal bar is often much darker and wider than Bonaparte's, and the secondaries lack a complete, uniform, black trailing edge. Note the paler tips to the innermost

20 Adult undergoing prebasic molt (old p10 with new incoming primaries). Dull orange-red legs in nonbreeding condition. AMAR AYYASH. MICHIGAN. SEPT.
21 Adult. Blackish hood, thin white eye crescents, and slender bill with tapered tip. An unusual amount of white seen on outer primaries here. FRED LELIÈVRE. QUEBEC. MAY.
22 Adult in basic plumage. Small earspot and gray wash on hindneck. Note white wedge on base of outer primaries and white on underside of far wing. AMAR AYYASH. MICHIGAN. NOV.
23 Adult in alternate plumage. White wedge on outer wing with black primary tips. No white on trailing edge. WAYNE SLADEK. WASHINGTON. APRIL.

primaries on Little. Little also has more black along the medial region of the outer primaries. The tail on Little often has a small divot at the innermost rectrices, with the tailband tapering off distally. On the folded primaries, Little Gull often shows a jagged, sawtooth pattern of black outer webs and white inner webs that isn't found on Bonaparte's. Ross's Gull lacks black trailing edge to secondaries, with inner tail feathers forming a long, pointed wedge. Black-legged Kittiwake is larger, with a continuous black M-pattern not breaking at the wrist. Black-legged has an all-white trailing edge.

2nd Cycle: This age group may be distinguished by one or more of the following: noticeable black markings on primary coverts, tail, tertials, and secondaries or black streaks along outermost primaries. More study needed, as no known-age 2nd-cycle plumages are on record.

MOLT Exhibits two partial molts, preformative and prealternate, in the 1st plumage cycle (CAS). Signs of preformative molt show as early as late August, first noted via juvenile scapulars being replaced with adult-like gray feathers (see plates 35–37 in the Introduction).

HYBRIDS None.

24 Adult. White underside to outer primaries, and gray underside to inner and mid-primaries. AMAR AYYASH. WISCONSIN. MAY.
25 Adult and 1st cycle in alternate plumage. Flight feather molt will soon begin. AMAR AYYASH. WISCONSIN. MAY.
26 Adults in alternate plumage. Orange-red legs, gray underwings, and white wedge on outer wing. EDWARD KROC. BRITISH COLUMBIA. APRIL.
27 Adult in basic plumage. Narrow wings with thin white leading edge. White flash on outer hand also found in Black-headed, but compare bill pattern and overall size. AMAR AYYASH. INDIANA. NOV.

9 BLACK-HEADED GULL *Choicocephalus ridibundus*

L: 14.5"–17.0" (36–43 cm) | W: 35.5"–39.0" (90–99 cm) | Two-cycle | CAS | KGS: 4–5

OVERVIEW Widespread hooded gull throughout Europe and Asia. Black-headed Gull is a fairly recent addition to North America, having first been recorded here in 1930 (Holt et al. 1986). First likely breeding record from Stephenville Crossing, on the west coast of Newfoundland (1977), but first actual nests found on Magdalen Islands, in the Gulf of St. Lawrence, Quebec in 1982 (Aubry 1984). Presence in North America believed to be correlated with an increasing population in Iceland and Greenland. Noteworthy is a banded bird from Iceland found in Rye, New Hampshire (2003), and historically, at least six individuals banded in Iceland and resighted in Newfoundland between 1943 and 1971 (Montevecchi et al. 1987). Usually found singly in the interior and Northeast, often in Bonaparte's or Ring-billed Gull flocks. Larger than Bonaparte's but smaller than Ring-billed. Feeding ecology and behavior often more similar to Ring-billed and Laughing Gulls than to Bonaparte's. Regularly visits landfills in the Old World, but rarely in North America. Can be very tame at times, taking handouts. Some are much more reserved, staying on large bodies of water away from humans. English name is a misnomer, as the species has a dark brown hood, not black.

TAXONOMY Black-headed Gull belongs to a well-supported clade of gulls known as "masked gulls" (as opposed to "hooded gulls"). Two subspecies. Birds in Newfoundland, Quebec, and along the Atlantic believed to be nominate *Choicocephalus ridibundus ridibundus*. Alaskan birds thought to be larger, E Siberian subspecies, *Choicocephalus ridibundus sibiricus*. Assigning subspecies in the field based on size is likely impossible. First-cycle *sibiricus* is said to have weaker, paler carpal bar and may average less black on outer tail feathers with a narrower tailband, but more study needed (Olsen 2018).

RANGE
Breeding: Eurasian breeder. Breeds in North America mainly in SW Newfoundland: nesting first noted there in 1977 (Finch 1978). Prefers lush vegetation, often near water's edge. Probably also breeds or has bred in S Labrador. Very small numbers (around 15 pairs) also breed or have bred in the Gulf of St. Lawrence, on the Magdalen Islands, since at least the early 1980s, and similar numbers at Pointe de l'Est National Wildlife Area on Iles-de-la-Madeleine in 2007 (Cotter et al. 2012). A juvenile being fed minnows by an adult on 2 August 1994 is strong evidence of nesting along the NW Iowa / SE Minnesota state line at Spirit Lake. Two adults were found in the region on this date, with multiple recurring sightings in subsequent years. Has nested or has attempted to nest in Massachusetts (1984 and 1986), Maine (1986), and possibly Nova Scotia; and one individual may have bred in NW Iowa or in adjacent S Minnesota during the late 1990s. From Europe, this species spread west, first breeding in Iceland in 1911 (where it is now common) and in W Greenland beginning in the early 1960s, although it is more elusive now. In Nuuk, 5–25 are observed annually, mainly from late April to May, and then again in August. Interestingly, only adults are seen (Lars Witting, pers. comm.). Breeding in small numbers suspected in S Greenland.

Nonbreeding: Uncommon locally in Newfoundland. At least 200 in St. John's Harbor is a North American high count (Dec. 2005). Recently decreased dramatically as a winter resident here, due to closure of sewage outfalls. Uncommon to rare in Gulf of St. Lawrence and Nova Scotia. Rare along Atlantic Coast south to New York, rare to New Jersey, North Carolina, Bermuda, and inland slightly to E Pennsylvania and E Ontario. Very rare farther south along the Atlantic Coast south of North Carolina, but reports to the south are increasing. Recorded in every month of the year, although most records are from between October and April, and it is strictly casual between late spring and early fall. Probably most often seen in mid-Atlantic and S New England in March–April. Most often found at coastal inlets, lakes, and ponds close to the ocean, sewage-treatment ponds and sewage outfalls, and even fast-food parking lots. Often associates with Bonaparte's and Ring-billed Gulls, less so with Laughing. Numbers along much of East Coast peaked during 1970s and 1980s but appear to have declined somewhat more recently, in part due to closure of some sewage outfalls and likely due to now-inaccessible sewage-treatment ponds. High count in the United States is 26, reported in Boston Harbor, Massachusetts (Dec. 1972).

Casual, mostly between late fall and early spring, in West Indies and south to Trinidad, as well as on Great Lakes west of Lake Ontario. Elsewhere in the interior, a casual migrant and winter visitor to E Great Plains and interior Southeast, south to South Padre Island, Texas. Casual north to Churchill, Manitoba, and west to SE Alberta and E Colorado. Accidental north to Bylot Island, Nunavut, and south to Veracruz, Mexico.

In the Pacific, Black-headed Gull is rare but regular in spring and early summer (primarily mid-May to mid-June) in W Alaska, mostly to W and central Aleutians and Bering Sea islands. Some representative high counts include 25 birds on Attu Island (May 1979), 5 on St. Paul Island (several dates), and 11 on St. Lawrence Island (17 May 2006). Very rare in E Aleutians and on mainland W Alaska coast. Casual in these same areas from midsummer to late fall (mostly Aug.–Oct.), and along the West Coast in migration and winter (Sept.–April) from S Alaska south to S California (over 50 records), especially inland on several occasions in Central Valley and Salton Sea. At least 8 state records in Oregon and over 20 in Washington. Accidental in Hawaii. In North America, eastern birds are presumably nominate *ridibundus*, whereas Bering Sea and Alaskan birds more likely *sibiricus*, strictly based on proximity. Those found farther south and in the interior are not understood, but some from the West Coast do appear relatively large when compared to accompanying Bonaparte's Gulls, so presumably *sibiricus*.

IDENTIFICATION

Adult: A small hooded gull with straight, slender bill. Dark brown hood in alternate plumage with dark, brownish-maroon bill. Dark eye with narrow white eye crescents more often joined at the rear than not. Head becomes white with dark earspot in basic plumage and often shows two vertical dusky bands: one above the eye and the other above the earspot. Lower neck may show faint, gray wash. Bill transitions to a brighter red with black tip as breeding condition diminishes. Leg color typically matches bill color throughout the year. The white leading edge on the upperside of the outer wing is distinctive. More distinctive is the black underside to the primaries (underside of two to three outermost primaries largely white). The primaries also have a narrow black trailing edge. Underbody and tail all white. Adults regularly show a pink blush to the underparts, although rarely the roseate color seen in Franklin's or Ross's.

Similar Species: Bonaparte's Gull is the primary confusion species to consider. Bonaparte's is smaller and more delicate, showing a petite black bill and dull pink legs. Black-headed is a shade paler (readily discernible in proper lighting). The underprimaries of Black-headed are largely black, not pale as in Bonaparte's (be aware that distant views and / or poor lighting conditions might give the impression of a darkish underwing on Bonaparte's). Voice can be useful in detecting a Black-headed Gull in a Bonaparte's

1 Juvenile. Similar to juvenile Bonaparte's, but slightly larger, with pale bill base. PETER HINES. ENGLAND. JULY.
2 Largely juvenile. Silvery greater coverts, paling bill base, and orange-brown legs distinctive. MARS MUUSSE. THE NETHERLANDS. JULY.

flock. Black-headed's call is a sharp but descending "*kreer*," on par with an emphatic Laughing Gull. Bonaparte's call is much more buzzy and tern-like, with a short, nasal "*kik*."

1st Cycle: Juveniles fledge with intense warm brown, cinnamon tones to crown, hindneck, mantle, and scapulars. These feathers are often mottled with white and show an admixture of contrasting gray and brown colors. Bill and legs are dull colored, soon becoming orange brown. Acquires adult-like gray mantle and scapulars in earnest, while head and neck become mostly white with dark earspot. The relatively weak

---

3   1st cycle. A paler individual with bolder white edging on upperparts. Preformative molt evident in neck and scapulars. RICHARD BOSNER. ENGLAND. AUG.
4   Typical 1st cycle in formative plumage, showing gray scapulars and whitish neck. Bill pattern distinctive. MAARTEN VAN KLEINWEE. THE NETHERLANDS. NOV.
5   1st cycle (*sibiricus*). East Siberian populations average larger and less brown on upperwing coverts. CHRISTOPHER TAYLOR. JAPAN. JAN.
6   1st cycle. Rather typical midwinter individual. Legs and bill base orange-brown. AMAR AYYASH. ENGLAND. DEC.
7   1st cycle with similar-aged Bonaparte's (right). Compare size, especially bill length and color. Black-headed slightly paler gray, with more dilute brown tertials. DELFIN GONZALEZ. SPAIN. FEB.
8   1st cycle. Dark chocolate hood. Some 1st alternate birds develop complete hood, but variable. Thin white eye crescents and dark reddish-brown bill. TONY ENTICKNAP. ENGLAND. MAY.
9   Juvenile. Dark carpal bar across wing coverts. Compare extensive pigment on inner primaries to Bonaparte's thinner trailing edge. BENGT BENGTSSON. SWEDEN. JULY.
10 Largely juvenile. Dark trailing edge and thin tailband. Largely unmarked outer-primary coverts with white wedge on outer wing. MARS MUUSSE. THE NETHERLANDS. AUG.
11 1st cycle in formative plumage. Brighter, orange bill. Inner-primary coverts well marked, unlike Bonaparte's. BERTIL BREIFE. SWEDEN. NOV.
12 1st cycle. Primary coverts largely unmarked. Pigment on inner primaries more extensive than Bonaparte's. AMAR AYYASH. PORTUGAL. DEC.
13 1st cycle (*sibiricus*). Upperwing with faint brown carpal bar. Reduced tailband on this individual. AMAR AYYASH. JAPAN. DEC.
14 1st alternate. Distinctive black on underside of primaries. Note extent of pigment across inner primaries and inner-primary coverts. MARS MUUSSE. THE NETHERLANDS. APRIL.

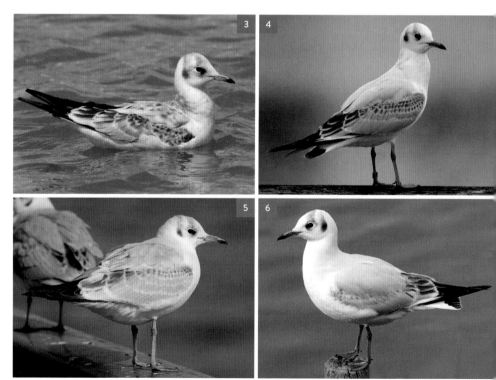

brown carpal bar across the upperwing coverts and the black trailing edge to the entire wing are distinctive. Similar to that of adults, the underwing shows blackish primaries, although at times not as bold as the adult's. The uppertail is white with a narrow black tailband. Not uncommon for outer web of outermost tail feather to be white, and some individuals have all-white outer rectrix. *Similar Species:* Again, similar-aged Bonaparte's is the only taxon expected to cause confusion in North America. Compare bill color and overall size. The underside to the primaries is blackish on Black-headed, although beware late-spring birds, which can often fade and cause confusion. Notice that the trailing edge on Black-headed has more pigment coming up the medial portion of the inner primaries. The carpal bar on Bonaparte's averages more pigment and more solid filling but is variable. Also, Black-headed typically has more white on outer-primary coverts, whereas in Bonaparte's these feathers are streaked with black, while the inner greater primary coverts are largely white (usually reversed in Black-headed, but there is considerable variation).

MOLT Adult *sibiricus* said to begin prealternate molt later than nominate. Partial preformative molt in hatch-year birds results in gray scapulars and mantle, and white head and body, and may replace a

15 2nd cycle, undergoing 2nd prebasic molt. Primary molt evident (new primaries emerging below tertials). Reddish-brown bill and legs, and faint brown hood. MARS MUUSSE. THE NETHERLANDS. JULY.
16 2nd cycle with adult-like p1–p4 (2nd basic). All other flight feathers juvenile. Trailing edge averages broader than Bonaparte's with cleaner outer-primary coverts. DAVID SANTAMARIA URBANO. SPAIN. JULY.
17 2nd cycle. Reddish bill and black underside to new primaries distinctive. MARS MUUSSE. THE NETHERLANDS. JULY.
18 Adult. Alternate plumage. Dark brown hood does not typically extend to the lower hindneck. Thin reddish bill and white on underside of far wing. MAARTEN VAN KLEINWEE. THE NETHERLANDS. JUNE.
19 Adult with similar-aged Laughing Gulls. Black-headed is smaller, with a finer bill, and noticeably paler. DIANNA LIETER. NEW JERSEY. APRIL.
20 Adult with incoming hood pattern (prealternate molt). Reddish bill base with blackish tip. Note white outer edge on primaries. MERIJN LOEVE. THE NETHERLANDS. FEB.
21 Adult (*sibiricus*). Basic plumage. Deep red legs and red bill. A large-bodied bird with relatively long bill. AMAR AYYASH. JAPAN. DEC.
22 Adult. Noticeable white wedge on outer primaries and all-white tail. Bill and legs dull orange-red. AMAR AYYASH. ENGLAND. DEC.
23 Adult with Common Gull (nominate *canus*). Obvious black on underside of primaries, white wedge on outer primaries, and reddish bill and legs. AMAR AYYASH. ENGLAND. DEC.
24 Adult (alternate). Dark brown hood may appear blackish when underexposed. Distinctive black on underside of primaries, with white on outermost primaries. MARS MUUSSE. THE NETHERLANDS. APRIL.

variable number of upperwing coverts; also a variable number of rectrices, but not common. Some tertials may be renewed to adult-like gray via preformative / prealternate. First prealternate molt also partial, resulting in irregular hooded pattern with white flecking. Head pattern varies from basic-like to complete brown hood. First prealternate may include a variable number of tail feathers (rarely all rectrices), typically restricted to a couple of inner pairs.

HYBRIDS Has hybridized with more species than any other "masked" gull. In North America, hybridization with Ring-billed Gull is most noteworthy—such as, for instance, an adult Black-headed × Ring-billed hybrid at Stephenville Crossing, observed bringing food into mixed colony (July 2005). Another putative adult hybrid was found on a nest in a Ring-billed Gull colony on Little Galloo Island on Lake Ontario, and that individual appeared to be paired up with a Ring-billed Gull in 1982 (Weseloh & Mineau 1986). This hybrid is reported once every two to three years in North America (see plates H8.28–H8.31). Hybridization with both Common Gull and Mediterranean Gull (*Ichthyaetus melanocephalus*) also reported (Taverner 1970; Adriaens et al. 2022). Copulation with Slender-billed Gull observed (Jones 1980). Purported to have hybridized with Relict (*Ichthyaetus relictus*) and Lesser Black-backed Gulls in captivity (McCarthy 2006).

25 Adult with freckled hood pattern at this date. Dull blackish-red bill. White outer wedge on wing and black on underside of far wing. AMAR AYYASH. ENGLAND. DEC.
26 Adult (basic). Maximum unmarked head pattern. Similar to other "masked gulls," such as Bonaparte's, Black-headed lacks white along the trailing edge of the wing. Dark reddish bill. AMAR AYYASH. PORTUGAL. DEC.
27 Adult with similar-aged Bonaparte's (left). Note differences in underside of primaries and bare-part colors. Bonaparte's averages slightly smaller. CHUCK SLUSARCZYK JR. OHIO. DEC.
28 Adult with similar-aged Little Gull (right). Note the brownish head and black limited to the underside of the primaries. MATS WALLIN. SWEDEN. JULY.

10 GRAY-HOODED GULL *Choicocephalus cirrocephalus*

L: 15.0"–17.5" (38–44 cm) | W: 39"–45" (100–115 cm) | Two / three-cycle | CAS | KGS: 5–6

OVERVIEW Gray-hooded Gull (or Grey-headed Gull) is a handsome southern-hemisphere species with a disjunct distribution. It is the only small gull resident to both South America and Africa. It is also unique in that it is the only species with a gray head, while other "hooded" species acquire brown or black heads in alternate plumage. Larger than Bonaparte's Gull and roughly the size of Laughing Gull, the species is increasing throughout its range, with some proclivity to wander, perhaps due to anthropogenic influences in the modern landscape. Colonial breeder. Although it is often found surface feeding and searching for exposed aquatic invertebrates, Gray-hooded has a rather catholic diet. It regularly scavenges, takes handouts, and has markedly increased at landfills in Argentina and South Africa. Otherwise, it is mostly found in coastal wetlands, harbors, beaches, and inland lakes and reedbeds.

TAXONOMY A "masked gull" belonging to the same genus as Black-headed Gull, which is its closest relative from the northern hemisphere. Two subspecies: nominate *Choicocephalus cirrocephalus cirrocephalus* from South America and *Choicocephalus cirrocephalus poiocephalus* from Africa. Nominate is slightly larger, with paler upperparts and a fainter gray hood. African race may average smaller p9–p10 mirrors, but larger dataset required to verify this. No transcontinental gene flow suspected between the two subspecies. Sometimes placed in *Larus* (del Hoyo et al. 1996).

RANGE Limited mainly to tropical and subtropical regions of South America and Africa. Breeds on both coasts of South America, where it is locally common from S Ecuador to S Peru. On the Atlantic Coast of South America, breeding occurs along waterways leading to Rio de la Plata and along the northern coast of Argentina. Nonbreeders move as far north as NW Argentina, Uruguay, and S Brazil. In Africa, distribution is patchy, but it also breeds along the coast and inland. Resides in the W Sahara, Madagascar, and south to South Africa. In Africa, no systematic migration known. The species wanders widely in the sub-Sahara, depending on available water sources inland. Found near tidal pools and saltpans along the coast.

Vagrant to Panama, where it is accidental, but there has been a spurt of records since 2000: approximately eight records, all northeast of Panama City, primarily July–October. Found associating with Laughing Gull in most instances. Singly to Punta Morales, Costa Rica (late April–May), Galapagos (Aug.), and Barbados (May–June).

Two records in the United States (origins unknown), in both instances found associating with Laughing Gulls. First in Franklin County, Florida (26 Dec. 1998)—alternate adult. Second in Kings County, New York (24 July 2011)—worn adult type, missing p1 on both wings, with tail feathers unusually frayed, unlike natural wear seen in free-flying birds. Lingered for nearly two weeks. Very likely to occur in the United States again.

Normal range is in South America and Africa. In South America, found near and along the Pacific Coast from S Ecuador to S Peru, rarely N Chile; and along the Atlantic Coast and well inland from central Argentina north to S Brazil, rarely or casually north to N Ecuador and to Paraguay and NE Brazil.

IDENTIFICATION
Adult: Pale gray upperparts similar to Bonaparte's Gull. Rich gray hood in alternate plumage, with darker outline toward the back of the "mask." White neck. Most adults have pale bone-white irises with a fish-eye stare, but some may have noticeable dark flecking. White eye crescents. Bill, orbital ring, and legs are a carmine red in high breeding condition. At other times of the year, the legs become a dark maroon or dark orange, with similar-colored bill showing a contrasting dark tip. Long, straight bill is proportionately thin. Underparts and entire tail white. Upperwing pattern is distinctive, consisting of gray inner primaries, black

outer primaries with p9–p10 mirrors, and a large white wedge extending to the leading edge of the hand. Adults in South Africa recorded with p8 mirror. Relatively broad wings with plumbeous gray underwing coverts getting progressively dark and fading into black underprimaries. No white trailing edge to the flight feathers. Outer secondaries and inner primaries may show dark shaft streaks, with dusky inner webs. In basic plumage, the head may show a diluted and faded hood with dark smudging or may be almost entirely white with an earspot. *Similar Species:* Distinctive, especially when sporting gray hood, but basic-plumaged birds should be compared to Black-headed Gull in the Nearctic. Black-headed does not have large black outer-primary triangle and does not show mirrors. Also, Black-headed maintains dark eyes.

1 Juvenile. Brown wash throughout with a blend of gray on the upperparts. Thin dark bill and faint earspot. Outer primaries not fully grown. AMAR AYYASH. PERU. NOV.
2 Same individual as 10.1, with 1st-cycle Franklin's Gull for size comparison. AMAR AYYASH. PERU. NOV.
3 1st cycle. Scapulars adult-like gray, whitish head with dusky markings, and reddish bill base. Tail pattern variable. AMAR AYYASH. PERU. OCT.
4 More advanced gray on wing coverts compared to 10.3. Similar to Black-headed Gull, but note lack of large white wedge on base of primaries. Gray-hooded averages darker gray (compare open wing). AMAR AYYASH. PERU. OCT.
5 Juvenile. Cinnamon-brown wash on head and neck. Secondaries all-dark, with white wedge on outer hand. MATT BRADY. GAMBIA. DEC.
6 1st cycle. Darker gray greater coverts and white wedge on outer hand. Thin tailband variable, at times reduced to a few dark segments. AMAR AYYASH. PERU. NOV.
7 1st cycle. Similar to Black-headed, but note entirely black webs to outermost primaries. AMAR AYYASH. PERU. OCT.
8 1st cycle. Dark, smoky gray underwing. All-dark underside to outermost primaries. AMAR AYYASH. PERU. OCT.
9 1st cycles. Dark secondaries, weak ulnar bars, and all-dark outermost primaries. Tail markings variable. AMAR AYYASH. PERU. NOV.
10 2nd-cycle type. Gray upperparts, white tail, and brighter red bare parts. Adult ruled out by dark eye and dark tips to primary coverts (below greater coverts). DAVID F. BELMONTE. PERU. JULY.
11 2nd-cycle type. Single small mirror (underside of far wing beyond tail tip) and dark eye suggest subadult. AMAR AYYASH. PERU. NOV.
12 Same individual as 10.11. Adult-like, but with single mirror and dark eye. Grayish hood, dark underwing, and white wedge on hand. AMAR AYYASH. PERU. NOV.

13 Adult. Gray hood, red bare parts, and pale iris. AMAR AYYASH. PERU. NOV.

1st Cycle: Juvenile has brown mottling on mantle and upperparts. Pale, buffy edging to upperparts quickly wears down. Extensive brown wash on head and breast in newly fledged birds soon becomes all white with a dark earspot. Juvenile upperwing coverts, and especially the greater coverts, may show contrasting silvery-gray centers. Bare-part colors variable in any one given flock. Blackish-brown bill transitions to dull orange with dark tip. Legs are orange brown to dull pink. The open wing presents a dark trailing edge to the flight feathers, with a brown carpal bar across the secondary coverts. Greater primary coverts show teardrop-shaped black tips with a flashy white wedge on the center of the hand. Underwing similar to adult's but with no mirror on the two outer primaries. Narrow tailband doesn't reach outer edge of outer tail feathers. Formative plumage acquired rather early and suddenly in many individuals, giving way to white, adult-like head with contrasting earspot, all-gray scapulars, light gray hindneck, and a reduced carpal bar. First alternate birds can show variable gray on head, with advanced birds acquiring a full hood. Some individuals may also replace all tail feathers at this time, taking on a white tail with no distal band. May also show renewed upperwing coverts and a variable number of replaced tertials, both of which are an adult-like gray. Eyes are brown at this time, but some are clearly a lighter brown, with a minority of birds taking on yellow irises similar to adult's. ***Similar Species:*** Should be compared to Brown-hooded Gull (*Chroicocephalus maculipennis*) in South America and with Hartlaub's (*Chroicocephalus hartlaubii*) in Africa. Although non-adults have not been recorded in the Nearctic, similar-aged Black-headed Gulls are likely to be the only confusion species. Outer-primary pattern from below and above is sufficiently different.

2nd Cycle: Overall resembles adult, but a combination of the following features may suggest a 2nd-cycle individual: light to dark brown eyes, smaller mirrors or p9 without mirror, duller bare parts, and noticeable dark wash across secondaries and tertials that contrasts with the rest of the upperwing.

MOLT Likely a three-cycle gull, as a somewhat distinct subadult form is often observed. Whether these are retarded adults or advanced 2nd cycles is not completely known. Data from known-age South

4 Adult in alternate plumage, with Brown-hooded Gulls (*Choicocephalus maculipennis*). Gray hood and darker upperparts distinctive. GUSTAVOE FERNANDEZ PIN. URUGUAY. JULY.

5 Adults. African race (*poiocephalus*) averages smaller, with deeper gray upperparts and fuller hood than nominate. GARY BYERLY. SOUTH AFRICA. JULY.

6 Adult. Similar to Black-headed, the hood does not typically extend to the lower head. Distinctive pale iris. AMAR AYYASH. PERU. NOV.

7 Adult. Basic head pattern, with similar-aged Franklin's Gull (left). Franklin's is a "hooded gull" with thick eye crescents, along with noticeable tertial and scapular crescents. AMAR AYYASH. PERU. NOV.

18 Adult type with 2nd-cycle Laughing Gull (left). Similar body size but with thinner bill, reddish bare parts, and pale iris. AMAR AYYASH. NEW YORK. JULY.

American birds is lacking, and determining whether some populations are two-cycle gulls and others three-cycle is still an open question. Some populations may breed year-round, such as those in Gauteng (Brooke et al. 1999). Birds in Senegal thought to undergo extensive to complete 2nd prebasic molts that result in adult-like aspect. Molt timing in many African birds is a function of breeding schedules, which invariably depend on peak rainfall periods, during which the availability of food increases. Some 1st cycles in Africa may replace all rectrices and secondaries in what is believed to be the 1st prealternate molt (Peter Adriaens, pers. comm.).

HYBRIDS Regularly hybridizes with closest relative, Hartlaub's Gull (*Choicocephalus hartlaubii*), in South Africa (Sinclair 1977).

19 Adult in alternate plumage. Large white wedge on outer hand, two large mirrors, and gray hood. Note darker secondaries. DAVID M. BELL. ECUADOR. NOV.
20 Adult. Similar to 10.19, with smaller mirrors and more wear. As in other masked gulls, gray secondaries lack white trailing edge. AMAR AYYASH. PERU. NOV.
21 Adult with faint gray hood. Consistently dark underwing, white wedge on outer hand, and two mirrors. AMAR AYYASH. PERU. OCT.
22 Adult type. Smoky gray wing linings and blackish underside to primaries. AMAR AYYASH. NEW YORK. JULY.
23 Adult types in various stages of molt, with alternate adult, 1st cycle, and Franklin's Gull. JOSHUA VANDERMEULEN. ECUADOR. NOV.

11 LAUGHING GULL *Leucophaeus atricilla*

L: 15.5"–17.0" (39–43 cm) | W: 38"–42" (97–107 cm) | Three-cycle | CAS | KGS: 8–9

OVERVIEW A slender-winged hooded gull with lead-gray upperparts. This three-cycle gull is locally abundant along the immediate coasts of the Atlantic and Gulf of Mexico. Also resident in the Caribbean and South America. Largest breeding colony in North America found in the intercoastal waterway surrounding Cape May, with an estimated 50,000 breeding adults. Common name derived from its boisterous, laugh-like call. Has generally benefited from human practices and is well adapted to people. Opportunistic and highly terrestrial. Found in salt marshes taking small crabs and crustaceans; also hawks for insects, forages for surface fish, follows shrimp boats, is a glutton for horseshoe crab eggs, and is no stranger to landfills. May move a short distance inland to work plowed or wet fields. Notorious beggar at beaches and tourist sites. Liable to turn up anywhere. Few gulls are prone to vagrancy worldwide as much as this species.

TAXONOMY The American Ornithological Society regards Laughing Gull as monotypic. However, two subspecies were proposed by Noble (1916) and were subsequently recognized by several workers, including Burger and Gochfeld (1996).

L. a. atricilla: Nominate race described from the West Indies, said to average smaller body and proportionately longer bill. Also more extensive black in wingtip, especially on p5–p6, with smaller apicals.

L. a. megalopterus: Ranges throughout the United States and Canada. Averages larger than nominate, with less black on wingtip and larger white apicals. Subterminal band on p6 thinner and more demarcated than in nominate (Pyle 2008). Age-related differences should be considered when assessing wingtip pattern.

RANGE
Breeding: Nests locally in colonies, mostly in marshes but also on rocky, vegetated, and sandy islands and beaches. Foraging birds occur in a wide variety of coastal habitats and in numbers well inland (up to 40 miles) at tidal rivers, animal-processing plants, landfills, farm fields, and shopping-center and fast-food parking lots. Numbers and range were substantially reduced by hunting and egging during the 1800s, but the species recolonized most areas lost by the 1970s and 1980s. Nests along Atlantic Coast from central Maine (common north only to Long Island) south to Florida and west and south along Gulf of Mexico, through Belize, to S Mexico. Also, locally in West Indies south to N South America (primarily French Guiana and parts of Venezuela). Former colony present on the Bird Islands in Nova Scotia until 1941, but no longer colonial there, with only several isolated records after 2001. There is a record from Machias Seal Island, New Brunswick. Has nested inland in SW Texas, at Lake Amistad and Falcon Reservoir. Along Pacific Coast, nests in Mexico from N Gulf of California south to Colima. Formerly bred in small numbers at Salton Sea, SE California, but not since 1950s. Nonbreeders widespread in all these regions during the breeding season. Single individuals have attempted breeding far inland, in S Great Lakes region, where they probably paired with Ring-billed Gulls, such as SE Chicago, where an adult was found on an egg in a Ring-billed colony (Bailey 2008).
Nonbreeding: Along Atlantic and Gulf Coasts and throughout most of the West Indies, nonbreeders are found in same habitats as breeders. Winter counts at some Florida landfills have reached highs of 100,000 birds. Also occur well out at sea, especially in Gulf of Mexico. Regular in summer and fall in small numbers in Nova Scotia and New Brunswick, very rare north to Newfoundland. In late summer (beginning in July) and early fall (after breeding), the numbers moving to foraging sites inland in E United States increase. Tropical storms and hurricanes may bring even larger numbers to the north (e.g., Atlantic Provinces) or inland. Many birds vacate the N and central Atlantic Coast by late October. In early winter

and well into December, small numbers linger regularly north to Delaware and S New Jersey, very rarely farther north, but by January they become difficult to find north of SE Virginia (during milder winters) or North Carolina. Spring migrants move north in March.

In the true interior, a rare but regular visitor to the Great Lakes region, as singletons (mostly in spring and summer), and also rare to very rare but widespread in many interior regions throughout the eastern half of the United States, progressively becoming rarer in W Great Plains, where it is very rare to casual. Accidental to S Hudson Bay. Rare in the interior West (away from Salton Sea, where it is a fairly common postbreeding visitor from NW Mexico).

Along the Pacific Coast, a rare visitor to S California at all seasons, very rare north of there; for example, there are only a handful of records to Oregon, with records north to SE Alaska; also to Hawaii (including a recovery of a bird banded in New Jersey!). South of the breeding range, nonbreeders are found regularly to northernmost Brazil and to Ecuador, and in small numbers to Peru. Casual south to SE Brazil and Chile, and well inland in South America. Also casual to Greenland, Iceland, and W Europe, and has records scattered elsewhere across many parts of the world, from E Europe to N and W Africa, and in the Pacific from Japan to Australia and Samoa.

Laughing Gull exhibits a disproportionate occurrence of aberrations in comparison to other gulls. One of these is a common bare-part color abnormality that results in a bright orange leg and bill color. This doesn't present any difficulties with respect to identification, but it is curious. A more problematic

1 Juvenile. Scaly, buffy brown upperparts and body, blackish primaries, and longish bill. AMAR AYYASH. DELAWARE. AUG.
2 Juvenile. Paler wing panel already showing wear, and bolder white edging to upperparts. Dark tail base, white underparts, and thin eye crescents. STEVE ARENA. MASSACHUSETTS. AUG.

aberration is melanistic individuals that recall much rarer species from outside the ABA Area. For example, in summer 2019, a melanistic Laughing Gull from Galveston, Texas, excited many people who were sure it was an adult Lava Gull (*Leucophaeus fuliginosus*). Closer examination of structural and plumage details revealed it was "just" a Laughing Gull. Another example is of a confusing gull from the 1987 Cameron Louisiana Christmas Bird Count. Independent parties who found the bird felt it was a Gray Gull, but the Louisiana Bird Records Committee rejected this account, since a melanistic Laughing Gull could not be eliminated.

IDENTIFICATION

Adult: Alternate adult has a black hood with white eye crescents which typically don't meet behind the eye. Dark gray upperparts. In breeding condition, the bill, legs, and orbital are dark red, approaching maroon at various times of the season. May show rosy-pink suffusion to head, neck, and underparts, especially in early spring. Legs are fairly long, as is the bill, which can appear heavy with a slight droop. At rest, the primary projection is noticeably long, with white apicals that are very prone to wear. The wingtip shows a prominent black wedge, which reaches the primary coverts. Black subterminal band typical to p6, with white tip to p5 (at times with thin or broken band found on nominate types). No mirrors. In flight,

3 1st cycle undergoing preformative molt. Whitening head, gray mantle and upper scapulars coming in, and molting tertials. White vent. DONNA POMEROY. VIRGINIA. OCT.
4 1st cycle. Dark hindneck and white loral region resembles Franklin's, but note flatter head, long droopy bill, and longer legs. SUZANNE ZUCKERMAN. FLORIDA. NOV.
5 1st cycle with adult-like scapulars, gray hindneck, and sides (formative). Lanky rear to the body, and longish bill. AMAR AYYASH. MEXICO. DEC.
6 Apparent 1st cycle, with rare complete black hood and white alternate neck pattern at this age. The reddish bill suggests hormonal asynchrony. JOHN GROSKOPF. FLORIDA. NOV.

7   1st cycle now with moderate molt in wing coverts and upper tertials, fairly common. Paler head and weak eye crescents, unlike similar-aged Franklin's. DARREN CLARK. TEXAS. MARCH.
8   1st cycle. Unlike Franklin's, white eye crescents thinner and do not typically meet at the back of the eye in most Laughings. Noticeably decurved culmen. AMAR AYYASH. FLORIDA. JAN.
9   Juvenile. Dark flight feathers with thin white trailing edge tapering off at inner primaries. Dark tail base and strong black bill. MICHAEL D. STUBBLEFIELD. NEW JERSEY. SEPT.
10 Juvenile. Note dark tail base and dark on outermost tail feathers (compare to Franklin's). Scaly brown upperparts and contrasting dark flight feathers. AMAR AYYASH. MARYLAND. AUG.
11 Juvenile. Long, pointy hand with dark underside. Marked wing linings. Note dark on outer edge of tail. AMAR AYYASH. NEW JERSEY. AUG.
12 1st cycle (formative). Darker underwing than 11.11, with gray wash on neck and sides. Note bill length and shape. AMAR AYYASH. FLORIDA. JAN.
13 1st cycle (formative). Typical long, narrow wings, long bill, and dark tail. Some postjuvenile wing coverts now grown in. AMAR AYYASH. FLORIDA. JAN.
14 Similar to 11.13 but with slightly more advanced molt in wing coverts. Variable number of tail feathers commonly replaced in 1st prealternate molt. AMAR AYYASH. FLORIDA. JAN.
15 1st cycle with juvenile primaries. All secondaries and tail feathers replaced (presumably via extensive 1st prealternate molt). ALAN KNEIDEL. MASSACHUSETTS. JUNE.
16 2nd cycle. Medium gray upperparts and long bill. Diffuse dark on head hints at hood. Visible primaries juvenile, but inner-primary molt has begun (2nd prebasic). TODD LEECH. WISCONSIN. JUNE.
17 2nd cycle (alternate). Commonly retains some white on the head at this age, with blacker bare parts. Note dark on tertials and absence of adult-like apicals. AMAR AYYASH. MASSACHUSETTS. APRIL.

the impression is of a long, narrow-winged bird with pointy wingtips. Tail entirely white. The underwing casts a darker shadow bar along the flight feathers. The underside to the outer primaries mirrors the prominent black triangle shown on the upperwing. In basic plumage, the head becomes largely white with dusky markings concentrated on the face, behind the eyes, and, to a lesser extent, on the crown. Bill and legs are decidedly black in nonbreeding condition but may show remnants of red. *Similar Species:* Franklin's Gull is more compact, averages shorter legs, and has a thinner bill that lacks the droopy structure found on Laughing. Eye crescents are noticeably thick and often touch behind the eye. The pink suffusion on Franklin's is more common and is typically more intense and vibrant. The wingtip on definitive adult Franklin's will usually show a white partition, or medial band, separating the gray bases of the primaries from the black subterminal bands. The underside of the wingtip on Franklin's shows a clearly demarcated black tip, not extending far beyond the wingtip. Basic adults average more black on their heads, with a semi-hooded appearance. Also note the signature dusky gray patch on the center of the uppertail on Franklin's. First alternate Franklin's Gull with 2nd-generation primaries can show a similar wingtip pattern, with no mirrors and extensive black. But the wingtip on these younger birds generally has

18 2nd cycle. Note new inner primaries (2nd basic). Retained juvenile outer primaries and secondaries. Medium-gray upperparts and dirty wing linings. WOODY GOSS. MASSACHUSETTS. JULY.
19 2nd cycle (alternate). Same individual as 11.17. Distinguished from adult by black on tail and primary coverts, and diffuse black on mid-primaries. AMAR AYYASH. MASSACHUSETTS. APRIL.
20 Adult (alternate plumage). Black hood and white eye crescents. Reddish bare parts in breeding condition. Long wing projection. AMAR AYYASH. MASSACHUSETTS. APRIL.
21 Adult (alternate). Bare parts duller than 11.20, becoming brighter soon. Note lanky rear to body and small apicals. Shape of hood altered by posture. AMAR AYYASH. VIRGINIA. MARCH.
22 Alternate adults in prime breeding condition. Variable pink suffusion (rightmost individual) sometimes found on underparts. AMAR AYYASH. MASSACHUSETTS. APRIL.

less black on the bases of p8–p9, at times not reaching the primary coverts on the inner web of p9 and the outer web of p8. Many 1st-cycle (and 2nd-cycle) Franklin's tend to show small black streaks on the primary coverts and often reveal a distinctive gray wash on the central tail feathers.

1st Cycle: Juvenile has a brownish-gray aspect with scaly upperparts showing pale edging. Head, breast, and sides have a brown wash. Thin white eye crescents. Black bill, dark legs, and dark flight feathers. On the open wing, note the broad white tips to the secondaries, becoming thinner and indistinct on the inner primaries. Uppertail coverts and rump largely white. Dark tailband is relatively broad and extends to the outer edges of the outer tail feathers. Underwings a patchy, smoky brown with a messy appearance. Underbody pale. Gray back of formative plumage may be limited to just a few feathers or may be extensive, resulting in all-gray scapulars. Similarly, wing coverts and tertials may remain all juvenile or could be extensively renewed with gray formative feathers before the end of the calendar year. At that time, the head, neck, and sides are mostly mottled with a mixture of gray, brown, and white feathers. Some take on a weak hooded appearance. First alternate birds develop a paler forehead, and a fair number of individuals take on a variable number of new white tail feathers. *Similar Species:* Franklin's Gull has an overall tidier appearance. Note the fuller semi-hood that appears more demarcated along its border. Paler neck and breast. Noticeably thicker and bolder eye crescents. Bill is shorter and skinnier, with no droop to the tip. Tailband averages thinner and does not reach the outer edges of the outer tail feathers. The underwing is whiter and cleaner than in Laughing. Also, on the open wing, note the paler inner primaries, which have broader white tips than in Laughing.

2nd Cycle: Overall appearance is adult-like, but with moderate to heavy gray wash on neck and breast. Upperparts can show strong brown tinge, black markings on tail and primary coverts, and few if any white tips to primaries. In alternate plumage, the hood can range from a lightly peppered look to virtually all black, resembling the adult's. Bare parts average darker and not as bright as adult's in breeding condition.

22

MOLT Apparently two partial molts in 1st cycle: preformative and prealternate. Some birds molt a variable number of wing coverts and tertials early in 1st cycle, which is associated with preformative. Others, with a later molt, presumably 1st prealternate, can molt a variable number of tail feathers, tertials, and head and body feathers, producing an adult-like hood pattern. Extreme individuals found with entire tail and all secondaries replaced, presumably in 1st prealternate before start of primary molt (Ayyash, pers. obs.). The extent of preformative and 1st prealternate needs study and may vary widely across populations.

The apicals on the outer primaries in adults can become completely worn down in spring (March–May), while the white tips to p5–p6 remain well intact. There are two reasons for this: first, by their nature, the white tips to the mid-primaries are generally larger than those on the outermost primaries in Laughing Gull; and, second, the tips to the mid-primaries don't suffer as much abrasion. This can give the impression of a genuine molt limit, and in fact there may be one. More study is needed to determine if some inner primaries are included in extensive 1st prealternate molt. It is not very rare to find adult types molting a variable number of primaries in December–February from Venezuela to Colombia. More study is needed to determine whether these are prebasic molts of birds on a southern-hemisphere molt schedule or complete prealternate molts similar to what we find in Franklin's Gull.

HYBRIDS With Ring-billed Gull (Henshaw 1992), possibly with Black-headed Gull (McKearnan 1999), and Gray-hooded in Senegal (Erard et al. 1984).

23 Adults. White head expected at this time of year (basic plumage). Hooded individual on left possibly underwent an early prealternate molt or is experiencing hormonal asynchrony. AMAR AYYASH. FLORIDA. JAN.
24 Adult (basic plumage). Head pattern shows maximum whiteness for this species. Black bare parts in nonbreeding condition. Forward-placed legs. ERIC ANTONIO MARTINEZ. MEXICO. NOV.
25 Adult (alternate). Black on outermost primaries reaches coverts on upperside and underside of hand. No mirrors. Worn apicals. STEVE ARENA. MASSACHUSETTS. MAY.
26 Adult (basic). Long-winged, thin white trailing edge, and all-black primary pattern, often to p6. AMAR AYYASH. MEXICO. DEC.
27 Adult with incoming hood pattern. Note all-black outermost primaries with contrasting gray underside to remiges. Black legs. AMAR AYYASH. FLORIDA. JAN.

12 FRANKLIN'S GULL *Leucophaeus pipixcan*

L: 13.5"–15.0" (34–38 cm) | W: 35"–38" (87–89 cm) | Three-cycle | CAS | KGS: 8–9

OVERVIEW A hooded gull of the northern prairie. Sometimes referred to as the "prairie dove" and also as "rosy dove," due to the pink flush commonly found in its plumage. Breeds colonially in marshy potholes and freshwater lakes. Fluctuating water levels can easily displace this species from one year to the next. Franklin's is a trans-equatorial migrant with one of the most notable migrations among North American gulls. Flocks move north and south through the central interior of the United States and Canada as they migrate to and from the nonbreeding grounds on the western shores of South America. Up to half of its annual cycle is near salt water, from which the nasal salt gland becomes enlarged (Burger & Gochfeld 1984). In the fall, migration coincides with the harvest, during which the species regularly follows the plow in search of earthworms, insects, and cropland pests, including voles and mice. Large, human-made lakes and inland reservoirs are used to roost and bathe in the evening, sometimes drawing tens of thousands of birds as staging arenas. At dawn, birds can be seen leaving their roosts to feed in agricultural fields. Swirling flocks can be found hawking insects at varying altitudes, and they're often heard before being seen. Despite its social and gregarious habits, Franklin's is not known to feed on the eggs or young of its own species and is seldom found engaging in kleptoparasitism. A future English name for this species could perhaps describe the unique gray wash on the uppertail of adults or highlight its annual journey through the Americas.

TAXONOMY Monotypic.

RANGE
Breeding: Breeds colonially in marshes in northern interior of western and central North America. Rare and irregular breeder south to NE California, Kansas, and NW Iowa. Nonbreeders may occur in summer at moderate distances away from nesting areas.
Nonbreeding: Along the coast, found most commonly in ocean waters and roosting on adjacent beaches. Also found in harbors and on mudflats. Inland, seen mostly at lakes, also in farm fields and pastures. Fairly rare at landfills. Migrants are widespread from late July or August well into November (rarely Dec.) and from March to May (a few into mid-June), often in large flocks, through Great Plains and Texas. It is not unusual to see concentrations in fall, less often in spring, of up to many thousands of birds (a few times up to hundreds of thousands, and a million or more prior to the 1970s) on the central and S Great Plains and in Texas. Many birds from Texas and E Mexico cross into the Pacific in vicinity of Isthmus of Tehuantepec. An estimate of 2.5 million individuals at Salt Plains National Wildlife Refuge, Oklahoma, would be a world high count (Burger & Gochfeld 2020). This species is partly pelagic during the nonbreeding season, occurring well offshore (up to 30 miles) on a regular basis. Much smaller numbers are found regularly west through the interior West to central British Columbia, Nevada, and E California, and eastward through the Midwest to the E Great Lakes (where it may linger in late fall, very rarely into Dec.).

Farther to the west and east, a very rare visitor to the West Coast from SE Alaska southward through W Mexico and from the mid-Atlantic region southward. Has occurred in moderate numbers in the Great Lakes region and along the East Coast on several occasions in late fall, following the passage of very strong low-pressure systems with strong westerly winds (598 on S Lake Michigan, Indiana, Nov. 2008). Casual north to the S Bering Sea region and across N Canada, New England, and the Maritimes to Greenland, Iceland, W Europe, and the West Indies. Casual or accidental in many other parts of the world, from E Europe to E and S Africa; in the south-central Atlantic Ocean, Middle East, and India; in the mid-Pacific; and from Japan to Taiwan and Australia. Noteworthy are at least 35 records in Spain!

Winters in the Humboldt Current region (also in agricultural fields) along the Pacific Coast from Peru to central Chile, where concentrations of many thousands may be seen. Rare migrant on Andean lakes.

12 FRANKLIN'S GULL

Uncommon to rare in winter north to Ecuador and Central America, rare south to S Chile. Very rare in winter north through Mexico to S United States, exceptionally as far north as S British Columbia and New York.

IDENTIFICATION

Adult: A complete black hood is acquired in alternate plumage, with contrasting white eye crescents. Eye crescents are considerably thick, giving the eyes a spectacled appearance. Dark eyes in all ages. The bill and legs are a deep red in breeding condition. The pink suffusion found in many adult types appears to be more bright and vivid in fresh alternate birds. The outer-primary pattern is highly variable. Some adults have extensive white dominating the wingtip, while others are mostly black, with a small p10 mirror and thin white medial band. Alternate adult types with full black hood, white medial band on the wingtip, and pink flush throughout have been recorded without p10 mirror (Ayyash, pers. obs.). It remains to be seen through known-age birds whether adults can be without a p10 mirror. When new, the apicals are rather large and conspicuous. The underwing shows a sharply demarcated black crescent limited to the wingtip. P10 commonly shows a thin black outer edge running all the way up to the primary coverts. A very helpful field mark is the signature gray patch found on the uppertail, which can range from a light, faint gray to a bold, prominent gray patch. Compact in size and evenly proportioned. Short legs, with a relatively short and straight bill. The wings are somewhat broad, and the hand can appear rounded at times. In basic plumage, the head retains a semi-hood that often reaches the back of the crown and envelops the eyes. Legs and bill become dark in nonbreeding condition. *Similar Species:* See Laughing Gull account.

1st Cycle: Juvenile scapulars are chestnut brown with paler edges. These feathers are soon replaced with adult-like gray formative feathers. Upperwing coverts with variable brown and gray tones. Greater

1 Juvenile. Dark wash on head with prominent eye crescents. Short bill. Grayish greater coverts. Note white underparts already at this age. White apicals. OLLIE OLIVER. WASHINGTON. AUG.
2 1st cycle. Some darker gray formative scapulars have grown in, as well as a whiter hindneck. White underparts. Short bill, dark legs, and prominent eye crescents. MASON MARON. WASHINGTON. SEPT.
3 1st cycle. Formative plumage, now with most scapulars adult-like. Gray wash on hindneck and white underparts. AMAR AYYASH. ILLINOIS. OCT.
4 1st cycle. Wing coverts moderately worn but still juvenile. Note white-tipped tail with narrow band. Bold eye crescents. JUSTIN PETER. GALAPAGOS. NOV.
5 1st cycle (right of center) with Laughing Gulls and Ring-billed Gull (upper left). Franklin's often keeps a semi-hooded appearance, bolder eye crescents, and a finer bill. But note variation in bill size in Laughing Gulls. MICHAEL BROTHERS. FLORIDA. NOV.
6 1st cycle. A number of replaced wing coverts and upper tertials, likely 1st alternate. Semi-hooded appearance expected at this age. JUSTIN PETER. PERU. FEB.
7 Juvenile. Narrow tailband, with black typically not reaching outer edge of outer tail feathers. White uppertail coverts. Relatively broad white trailing edge with grayish inner primaries. RYAN MERRILL. WASHINGTON. AUG.
8 1st cycle. Generally shows clean, white undertail coverts and underparts (compare to Laughing). Semi-hooded look with bold eye crescents. AMAR AYYASH. INDIANA. OCT.
9 1st cycle (formative). Compared to Laughing, wing linings are largely white, hand is paler, and black does not typically reach outer edge of tail. AMAR AYYASH. ILLINOIS. SEPT.
10 1st cycle. Formative plumage. More brownish upperwing than 12.7, but note narrow tailband, semi-hood with pale lower neck. MIKE BOURDAN. MICHIGAN. NOV.

coverts sometimes predominantly with dark shaft streaks. Eye arcs are visible in youngest fledglings, against a dark head with dusky neck and breast markings. These hindneck feathers soon become mostly white, contrasting with a black semi-hood. The lower border to the hood is similar to that of basic adult, but hindneck can show some dusky spotting or smudging. Bill black. Legs can be a dark, dirty pink color when fledging, but most are dark, approaching black. Primary tips often show white, chevron-shaped apicals. Tertials are dark centered with pale edging. Body becomes increasingly white on most birds during southbound migration in the fall. On the open wing, extensive white can be seen on trailing edge to secondaries and inner primaries. Some individuals can have surprisingly adult-like gray inner primaries

11 1st-cycle molting primaries in 1st prealternate molt on wintering grounds (p6 A1, p7 dropped, p8–p10 juvenile). Visible secondaries and several lesser coverts are juvenile. MANUEL RABANAL. PERU. MARCH.
12 Similar to 12.11 but fuller hood, 1st alternate greater coverts and new p6–p7 (p9–p10 juvenile). GRAHAM WILLIAMS. FLORIDA. MARCH.
13 1st cycle with Laughing Gull (right) and American Herring. Franklin's averages more delicate proportions and often has connected eye crescents at the back of the eye. See 12.16 for open wing. LEO MCKILLOP. NEW HAMPSHIRE. MAY.
14 1st alternate. All primaries replaced in 1st prealternate molt. Extensive black on primaries, nonuniform gray upperparts, and incomplete hood, unlike adult. LEROY HARRISON. ILLINOIS. MAY.
15 2nd cycle. The date and limited primary molt suggest 2nd prebasic molt. Thus no remiges were replaced in the 1st prealternate molt, a pattern found in some 1st cycles, often in those that remain north of the wintering grounds. CHRISSY MCCLARREN. MISSOURI. JULY.
16 1st cycle. Same individual as 12.13. 1st prealternate molt commonly incomplete, as seen here (suspension at p6/p7). Tail feathers replaced (1st alternate). Most secondaries and p7–p10 juvenile. LEO MCKILLOP. NEW HAMPSHIRE. MAY.
17 1st cycle in alternate plumage. Aged by black streaks on primary coverts and tail, extensive black on wingtip, and incomplete hood. CHRISSY MCCLARREN. MISSOURI. MAY.
18 2nd-cycle type. Commonly without p10 mirror and subtle medial band. Southbound migrants with such a wingtip at this time of year are relegated to 2nd basic. AMAR AYYASH. INDIANA. OCT.
19 2nd-cycle type, aged by time of year and primary pattern. Compared to adult in 12.33, shows much black on wingtip, reduced medial band, and duskier wing linings. JAMES PAWLICKI. NEW YORK. NOV.
20 2nd-cycle type. Presumably undergoing 2nd prealternate molt (see 12.21 for open wing). Unlike adult, extensive black in wingtip (worn apicals). RICARDO RODRIGUEZ. SPAIN. MARCH.
21 2nd-cycle type. Active 2nd prealternate molt. New secondaries, innermost primaries, and several tail feathers (2nd alternate). Retained p6–p10, presumed 2nd basic. RICARDO RODRIGUEZ. SPAIN. FEB.
22 2nd alternate type (see 12.23 for open wing). Extensive black on wingtip, incomplete hood, and black bare parts point away from adult. WOODY GOSS. WISCONSIN. JUNE.

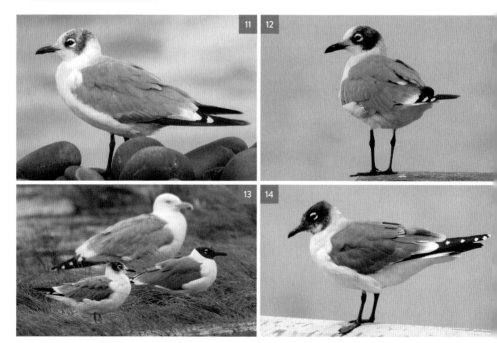

with black subterminal bands on p4–p5. Black tailband fairly thin, with little to no pigment on outermost rectrices. Uppertail coverts mostly white, with occasional dark shaft streaks coming up the tail feathers. A radical transformation occurs in many 1st cycles when they undergo their 1st prealternate molt, which can be complete and typically occurs on the wintering grounds, mostly in South America. In the most advanced birds, adult-like flight feathers are acquired with a broad white trailing edge, but wingtip pattern is generally without p10 mirror and may show brown tinge throughout upperparts. Medial band often difficult to detect and is reduced to small white tongue tips. Primary coverts often with some black streaking, and outer primaries at times with thick black shafts. First alternate tail becomes white, or mostly white. Some birds suspend 1st prealternate molt and retain a variable number of juvenile outer primaries and their respective greater coverts. In these individuals, secondaries are largely juvenile as well and are often worn. The hood in 1st alternate plumage is usually less than full, with white freckles around the lores, chin, and forehead. *Similar Species:* See Laughing Gull account.

2nd Cycle: Separation of 1st alternate and 2nd basic Franklin's is based on time of year, hood pattern, and primary molt. Much like adult, but 2nd cycle averages more black on wingtip and typically lacks mirrors (more study needed). An unknown percentage does appear to show a small p10 mirror (usually restricted to inner web). White medial band separating black from gray on wingtip may be absent or limited to small white tongue tips on p5–p7. Black markings on the primary coverts are not unusual on presumed 2nd basic birds (Ayyash, pers. obs.). Also averages more-dusky markings on the underwing coverts, and the bare parts average darker and duller than adult's.

23 2nd alternate type with Bonaparte's Gull. This individual appears to have replaced all secondaries in an incomplete 2nd prealternate molt, with retained primaries (2nd basic). WOODY GOSS. WISCONSIN. JUNE.
24 Adult (alternate). The wingtip pattern is highly variable in adults. This individual shows minimal black. RICHARD BONSER. CHILE. APRIL.
25 Adult (alternate). Relatively extensive black on the wingtip, but note large mirror on p10 (underside of far wing). RICHARD BONSER. CHILE. APRIL.
26 Adult with Laughing Gull (right). Franklin's generally shows larger apicals and a smaller bill, and more commonly acquires pink suffusion on the body. BENJAMIN VAN DOREN. TEXAS. APRIL.

12 FRANKLIN'S GULL

MOLT Our only gull with two complete molts per cycle. Complete prebasic and complete prealternate molt typical in adults. Outer primaries on adult types rarely show heavy wear, due to flight feathers being replaced twice a year. First cycles begin preformative molt in earnest after fledging, typically limited to head, neck, scapulars, and underparts. Variable 1st prealternate molt (on wintering grounds) may be

27 Adult with 2nd-cycle type (left). Alternate adults typically have complete hood by this date. Note brighter bare parts and more extensive white on primaries in adult. AMAR AYYASH. WISCONSIN. MAY.
28 Same adult as 12.27 (right). Prominent white medial band and gray wash on central tail. It's not unusual for adults to retain 1–3 basic outer primaries, as seen here (p10). AMAR AYYASH. WISCONSIN. MAY.
29 Adult (alternate). A few Franklin's arrive on the breeding grounds unusually early. Seen here on ice with California Gulls. ALAN KNOWLES. ALBERTA. MARCH.
30 Adult (basic). Defined border to back of hood in basic plumage. Short bill and thick eye crescents. AMAR AYYASH. MICHIGAN. NOV.
31 Adult (alternate). White medial band (and mirror on p10) not found in Laughing Gull. Note distinctive gray on central tail often shown by this species. RICHARD BONSER. CHILE. APRIL.
32 Adult undergoing prebasic molt (p1–p2 new). Old primaries worn, but white medial band apparent. All-red bill in high breeding condition. MARK CHAPPELL. IDAHO. JULY.

partial, with no flight feathers replaced, or incomplete, with variable number of rectrices, secondaries, and primaries replaced. Others have complete 1st prealternate. It's quite common to find fully grown 1st alternate p1–p6 / p7 and juvenile p7 / p8–p10 with no molt gap in May–July (Ayyash, pers. obs.). Second basic plumage perhaps more delayed in appearance on birds which had absent to incomplete 1st prealternate molts. This can be observed in June–July in 1st-cycle individuals with retained 1st basic primaries, molting into 2nd basic. Second basic p10 typically without a mirror (although some may have smaller mirror on inner web), with reduced or entirely absent white medial band, and increased black on wingtip. These individuals often show black shaft streaking on primary coverts and alula. Second prealternate molt on wintering grounds appears to begin at inner secondaries / tertials in many individuals and works distally to outer secondaries, continuing to inner primaries. Others suspend 2nd prealternate at s1 / p1; determining start of 2nd prebasic molt at inner primaries is not straightforward (Ayyash, pers. obs.). Adult types are also recorded with incomplete prealternate molts, but less commonly. This may be due partly to an observer bias in which molt contrast is more difficult to detect. For instance, from a sample of 52 alternate adults migrating north in the interior United States, 7 appeared to have retained 2nd basic p10 (Ayyash, pers. obs.). Data from known-age birds is surprisingly sparse, which is unfortunate, especially considering the unique molt and migration patterns of this gull.

HYBRIDS A single individual was found incubating eggs over two consecutive years at the same site in a Ring-billed Gull colony, paired with Ring-billed. Two chicks successfully hatched (Weseloh 1981). A convincing example of this pairing is an adult individual, believed to be the same bird, returning to Hughes County, South Dakota, in March–April 2006 and 2007 (Ricky Olson, pers. comm.), and an individual from Alberta (Weseloh 1981). Vagrants to Africa also recorded hybridizing with Gray-hooded Gull in Senegal (Borrow & Demey 2001).

33 Adult (alternate). Compared to Laughing, extensive white in wingtip, including distinctive medial band (white partition between black wingtip and gray). KEN HANSEN. ALBERTA. MAY.
34 Adult in basic plumage. White mirror on p10 rules out Laughing, and also note sharp demarcation to semi-hooded head. Bold eye crescents. ALEX LAMOREAUX. TEXAS. NOV.
35 Adult and 2nd-cycle type. The bottom bird shows no p10 mirror, reduced white medial band, and increased black on the underwing. Such birds are aged as 2nd cycles at this time of year, believed to be in 2nd basic plumage. ROB WILLIAMS. PERU. NOV.

SECTION 2

LARUS GULLS

13 HEERMANN'S GULL *Larus heermanni*

L: 18.0"–20.5" (46–52 cm) | W: 50"–51" (127–130 cm) | Four-cycle | SAS | KGS: 10–11

OVERVIEW A striking and evenly proportioned four-year gull belonging to *Larus*. Essentially a Mexican species, with most Heermann's Gulls breeding colonially in the Gulf of California, on Isla Rasa. Unlike any other North American gull species in all plumages. Migration pattern is also unique, in that a significant "reverse" migration northward occurs after breeding. North of breeding grounds, restricted primarily to the Pacific Coast from S California to Vancouver Island. Found feeding offshore as well as in intertidal waters. Readily scavenges, and often found pirating other gulls and terns. This behavior, coupled with the sooty, chocolate-brown plumages of younger birds, often recalls a jaeger from afar. Its future English name may highlight these jaeger-like elements, emphasize the dark plumages found in all ages, or perhaps stress its affinity to the Gulf of California. Also commonly steals food from Brown Pelicans, alighting on their backs and snatching fish directly from their pouches. An interesting anecdote that highlights this species' potential for site fidelity is of a one-legged individual that returned to a small marina in San Rafael, California, for at least 16 years (George 2005). Adult population estimated at approximately 150,000 pairs. With the overwhelming majority of Heermann's Gulls nesting on one small island, the International Union for Conservation of Nature lists it as a near-threatened species (BirdLife International n.d.-b). Formidable threats to Heermann's Gull and other species in the Gulf of California are overfishing and unpredictable water temperatures (Islam & Velarde 2020; Elias-Valdez et al. 2023).

TAXONOMY Monotypic.

RANGE
Breeding: Breeds on rocky islands in the Gulf of California, NW Mexico, with approximately 95% of the species on the volcanic island Isla Raza. Small numbers also nest south to Jalisco and along the west coast of the Baja California peninsula. Several nesting attempts, all since 1979, along the California coast between San Luis Obispo (Shell Beach) and San Francisco (Alcatraz Island) Counties, as well as a lesbian pair inland at Salton Sea, California. A small but resilient colony in Seaside, California, since 1999 is the only known nesting site for the species in the United States (Chin 2020). Attempts to protect and bolster this colony with a human-made floating island on Roberts Lake starting in 2019 had proven unsuccessful up to the time of this writing.

Nonbreeding: Whereas almost all Heermann's Gulls nest in Mexico, very large numbers of postbreeders disperse north, mainly June–November, along the Pacific Coast, as far north as extreme SW British Columbia. Found commonly on sandy beaches and rocky shores, as well as offshore; somewhat rare on coastal mudflats, at most coastal lakes, and in parking lots not close to the beach. Also moves south as far as Colima, in central Mexico; very rarely south to Guatemala and Honduras, exceptionally Costa Rica. Northbound movement along California coast between mid-June and mid-July can be very evident, with constant procession of individuals and very small flocks. Peak numbers along California coast typically reached by mid-July and last into November, when birds start to move back south. Farther north, the first birds typically arrive at S Vancouver Island by early or mid-July, and peak numbers in Washington are from August to October (high count of 20,000 off the Elwha River mouth in Sept. 2009). Southward withdrawal is fairly rapid in the north, and very few remain after early November north of central California, but moderate numbers may remain in coastal S California through midwinter, with fewer still through spring, until the postbreeding influx arrives in early summer. Also occurs offshore regularly out to a distance of about 30 miles. Small to moderate numbers of nonbreeders remain in S California, north to Monterey Bay, through the breeding season.

A rare but regular visitor inland, mostly in spring and summer at the Salton Sea, and mostly in fall in Arizona. Casual or accidental in interior Northwest and well to the east, where it is recorded in a number of states, from NM, UT, TX, WY, and OK east to Great Lakes region (four state records in MI), VA, and

FL. Most recently, a single, wayward bird first detected as a juvenile in Florida in August 2019 spent at least to January 2024 moving between GA, SC, NC, NJ, RI, VA, and MA. A similar story unfolded in Baltimore, Maryland, with that state's first state record believed to have returned as a 2nd cycle and then migrated to Ohio (April 2023) and New York (May 2023). Vagrants in E United States are typically young individuals that have the tendency to linger for months. A single sight record in Costa Rica.

IDENTIFICATION

Adult: Distinctive dark gray upperparts with lighter gray neck, breast, and underparts. At rest, and especially in fresh plumes, a contrasting white tertial crescent is seen, and to a lesser extent a white scapular crescent. Clean-white head in alternate plumage, with bright red bill and red orbital in high breeding condition. Reduced black tip on bill becomes larger in nonbreeding condition. The only member of *Larus* with black legs. Generally dark eyed, but some individuals show a lighter, bronzed-gray coloration to the iris. In basic plumage, a dusky gray head pattern often develops and may become densely marked, forming a quasi hood. Open wing reveals a contrasting white trailing edge to the secondaries, which is sometimes visible as a drooping white secondary skirt at rest. Primaries with indistinct, thin, white tips, usually out to p6/p8 (unlike 3rd cycle). Outer primaries all dark without mirrors or the prominent apicals found in many *Larus* species. Dark tail is white tipped with paler gray uppertail coverts. An uncommon but regularly occurring aberration found in this species is a small leucistic patch on the primary coverts, often symmetric in nature on both wings. *Similar Species:* Distinctive at this age. See 3rd-cycle Heermann's for discussion on aging.

1 Juvenile. Sooty brown with pale edging throughout gives scaly appearance. Dark tail. Black legs. PAUL FENWICK. SEASIDE, CA. JULY.
2 Juvenile. Paler than 13.1. Both individuals presumed from local breeding colony in Seaside. Migrants from Mexico typically arrive with more wear and some postjuvenile scapulars. BYRON CHIN. SEASIDE, CA. JULY.
3 Juvenile now with signs of wear. Paler underparts approach California Gull, which is not as uniformly brown on the head and neck, and has pale legs. ALEX A. ABELA. CALIFORNIA. AUG.
4 1st cycle with faded and worn juvenile coverts and tertials. Dark brown head, scapulars, and body feathers are postjuvenile (formative or 1st alternate). ALEX A. ABELA. CALIFORNIA. SEPT.

13 HEERMANN'S GULL

1st Cycle: Fresh juveniles are a chocolate brown throughout, with pale, scaly edging to their upperparts. Dark legs. Black bill becomes pink based very soon after fledging, and later in 1st cycle some birds develop a grayish-green hue around the bill base before the gape. Postjuvenile feathers are leaden brown and contrast with juvenile feathers, which often become faded and worn by midwinter. Some birds replace a variable number of tertials in their 1st cycle, which will often reveal contrasting white tips against an entirely dark bird. In flight, underside of remiges is lighter than upperside. Flanks and undertail can show a scaly, neatly barred pattern at times, and this is often seen on the rump and uppertail coverts of juveniles. Uppertail dark. *Similar Species:* Commonly accused of being a jaeger from a distance, but note that Heermann's flight pattern is more sluggish, and it lacks visible white flashes on the underside of

5   1st cycle. This vagrant lingered for several months, replacing most juvenile upperparts, rectrices, and some inner secondaries in an advanced 1st prealternate molt. Resembles 2nd cycle, but with less gray on body. Note bill pattern and pointed primary tips. DANIEL LEBBIN. VIRGINIA. NOV.
6   1st cycle with Laughing Gulls (slightly smaller). This individual first appeared in Florida as a juvenile and spent a few years sporadically moving throughout the Eastern Seaboard. AMAR AYYASH. FLORIDA. JAN.
7   Juvenile. Uniformly brown with scaly appearance. Note dark uppertail coverts. Long bill already pale at this age. MIKE DANZENBAKER. CALIFORNIA. AUG.
8   Juvenile. Jaeger-like gull with kleptoparasitic habits. Lacks white flashes on underside of primaries, with plain wing linings and long pale bill. PAUL FENWICK. CALIFORNIA. JULY.
9   1st cycle. Postjuvenile body feathers uniformly brown, now without scaly appearance. Pale bill base and dark legs. BRIAN SULLIVAN. CALIFORNIA. OCT.
10 1st cycle. Dark gray postjuvenile scapulars and uppertail coverts. It's not rare for some tail feathers to be replaced at this age. AMAR AYYASH. CALIFORNIA. JAN.
11 1st cycles. Sooty-brown birds with pale bill. All with postjuvenile scapulars and some inner-wing coverts replaced. BRIAN SULLIVAN. CALIFORNIA. OCT.
12 2nd cycle. Incoming 2nd basic primaries (black below tertials). Bill pattern highly variable at this age, but generally becomes orange, then red, with dark tip. ALEX A. ABELA. CALIFORNIA. JULY.
13 2nd cycle. The bill pattern, uniform upperparts lacking scaly appearance, and black primaries (still growing) with rounded tips differ from juvenile at this time of year. AMAR AYYASH. WASHINGTON. AUG.

the primaries. At a closer distance, note the lack of white shafts on the upperside of the primaries. Also compare to dark juvenile California Gulls, which could show a similar plumage aspect, but legs are pale in California Gull, and bicolored bill takes on a sharper demarcation at the tip. Even the darkest juvenile California Gull should display a paler ground color to the uppertail coverts, wing linings, and white-tipped

14 2nd cycle. Dark grayish-brown 2nd alternate scapulars, median coverts, breast, and underparts. Brighter bill. Delayed head pattern often becomes mottled white. ALEX A. ABELA. CALIFORNIA. MARCH.
15 2nd cycle (basic). Plain grayish-brown upperparts without pale edging of 1st cycle. Note rounded primary tips, gray uppertail coverts, and orange bill. STEVE HAMPTON. WASHINGTON. AUG.
16 2nd cycle with 1st cycle (top). 2nd cycle often with prominent white tips to lower scapulars and tertials. Also, note plainer and darker wing coverts, and grayer tail coverts on 2nd cycle (although see 13.10). LARRY SANSONE. CALIFORNIA. JAN.
17 3rd cycle. Reddish bill. White head, upper scapulars, many lesser and median coverts, and central tail feathers acquired by 2nd prealternate molt. Visible secondaries and primaries 2nd generation (3rd basic p1–p2 concealed). BYRON CHIN. CALIFORNIA. JUNE.
18 3rd cycle (basic). Bill becomes dull orange. Brown tone to fresh upperparts. Head more densely marked and muddier-brown compared to adult. Also lacks white tips to visible primaries. TIMOTHY KEYES. GEORGIA. OCT.
19 3rd cycle. Alternate plumage. White head, gray neck, and gray underparts. Red bill with black tip. Same individual as 13.6. Would likely be aged as an adult without known life history. DAVID BAKER. VOLUSIA COUNTY, FLORIDA. MARCH.
20 3rd-cycle type. Basic plumage. Brown head can form quasi hood appearance. Bill orange-red. Distinguished from 2nd cycle by thin white tips to inner primaries, secondaries, and tail feathers. ALEX A. ABELA. CALIFORNIA. OCT.
21 3rd cycle. Thin and irregular white trailing edge, thin white tips to rectrices, and brown wash throughout. White trailing edge on adult bolder and more defined. White primary covert patches expected aberration in this species, found most commonly on adults. PAUL FENWICK. CALIFORNIA. JAN.
22 4th cycle undergoing 4th prebasic molt (same individual as 13.6, 13.18, 13.19). Note how the primaries and secondaries have faded from black to brown, and that the tail feathers have lost their white tips due to wear. Resembles typical adult undergoing prebasic molt. DENNY SWABY. FLORIDA. AUG.
23 Adult type. Aging difficult. Brown wash across upperparts not uncommon in adult types wintering farther south, especially from March to Aug. Thin white tips to rectrices may be due to wear (or age-related). "Dirty" rump and lack of white tips to mid-primaries (unabraded) suggest subadult. CHRIS GIBBINS. MEXICO. MARCH.
24 Adult type. Perhaps not safely aged. Many adults are largely white-headed by now, with bright red bill. Clean gray upperparts, prominent white on exposed secondaries, and tail feathers suggest adult, but this may be due to extensive 3rd prealternate molt. Note growing central tail feathers. ANDREW BIRCH. CALIFORNIA. JAN.
25 Adult in alternate plumage. Unmistakable. Clean dark gray upperparts. White head gradually fading to gray neck. White tips on fresh outer primaries, although regularly giving way to wear. BLAKE MATHESON. CALIFORNIA. JAN.

tail feathers. Some darker juvenile Black-tailed Gulls may show similar resemblance, but note their pinkish-colored legs; long tubular bill, which is typically paler; and paler uppertail and undertail coverts.

2nd Cycle: Second basic plumage similar to that of 1st cycle, but generally more solidly dark throughout, with paler, gray uppertail coverts and dusky gray body. Not as brown as 1st cycle and lacks pale edging on upperparts. In 2nd alternate, the upperparts may begin to resemble gray coloration found in adults, especially on upper scapulars, median coverts, lower neck, and sides. Bill becomes more orange now, and in some advanced birds, a dirty white head may come in an alternate plumage (some are entirely white headed) and may also show thin white eye crescents. A few individuals show thin white tips on new, 2nd alternate tail feathers (typically limited to some central tail feathers).

3rd Cycle: Upperparts uniformly dark gray, but often with brownish hue throughout. Paler gray underparts. Now with bolder white tertial and scapular crescents than 2nd cycle, but often less defined than adult's. Head pattern varies from white (alternate) to dark (basic) mottled hood appearance similar to adult's but generally more solidly brown and smudged, forming a quasi hood. In flight, the secondaries and tail feathers show white tips (limited to innermost secondaries and tertials in 2nd cycle). Uppertail coverts gray, sometimes with brown wash, unlike adult's. This age can be confusingly variable, with some birds resembling advanced 2nd cycles and others very adult-like. Compared to those on fresh adults, outer primaries on 3rd cycles often show a dark brown hue, and thin white tips restricted to inner primaries (p1–p4), with all-dark tips to the mid-outer primaries. The white tips to the tertials, secondaries, and tail are not as prominent as adult's, although a known-age 3rd alternate individual from North Carolina seemed to show features consistent with an adult except for white tips on the mid-outer primaries (Feb. 2022). Third-cycle head markings also more clouded and denser in basic plumage—often brownish—while adult head shows more of a gray base color with finer pattern. Third cycle should average earlier flight feather molt than adult, with outer primaries generally grown by mid- to late September. Adult types can show brownish secondaries and outer primaries in late summer (July–Sept.), due to fading, and this may suggest 3rd-cycle type, but note retained secondaries and tail feathers with noticeable white tips. Some individuals will defy proper aging.

MOLT More study is needed to determine whether the inserted molt in 1st cycle is a single protracted molt (1st prealternate) or two separate molts (Pyle 2008). Heermann's is partly resident and/or a mid-distance migrant, and the extent of its inserted molt in 1st cycle varies greatly (quite evident in several vagrants to the East Coast; Ayyash, pers. obs.). Inserted molt in 1st cycle may be complete (Howell & Wood 2004; Howell 2010) but is typically partial to incomplete. Some at the southern limits in Mexico have been documented with eccentric primary molt (Howell & Wood 2004). A vagrant 1st cycle to Texas is the only known example from the United States in which juvenile primaries appear to have been replaced in an eccentric pattern (Ayyash, pers. obs.). Careful consideration should be given to individuals

26 Adult completing prebasic molt. Note noticeable white tips to primaries (unlike 3rd-cycle type). Irregular white secondary skirt due to feathers still growing. Juvenile Thayer's (left) for size reference. LIAM SINGH. BRITISH COLUMBIA. OCT.

with apparently "false" molt limit. Sun-blazed outermost primaries appear to have worn, bleached tips with a dark and fresh-looking subterminal region and faded bases to the outer webs, giving the impression of multigenerational primaries (Peter Adriaens, pers. comm.).

Examination of molt in vagrants to the East has been eye opening (although vagrants can exhibit atypical extent and timing in molt). The molt pattern of a vagrant 1st cycle to Toronto was well documented over a 10-month period from November 1999 through September 2000 (Iron & Pittaway 2001), and in August 2019, a juvenile found in east-central Florida became the most well-documented Heermann's Gull on record, roaming up and down the Florida coast and eventually to SC, NC, GA, VA, NJ, RI, and MA. This individual lingered until at least January 2024, becoming a great case study for plumage succession in this species, especially when comparing 3rd and 4th cycles.

HYBRIDS California Gull. The only known record of Heermann's Gull hybridizing with another species comes from Lahontan Reservoir in Churchill County, Nevada, 19 May–3 June 1990 (Chisholm & Neel 2002). Two eggs were found, and one presumably fledged. A wonderful photo of the presumed-hybrid chick shows large, dark head and hindneck spotting typical of California Gull and a creamy breast with completely jet-black legs, as found in Heermann's Gull (Martin Meyers, pers. comm.). Interestingly, an adult Heermann's was reported at the same site the previous year but with no evidence of hybridization.

27 Adult type. Broad white tips to visible secondaries and black primaries with worn tips. Noticeable wear on tertials. CHRIS GIBBINS. MEXICO. MARCH.
28 Adult type in basic plumage. White trailing edge well-defined, with noticeable white tips to mid- and outer primaries, and bold white tips to rectrices. Note, however, that these are fresh, unabraded, feathers. BRIAN SULLIVAN. CALIFORNIA. OCT.
29 Adult (alternate). Gray neck and underparts, dark underwing, and black legs. Prominent white trailing edge and bold white tips to tail feathers. PAUL FENWICK. CALIFORNIA. JAN.
30 Adult (alternate). White head often acquired around this time of year. Noticeable white tips to p7 on this individual, and dark gray upperparts (compare to 13.23). GREG GILSON. CALIFORNIA. JAN.

14 BELCHER'S GULL *Larus belcheri*

L: 18.0"–21.5" (46–55 cm) | W: 48"–49" (122–124 cm) | Three-cycle | SAS | KGS: 14–16

OVERVIEW A medium-sized black-backed gull with a conspicuous tailband in all plumages. Roughly the size of Heermann's Gull. Generalist found primarily on the arid shores of Peru and N Chile. This gull is endemic to the Humboldt Current littoral and does not move far offshore or far inland, although it is regularly found on inshore waters, where it breeds colonially in small groups. Belcher's is one of the darkest gulls in the world, with a powerful, stout bill that appears disproportionately heavy. The wings are relatively broad, and the somewhat short tail appearance is accentuated by the large black subterminal band. Once considered conspecific with the similar-looking Olrog's Gull (*Larus atlanticus*), which resides on the southeast coast of South America. Some authors employ the name Band-tailed Gull for Belcher's, once a well-established name for this taxon, before its split with Olrog's. Also known as Peruvian Gull locally, which would be a suitable English name in the future.

TAXONOMY Monotypic.

RANGE
Breeding: Generally restricted to the coasts of N Peru to N Chile.
Nonbreeding: Nonbreeders may be found rarely north to Ecuador and south to central Chile; casual or accidental north to Colombia (Ellery & Salgado 2018), Panama, and Costa Rica. In North America, there are just a few reports: a well-documented, accepted record from S California (San Diego County, Aug. 1997–Jan. 1998) and, surprisingly, up to four reports from Florida, from the "wrong" ocean, along the Gulf Coast, between 1968 and 1976 (Stevenson 1980). None of these has gone before the records committee in Florida, as the sightings were before the Florida Ornithological Society Records Committee's time, and they have yet to be reviewed. Questions regarding provenance and confusion with the then conspecific Olrog's Gull should be addressed. As unusual as it is to keep gulls in captivity, Miami-based bird keeper Charles P. Chase admitted to importing several "Band-tailed Gulls" to Miami in 1968 (Olson 1976). (Coincidentally, this was the same year in which *L. belcheri* was recorded in Florida.) Chase reported that none of his birds had escaped from his collection, nor were any sold in Florida.

IDENTIFICATION
Adult: The blackish upperparts sharply contrast with an all-white head in alternate plumage. In favorable lighting, the gray wash on the upper neck, breast, and belly of alternate adults is quite noticeable. Dark eyes, with yellowish orbital. Thick, yellow legs are mustard colored in high breeding condition. The bill is a similar yellow, stocky, and parallel edged from its base to the gonys. The black on the bill tip often forms a thick "Ring-billed" band, encircled with a variable amount of red, particularly on the very tip. Often seen on the perched bird is a thin white secondary skirt. Insignificant tertial and scapular crescents. The impression of the standing bird is a somewhat long-legged gull, long winged, with thin structure to the body behind the legs and throughout the vent region.
  In basic plumage, adults typically acquire variable black on the head, much more than most other white-headed gulls. The patchy hooded appearance has an undefined border, and the head often shows some white in the loral region. A brownish hue to the black upperparts is more noticeable in late basic plumage, perhaps due to external factors. White eye crescents become noticeable in basic plumage. A dusky wash develops on the upper breast, with variable dark markings on the upper neck.
  On the open wing, the most distinct feature is the broad black tailband against an otherwise-white tail. No mirrors on the outer wing. The white trailing edge to the secondaries is relatively broad but quickly tapers off to insignificant white tips on the innermost primaries. The underside to the remiges is a smoky

14 BELCHER'S GULL

1   Juvenile. Brown hood extends to upper breast; thick bill with black tip and extensive pale edging to upperparts. RICHARD BONSER. CHILE. APRIL.
2   1st cycle showing some dark, grayish-brown postjuvenile upperparts. Heavy yellow bill with black tip and red on nail. AMAR AYYASH. PERU. NOV.
3   1st cycle. A paler individual with several dark gray postjuvenile coverts and tertials. Brighter bill with noticeable red tip. AMAR AYYASH. PERU. NOV.
4   1st cycle with extensive 1st prealternate molt. Compare to Olrog's, which has paler head and neck, with whiter loral region and chin. AMAR AYYASH. PERU. NOV.
5   1st cycle with similar-aged Kelp Gull (left). Belcher's is smaller, with a relatively thick, strong bill for its size. Both individuals have undergone extensive 1st prealternate molts, replacing most wing coverts. AMAR AYYASH. PERU. OCT.
6   Largely juvenile. Extensive pale edging to upperparts, dark tail and already a bright bill at this age. Olrog's averages paler uppertail coverts and whiter forehead. AMAR AYYASH. PERU. NOV.

14 BELCHER'S GULL

gray black, with white wing linings. *Similar Species:* A fairly straightforward gull in its range, but compare to adult Kelp Gull where the two overlap. Kelp Gull is noticeably larger, typically lacks large amounts of black on the bill tip, and has a thicker white tertial crescent and duller legs year-round. On the open wing, adult Kelps do not show a tailband. For out-of-range individuals, such as those reports from Florida, consideration should be given to Olrog's Gull, which is generally larger, thicker billed, and with an overall

7   1st cycle. Dusky grayish-brown underwing and whitish vent. Dark brown head and neck, and thick yellow bill with black tip. ROBERT TIZARD. CHILE. OCT.
8   1st cycle. Well-marked underparts extend down to the lower belly on this individual, as in some Olrog's. Belcher's averages darker wing linings, with darker forehead and lores. SASKIA HOSTENS. CHILE. APRIL.
9   2nd cycle. Basic plumage. Olrog's has noticeably thicker bill and brighter white lower hindneck. Aged by grayish-brown belly and lower neck, and extensive brown on upperparts. RICHARD BONSER. CHILE. APRIL.
10   2nd cycle undergoing prealternate molt. Cleaner head and breast than 14.9. Bright bare parts similar to adult's. A thick-billed individual. Distinctive gray wash on body not found in Olrog's. AMAR AYYASH. PERU. NOV.
11   2nd cycle. Alternate plumage. White head and neck, with subtle gray wash on lower neck and breast. Solid black scapulars and many 2nd alternate wing coverts. AMAR AYYASH. PERU. NOV.
12   2nd cycle. 2nd prebasic molt well underway (new s1/p1–p6). Tail feathers possibly 1st alternate. Darker forehead and dingier lower neck compared to Olrog's. DOUGLAS FAULDER. PERU. JAN.
13   2nd alternate. Broad black tailband with white tips resembles Olrog's, but with gray wash on lower neck and body. This individual has replaced primaries nonsequentially, likely in 2nd prealternate molt. AMAR AYYASH. PERU. OCT.
14   2nd cycle in alternate plumage. Distinctive broad black tailband. White head and noticeable gray wash to lower neck. p9–p10 oddly fresh. AMAR AYYASH. PERU. NOV.
15   Adult type. Basic plumage. California's only Belcher's Gull, which lingered for five months in San Diego County, just a couple of miles from the Mexican border. ANTHONY MERCIECA. CALIFORNIA. AUG.
16   Adult with basic head pattern. Hooded appearance and extensive black on bill tip expected. Solid black upperparts, with noticeable white tips to secondaries. AMAR AYYASH. PERU. NOV.
17   Adult in alternate plumage. White head and bright yellow bare parts, with much red on bill tip. Note black tail. Olrog's averages more prominent white secondary skirt, but variable at rest. AMAR AYYASH. PERU. NOV.
18   Adult. Saturated-yellow bare parts. Belcher's is relatively long-legged, with a long bill that is apparent here. This individual has extensive gray wash on body, which can be found year-round (compare to 14.17). AMAR AYYASH. PERU. NOV.

stockier body. Olrog's lacks the gray wash found on the body of Belcher's and averages lighter and less pigment on the head in basic plumage. If visible, the orbital ring on Belcher's is more of a yellow, compared to the reddish orbital of Olrog's. Black-tailed Gull shares a similar tailband but is much paler gray on the upperparts, with a pale eye and thinner bill.

1st Cycle: Fresh juveniles have a sooty-brown head and neck with variable mottling on the breast and flanks. The plumage aspect of juveniles ranges from a ghostly white to a dark chocolate brown, but overall, the upperparts of juveniles appear long feathered, with brown centers. Dark eyes. Pinkish legs become a dull yellow throughout 1st cycle. Pale bill base has a black tip. Young birds will often show an indistinct red spot on the nail. Dark primaries sometimes show small white tips. Vent region behind the legs is typically pale, but some barring and chevrons are often present on the undertail coverts in fresh juveniles.

On the open wing, note the scaly appearance to the upperwing coverts with pale edging. The dark tail contrasts with mostly white ground color to the uppertail, which has variable brown smudged marks and barring. The secondaries and tail have fine white edging. Smoky-brown underwing. Variegated postjuvenile upperparts are generally boldly dark, some appearing adult-like but with a patchy brown hue. The mixture of worn juvenile feathers and newer 1st alternate feathers makes for a very blotchy and scraggy appearance on most individuals. *Similar Species:* Compared to similar-aged Olrog's Gull, Belcher's is thinner billed, with a thinner bill base and a slimmer body. Belcher's generally has a much

19 Adults and 2nd cycle (right). Lanky body, long legs, and long thick bill. Solid black upperparts. Contrasting white eye crescents seen well in basic plumage. ROHAN VAN TWEST. PERU. APRIL.
20 Adults and 1st cycles. Belcher's is roughly the size of a Ring-billed Gull, but with longer wings and legs, and a long, heavy bill. Adults are among the darkest gulls in the world. AMAR AYYASH. PERU. NOV.
21 Two adults with Kelp Gulls. Note size difference. Adult Kelp Gull (center) shows dull greenish-yellow legs and limited red on bill tip, and lacks gray wash on body. AMAR AYYASH. PERU. NOV.
22 Adult concluding prebasic molt. Solid-black hood pattern not shown by adult Olrog's. Upperwing similar to Olrog's, but Belcher's has noticeably thinner white trailing edge to secondaries and averages a broader black tailband. DAVID F. BELMONTE. PERU. MAY.

more solidly filled-in head pattern at this age, whereas Olrog's has a distinctly pale forehead, pale feathering around the bill base and chin, and a more patchy appearance throughout the head. The uppertail coverts on Olrog's are also largely pale, with fewer markings than Belcher's, but consideration should be given to the extent of molt of uppertail coverts (i.e., some Belcher's replace juvenile uppertail coverts early on, and these feathers can be predominantly white).

2nd Cycle: Approaches adult aspect, but with variable brown and pale feathers throughout the upperparts. Alternate birds with mostly white head may show a light gray wash to the breast, but not as vivid and intense as adult's. In basic plumage, the head is mottled brown and takes on a patchy appearance, with a poorly defined hooded look. Some individuals in basic plumage can show extensive brown wash down the neck and breast, as in 1st cycle. Legs are generally yellow but with a greenish tinge. The open wing reveals primaries that are browner than adult's, dark-centered secondaries, and a narrow white trailing edge. White uppertail with a wide black tailband. Tailband may average more pigment on outer tail feathers than is found in adult. Dark eye, and bill a dull to bright yellow, averaging more black than older birds. Variable red on bill tip. *Similar Species:* Compared to similar-aged Olrog's, note Belcher's gray buffer between the head and mantle along the hindneck (in basic plumage). Olrog's should generally show a thicker white trailing edge on the secondaries and average larger proportions, especially throughout the bill.

MOLT Complete molt worked in around breeding, primarily in late October through March, but some year-round breeding possibly takes place in a small percentage of the species (more study needed). Complete molt in southernmost populations appears to be slightly earlier in the calendar year, while northern populations average later molts (e.g., compare those wintering in central Chile with those in far N Peru). Thought to have one inserted molt in 1st cycle (1st prealternate) and a prealternate molt in subsequent cycles, so relegated to SAS. A variable number of upperwing coverts may be replaced in 1st prealternate molt. Individuals replacing some wing coverts in 1st prealternate are more likely to replace several tertials as well. A variable number of rectrices may be replaced in 1st prealternate (Ayyash, pers. obs.). One individual apparently replaced 10 rectrices in 1st prealternate molt. I suspect a variable number of secondaries are also replaced in prealternate molt in this species, with some birds apparently replacing all secondaries (Ayyash, pers. obs.). One 2nd-cycle-type bird recorded with atypical, eccentric-like, primary molt sequence. Records from North America pertain to adult-type birds. A well-documented Belcher's Gull from California was observed with a somewhat asynchronous molt schedule.

HYBRIDS None.

---

23 Adult in alternate plumage. Black tailband bordered by white tips. No mirrors. White trailing edge tapers off to thin white tips at inner primaries. ROGER AHLMAN. PERU. SEPT.
24 Adult (alternate plumage). Wing linings commonly have dusky gray appearance (not pure white), becoming darker on underside of primary coverts. AMAR AYYASH. PERU. OCT.

15 BLACK-TAILED GULL *Larus crassirostris*

L: 18.0"–21.5" (46–55 cm) | W: 48"–49" (122–124 cm) | Four-cycle | SAS | KGS: 8–10

OVERVIEW A medium-sized four-year gull of slender proportions. Widespread throughout various coastal habitats in NE Asia. Found in estuaries and fishing harbors, scavenges in landfills, and will readily take handouts. Epipelagic. Vagrant to North America, with sporadic and remarkable records from coast to coast. Liable to turn up anywhere other gulls are found, from human-made reservoirs to farmland! Ring-billed size, but proportionately longer billed, with a noticeable long-winged appearance. Upperparts are most similar to those of Laughing Gull or a pale Lesser Black-backed Gull in coloration. Broad black tailband against an otherwise all-white tail is unique among large gulls in the northern hemisphere. Its cat-like sounds have earned it the name Sea Cat in Japan, where it is also revered and protected, nesting colonially in the thousands at several shrine sites.

TAXONOMY Monotypic.

RANGE
Breeding: Native to coastal NE Asia, where it breeds colonially on vegetated islands and rocky shores. Abundant in Japan, Korea, and China. Ranges as far north as N Sakhalin and northeast to Kuril Islands (Brazil 2009). Vagrant to Kamchatka, Malaysia, and as far south as Australia.
Nonbreeding: In North America, this species has turned up at a variety of habitats frequented by flocks of other mid-sized and large gulls. As would be expected, given the species' normal range, records are concentrated in Alaska, where it is a casual visitor to W Alaska in spring and early summer (May–July) and to south-central and SE Alaska in summer and fall (June–Oct.); less so farther south along Pacific Coast as far as S California, where most records are from fall and winter (Aug.–Jan.). Very surprising are the scattered records (30+) from various seasons and many states and provinces throughout North America, from Northwest Territories, SW Hudson Bay, Newfoundland, Nova Scotia, S Ontario, Vermont, Virginia, Wisconsin, Illinois, Indiana, Ohio, and Iowa south to Texas, Florida, SE New Mexico, Sonora, Mexico, Belize, and Bermuda, as well as several other states along the mid-Atlantic.

IDENTIFICATION
Adults: Upperparts dark gray. Mustard yellow legs. Bill yellow with black subterminal mark surrounded in red out to tip. Pale yellow eye with red orbital. Structurally, the wings sit far behind this bird at rest, and the vent region is sleek and thin. The open tail reveals a noticeable black tailband from above and below, with prominent white tips to the rectrices. Sides of black tailband also easily seen at rest. White head in alternate plumage. In basic plumage, the face often takes on dusky markings—heavy at times—especially in the postocular region, back of the head, and down the nape. Trailing edge to secondaries is white, tapering off along the inner primaries. Outer primaries show variable amount of black on centers from p6 to p10, with indistinct white tips. Weak subterminal band regularly found on p5, broken band on p4 and at times even out to p3. A notable percentage of adults apparently show black-marked primary coverts that are often associated with subadult plumages in other species. For instance, a known-age 30-year-old adult in Japan displayed large black centers to its outer greater primary coverts. Sometimes shows a small p10 mirror, often restricted to a single web and asymmetric. *Similar Species:* Similar in size to Ring-billed Gull but distinctly darker. In North America, this age group should be straightforward, but there is superficial resemblance to Lesser Black-backed and California Gulls. Smaller than Lesser Black-backed Gull, although somewhat similar structurally. Bill on Black-tailed is longer, thinner, and parallel edged, with no noticeable expansion on the gonys, and without red out to the tip. Compared to California Gull, Black-tailed is pale eyed and darker above. Black-tailed lacks mirrors on the outer wing. Tailband readily settles this identification. Olrog's and Belcher's Gulls from South America have similar tail patterns but are considerably darker, with broad white trailing edge and much thicker bill.

15 BLACK-TAILED GULL

1. Juvenile. Brown head, neck, and sides rather typical. Dark tail and already a sharply bicolored bill with bright pink base. AKIMICHI ARIGA. JAPAN. JULY.
2. 1st cycle. Postjuvenile scapulars highly variable, here showing much white. Long bill can show a slight droop near the tip. Paling forehead. MARK STEPHENSON. CALIFORNIA. JAN.
3. 1st cycle. A paler individual, now with all juvenile scapulars replaced. A smooth, velvety appearance is maintained on the hindneck, breast, and sides. AMAR AYYASH. JAPAN. JAN.
4. 1st cycle. A large-billed individual, perhaps male. Paling head and considerable wear now evident. RICHARD BONSER. JAPAN. DEC.
5. 1st cycles. Variable head patterns. Underparts more white now, with moderate wing covert molt. MASAMI YOSHIMURA. JAPAN. MAY.
6. Juvenile. Even, brown tones, with scaly appearance to fresh upperparts. Dark tail contrasts with white tail coverts. Pink bill base and brown head. YANN MUZIKA. JAPAN. JULY.
7. 1st cycle. Dark remiges lacking pale inner-primary window. Pale postjuvenile scapulars. Paling forehead and smooth brown neck pattern. AMAR AYYASH. JAPAN. JAN.

15 BLACK-TAILED GULL

1st Cycle: Juvenile plumage is dark brown with buffy edging throughout the upperparts. Bill becomes bicolored soon after fledging, with pink base and large black tip. Pink-colored legs. White eye crescents often visible. Pale forehead and chin develop early on. Undertail coverts and vent region are generally white, while underbelly is a soft brownish gray. Uppertail coverts have few if any markings, with a

8 1st cycle. Recalls California Gull, but Black-tailed averages whiter vent, has smooth, solid brown flanks and neck, with plain greater coverts. HIROSHI SATO. JAPAN. OCT.
9 2nd cycle. Similar to 1st cycle, but note new, jet-black primaries (not fully grown) and black tail. Some red on nail. MIKIYA OIKAWA. JAPAN. AUG.
10 2nd cycle. Black tail with noticeable white border. Iris now paling. Leg and bill color yellowing. AMAR AYYASH. JAPAN. JAN.
11 2nd cycle. Plain, dark gray scapulars (2nd alternate) and long greenish-yellow bill now. Hindneck and breast often regress to smooth brown wash. Wing coverts are neatly fringed with pale edging. AMAR AYYASH. JAPAN. JAN.
12 2nd cycle in alternate plumage. Advanced bare parts with black-and-red pattern on bill tip. Underparts largely white. Wing panel can become whitish, especially when faded in spring. MASAMI YOSHIMURA. JAPAN. MAY.
13 2nd cycle. Plain, all-dark tail, and white coverts. Paling iris. Dark wash on head appears smudged and not streaked, as in California or Lesser Black-backed. AMAR AYYASH. JAPAN. DEC.
14 2nd cycle. Paler lesser and median coverts on this individual. Advanced tail pattern recalls 3rd cycle (perhaps replaced in 2nd prealternate molt). Bill pattern also advanced. SIMON COLENUTT. SOUTH KOREA. JAN.
15 2nd cycle. A darker individual (and slightly underexposed). Note smooth head pattern and plain greater coverts. MIKIYA OIKAWA. JAPAN. MARCH.
16 2nd alternate. 1st cycle ruled out by round primary tips, pale underwing and body, pale eye, and yellowish legs. Some can show duskier brown underwing and body at this age. Note dark tail. AMAR AYYASH. JAPAN. JAN.
17 3rd-cycle type. Compared to adult, note brown wash across upperparts, primaries lacking white apicals, and dull bill pattern with extensive black tip. MICHAEL TODD. ILLINOIS. JAN.
18 3rd-cycle type. Same individual as 15.17. Much black on primary coverts and alula, brown wash across upperparts, and irregular black tailband. CHRISSY MCCLARREN. ILLINOIS. JAN.
19 3rd cycle. Broader tailband than adult, with irregular black along proximal edge. Black on secondaries. Difficult to age when perched, although white apicals often reduced or absent. AMAR AYYASH. JAPAN. DEC.

15 BLACK-TAILED GULL

20 Subadult. Overall adult-like, but with brown hue across upperparts, and unusually dull bill for this date, with extensive black on tip. Presumed 4th basic incoming primary beneath tertials. MIKIYA OIKAWA. JAPAN. AUG.

white base color that contrasts heavily with a wide, dark tailband. Rectrices have indistinct white tips. Upperwing uniformly dark. On the open wing, note the lack of any window on the inner primaries. Upperwing coverts, especially greater coverts, are plain and lack patterning. Underwing coverts are a paler smoky brown. Postjuvenile scapulars sometimes have a silvery white or a dark sooty gray, with small, dark centers. *Similar Species:* California Gull is generally much more patterned on the upperwing coverts, tertials, and uppertail coverts, and it lacks pale eye crescents. Heermann's Gull is more uniformly dark, with dark uppertail coverts, noticeably shorter bill, and black legs. Black-tailed has a mostly white vent region, which these two lack.

2nd Cycle: Many individuals with solid gray back at this age, but with a brownish hue. Some with noticeable pale regions on the scapulars. Pale bill base is yellow and often shows a grayish-blue color. Some red is often present close to the black bill tip, especially later in the season (starting around late April–May). Eyes variable, from dark to yellow, but note the white eye crescents, which typically contrast with a dusky head. Dusky hindneck, heavily marked at times, but head may be entirely white with clean neck in alternate plumage. Leg color varies from pink colored to greenish yellow and pale yellow. White uppertail coverts with few if any markings, against a wide black tailband. Plain upperwing with variable adult-like gray at this age, especially median coverts. On the spread wing, the inner primaries and secondaries may show pale edging—bold white at times—but compared to 3rd cycle, note the absence of an adult-like trailing edge on the secondaries. *Similar Species:* California Gull is noticeably paler gray above, with a shorter bill. Red markings on the bill are not expected in 2nd-cycle California but are often present in similar-aged Black-tailed. Head markings on California Gull are streakier and spotted, as opposed to the uniformly smudged look on Black-tailed, which often develops a dense necklace. Note the lack of white eye crescents on California Gull. Upon close inspection, Black-tailed typically shows contrasting white eye crescents, which are seen well in basic plumage. Most Black-taileds will show a paling iris by now, while California is dark eyed. On the open wing, Black-tailed averages a darker tail base with cleaner uppertail coverts and vent and maintains darker inner primaries with no p10 mirror. Ring-billed Gull is significantly paler above and lacks any red on the bill.

3rd Cycle: Overall, similar to adult, but some features that set it apart include variable black on secondaries (sometimes reduced to thin black shaft streaks), more-extensive black markings on primary coverts and alula, and wider tailband with irregular proximal edge showing black up the tail feather shafts (some adult types appear to have this irregular tail pattern as well). Outer primaries average more pigment with brown tones, smaller apicals are reduced or may be absent entirely, and black is commonly found on outer webs of p2–p3. A dusky brown hue across the upperwing is not uncommon. Third cycles average more head markings in basic plumage. Black tip to the bill may be larger with less red, and overall duller coloration to bare parts. Wing linings sometimes with a "dirty" wash compared to clean white underwing of the adult. *Similar Species:* Lesser Black-backed lacks a wide black tailband at this age, is larger with a heftier bill, and typically shows mirrors. Also note the differences in head patterns: streaked in Lesser Black-backed and smudged in Black-tailed.

21 Adult in alternate plumage. Long, pencil-shaped bill, with red often on both edges of black band. Pale iris and red orbital. Black on tail exposed here. ERIC VANDERWERF. JAPAN. FEB.
22 Adult with adult-type Black-legged Kittiwake (right). Long wing projection, yellow legs, and relatively dark upperparts. Note bill shape and pattern. RICHARD MACINTOSH. BERING SEA; RUSSIA-ALASKA BORDER. JULY.
23 Adults. Slightly overexposed. Similar to Ring-billed Gull in size, but Black-tailed has darker, slate-gray upperparts with longer bill that always shows red. MASAMI YOSHIMURA. JAPAN. JAN.
24 Adult in basic plumage. Smudged head pattern often restricted to back of head. Note bill pattern, weak white tertial crescent, and extensive black on tail. CHRISTOPHER TAYLOR. CALIFORNIA. NOV.

MOLT Partial 1st prealternate molt regularly includes a variable number of upperwing coverts and may include some tertials. On rare occasions, a variable number of tail feathers are replaced (Ayyash, pers. obs.).

HYBRIDS The only reported hybridization was in captivity, at Kyoto Zoo (1939–1940; Kuroda 1941). A female Black-tailed and male Silver Gull (*Chroicocephalus novaehollandiae*) bred over two years. One chick from two broods survived and reached maturity; the hybrid was said to resemble Black-tailed Gull. No other details were given.

25 Adult. Faint dusky markings on back of head and face. Apparently undergoing prealternate molt with new scapulars, lesser coverts, and lower tertials. JOSH JONES. JAPAN. JAN.
26 Adults (lower left). Noticeably smaller and paler than adult Slaty-backed Gull (right). 1st-cycle Slaty-backed Gulls above. MARTEN MULLER. SOUTH KOREA. FEB.
27 Adult. Long-winged, with relatively thin, white trailing edge. Wingtip recalls Laughing Gull. Pale eye. Broad black tailband, with prominent white tips distinctive. RICHARD BONSER. JAPAN. DEC.
28 Adult. A small percentage of adults show a small p10 mirror, usually on the inner web. Also, it's not uncommon for apparent adults to show a light black wash on primary coverts. AMAR AYYASH. JAPAN. JAN.

16 SHORT-BILLED GULL *Larus brachyrhynchus*

L: 16"–17" (41–43 cm) | W: 41"–44" (104–112 cm) | Three-cycle | SAS | KGS: 6–8

OVERVIEW Long known as Mew Gull in the ABA Area, Short-billed Gull is a three-cycle species with a superficial resemblance to Ring-billed Gull, to which it is closely related. Thrives in several biomes throughout Alaska and NW Canada, including the boreal forest and tundra. An impressively adaptable gull, breeding in both marine and freshwater habitats, from coastal cliffs and rocky islets to rivers, marshes, and grassy meadows. Regularly nests in trees located near water, mostly conifers, but also reported on limbs of deciduous trees, as well as in large cavities in tree stumps and on poles. Nests singly, in small groups, and less commonly in colonies exceeding 50 pairs.

Omnivorous. As with many other gulls, its diet in the breeding season differs from that in migration and at the wintering grounds. Commonly feeds on aquatic and terrestrial invertebrates and frequents herring spawn events, and plowed fields for exposed worms and larvae. Customarily found working mudflats and exposed substrate at low tide, and patrols the surf in between resting at beaches. Can be numerous at sewage-treatment ponds and outfalls. Often picks insects from the water's surface, where feeding behavior is phalarope-like. Less commonly feeds in landfills or pirates from other birds. Accepts handouts throughout parts of its range, but in other areas is much shyer and more reserved.

TAXONOMY Monotypic. No recognized geographic variation in North America, although studies are lacking. Type specimen from Great Bear Lake, in the Northwest Territories (Swainson & Richardson 1831). In 2021, the American Ornithological Society split *brachyrhynchus* from a larger complex of similar-looking old-world taxa known collectively as Mew Gulls or the Common Gull (*canus*) complex (Chesser et al. 2021). Mew Gull was also the English name used widely for *brachyrhynchus*. Prior to being allied with the old-world taxa, *brachyrhynchus* was treated as a separate species by the American Ornithological Society through the third edition of its *Checklist of the Birds of the World*, in which it was referred to as Short-billed Gull (Peters 1934). This name was resurrected in 2021, when it was split. Old-world taxa all remained under Common Gull (*Larus canus*; see that species account). The split was said to be supported by genetic work that suggested *canus*, *heinei*, and *kamtschatschensis* clustered together, with notable genetic differences detected in *brachyrhynchus* (Sternkopf 2011; Sonsthagen et al. 2016). In addition to their being geographically isolated, their distinct morphological and vocal differences bolstered the split (Adriaens & Gibbins 2016b).

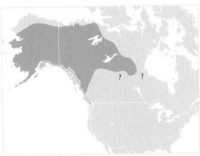

RANGE
Breeding: Breeds in mainland Alaska, throughout the interior lowlands and southeast region of the state. The most widespread gull species in Alaska, although it becomes rare farther north in North Slope County. Also breeds from British Columbia, including Vancouver Island, east to Yukon, mostly S Northwest Territories, but locally to Inuvik and surrounds. Recently found in small numbers to N Saskatchewan and Alberta, and possibly in extreme N Manitoba, although unconfirmed. Two unsuccessful nests were found in Churchill in 1980 (Godfrey 1986).

Nonbreeding: Vacates NW Alaska by the end of September, but small numbers linger in south-central Alaska into November. Small numbers winter as far north as Aleutian Islands, Kodiak Island, and east to Prince William Sound region. Generally considered a medium-distance migrant, although it is unknown if any southern breeders from British Columbia are year-round residents. The first migrants (juveniles) may appear south of the breeding range as early as mid-July and especially during August in coastal Pacific Northwest and by early or mid-September in N California. But the species is not numerous in these areas until mid-October, and even larger numbers arrive during November, when it becomes common to locally abundant at a variety of coastal habitats, as well as appreciably inland, west of Cascades and Sierra Nevada, in wet farm fields and pastures from SW British Columbia south through central Oregon (e.g., Willamette Valley) to Sacramento Valley, California. Does not arrive until late October in most of

S California. Numbers in winter in much of S California have declined in recent decades. In winter, it is most numerous in Pacific Northwest and northern and central California, then fairly common south to Santa Barbara or Ventura Counties and uncommon to rare south to N Baja California. Casual south to Baja California Sur. Relatively scarce offshore but found in small numbers out to about 50 miles.

Most interior records are from large lakes and riverways. Rarer migrant and winter visitor farther into interior from south-central British Columbia and in E Washington and Oregon south to W Nevada, Salton Sea, Lower Colorado River in California / Arizona, and N Gulf of California in NW Mexico. Very rare to casual west to Bering Sea islands. A handful annually in E Colorado, with widespread records, mostly fall and winter. Casual from Nunavut to Great Lakes, Great Plains, and south to Texas and Louisiana, with scattered records throughout the interior. Exceptional to the East Coast, where a spike in records began in 2013, and there is a continuous series of sightings from Kings County, New York, beginning in 2015. Still unrecorded in several eastern states and far NE Canadian provinces. A few are found annually in Japan, mainly in the east of the country (Michiaki Ujihara, pers. comm.). Extralimital records to Azores (Alfrey & Ahmad 2007) and Hawaiian Islands (2015).

Departs winter range mostly in March, with some still present through mid-April (Pacific Northwest) and a few lingering into May. Nonbreeders are uncommon but regular in NW Washington, and random individuals, usually 1st cycles, are rare to casual in summer farther south of the breeding range, as far south as S California.

IDENTIFICATION

Adult: A small white-headed gull with medium gray upperparts, short legs, and a long-winged appearance. The head is generally rounded and is complemented by a short, thin bill with a tapered tip, showing no appreciable gonys expansion. Variable eye color: some with all-dark eyes, but many have a clouded, amber-colored iris, which may appear all dark from a distance; others have paler olive-gray iris with dark flecking or rarely bright, all-yellow iris (as in Ring-billed). Red orbital. Mustard-yellow legs and unmarked yellow bill in breeding condition. In nonbreeding condition, the bill is a duller yellow with weak, grayish smudges near the tip, sometimes forming a faint band, but typically not defined and not boldly black. Does not show red on the bill. At rest, note the bold white tertial crescent and frequently seen gray base to the folded primaries. The head markings are distinctive, showing dusky mottling on the crown and face, converging to a densely smudged brown lower hindneck. This smooth brown wash often extends down to the sides and breast in the most heavily marked birds.

On the open wing, a broad white trailing edge is conspicuous, often extending to the tips of the inner primaries. Large mirrors on p9–p10 (rarely ever without p9 mirror). A small mirror on p8 is quite rare. Typically has a full subterminal band on p5, but some have a broken band, and in extreme birds just a small mark or all white. Many individuals show some marks on the outer edges of p4 (very rarely a small mark to outer web of p3). Often, there are prominent white tongue tips on p5–p8, with longer gray tongues

---

1 Juvenile. Small bill, long wings, and overall plain tertials and greater coverts. Warm, even, brown underparts. AMAR AYYASH. ALASKA. AUG.
2 1st cycle. Some gray, postjuvenile scapulars have grown in. Pale bill base common by fall. Low-contrast plumage and docile appearance recalls a mini Thayer's. AMAR AYYASH. CALIFORNIA. JAN.
3 1st cycle. Short, petite bill, sometimes with pinkish base, as in Ring-billed, which lacks this smudgy appearance to the neck and breast and shows more contrasty wing coverts. AMAR AYYASH. BRITISH COLUMBIA. MARCH.
4 1st cycle. Prominent pale edging on primaries not uncommon. Extensive pale edging on upperparts and white greater coverts recall Kamchatka, but note short bill and smudged neck and sides. AMAR AYYASH. WASHINGTON. JAN.
5 1st cycle with Ring-billed Gull (back). Remarkable differences in bill and body size, not always this obvious. Smudged neck and marked undertail coverts point away from Common Gull. ROBERT RAKER. COLORADO. APRIL.
6 1st cycle. Unusually pale bodied for this date. Broad white tertial tips and lower scapulars not very common in this taxon, but does occur (compare to Kamchatka). AMAR AYYASH. WASHINGTON. JAN.
7 1st cycle. Note long wings, lanky rear to body, and tapered bill tip with dull bill base. Can become extremely bleached and worn late in the season, at times with contrasting, adult-like scapulars. MARLIN HARMS. CALIFORNIA. APRIL.
8 Juvenile. Even, sandy tones with low-contrast brown remiges. Dark tail at times, with paler base. Considerable barring on uppertail coverts. Pale inner-primary window, with distinct dark tips. AMAR AYYASH. ALASKA. AUG.
9 1st cycle. Short bill obvious. Whitish appearance across upperparts recalls Kamchatka, but note pale and plain inner primaries with dark tips. Short-billed averages darker tail base as well. AMAR AYYASH. BRITISH COLUMBIA. MARCH.

on p8–p9. Importantly, p9 sometimes shows a thin white tongue tip. The inner webs to p9–p10 also average relatively long gray tongues. This, in addition to the restricted black on the outer webs of p6–p7 (short wedges), makes for a unique primary pattern. The wingtip is appreciably variable, however, with some birds impressively white winged and others averaging more black on the inner webs of p8–p9 with

---

10 1st cycle. Some individuals show darker outer webs to the inner primaries. Note smudged neck and dark tail. LIAM SINGH. BRITISH COLUMBIA. FEB.
11 1st cycles. Underwing generally a dusky brown, as seen on the individual at left, but somewhat variable. Barring on axillaries not expected. Dark belly patch found in some 1st cycles (left). AMAR AYYASH. WASHINGTON. JAN.
12 2nd cycle. Smudged neck, short, petite bill, long wings, and lanky body. Brown wash across upperparts and black tertial spot common at this age. AMAR AYYASH. CALIFORNIA. JAN.
13 2nd cycle with California Gull. Compact gull with thin bill and short legs. Yellowing bill with dark smudges. Delayed fleshy leg color. Lacking white apicals with markings on primary coverts, unlike adult. AMAR AYYASH. BRITISH COLUMBIA. MARCH.
14 2nd cycle (right) with adults. Frosty-white wing covert pattern not uncommon (compare to Kamchatka). Black on tail and tertials expected. AMAR AYYASH. BRITISH COLUMBIA. MARCH.
15 2nd cycle with adult Ring-billed (right). A larger individual with advanced, adult-like upperparts. Open wing used for aging, but note dull bare parts and absence of white apicals on fresh primaries. CALEB PUTNAM. MICHIGAN. OCT.
16 2nd cycle. Smudged neck distinctive. Wingtip variable, but note gray base to p8 and subterminal bands to p3. Sometimes with complete tailband and, more rarely, thin dark streaks on secondaries. LIAM SINGH. BRITISH COLUMBIA. MARCH.
17 2nd cycle. Cleaner upperparts and tail than 16.16, and larger mirror on p9, which Ring-billed lacks entirely at this age. Reduced black on p8, and black on p3–p4 expected. Black on primary coverts restricted to feather tips here. BRUCE MACTAVISH. BRITISH COLUMBIA. DEC.
18 2nd cycle. Dusky wing linings with dark fringes fairly common. Smudged neck and short bill. Note gray bases to outermost primaries. AMAR AYYASH. ALASKA. MARCH.
19 Adult in alternate plumage. Clean yellow bill without a red gonys spot. Eye color variable, but typically darkish. Red orbital. Much gray on exposed bases of primaries. STEPHAN LORENZ. ALASKA. MARCH.
20 Adult. Extensive gray base to primaries just below the tertials, with bold white tongue tips to p5–p7. Broad tertial crescent. Saturated bare-part colors expected in late winter. AMAR AYYASH. BRITISH COLUMBIA. MARCH.
21 Adult (basic). Sometimes with a dark smudge near the bill tip, recalling Common Gull. Unlike Common, note the paling iris and smooth wash on the face and neck (more spotted and streaked in Common). LIAM SINGH. BRITISH COLUMBIA. JAN.

a complete subterminal band on p4. These patterns are likely age related overall, but some may simply be individual variation (Liam Singh, pers. comm.). *Similar Species:* Ring-billed and California Gulls are both yellow legged and are sometimes confusion species, but the resemblances are superficial and can quite readily be sorted out (see those species accounts). Also, some confusion with Black-legged Kittiwake may arise where the two overlap (especially in coastal Alaska), but wingtip pattern and leg color readily settle the matter.

The most formidable identification challenge involves separation from old-world Common Gulls, often in vagrant scenarios. Comparing wingtip patterns is essential. It must be emphasized, however, that wingtip patterns can belie expectations at certain times, due to age-related factors, inherent variation, and general overlap in some features. A collection of field marks should be employed, including bare-part patterns, darkness of gray upperparts, head and neck markings, and overall size and structure. Common Gull (hereafter implied as nominate *canus*) averages paler gray upperparts with distinct spotting on the head and finer neck markings usually ending abruptly and not coalescing into a dense, brown neck band. In Short-billed, the lower hindneck is especially densely marked, forming a brown boa, with a smudged appearance. The bill tip on Common Gull frequently shows a rather defined black band in nonbreeding condition (recalling some Ring-billeds), unlike the weak, gray markings on Short-billed. Eye color in

22 Adult (basic). Heavy, dusky wash on neck. Dull bare-part colors. Lanky body shape, with thin bill and short legs. AMAR AYYASH. CALIFORNIA. JAN.
23 Adult (center) with Ring-billed (back) and California Gull (front). Smallest of the three with petite, unmarked bill and darker upperparts matching California. Bolder and more contrasting tertial crescent on darker taxa. AMAR AYYASH. COLORADO. NOV.
24 Adult with Thayer's Gull (right). Fairly distinctive. Our smallest white-headed gull. Compare to Common and Kamchatka Gull. AMAR AYYASH. BRITISH COLUMBIA. MARCH.
25 Adult. Classic wingtip pattern showing string-of-pearls. Extensive gray on p8–p9 and bold, white tongue tip on p8. Short black wedge on p7 results in subterminal band similar in size to p6. AMAR AYYASH. WASHINGTON. JAN.

Short-billed is variable but is systematically dark in Common. The white trailing edge to the secondaries is strikingly broad on Common, abruptly thinning out on the tips of the inner primaries. In Short-billed, many adults show relatively broad white tips to the inner primaries too. Both taxa show large p9–p10 mirrors, but on average the p9 mirror on Common is larger, or at least closer in size to the mirror on p10. Quite distinctive is the extensive amount of black often found on the three outer primaries on Common, in contrast to the rather long gray tongues that are often found on Short-billed. On Short-billed, p8 typically has an extensive gray tongue with a prominent white tongue tip. Overall, the string-of-pearls effect on Short-billed is more prominent and eye-catching and commonly extends to p8, with shorter black wedges on p6–p7. A p8 mirror is very rare in Short-billed but not uncommon in Common. Also, Common is much more likely to show no markings on p4 and can have limited black on p5 (at times completely unmarked). Short-billed typically shows a complete, symmetrical band on p5 and often has some markings on p4, although it very rarely has unmarked p5; but the extensive white in the outer wingtip is distinctive and does invite confusion with Common. Finally, any individual with a long, pale tongue on p10 covering more than half the length of the feather, or with a white tongue tip to p9, should invariably be a Short-billed Gull (Adriaens & Gibbins 2016b). Kamchatka Gull is larger billed, with overall larger proportions, and averages darker gray upperparts (although with considerable overlap). See that account for details.

1st Cycle: Note the compact head and shorter bill, with a long-attenuated look to the folded wings. Overall, the head and neck have a soft, uniform brown coloration, extending to the breast and underparts. The juvenile greater coverts are largely pale and plain, while the lesser and median coverts vary from broad, dark centers to paler centers showing dark, elongated-diamond tips. The tertials are a medium, plain brown and generally show thin, pale fringes. The folded primaries often have prominent pale edging, which makes for a rather distinct appearance. The elegant look can be very reminiscent of a "mini" Thayer's. The bill usually develops a paler base by early fall. Pinkish legs. Dark eye, which often appears disproportionately large. As the postjuvenile molt progresses and the upperparts start to fade, a more contrasting grayish-brown aspect develops. The head can become pale with slight mottling, but the lower neck and breast often remain uniformly smudged. Some individuals develop a contrasting belly patch. The postjuvenile scapulars are generally weak gray with marbled brown tones and whitish fringes, especially on individuals molting earlier in the fall.

On the open wing, a pale inner-primary window is found, often with distinct dark tips. The greater coverts are plain and the palest tract on the upperwing, seldom with darker diamond tips. The tail is largely dark with much barring on the coverts, although some may have paler outer tail feathers and sparsely barred coverts. The undertail coverts are also barred, with some dark patterning often extending to the vent. The underwing coverts are smoky brown and dusky, generally matching pigment tones on the belly. *Similar Species:* Compare to Ring-billed and Common Gulls (later in this section) and Kamchatka Gull (see that account). Ring-billed averages a larger body, longer legs, and bigger bill, with an overall more rugged appearance. Be aware that some smaller, female-type Ring-billeds can show a petite bill with a rather tapered tip. The bill base on Ring-billed is often a more vivid pink. Plumage differences should settle this identification. The head and neck markings show uneven streaking and blotches on Ring-billed, lacking the softer, velvety texture of those of Short-billed. Short-billed has a more uniform, often sand-colored appearance throughout, often extending to the belly, while on Ring-billed, the base color to the body is largely white, showing more contrast with the upperparts. The tertials and inner coverts can show notching on Ring-billed, and this is generally not found in Short-billed. Also, Short-billed is more likely to show extensive pale edging to the folded primaries. On the open tail, note the largely white uppertail coverts on Ring-billed, showing weak barring or sometimes entirely white with a relatively narrow tailband. The uppertail coverts on Short-billed show more uniform and consistent dark barring and a darker tail base. Ring-billed averages a whiter underwing and axillaries.

Common and Short-billed Gulls have disjointed ranges, but they do present something of an identification challenge from the Niagara River corridor east to the Atlantic, especially in early juvenile plumage. The smaller bill and more elongated look to the wings on Short-billed are supportive, though subjective without direct comparisons. The key to separating these two is strictly based on plumage, with little to no attention given to size, as there is much overlap. Short-billed averages a more grayish-brown plumage aspect and retains a darker, smoother pattern on the neck and underparts. Note that many Common Gulls take on a rather pale body (more white, like Ring-billed) and show a paler vent region. Also, the postjuvenile scapulars in Common are often solid gray and adult-like, lacking the marbled brown tones and pale fringes seen in many Short-billeds. Common has uppertail coverts that are generally white with limited dark markings and a rather defined tailband. Short-billed shows consistent barring on the tail coverts and darker bases to the outer tail feathers. The underwing coverts on Short-billed are smoky and washed dark, whereas Common has largely white centers to these feathers, with dark fringes. A small percentage (3%–4%) of 1st-cycle Common Gulls can show a noticeable p10 mirror, which has not been recorded in Short-billed (although, on very rare occasions, a faint pinhole mark is found on *Larus brachyrhynchus*).

2nd Cycle: Many 2nd-cycle Short-billed Gulls have a rather dusky gray appearance. Solid, adult-like scapulars are typical, but often with extensive brown wash throughout the wing coverts. Commonly show black patches on the tertials. Can regularly have heavier head markings, with a pattern quite similar to the adults', showing a mostly smooth or smudged texture and not finely streaked or spotted. This wash has the tendency to extend farther down the breast and sides than on adults. Most birds retain a darkish bill band and have a grayish-yellow bill base, typically dull, not bright. Leg color variable, from dull pink to greenish yellow, although seldom an intense yellow, as in the adults. Most have dark eyes at this age, but upon close inspection some show a paling iris. On the perched bird, note the lack of white primary tips. About half the time, a mirror is found on p9. Also, p9 can quite expectedly show gray on the base of the outer web. Longer gray tongues on p8–p9 are not uncommon, and black on outer webs of p7–p8 often does not reach primary coverts. White tongue tip on p7 is rather expected and sometimes shows on p8. Complete subterminal bands on p5–p6, and very often on p4. Not rare for p3 to show some markings, and at times to p2 in heavily marked individuals. The greater primary coverts often show restricted black markings toward the tips, with cleaner gray bases. The secondaries are generally white, with some birds occasionally showing diffuse streaking around the shafts (typically limited to a short segment of the secondaries, and not the entire tract). Commonly shows markings throughout the tail, often forming a complete black tailband. The tail coverts are predominated by white, although some may show faint markings and light barring, particularly the undertail coverts. Many individuals show moderate brown edging on the underwing coverts. *Similar Species:* Second-cycle Common and Short-billed Gulls have several real similarities, in that both are medium-sized, three-cycle gulls with relatively small bills and are generally compared to Ring-billed Gull. But overall, their plumages are quite distinct. Common Gulls have rather adult-like upperparts with little brown wash to the coverts, usually unmarked secondaries, minimal tail markings, cleaner underwing coverts, and no smooth wash to the head and neck patterns. Their wingtip patterns also have some noticeable differences at this age, and they typically show more black coming up the bases of p7–p9. An individual with a gray base to the outer web of p9 should be a Short-billed, and a gray base to the outer web of p8 is very indicative of Short-billed. Also note the longer gray tongues on p8–p9 on Short-billed. A mirror on p9 is rather standard in Common Gulls but is found in only about half of 2nd-cycle Short-billeds. More primaries are marked in Short-billed, with a complete subterminal band on p4 being the norm, and marks commonly extend to p3. On Common, p4 rarely has a complete subterminal band; more often, only the outer web is marked, or p4 is completely unmarked. See also Ring-billed and Kamchatka Gulls accounts.

MOLT The partial inserted molt in 1st cycle is attributed to a prealternate molt and limited to scapulars, head, neck, and underparts. Does not typically include upperwing coverts or tertials, although three exceptional individuals have recently been recorded with a few replaced coverts, all in February 2023. Juvenile wing coverts are often retained into May, when the entire wing panel can be uniformly bleached to a sullied white. Uncommon but routinely found are 2nd cycles with a variable number of fresh, gray upperwing coverts in late winter and early spring, which are presumably 2nd alternate. These fresher coverts (and variable scapulars) coincide with an overall alternate aspect well before primary molt commences.

16 SHORT-BILLED GULL

26 Adult. Large white tongue tips on wingtip, including on p9, which is distinctive. Limited black on outer web of p8, full p5 band, and spots on outer edge of p4. AMAR AYYASH. BRITISH COLUMBIA. MARCH.

27 Adult. Dark bill band rare; recalls Common Gull. Note extensive gray base on outer web of p8–p9, bold white tips to inner primaries, and relatively long tongues on underside of far wing. AMAR AYYASH. BRITISH COLUMBIA. MARCH.

28 Adult. Smaller mirrors, extensive black on outer primaries, with diffuse black on bases of p7–p8, and complete band on p4 likely age-related. Smudged neck and broad white tips to inner primaries typical. AMAR AYYASH. WASHINGTON. JAN.

29 Adults. Large p9–p10 mirrors, broad white tips to inner primaries, and flashy string-of-pearls distinctive. White tongue tip on p9 (right) and long gray tongue on p9 cutting into mirror (left) diagnostic Short-billed features. LIAM SINGH. BRITISH COLUMBIA. MARCH.

30 Adult. Extensive gray tongues to p7–p8, with relatively little black. Contrasting, darker gray to underside of remiges. DONNA POMEROY. CALIFORNIA. DEC.

HYBRIDS Possibly with Ring-billed Gull. Several unconfirmed reports include a 1st cycle from Clallam County, Washington (Mlodinow et al. 2008); an adult from Worcester County, Maryland, in which it is uncertain if Short-billed Gull or Common Gull is involved (Howell & Dunn 2007); and a suspected adult from Santa Clara County, California, whose vocalizations were intermediate in sound (Alvaro Jaramillo, pers. comm.). These two species have never been reported paired up or nesting.

17A COMMON GULL *Larus canus canus*

L: 16.0"–19.5" (41–50 cm) | W: 39"–51" (100–130 cm) | Three-cycle | SAS | KGS: 5–7

OVERVIEW Common Gull comprises three subspecies, with a breeding range encompassing all of N Eurasia. It is often compared to Ring-billed Gull, which it superficially resembles. In particular, the name Common Gull is often used to denote only the nominate subspecies, *Larus canus canus*. Nominate *canus* breeds in N Europe and has long been the default Common Gull in E North America. It is recorded annually in small numbers in Newfoundland, regularly to Nova Scotia, and less commonly farther south to New England and New York. However, the E Siberian race, *Larus canus kamtschatschensis* (Kamchatka Gull), has very recently made a sweeping entrance to the Northeast, with no fewer than 28 credible sightings as of this publication. Although several are believed to be returning individuals, the odds of seeing a Kamchatka Gull in some places throughout New England have just about leveled out with, if not surpassed, those of seeing a Common Gull. Prior to this phenomenon, Kamchatka Gull was known only as a rarity from several Bering Sea islands in Alaska. This subspecies is overall quite distinctive and is detailed in a separate account.

A number of banded Common Gulls have been recorded in the Northeast in recent years, most of which had their origins in Iceland (where Common Gulls began breeding in 1955). Exceptionally, one individual taken in Newfoundland, and believed to be the first confirmed *canus* documented in North America, was apparently banded as a chick in the W White Sea region of Russia (Tuck 1968). This is on the eastern edge of nominate's breeding range (Peter Adriaens, pers. comm.). A third subspecies, *Larus canus heinei* (Russian Common Gull), breeds to the east of this area. Although *heinei* is geographically situated farthest from North America, and there have yet to be any confirmed reports of this taxon in the western hemisphere, avid gull-watchers have begun cultivating a search image for its key features (Garvey & Iliff 2012; see this account's "Taxonomy" section).

TAXONOMY Polytypic. Three subspecies spanning from the N Atlantic of Europe (nominate *canus*), east through N Asia (*heinei*) and NE Siberia to the Pacific (*kamtschatschensis*). Until recently included Short-billed Gull / Mew Gull (*L. brachyrhynchus*), which was given species status in 2021 (Chesser et al. 2021). See also Short-billed Gull account.

L.c. canus (KGS: 5–7). Common Gull sensu stricto. Nominate *canus* stretches from Iceland east to the White Sea region. A comparison of adults sampled from Scotland and Estonia revealed eastern birds show more black in the wingtip and perhaps average larger bodies and longer wings (Adriaens & Gibbins 2016b; Adriaens et al. 2022).

L.c. kamtschatschensis (KGS: 6–9). Kamchatka Gull. See that account.

L.c. heinei (KGS: 5–9). Russian Common Gull. Situated between the ranges of nominate *canus* and *kamtschatschensis*. Breeding range spans approximately 2,500 miles, from the western boundary, believed to be the Moscow region, east through the Lena Basin to N Mongolia. Precise boundaries are still being investigated, although presumably complicated by contact zones (in the west with nominate *canus* and in the east with *kamtschatschensis*). Winters primarily in the Black Sea region, S Caspian Sea, and parts of the Middle East. Also to E China, but exact status there unknown. Small numbers said to reach Korea and Japan, where it is scarce. Banding data confirms movements to W Europe in the winter, where it is being recorded with more frequency (Adriaens et al. 2022).

Remarkably, *heinei* is not morphologically intermediate between Common and Kamchatka Gull despite being geographically situated "between" these two subspecies (Adriaens & Gibbins 2016b). Characteristically, *heinei* overlaps primarily with nominate *canus*. Note that *heinei* is not fully detailed in this guide, but some suggestive field marks that should warrant a closer look include a clean white head with thin, fine streaks forming a necklace-like pattern on the lower hindneck (in basic plumage), an orange-yellow bill which may show a black band near the tip, and a paler iris. Overall, the wingtip shows considerable black and less white than nominate. Some wingtip features that signal a *heinei* candidate (instead of nominate) include black reaching the primary coverts across the entire base of the outer web of p8, a long black wedge on the outer web of p6 combined with a small p9 mirror, a broader black subterminal band, and black marks on p4. Also, the absence of a white tongue tip on p7, or a relatively thin white crescent, is expected in *heinei*, although some individuals seem to defy this rule with noticeable pearl (more study needed). Note that these field marks are not exclusive and may be refined following future studies. For more comprehensive details on this form, see the colossal paper on the *canus* complex by Peter Adriaens and Chris Gibbins (2016b). This study is the principal source on the subject and indeed has been an invaluable resource in shaping the accounts relating to this group.

RANGE
Breeding (L.c. canus only): Atlantic Europe from Iceland eastward, through N Europe, east to Moscow region. Does not breed in Greenland, although one pair likely nesting in S Greenland in 2021 and 2022 (David Boertmann, pers. comm.). No breeding records in Canada or United States. ***Nonbreeding:*** In North America, recorded annually in small numbers in Newfoundland and regularly in Nova Scotia in winter. Associates primarily with Ring-billed and Black-headed Gulls. Very rare to casual farther south, where in recent years reports have increased in S New England and New York (Long Island). The increase in reports in this region of both Common Gull and Kamchatka Gull, as well as Short-billed Gull, has added much excitement to looking through flocks of Ring-billed Gulls. Most records come from winter and early spring. Casual to North Carolina, where several records comprise the earliest Atlantic Coast reports for this species (first in 1980). Casual to Quebec, but reports increasing along the St. Lawrence River. Casual to the Niagara River and Lake Ontario region, with a surge in records between 2013 and 2023 (4–5 records). An adult from Brant County, Ontario (Riley et al. 2021), and another intermediate adult from Cuyahoga County, Ohio, are noteworthy inland records. Unrecorded on the remaining Great Lakes. Casual to Greenland, with fewer than 10 sightings around Nuuk from 1997 to 2022 (David Boertmann, pers. comm.).

IDENTIFICATION (*L.C. CANUS* ONLY)
Adult: A smaller white-headed gull with gray upperparts generally darker than those of Ring-billed Gull and paler than those of Short-billed Gull, although can overlap with either at their extremes. Short legs, situated rather forward, with much body behind the legs, adding to the slender, long-winged shape of the bird. The head is typically nicely rounded and is complemented by a short, thin bill that is somewhat tapered at the tip, showing no appreciable gonys expansion. Dark eyes, but on rare occasions can show a slight paling to the iris (seen at close range in favorable lighting). Red orbital. Mustard-yellow legs and unmarked yellow bill in breeding condition. Does not show red on the bill. In nonbreeding condition, the bill base is a duller yellow with noticeably grayish-green tones, often with a thin blackish bill band near the tip. The head markings are distinctly spotted around the forehead and crown, with finer streaking on the face and back of the head, which may become more concentrated on the lower neck. At rest, note the broad white tertial crescent and large, conspicuous p10 mirror on the underside of the far wing.

On the open wing, a broad white trailing edge to the secondaries is striking; it abruptly tapers off to thinner white tips on the inner primaries. Large p9–p10 mirrors. At times, the mirror on p9 is exceptionally large, appearing almost equal in size to that of p10. A mirror on p8 is not rare. Relatively extensive black running up the bases to p8–p10, with moderate black on outer web of p7. White tongue tips on p6–p7 generally form a lackluster string-of-pearls pattern. Pattern on p5 is variable, ranging from completely unmarked to a full p5 band, with a black spot sometimes on p4 (believed to be more common in eastern populations). *Similar Species:* Ring-billed Gull has pale yellow eyes, shows a bold black ring on the bill that is well defined, and averages paler gray upperparts (although often difficult to appreciate in various lighting scenarios). Structurally, Ring-billed shows longer legs, a flatter crown, and less wing projection. Also, Ring-billed has a thinner trailing edge to the secondaries and a smaller p9 mirror. In

17A COMMON GULL

1. Juvenile. Compact gull with grayish-brown aspect and scalloped upperparts. Petite bill with base beginning to pale. White base color to underparts. Note white vent. MAARTEN VAN KLEINWEE. SWEDEN. AUG.
2. 1st cycle. Mostly juvenile upperparts with some 1st alternate scapulars. The head and underparts become rather white, and the bill base pinkish. MARS MUUSSE. THE NETHERLANDS. DEC.
3. 1st cycle. An individual with more typical dull bill base. Noticeably gray greater coverts recall Ring-billed, but note the more refined and shorter bill, and neater pattern to the upperparts (namely the plain lesser and median coverts). ANDREW BAKSH. KINGS COUNTY, NEW YORK. DEC.
4. 1st cycle. Postjuvenile scapulars plain gray and adult-like. Small, fine bill has a tapered tip. Spotted head. Lesser coverts are even brown with darker shaft streaks, and the tertials are dark with broad white tips. Note white tail base. AMAR AYYASH. ENGLAND. DEC.
5. 1st cycle. An individual with a rather dark and uniform wing panel. Fairly long wing projection and overall slighter than Ring-billed. MARS MUUSSE. THE NETHERLANDS. MARCH.
6. 1st cycle. Worn wing panel. Small round head, short bill with dull base color, and spotted head are helpful. Direct comparison with Ring-billed should reveal smaller proportions and perhaps darker gray scapulars. Short-billed shows smooth wash to the neck and breast, with warmer underparts at this date. RICHARD BONSER. PORTUGAL. DEC.
7. 1st cycle. A small percentage replace a variable number of wing coverts. Note plain, spade-shaped tertial centers with broad white tips, which differ from Ring-billed. AMAR AYYASH. ENGLAND. DEC.
8. 1st cycle. A fairly typical tail pattern with defined tailband (more diffuse and patterned borders on Ring-billed). Pale window. Compact head and bill, with dull base color. AMAR AYYASH. PORTUGAL. DEC.
9. 1st cycle. Similar to 17a.8, but with more head streaking and obvious barring on the uppertail coverts. Note white tail base, however. Impressive plumage condition for this date. FRODE FALKENBERG. NORWAY. APRIL.
10. 1st cycle. An individual with uniform brown upperwing, lacking a primary window. Light spotting on uppertail coverts, but with clean, white tail base. Some juvenile scapulars retained. MARS MUUSSE. THE NETHERLANDS. MARCH.
11. 1st cycle. Underwing whitish, but with dark edging throughout. The greater coverts, in particular, have distinct dark chevron tips. Largely white body, vent, and undertail coverts. RICHARD SMITH. ENGLAND. APRIL.
12. 1st cycle. Some patterning on the uppertail coverts. Note dark chevron edging to tips on underwing linings, especially the greater coverts. MARS MUUSSE. THE NETHERLANDS. MAY.
13. 2nd cycle with adult Black-headed. Overall adult-like at this age, often with brownish cast to upperparts. Primaries lack bold white apicals and more extensively marked neck. GARY THOBURNS. ENGLAND. FEB.

addition to Ring-billed, observers in North America may have an initial impression of something that recalls California Gull, as the two have some commonalities, including dark eyes, throat, and front of the neck, often with limited markings; grayish-green bare parts; and an attenuated look to the rear of the body. California is a noticeably larger bird with obvious red on the gonys and a deeper bill tip.

In recent years, separation from Kamchatka Gull has become a legitimate identification issue, particularly in Eastern Canada and New England. The largest and darkest Kamchatka Gulls are unmatched in appearance and should be straightforward to separate. However, some do not have the strikingly dark upperparts they're often associated with, and color comparisons in the field can be somewhat ambiguous, especially in cases of vagrancy. Smaller Kamchatka Gulls are routinely recorded, but despite this, *canus* typically appears more compact, with a rounder head and smaller bill. Kamchatka shows larger proportions with an oblong body, a longer and more robust bill, a flatter crown, and a more sloping forehead. Head markings on *canus* are generally finely spotted and streaked and mostly limited to the head, at times being neatly demarcated from the lower neck. Kamchatka often has a messier mix of blotchy streaks and mottling, which commonly extends to the lower neck and sides, with markings that generally appear darker, coarser, and more dense. Common Gull regularly shows a dull grayish-green bill base with a dark bill band, unlike the largely unmarked yellow bill of Kamchatka. Note, however, that some Kamchatka Gulls do show a faint bill band near the tip, but it is not as consistently dark and not as defined as in many *canus*. Also, nominate is a dark-eyed bird; only a small minority show some hints of a paling iris, whereas eye color in Kamchatka is variable, with approximately 36% showing an obviously pale iris in the Adriaens and Gibbins (2016b) sample. Be aware that distance and unfavorable lighting can render a pale-eyed bird dark eyed. An appreciable number of adult Kamchatka Gulls share wingtip characteristics with nominate *canus*. Some fairly reliable average differences suggest *canus* includes shorter gray tongues on the three outer primaries; an obviously larger p9 mirror, which can appear to form a solid block of white with adjacent p10 mirror; little to no white on the tongue tip of p8; and a broken p5 band (as opposed to a complete p5 band found in most Kamchatka Gulls). It is not rare for *canus* to have no black on p5, and it's quite expected for it to show an unmarked p4. Nominate is also more likely to show a p8 mirror. Given its tendency to sometimes have a prominent white tongue tip on p8, the primary pattern on Kamchatka Gull can show an appreciable string-of-pearls effect on p5–p8 that is generally not expected in Common Gull. See also Short-billed Gull account.

1st Cycle: Juveniles are grayish brown with scalloped upperparts. The bill soon goes from black to a rather dull pale base, and the body begins to whiten, which persists through the winter. The head often shows distinct small spotting on the crown with an obvious white base color to the neck and underparts. The coverts appear tidy with long, pale edges. The lesser and median coverts have dark centers, often with contrasting darker brown shaft streaks. The greater coverts are largely plain, being somewhat dull grayish brown. The tertials also show pale edging, becoming significantly broad at the tips, to the extent that on the resting bird, the tertial tips can be the palest region on the upperparts. Whitish vent. Postjuvenile scapulars

14 2nd cycle. Dull greenish bare parts, some with extensive black bill tip. Indistinct white apicals, but noticeably large p10 mirror. Unmarked tail. MARS MUUSSE. THE NETHERLANDS. OCT.

15 2nd cycle. Upperparts rather adult-like at this age. Dark eye, spotted head pattern, grayish-green bill base. Small p9 mirror unexpected in Ring-billed. AMAR AYYASH. ENGLAND. JAN.

16 2nd cycle. Commonly shows white trailing edge to secondaries and all-white tail (unlike Kamchatka). Extensive black on three outermost primaries with standard p9 mirror. Short-billed and Kamchatka regularly show dark tip to p4 (often unmarked in Common). AMAR AYYASH. ENGLAND. DEC.

17 2nd cycle. Largely white underwing and white trailing edge. Stray black markings on the tail, rarely extensive. Primary coverts commonly marked, and a p9 mirror is standard at this age. MARS MUUSSE. THE NETHERLANDS. OCT.

18 2nd cycle (advanced) or delayed 3rd-cycle type. More defined wingtip and lightly marked primary coverts. Large p9 mirror, noticeable white tongue tip to p7, thin p5 band, and unmarked p4. AMAR AYYASH. ENGLAND. DEC.

19 Adults. Alternate plumage. Dark eye, red orbital, and unmarked yellow bill. Note variation in bill size (closer bird presumably female). Much body behind legs. BERTIL VERRA BREIFE. SWEDEN. JUNE.

20 Adult. Basic plumage. Thicker bill and overall larger proportions (male type). The darker bluish upperparts, dark eye, and long wing projection can recall a small California Gull in some individuals. AMAR AYYASH. ENGLAND. DEC.

21 Adult. More delicate proportions, with rounder head and finer bill. Dull bill with dark band, typical in nonbreeding condition. Head markings consist of small fine streaks and spotting. HARRY HUSSEY. IRELAND. FEB.

22 Adult (center) with Ring-billeds. Dark eye immediately stands out. Finer bill with tapered tip, showing faint residual marks late in the season. Gray upperparts generally darker than Ring-billed, but some very similar. DAVE CURRIE. NOVA SCOTIA. MARCH.

23 Adult. Large p9–p10 mirrors often form a contiguous swath of white. Broad white trailing edge tapers to thin white tips on inner primaries. Lacks bold white tongue tips on p6–p8, as in many Short-billeds. Paler gray upperparts. AMAR AYYASH. ENGLAND. DEC.

24 Adult. Streaked necklace and white head recall the Russian subspecies, *heinei*. But note dull bill color, gray on base of outer web of p8, white tongue tip on p7, reduced black on outer web of p6, and broken subterminal band on p5. AMAR AYYASH. ENGLAND. DEC.

25 Adult. Large p9–p10 mirrors, much black on p8–p10, although p8 has gray base to outer web here. Completely unmarked p5 is expected in some nominate *canus*. Note broad trailing edge, which tapers to virtually no white on inner-primary tips. AMAR AYYASH. ENGLAND. DEC.

26 Adult. Spotted head with clean lower neck. Full bill band. Large p9–p10 mirrors. Mirror on p8 found regularly. Noticeable white tongue tips on p5–p7 don't typically extend to p8, which has all-gray tongue here (compare to 17b.26). AMAR AYYASH. ENGLAND. DEC.

27 Adult. Wingtip pattern matches that of Russian Common Gull, *heinei*. Namely, p7 is without a white tongue tip. Also, the extensive black on the base of p8, small p9 mirror, broad black band on p5, and small mark on p4 support that taxon as well. But the head markings, bill pattern, dark eyes, and general structure seem more typical of nominate *canus*. This individual may be attributed to a far-eastern *canus* or perhaps an intergrade, highlighting the importance of using a suite of features for identification. DAVE BROWN. NEWFOUNDLAND. FEB.

28 Adult (ssp. presumed *heinei*). Note paling iris, white head with faint streaks limited to lower neck, small p9 mirror, no white tongue tip to p7, broad p5 band, and marks on outer edges of p4. ADAM RACIBORSKI. POLAND. DEC.

often create an entirely solid gray back that appears rather adult-like. On the open wing, note again the plain and paler greater coverts tract, pale primary window, and rather tapered black tailband with all-white uppertail coverts. Some individuals may show slightly darker tail bases, with scattered spotting and barring on the uppertail coverts, although this is less common. The underwing is generally pale with darker fringes. **Similar Species:** Ring-billed is slightly larger, with a bigger head and deeper bill, although overlap exists, particularly if one encounters a small, female-type Ring-billed. The bill on Common has a duller base and is not as bright, showing a somewhat diffuse pattern. Compare bill size and structure, with Common showing a shallower and more tapered tip. On average, the crown markings on Common appear cleaner and more distinctly freckled, whereas Ring-billed shows more short streaks. Ring-billed tends to show a more jagged pattern to the lesser and median coverts, often with pointed tips, and overall displays a more disorganized appearance to the wing panel. The lesser and median coverts on Common appear more even, dusky brown, with blunter tips and neater pale edging, often with distinct dark shaft streaks. The greater coverts are sometimes distinctly silver and barred in Ring-billed, and at times with notched tertials. On the open wing, Ring-billed averages more-silvery greater coverts, and the tailband is more intricately patterned along its proximal edge (more demarcated and defined in Common). On the underwing, Common tends to consistently show deeply scalloped edges on the greater coverts, which makes for a neater effect.

On Ring-billed, the pattern is not as strong, and it is often entirely unmarked here. This identification becomes increasingly challenging in late winter and early spring, when wing panels have faded. Compare overall size, bill patterns, tailbands, and underwing pattern, and when possible compare gray coloration on scapulars (Ring-billed's being slightly paler). See also Short-billed and Kamchatka Gulls accounts.

2nd Cycle: Overall, 2nd-cycle Common Gull has an adult-like plumage. The upperparts are predominated by solid gray with minimal signs of immaturity. Sometimes with faint black tertial spots or faint brown wash on some coverts. On the open wing, most subadult markings are found on the primary coverts, alula, and marginal coverts. The head averages more markings than the adult's and is rather variable, at times extensively. The general pattern is much spotting but can also be streaked or even blotchy around the lower neck. Dark eyes. The bill very often has a dark band with a dull, yellow-green base. The legs, as well, appear dull yellow green. On the perched bird, note the lack of white primary tips. Generally shows the least black on the mid-primaries when compared to other members of the Common Gull complex. Not uncommon for p4 to be unmarked or to show black only on the outer web. A mirror on p9 and p10 is a standard feature, although on rare occasions the mirror on p9 may be absent. The secondaries are typically unmarked and adult-like. Usually has an all-white tail, but a smaller percentage sometimes show isolated black markings. Underwing is generally advanced and minimally marked. *Similar Species:* Second-cycle Kamchatka Gull very commonly shows brown wash throughout its wing covert panel with extensive white edging, large black tertial patches, black on the secondaries, and black on the tail. The aspect on Common Gull is much more adult-like, and this carries over to the wing linings, on which Kamchatka averages more dark edging and Common generally has a much cleaner look. Head patterns may overlap at this age, but Kamchatka averages a denser shawl and an overall messier appearance. Kamchatka averages a longer bill with brighter yellow base and can show a paling eye at this age. The wingtip patterns also show some noticeable differences. A mirror on p9 is rather standard in Common but is found in only about half of 2nd-cycle Kamchatka Gulls. More primaries are marked in Kamchatka, with a complete subterminal band on p4 being the norm and marks commonly extending to p3. On Common, p4 rarely has a complete subterminal band; more often, only the outer web is marked, or p4 is completely unmarked. See also Short-billed and Ring-billed Gulls accounts.

MOLT (*L.c. canus* only) The partial inserted molt in 1st cycle is attributed to a prealternate molt, which includes scapulars, head, neck, and underparts. A sizable percentage of nominate *canus* replace most to all of their scapulars in the 1st partial molt (Adriaens & Gibbins 2016b). Postjuvenile molt in northern populations can be limited, with most scapulars retained through the winter. Less commonly, birds may replace a variable number of upperwing coverts, typically lesser and median coverts (approximately 4% of a sample of 175; Ayyash, pers. obs.). Some 1st cycles show juvenile innermost secondaries and outer tertials (lowest tertials on perched bird) that appear advanced and adult-like, which may give the impression of replaced feathers when in fact they're not (Peter Adriaens, pers. comm.). A few (very rarely) may replace a random tertial or two in 1st prealternate (needs more study), although more likely adventitiously.

HYBRIDS A banded adult found in Ireland in 2008 was initially identified as a Ring-billed Gull, but records revealed it was banded in a Common Gull colony in 2004, where Ireland's first Ring-billed Gull was confirmed breeding that year (Charles 2008). Closer examination of this individual began to reveal its intermediate features, and interestingly it remained faithful to this area until at least January 2022 (see "Other Hybrids" section). An adult Ring-billed known as Kajzerka, fitted with a GPS logger in Poland in December 2021, successfully hybridized with a Common Gull 250 miles east of Moscow in the breeding season of 2022 (Marcin Faber, pers. comm.). An adult from Worcester County, Maryland, was believed to be a hybrid with Ring-billed Gull, although it is uncertain whether *Larus brachyrhynchus* or *Larus canus* was involved (Howell & Dunn 2007). In 2016, an adult from Massachusetts showed features that suggested Common × Ring-billed, although it was unconfirmed (Nathan Dubrow, pers. comm.). There are also reports of hybrids with Mediterranean and Black-headed Gulls (Olsen 2018). Intergrades with *heinei*, presumably in contact zone near Moscow, although boundaries are obscure (Adriaens et al. 2022).

17B KAMCHATKA GULL *Larus canus kamtschatschensis*

L: 17.0"–20.5" (43–52 cm) | W: 41"–53" (105–135 cm) | Three-cycle | SAS | KGS: 6–9

OVERVIEW A three-cycle, medium-sized gull. Kamchatka Gull is the largest and darkest of the Common Gulls, breeding in E Russia, and is a rarity anywhere in North America. It is a curious novelty in the eastern part of the continent, where a recent wave of records has heightened observer awareness and interest in this complex. In some ways, the occurrence of Kamchatka Gull in E North America might mirror that of Slaty-backed Gull, whose breeding range largely overlaps with Kamchatka's. However, Kamchatka Gull has a more restricted diet and at present appears to be detached from landfill foraging. Feeds primarily on small fish, such as sand eel, capelin, and stickleback, and on crayfish and mollusks (Flint et al. 1984). Vagrants in E North America are strongly associated with coastal-feeding opportunities, such as the plankton-rich spreads found off S New England.

Kamchatka Gull can easily match and sometimes exceed the size of a Ring-billed Gull, but some are quite small too. There is also considerable variation in its upperpart coloration and wingtip pattern. Ordination plots in Adriaens and Gibbins's (2016b) work illustrate this well, with Kamchatka Gull occupying a rather central position. That is, Kamchatka presents a number of variable features, overlapping with nominate *canus* in some traits and with *L. brachyrhynchus* in other respects. This raises an important question: With the multitude of reports of Kamchatka Gull throughout New England and Eastern Canada in recent years, why aren't there as many—or even one-tenth as many—being recorded throughout the Pacific Northwest and West Coast? Given the geographic proximity of its breeding range, one naturally expects there to be more reports from there. It may be that some are overlooked in the bevies of Short-billed Gulls on the Pacific Coast. As of this publication, there is only a single pending record in the entire W Palearctic, which, interestingly, comes from Ireland (Adriaens et al. 2022).

TAXONOMY Subspecies in the *canus* group (see Common Gull account). Kamchatka Gull (*L.c. kamtschatschensis*) may prove to be a separate species following enhanced molecular studies and fieldwork from the breeding grounds. Currently believed to intergrade with *L.c. heinei*, which, curiously, it is not very similar to morphologically (Adriaens & Gibbins 2016a).

RANGE
Breeding: No breeding records anywhere in North America. Breeds throughout E Russia, primarily from Kamchatka, S Chukotka, and the Sea of Okhotsk region. Also into Yakutia, possibly extending farther west than previously stated in the literature. Zone of intergradation with *heinei* suspected there, although boundaries are unclear (Adriaens et al. 2022). Breeds in coastal areas, wetlands, and river deltas, on both fresh and salt water. Nests singly, in small colonies, and more rarely in colonies reaching 1,000 pairs (Kondratyev et al. 2000).

Nonbreeding: Winters primarily between Japan and Korea, with smaller numbers to E China. In North America, a rare to very rare spring visitor to the W Aleutians and Bering Sea islands, casual in fall. Reported with the most frequency on St. Paul Island in late spring, likely due to an increase in observer presence. Reports there are typically of singletons, rarely exceeding 2. Approximately 15 records from St. Lawrence Island—all from Gambell—with two-thirds of these from late spring. At least 5 records from Hawaiian Islands, including Midway.

Extralimital records to Newfoundland, Nova Scotia (returned for three winters), Quebec (5 accepted records through 2022, of which 4 from St. Lawrence River), Connecticut (Bonomo 2017), and Massachusetts (multiple records—more than any state on the Atlantic; some likely returning individuals). Also reports from RI, PA, and DE, although lacking details. An adult from Maine in 2022 showed much promise and is likely this subspecies). Away from the Atlantic Coast, there are several convincing reports from Ontario, and at the time of writing, the farthest inland was a suspected 3rd-cycle type from Illinois (Stotz 2008). Recently, an adult from Ohio in 2022 showed intermediate characteristics of something

between *kamtschatschensis* and *heinei*. Along the Pacific Coast, there are no accepted records outside the Bering Sea region, but several plausible reports exist from mainland AK (Nome, Kodiak Island, Juneau), BC, WA, and OR. Most of these somewhat lack critical analysis of wingtip patterns or have simply not been reviewed by regional records committees. The identification obstacle posed by Short-billed Gull has likely obscured the true status of Kamchatka Gull in W North America.

IDENTIFICATION

Adult: A moderately long-winged gull with medium to dark gray upperparts. Upperpart coloration is usually the first tipoff with vagrants, appearing noticeably darker than Ring-billed Gull, with extreme birds approaching a pale Lesser Black-backed Gull. Others can overlap with Common Gull in upperpart coloration. Can easily match or exceed Ring-billed Gull in size; at the other extreme, a few are smaller and more compact, with rounded heads and shorter bills. On average, the bill is relatively long and straight, often with a tapered culmen and no appreciable gonys expansion. Bill color varies from a dull greenish yellow to bright saturated yellow with an orange glow (especially near the tip). Some diffuse markings can be found near the bill tip, forming a faint band, but typically neither defined nor dark black. Eye color is variable: many show obvious paling in the eyes; others are dirty yellow. Completely dark-eyed birds are much less common. Red orbital. Head markings can be rather messy, sometimes beginning as finer spots and streaks on the crown, becoming more mottled around the face and back of the head, and then transitioning to dense, concentrated blotches on the lower neck. At times, the head is weakly marked and remains largely white but with a heavily marked shawl around the lower neck (known as a "tidemark"), which may extend to the breast and sides. Markings on breast and sides can often appear scaly.

On the open wing, the trailing edge to the secondaries is moderately broad, usually tapering to thinner tips on the inner primaries. But a minority can show broad white tips to the inner primaries, which invariably invites confusion with Short-billed Gull (see following section, "Similar Species"). The wingtip pattern is doubtlessly variable, overlapping with Short-billed at times and with Common Gull in other instances. Some commonly expected features include a large p10 mirror, a relatively small p9 mirror, a complete subterminal band on p5, and white tongue tips on p5–p7—noticeably large on p7. It's not uncommon for p8 to also have a longer gray tongue and noticeable white tongue tip. These types usually display a showy string-of-pearls pattern. Others have more black on outer web of p8, with a shorter gray tongue and showing no white tongue tip. Marks on p4 are not uncommon, although largely limited to the outer web or edges (i.e., broken band). A small percentage of birds can show a mirror on p8 that is generally smaller in size and restricted to the inner web. *Similar Species:* Foremost confusion species is Short-billed Gull (but see also nominate Common Gull account). Short-billed and Kamchatka Gull are a particularly problematic pair but can usually be distinguished using a collection of average differences. The largest and darkest Kamchatka Gulls are unmatched in appearance and should be straightforward

1 Juvenile. Upperparts commonly have broad white edging and various tones of light brown centers. Pink bill base and long, sloping forehead. Some routinely retain much of their juvenile scapulars late into the winter season. AMAR AYYASH. JAPAN. JAN.
2 1st cycle. Coarse blotching and spotting on neck and breast extend to vent region. Several dark gray postjuvenile scapulars have grown in. Note paler base to tail with diffuse pattern. OSAO UJIHARA. JAPAN. JAN.
3 1st cycle with similar-aged Black-tailed Gull (center) and adult Vega Gull. Note shorter bill and overall plump body shape at rest. Blemished head pattern sometimes shows fine dark spots. AMAR AYYASH. JAPAN. DEC.
4 1st cycle. Duller bill base and light, smooth wash to hindneck. Extensive wear across wing coverts and overall sandy-white aspect. Upper scapulars postjuvenile and show low contrast. OSAO UJIHARA. JAPAN. FEB.
5 1st cycle. White plumage aspect resembles Ring-billed. Brown, spade-shaped tertial center with broad white edging, carrying over to the lowest scapulars, unlike Ring-billed. Uniform-colored covert centers and frosted postjuvenile scapulars distinctive. AMAR AYYASH. JAPAN. DEC.
6 1st cycle. Wing panel heavily bleached. Darker postjuvenile scapulars (unlike Ring-billed). Pinker bill base and blotchier neck shawl compared to Short-billed. Also note heavier bill, long, sloping forehead, and plump body. AKIMICHI ARIGA. JAPAN. FEB.
7 Juvenile (same as 17b.1). Noticeable inner-primary window and broad pale edging to lower scapulars. Tail pattern variable (can show darker base than this), often with noticeable, chevron-shaped barring on coverts. AMAR AYYASH. JAPAN. JAN.
8 1st cycle. Silvery sheen to the greater coverts and inner primaries recalls Ring-billed. That taxon typically lacks broad white edges to the lower scapulars and shows anchor-shaped tips to the median and greater coverts. Note dense neck shawl. CHRIS GIBBINS. SOUTH KOREA. JAN.

to separate. There is, however, considerable overlap in their gray upperpart coloration, making this a nonstarter field mark. More importantly, evaluate overall size and shape differences, head markings, and wingtip patterns. Most Short-billed Gulls appear smaller and compact, with a more rounded head and a daintier bill. Kamchatka shows larger proportions, with an oblong body, a longer and more robust bill, a flatter crown, and a more sloping forehead. Average differences in head markings can be useful too. Short-billed shows a smooth wash across the lower neck, which appears more uniformly smudged. Kamchatka can show a dark wash on the lower neck as well, but the markings appear blotchier and coarser, sometimes scaly, and generally messier. Both very regularly show unmarked yellow bills, but Kamchatka commonly shows an orange glow near the tip. Extensive dark bill bands are rare in both, perhaps slightly rarer in Short-billed. Although eye color varies in both, Kamchatka is more likely to show an obviously pale iris, while Short-billed is more likely to show an all-dark eye. As for key differences in wingtip patterns, the importance of acquiring clear and detailed views cannot be stressed enough. Ideally, this is done using still photographs or video recordings. Kamchatka shows more black coming up the bases of p8 and p9, very often with an entirely black base to the outer web of p9. Short-billed averages less black on the base of the outer web of p9, as well as a longer gray tongue on the inner web. The black wedges on the outer webs of p7–p8 average shorter on Short-billed, especially on p7, while Kamchatka shows more black here. Both can show a white tongue tip on p8, but it's more commonly found in Short-billed. This white tongue tip is typically bolder and larger on Short-billed, combining for a flashier string-of-pearls effect. The white tips to the inner primaries are often noticeably broad on Short-billed, approaching the size of the tips to the outer secondaries. This is much less expected in Kamchatka; many Kamchatkas seem to show thinner white tips to the inner primaries. A mirror on p8 favors Kamchatka and is found in around 10% of adults, whereas in Short-billed, it is considered very rare at best. The subterminal band on p5 appears more even and symmetric in many Short-billeds. Indeed, adult Kamchatka Gulls found on the Bering Sea islands in Alaska commonly show asymmetric p5 bands. An incomplete p5 band is rare and not typical in either taxon, but rarer in Short-billed (more study needed). Finally, any individual either with a long, pale tongue on p10 that covers more than half the length of the feather or with an obvious white tongue tip to p9 should be Short-billed (Adriaens & Gibbins 2016b). See also nominate Common and Ring-billed Gulls accounts.

1st Cycle: Note the oblong body and relatively long bill. The bill base is often a brighter pink with demarcated black tip, and some retain diffuse black throughout most of the bill. Pink-colored legs. The head markings are rather coarse and often give a "dirty-headed" appearance consisting of heavy spotting and brown blotching. Variable plumage aspect: some an even mid-brown; others with distinct white ground color to the body (recalling Ring-billed). Importantly, the wing coverts, lower scapulars, and lower tertials often show extensive pale fringes and tips, creating a very frosted look to the upperparts—a

9   1st cycle. Late-season fading evident. Retained lower scapulars show broad white edges and frosted edges to postjuvenile scapulars. Relatively pale tail, but note diffuse proximal edge and distinct, rust-colored spotting throughout coverts (compare to nominate *canus*). MICHIAKI UJIHARA. JAPAN. MARCH.
10 1st cycle. Variable uppertail coverts with virtually same tail patterns. Darker underwing common. Frosted edges to upperwing coverts, and some visible pigment on outer webs of inner primaries. MICHIAKI UJIHARA. JAPAN. JAN.
11 1st cycle. Darker underwing resembles Short-billed, but Kamchatka has more coarsely marked head and blotchier neck. Barred axillaries fairly common, which is not expected in Short-billed. Dark belly patch. CHRIS GIBBINS. SOUTH KOREA. JAN.
12 2nd cycle. Long bill, plump body, and dense neck shawl. Medium gray scapulars. Dark tertial marking a standard feature in this age group. YANN MUZIKA. JAPAN. JAN.
13 2nd cycle. Expected dark marks on tertial bases with broad white tips. Commonly has frosted white wing panel with sandy-brown covert centers. Blotchy head and neck pattern. Noticeably sloped forehead. CHRIS GIBBINS. SOUTH KOREA. JAN.
14 Presumed 3rd-cycle type (left) with 1st- and 2nd-cycle Ring-billed Gull. Many features consistent with *kamtschatschensis*, although wingtip pattern in this age group may overlap with that of nominate *canus*, or even Short-billed, in which case size, structure, and bare parts should be weighed. TOM BORMANN. ILLINOIS. FEB.
15 2nd cycle. Extensive dark on outer primaries, with complete dark tips to p4. Black patches on secondaries are fairly common. Note frosty-white aspect to wing coverts and faintly marked uppertail coverts. CHRIS GIBBINS. SOUTH KOREA. JAN.
16 2nd cycle. Slightly more advanced than 17b.15, with all-white secondaries, small p9 mirror, and white tongue tips on p4–p7. Commonly shows black markings on tail. MARTEN MULLER. SOUTH KOREA. FEB.
17 2nd cycle. Complete tailband not very rare. Note black markings on secondaries and dark edging to underwing coverts. Mirror on p9 often reduced or absent altogether at this age. OSLO UJIHARA. JAPAN. MARCH.

18 Adult in basic plumage. Dark gray upperparts matched by Short-billed, which is very similar, but that species averages a more compact body with a thinner bill and smudgier lower-neck pattern. Nominate *canus* averages more spotted head pattern, paler upperparts, and defined bill band. Compare open wings. YANN MUZIKA. JAPAN. OCT.
19 Adult. Relatively long, sturdy tubular bill, often bright yellow in breeding condition and unmarked. Averages more sloping forehead than Short-billed and Common. Medium-gray upperparts, red orbital, and paling iris. Active primary molt with p5/p6 dropped. PETER HARRISON. KAMCHATKA, RUSSIA. JULY.
20 Adult with Ring-billed Gull (left), which is commonly matched in size. Darker gray upperparts and, on this individual, a somewhat thinner and shorter bill. The bill typically lacks the thin dark band found in nonbreeding Common Gulls. Dense lower neck and breast markings expected. NICK TEPPER. MASSACHUSETTS. FEB.
21 Adult with Ring-billed Gull (right). A large individual with noticeably long wings approaching a small Lesser Black-backed. Gray upperparts show maximum darkness for this taxon. Sturdy, plain, yellow bill, and coarse head markings becoming denser and blotted on the lower neck. ALEX LAMOREAUX. CONNECTICUT. FEB.
22 Adults showing variable gray upperpart coloration, eye color, and bill size, demonstrating the importance of using a suite of field marks when identifying vagrants. MASAMI YOSHIMURA. JAPAN. MARCH.

rather interesting juxtaposition, given the otherwise untidy appearance of the dirty head and coarsely marked body. The underparts are variably marked but generally show a patchily marked belly and sides. The pattern on the undertail coverts commonly extends to the vent region, although it's not uncommon for the vent to be largely unmarked with a pale base color. Some individuals retain all juvenile scapulars into late winter; others have a fair number of upper scapulars already replaced by early November. The postjuvenile scapulars can be solid adult-like gray in advanced birds or show brownish shafts and paler fringes in less advanced individuals. On the open wing, the wing panel can appear bleached, due to the rather bold white fringes on the coverts. Overall, the greater coverts are generally the least patterned tract on the upperwing and can be excessively pale—virtually white. The inner primaries can show a gentle two-toned look in some individuals, with darker outer webs and paler inner webs. Others simply show an all-pale window with darker tips. Tail pattern is appreciably variable, with many showing a relatively dark tail with diffuse pattern on the tail base, with barring on outer web of outermost tail feather. The uppertail coverts generally have a white base color with considerable barring, often appearing as crescent-shaped segments or chevrons. The underwing coverts, especially the lesser coverts, can be smoky brown with extensive dark fringes. The axillaries, as well, have dark fringes and can often show variable barring.

Similar Species: Compared to Short-billed Gull, Kamchatka is longer billed with a pinker bill base, shows a squarer head with a sloped forehead, and averages a more oblong and thickset body. But in the context of vagrants, size alone will not be a deciding factor. Average plumage differences should be considered, and even here, much overlap exists. Many 1st-cycle Kamchatka Gulls show largely frosted white edges to the wing coverts and juvenile scapulars, but this is also found to some extent in some Short-billeds; the pattern is typically not as bold and pronounced as we find in Kamchatka. The greater coverts can be unpatterned and pale in both taxa but average whiter on Kamchatka. Consider the smoother, more smudged texture to the lower neck, breast, and sides on Short-billed (not as coarsely patterned as on Kamchatka). Head patterns also differ, with Kamchatka often showing large, dark, isolated spots. On the folded primaries, Short-billed is more likely to show extensive pale edging. The tertials on Kamchatka average broader white fringes, while these are rather thin on most Short-billeds. Tail patterns may overlap, but Short-billed generally shows a darker tail base with more extensive barring on the coverts. A thin, well-defined tailband with all-white uppertail coverts is not expected in either taxon (as in Common Gull). But the pattern on the uppertail coverts may be almost entirely lost late in the season, due to wear and fading, especially in Kamchatka. A p10 mirror is not expected in either taxon. Barring on the axillaries is expected on Kamchatka but not expected in Short-billed, although some may show faint bars upon detailed inspection, and rarely with obvious barring (on one wing and not the other more likely). Both can show a contrasting dark belly patch, although more pronounced and scruffy in Kamchatka. The appearance to the postjuvenile scapulars averages more uniform on Short-billed, while in Kamchatka, there is a tendency to see diffuse brown around the shafts and more marbled-white tips.

In some instances, Ring-billed may be just as much a confusion species in North America, especially late in the season, when wear and bleaching have set in. Compare bill size and shape. Ring-billed has a deeper bill tip with more of a gonys expansion. The upperwing coverts on Kamchatka are plainer, without the distinct anchor pattern often seen on Ring-billed. The tailband also lacks patterning along the tips. If present, the postjuvenile scapulars on Kamchatka are darker gray, and the white tertial tips are broader. Ring-billed averages more-silvery inner primaries and a paler underwing. See also Common Gull account.

2nd Cycle: Many 2nd-cycle Kamchatka Gulls combine advanced and immature features. Typical for individuals to have solid adult-like scapulars but with extensive brown wash throughout the wing coverts, with distinct white frosted edges. Very often show large, black patches on the tertials and can regularly show a darker neck shawl and extensive markings throughout the head. The head markings are similar to the adult's, being somewhat coarse, making for a messy appearance. The bill base can become yellow in some individuals, at times surprisingly bright, and in others a dull greenish yellow. A dark bill band is still expected at this age. Leg color is generally dull grayish yellow, although variable. Eye color is typically dark, but upon close inspection some show a paling iris. On the perched bird, note the lack of white primary tips. On the open wing, very often has a complete subterminal band on p4 and commonly shows a relatively large p10 mirror and a small p9 mirror, although some only have a p10 mirror. Regularly shows markings on the secondaries: sometimes quite extensive segments of black. Often shows markings throughout the tail, and it is not uncommon to see a complete black tailband. The uppertail coverts are predominated by white, although there may be obvious spotting on the central tail coverts or light barring on the undertail coverts. Many individuals show moderate brown edging on the underwing coverts.

Similar Species: Second-cycle Kamchatka and Short-billed Gulls share some striking plumage similarities at this age. Relying on size will be rather subjective, but, just as with other ages, consider the elongated body, longer bill, and more sloped forehead on Kamchatka. Kamchatka also averages a brighter yellow bill base at this age. The head pattern on Short-billed is more smudged, with smoother wash across the lower neck. Both can show considerable brown wash across the wing coverts, but Short-billed averages fewer white fringes on the lesser and median coverts. On the open wing, Short-billed typically lacks the dark blocks of black on the secondaries that are often found in Kamchatka. Both commonly show complete bands on p4 and average similar tail patterns. Short-billed has less black on the bases of the outer webs of p7–p9. An individual with a gray base to the outer web of p9 should be Short-billed, and a pale base to the outer web of p8 is very indicative of Short-billed. Importantly, Short-billed sometimes shows a white tongue tip on p9, which is not expected in Kamchatka. See also Common and Ring-billed Gulls accounts.

MOLT The partial inserted molt in 1st cycle is attributed to a prealternate molt. Limited to scapulars, head, neck, and underparts. Does not appear to include upperwing coverts or tertials. Extent of scapular

23 Adult. Plain bill with orange glow to tip. Wingtip rather variable, but note black on outer web of p9, reaching the primary coverts, and broad p5 band. Mirror on p9 generally averages small. Considerable black on outer webs of p7–p8 on this individual. MARTEN MULLER. SOUTH KOREA. FEB.

24 Adults, obviously paler than Black-tailed Gull (right), although matched in body size. Note steep forehead, deep breast, relatively long, tapered bill, and yellow legs. MARTIN MULLER. SOUTH KOREA. FEB.

25 Adult type with well-marked head (coarser than Short-billed). Paler iris not uncommon, and mark on p4 fairly common. White tongue tips on p5–p7 insignificant on this individual. YANN MUZIKA. JAPAN. FEB.

replacement in winter slightly more variable than in Common Gull, although timings have not been provided (Adriaens & Gibbins 2016b). Numerous individuals wintering in Japan retain many of their juvenile scapulars into February and even March, although percentages have not been provided (Ujihara & Ujihara 2019). Other 1st cycles have already replaced many of their scapulars by early November, perhaps relating to geographic origins (study needed).

HYBRIDS None. Believed to intergrade with *heinei*; however, exact regional boundaries need further study (Adriaens & Gibbins 2016b).

26 Adult. Mirror on p8 and larger p9–p10 mirrors recall Common Gull. Noticeable white tongue tips, especially out to p8, point away from Common, as does the long, bright bill and thinner trailing edge. AKIMICHI ARIGA. JAPAN. MARCH.
27 Adult. String-of-pearls on p5–p8 recalls Short-billed, which averages less black on outer web of p8–p9, shorter black "wedge" on outer web of p7, broader white tips to inner primaries, and more smudged lower neck markings. CHRIS GIBBINS. SOUTH KOREA. JAN.
28 Adult with relatively broad white tips to inner primaries and bold string-of-pearls out to p8, recalling Short-billed. Black on outer webs of p7–p8 averages shorter on Short-billed, and p9 typically shows more gray on that species, especially on individuals with incomplete p5 band. PETER ADRIAENS. JAPAN. MARCH.
29 Adult. Much bill length protruding beyond head. Dense lower neck shawl fairly distinctive (smoother and more smudged in Short-billed). Darker gray remiges apparent. Fairly average underside to wingtip. YANN MUZIKA. JAPAN. JAN.

18 RING-BILLED GULL *Larus delawarensis*

L: 16.5"–20.5" (43–53 cm) | W: 44.5"–49.0" (113–124 cm) | Three-cycle | SAS | KGS: 4–5

OVERVIEW Medium-sized gull with even proportions. This three-cycle, large white-headed gull is the most widespread gull in the United States. Highly terrestrial. Seen anywhere gulls are expected, from beaches and rivers to parking lots, landfills, and agricultural fields. Has been recorded feeding in date palms and in crab apple, Russian olive, and cherry trees. Strong tendency to nest near human-populated areas. Overall, prefers fresh water for both breeding and migration, but readily breeds and associates on salt water edges. Seldom found far offshore.

TAXONOMY Monotypic. Western Canadian populations average larger than breeders from Eastern Canada.

RANGE
Breeding: Breeds in colonies on sparsely vegetated, low-elevation islands in lakes, rarely in rivers, from S Northwest Territories, central British Columbia, and NE California east across N North America to James Bay and Labrador, and south to N Nevada, Wyoming, S Great Lakes region, Prince Edward Island, New Brunswick, N New York, and N Maine. Rooftop and dry-land nesting apparently increasing, such as the 300-odd adults found nesting on the ground in a very busy commercial-development area in Simcoe County, Ontario, well away from any large body of water (Wukasch 2014). In 2021, small numbers successfully nested and fledged approximately 40 young in Jackson County, Colorado, after a breeding absence in the state since 1900 (Nicholas Komar, pers. comm.). Declined during 1800s and early 1900s due to hunting and egging but recolonized Great Lakes from 1920s onward. Populations in many areas, particularly in the Great Lakes region and Atlantic Provinces, increased substantially during the 1960s–1980s. Nesting spread into the Northwest Territories and N Ontario during the 1980s. A few pairs have nested sporadically in coastal Washington since the 1970s.

A wing-tag study of 763 birds in Massachusetts from September to March (2008–2010) discovered that 13 of the 461 resighted birds were found in Ontario, 26 in Newfoundland, 23 in Quebec, 14 in New Brunswick, and 3 in Nova Scotia. Of most interest is the resighting of individuals on the Great Lakes during the breeding season, as this population was historically thought to winter strictly in the interior, Florida, and the Gulf Coast, and not along the NE Atlantic (Weseloh & Clark 2011).

Nonbreeding: Found commonly to locally abundantly at a variety of coastal and interior habitats, including beaches, mudflats, lakes, rivers, marshes, agricultural fields and pastures, cattle feedlots, golf courses, parks, parking lots, and fast-food restaurant dumpsters. Found throughout virtually all of North America from central Canada southward during migration. The earliest dispersing birds may be found well away from breeding sites already by the very end of June, and numbers of such birds, including juveniles, are routine by mid-July. Arrive in numbers in southern part of winter range, such as along Gulf Coast and N Mexico, beginning in October. Northbound birds begin moving already in mid-February and are widespread in March.

Northern limit of winter range determined by severity of the winter. For example, large numbers are found in Great Lakes region through December, but abundance thereafter is variable. Casual in winter north to SE Alaska, across much of interior S Canada away from Great Lakes, and in Newfoundland. In the south, uncommon to rare in Central America, rare in Caribbean south of Bahamas, and casual to extreme in N South America.

Summering nonbreeders are found regularly in small to locally moderate numbers well south of the nesting range. But in some regions, such as along the West Coast, such birds are regularly overreported and misidentified at habitats like sandy beaches and rocky shores, where they are very rare at this season, and where young, nonbreeding California Gulls are fairly numerous. The Ring-billeds are most likely to be found in small numbers at nearby freshwater ponds, parks, and parking lots.

18 RING-BILLED GULL

This species does not occur regularly offshore more than 1–2 miles, with just a few records out to perhaps 10–15 miles, farther only exceptionally.

Very rare to casual visitor north to Bering Sea, Hudson Bay, and Greenland (none between 2010 and time of writing); west to Hawaii, Japan, and Korea; east to Europe (mostly W Europe) and W Africa; and south to Amazonian Brazil.

IDENTIFICATION

Adult: Mustard-yellow legs and pale yellow eyes. Sharply demarcated black ring on bill is distinctive, never showing any red. The bill is relatively straight, with little expansion at the gonys. Gray upperparts are pale and nearly identical in shade to those of American Herring Gull. Orbital ring and gape a fiery red when in breeding condition. All-white head in alternate plumage. Wingtip pattern highly variable, with exceptional birds showing clean *thayeri* pattern on p9. Others have increased black in wingtip with messy, isolated markings on inner web of p8 and p7. Many adults show black subterminal band or broken band on p5, but some are unmarked. Commonly show two variably sized mirrors on the two outermost primaries or only a p10 mirror (all-white tip to p10 is exceptionally rare). It's not uncommon for the primary pattern to show messy and irregular black markings along the proximal edges, as if a printer error occurred while in production. White tertial crescent indistinct and scapular crescent even less so. In basic plumage,

1 Juvenile. Plumage highly variable, but note thin, pink-based bill and grayish greater coverts. AMAR AYYASH. ILLINOIS. AUG.
2 Juvenile. Commonly shows bicolored bill early in the season, recalling California Gull, but note plain, silver-gray greater coverts. AMAR AYYASH. ILLINOIS. OCT.
3 Juvenile. A more boldly patterned bird with whiter base color. White-tipped primaries are not uncommon at this age. AMAR AYYASH. INDIANA. AUG.
4 1st cycle. A compact individual with petite bill, domed head shape, shorter legs, and long wings, reminiscent of Short-billed and Common Gull, both of which show simpler and neater pattern on the upperwing coverts (particularly the greater coverts). MAX EPSTEIN. NEW YORK. DEC.

head streaking is made up of short, fine, thin streaks that can gradually appear spotted on back of head and hindneck. Some adults acquire heavily marked and dusky heads, which may take on a pseudo-hood appearance. Others form a pseudo-necklace. Legs and bill often a dull green yellow in nonbreeding

5   1st cycle. Ghost type with cinnamon wash across head and neck. Note scaly chevron pattern on breast and sides. AMAR AYYASH. ILLINOIS. OCT.
6   1st cycle with adult. Most scapulars replaced with adult-like gray. Commonly develops sharply bicolored bill with pink base. Pink legs. CHUCK SLUSARCZYK JR. OHIO. NOV.
7   1st cycle. Some hatch-year birds have a somewhat extensive postjuvenile molt (1st prealternate?), which can include many wing coverts and a variable number of tertials. AMAR AYYASH. ILLINOIS. AUG.
8   1st cycles. Variable at this age, but the bill pattern and overall size and structure help separate Ring-billed from other taxa. Note the differences in scapular and covert patterns in these individuals. AMAR AYYASH. FLORIDA. JAN.
9   1st cycle (foreground). 2nd-cycle American Herring Gull (directly above) has superficial resemblance, but is much larger, with marbled coverts and tertials, and rounded primary tips (2nd generation). AMAR AYYASH. FLORIDA. JAN.
10 1st cycle. Late in the season, some take on a very advanced appearance, with yellowish bill and legs, white head, and white body. Fresh gray upperparts are likely from postjuvenile molt (1st prealternate?). Note faded brown juvenile primaries with pointed tips. All primaries still visible on open wing. AMAR AYYASH. ILLINOIS. APRIL.
11 Juvenile. Small black bill still developing. Silver-gray inner primaries and greater coverts. Commonly shows white tail base and patterned white tips to tail feathers. AMAR AYYASH. WISCONSIN. JULY.
12 1st cycle. Note dark diamond tips to median and greater coverts. Diffuse pattern on tail, with white base color to uppertail coverts. AMAR AYYASH. INDIANA. AUG.
13 1st cycle. White uppertail with well-defined tailband recalls Common Gull. Ring-billed averages messier and more patterned upperparts. Note distinct anchor tips to greater coverts and dark chevrons on postjuvenile scapulars (which are typically plain gray in Common). AMAR AYYASH. OHIO. DEC.
14 1st cycle. A rather typical late-winter appearance. Pale gray postjuvenile scapulars and heavily faded wing panel. Note plain greater coverts and grayish inner-primary window. AMAR AYYASH. INDIANA. MARCH.
15 1st cycle. An individual with entirely dark upperwing, including inner primaries and greater coverts. Such Ring-billeds are rare but encountered annually. In addition, some can show a more extensively dark tail than this. Compare to 17a.10, which shows a narrower tailband. CALEB PUTNAM. MICHIGAN. APRIL.
16 1st cycle with similar-aged American Herring Gull (above). Note size difference, white aspect of Ring-billed with paler belly, pale tail base and vent region, and dark secondary bar. AMAR AYYASH. FLORIDA. JAN.

condition, with orbital ring and gape becoming dark. ***Similar Species:*** Adult leg color limits choices to California, Short-billed, Common (*L.c. canus*), and Kamchatka Gulls. Ring-billed is paler gray than these taxa but comes quite close to Common at times. Averages smaller than California Gull but larger than Short-billed Gull. Adult California Gull is dark eyed, with red on gonys, and is longer billed, shows longer wings at rest, and is larger bodied than Ring-billed. Wingtip on California Gull typically shows a fuller black triangle. Short-billed has a more petite, plover-like bill with a pointier tip and lacks a demarcated black ring. Short-billed is noticeably darker gray with contrasting white scapular and tertial crescents. The wingtip usually has larger white mirrors on p9–p10, with distinct tongue tips on p7–p8. Head markings appear more smudgy than streaky. Structurally, Short-billed often shows a rounder head and appears bug-eyed, with the eye more centered on the face. The body shows considerable wing projection, with skinnier lower belly and ventral region. In the Northeast, Common Gull, albeit rare, has slightly darker upperparts and may show more of a complete band on the bill but is often dark eyed. Surprisingly, Kamchatka Gull is increasing in the Northeast, but its noticeably darker upperparts with cleaner bill make it difficult to confuse as Ring-billed.

1st Cycle: Juveniles are highly variable with mottled mix of brown to upperparts. Fresh feathers have pale edges and soon lose their pattern due to wear. Underparts are pale with dark spotting and streaks, sometimes scaly on the breast and flanks. Overall ground color to body is whitish, but some birds with chocolate aspect and others with pale ghost aspect. Many 1st cycles have plain, gray-brown greater coverts showing dark, spade-shaped tips. Small black bill becomes pale at base soon after fledging and regularly develops into an all-pink base with sharply demarcated black tip. A few show diffuse black on the entire bill late into the winter season. Dark iris. Legs pink colored. The primaries are dark, contrasting with the rest of the body, and may have pale edges. On the open wing, the inner primaries often form a pale, silvery window with brown subterminal diamond tips. A dark secondary bar is seen well in flight. Uppertail highly variable, but coverts are often speckled and weakly barred with a white ground color. Tailband variable in width, with fine internal markings and pale terminal border. Undertail is largely white with variable barring and small chevrons. Resident birds often take on a gray back by late September and begin to look more like 2nd-cycle four-year gulls. Wing coverts by then become more weakly patterned, due to wear and fading, and body takes on a mostly white appearance with diffuse head markings. ***Similar Species:*** Juvenile California Gull is darker with a browner ground color. California Gull is larger and consistently shows a dark tail with outer rectrices often dark to their bases. The inner primaries on California Gull lack the silvery-gray window shown by Ring-billed. Short-billed Gull is smaller bodied, with a noticeably smaller and more petite bill. Short-billed plumage appears finer, with a smoother texture and milky-brown coloration. The uppertail coverts on Short-billed are well marked and barred, and the folded primaries often show extensive pale edges, recalling a small Thayer's Gull. Compared to Common Gull, Ring-billed has bulkier proportions, averages a shorter wing projection, and is less uniformly patterned above. Ring-billed has noticeably larger, pink-based bill, while Common regularly displays a

---

17 1st cycle. Ghost type. Silver-gray greater coverts and inner primaries expected. Largely plain underwing not uncommon. Spotted body, often becoming all white by winter. AMAR AYYASH. INDIANA. OCT.
18 1st cycle. An individual with darker inner primaries (far wing) and darker underwing coverts (greater coverts largely white, however). Dark tail base on this individual. AMAR AYYASH. FLORIDA. JAN.
19 2nd cycle. Blackish primaries lack white apicals found on adult. Tertial centers, sometimes with black blots. Paler wing coverts, pink bill base with large black tip commonly seen at this age. Eye color advanced. Leg color typical. ALEX A. ABELA. CALIFORNIA. OCT.
20 2nd cycle. Advanced bill pattern, but darker eye than adult, and lacks bold white apicals on primaries. Also note blemishes throughout gray upperparts and black on tail. Adults don't typically have such heavy markings on lower breast and flanks. AMAR AYYASH. MICHIGAN. DEC.
21 2nd cycle and adult in alternate plumage. Compare eye color, primary tips, and tail. Also, note less-advanced pattern on outer lesser coverts and greater coverts. AMAR AYYASH. INDIANA. APRIL.
22 2nd cycle. Advanced (or 3rd-cycle type). Compact appearance and fine head spotting recall Common Gull, but note thicker yellow bill base, with sharply demarcated black ring, weak tertial crescent, and absence of large p9–p10 mirrors. AMAR AYYASH. FLORIDA. JAN.
23 2nd cycle. Narrow tailband can be complete, heavy, dark wash on secondaries and dirty primary coverts. Commonly without p10 mirror and extensive black on base of wingtip. AMAR AYYASH. ILLINOIS. SEPT.
24 2nd cycle. More advanced than 18.23, with largely white tail and thin black shaft streaks on secondaries. Small mirror on p10 fairly common. JANET HIX. TEXAS. NOV.
25 2nd cycle. Advanced bill pattern and pale eye appear adult-like. New 2nd alternate scapulars and median coverts contrast with older 2nd basic feathers. AMAR AYYASH. INDIANA. MARCH.

greenish-gray cast on its thinner and pointier bill. On fresh feathers, Common shows more blunt tips to the lesser and median coverts (Ring-billed covert centers often have pointy, dark tips). The tertials on Common can show more extensive white fringes. Wing linings on Common are more consistently tipped with dark edges throughout. The wing linings can be almost entirely white on some Ring-billeds. Uppertail on Common often paler, with defined tailband, lacking the variegated proximal tailband markings

---

26 2nd cycle. Typical underwing at this age. Obvious black on underside of tail and on a few secondaries. Yellowish legs and bill similar to adult, but dark eye. AMAR AYYASH. WISCONSIN. NOV.

27 3rd cycle. Flight feathers 2nd basic, but note active primary molt (p1–p3 dropped); thus, 3rd prebasic molt has commenced. White head and contrasting darker gray scapulars and lesser/median coverts are presumed 2nd alternate. AMAR AYYASH. ILLINOIS. JUNE.

28 Adult. Alternate plumage. Pale gray upperparts. Bright yellow legs and bill, with demarcated black tip. Pale eye and red orbital. White apicals average smaller than other white-headed gulls. AMAR AYYASH. ILLINOIS. MARCH.

29 Adults. Alternate plumage. Body-size variation obvious in male (right) and female pairing. Note significant difference in bill size. AMAR AYYASH. ILLINOIS. APRIL.

30 Adult. Basic plumage. Typical head markings consist of short thin streaks, sometimes extending to denser wash on lower neck. Duller bare parts in nonbreeding condition. ALEX A. ABELA. CALIFORNIA. JAN.

31 Adults in basic plumage. An individual with maximum head markings. May appear "hooded" from a distance. AMAR AYYASH. OHIO. DEC.

32 Adult with California Gull (left). Ring-billed Gull never shows red on the bill, and adults have a pale eye. Ring-billed generally shows a thin tertial crescent with weak contrast, due to its paler gray upperparts. LIAM SINGH. BRITISH COLUMBIA. MARCH.

33 Adults in alternate plumage. One of our only white-headed gulls regularly reported with pink suffusion throughout the body. KRZYSZTOF KURYLOWICZ. ILLINOIS. MARCH.

34 Adult. An average wingtip in terms of black pigment. Two mirrors typical (relatively small p9 mirror here). Commonly shows stray, irregular black markings on primary pattern, as seen here on p7. AMAR AYYASH. INDIANA. MARCH.

35 Adult. Large p10 mirror on this individual and thin p5 band. Note messy pattern on base of p7–p8. AMAR AYYASH. WISCONSIN. MARCH.

36 Adult. Limited black on inner web of p8 and unmarked p5. Broader white trailing edge to secondaries on this individual. AMAR AYYASH. WISCONSIN. MARCH.

37 Adult. Dull bill with average basic head pattern. Broken p9 mirror. Black on outer edge of primary coverts is base of p10, which is commonly exposed. AMAR AYYASH. FLORIDA. JAN.

commonly seen on the outer tail feathers of Ring-billed. First-cycle Ring-billed Gulls with renewed gray scapulars take on an aspect shown by 2nd-cycle Herring and California Gulls. Ring-billed is smaller, with more refined proportions, and generally paler throughout the body and underparts.

2nd Cycle: Similar to adult. Most individuals have largely gray upperparts, showing brown tones and blemishes. More weakly defined tertial crescent than in adults. Primary coverts are usually streaked with black markings, and, less often, there are black markings on a variable number of secondaries and tertial centers. Primary pattern may show a small mirror on p10 or no mirror, and rarely a small mirror on p9. Primary tips average smaller white apicals than adults' or may be completely absent. Wingtip in general has extensive black to the feather bases. Uppertail can range from complete narrow black band to all white. Legs and bill may be a dull yellow green, and the bill's ring not as well defined. Iris color variable, but commonly dark or with dusky coloration; some adult-like yellow. Head streaking often more prominent than adult's in all seasons and may extend farther down the breast and flanks. Underwing largely white with stray dark edging on wing linings. *Similar Species:* May be separated from similar-aged Short-billed Gull by size and, mostly, bill size and shape, as well as the black pattern on the bill tip. Short-billed averages larger mirror(s), with a broader white trailing edge and darker gray upperparts. See also Common Gull account.

MOLT Many juveniles quickly begin molting scapulars almost immediately after fledging. Juvenile scapulars are commonly replaced by fall with an adult-like gray aspect. It is thought the rapidity of this molt is due to a need to leave the nest in earnest and replace weak juvenile feathers. Although juvenile and postjuvenile feathers may have a two-toned pattern, with brown, spade-shaped centers or edges, they regularly show a mishmash of adult-like gray and juvenile-like brown patterns on a single feather. Local resident populations are thought to have more extensive molt, while northern populations with longer migrations retain more juvenile feathers. Hatch-year birds may replace tertials and a variable amount of wing coverts in fall, giving them a 2nd-cycle appearance, but note that flight feathers are not replaced until 2nd prebasic molt in the following calendar year (usually beginning in early May).

HYBRIDS A remarkable list of putative (mostly unconfirmed) hybrids with both hooded and white-headed gulls: Laughing, Franklin's, Black-headed, Common, Short-billed, California, and, most recently, Lesser Black-backed in Spain. Two confirmed incidents of hybridization with Common Gull recorded in Ireland and Russia (see that species account for details). Possible adult Common × Ring-billed from Massachusetts in 2016 showed features that suggested this combination (Nathan Dubrow, pers. comm.). One to two hybrids with Laughing Gull are now encountered on the Great Lakes annually, and hybrids with Black-headed Gull are encountered in the Northeast every few years.

---

38 Adult type. Some adults have only a p10 mirror, but one that is typically larger than this. The extensive black on the underwing and smaller apicals suggest a young adult or 3rd-cycle type. AMAR AYYASH. ILLINOIS. JAN.

39 Adult. A small minority show a *thayeri* pattern on p9, with limited black on the wingtip. Note limited black on underwing, with long gray tongue on p10 (and pseudo mirror). AMAR AYYASH. ILLINOIS. JAN.

19 CALIFORNIA GULL *Larus californicus*

L: 18"–23" (45.5–58.0 cm) | W: 48"–55" (122–140 cm) | Four-cycle | SAS | KGS: 5.0–7.5

OVERVIEW A gull of "inland seas," breeding predominantly from the Great Basin east into the N Great Plains of the United States and S Canada. Patchy distribution throughout arid regions of the West. It has been suggested that California Gull is more adept at thriving in habitats other large gulls cannot survive in, filling an interesting niche in the interior. It is the most common large gull at the northern limits of its breeding range and sometimes the only gull at the southern end of its range. A colonial breeder which commonly nests on inland islands free of predators, including hypersaline lakes. Also nests alongside Ring-billed Gull and other waterbirds on marshes and rivers. California Gull is among the smallest of our four-year gulls, averaging smaller than Herring Gull, but larger than the three-year Ring-billed Gull. There is, however, considerable variation in size, both sex based and geographic. It has an incredibly diverse diet, ranging from benthic prey, insects, and plant matter to small mammals, eggs and chicks of other birds, fish, and carrion. Frequently follows the plow, where it ambitiously feeds on cropland insects and rodents. No stranger to landfills. Opportunistic. As with many other large gulls, its diet is shaped by availability, both seasonally and even daily. Known to pick cherries at cherry orchards, and gorges on massive brine fly hatches, such as those on the Great Salt Lake. An interesting phenomenon noted with this species is what appears to be a reverse northward migration. In mid- to late summer, thousands stage from coastal Washington to coastal British Columbia. By September and through October, they virtually disappear, with only a few overwintering. Whether these are individuals that have moved directly to the coast from interior Canada or birds that originate in the West with a postbreeding dispersal north remains in question. Perhaps both populations contribute to this phenomenon, as birds are recorded migrating over the mountains, as well as along the Pacific Coast. Data from banded birds suggests a continual movement along the Pacific Coast from fall through spring (Winkler 2020). A California Gull banded in Greeley, Colorado, in June 1977 and found dead in June 2021 is the oldest known gull recorded in North America, living a whopping 44 years (Scott Rashid, pers. comm.). State bird of Utah.

TAXONOMY
L.c. californicus (KGS: 6.0–7.5): Nominate, Great Basin race. Averages smaller size, darker upperparts, and more extensive pigment on the wingtip. Black on base of p8, often reaching primary coverts.

L.c. albertaensis (KGS: 5–6): Great Plains race. Assigned by Jehl (1987). Etymology of trinomial refers to the Canadian province Alberta. Larger body and larger bill dimensions. Paler upperparts (many approaching or even matching those of Ring-billed and American Herring). Wingtip averages less black on bases of p7–p8, often not reaching primary coverts on p8, with deeper gray tongue on p7. Averages larger p9–p10 mirrors; p10 may more often have an all-white tip when compared to nominate; and more likely to show thin white tongue tips on p5–p7 (King 2000).

Jehl, Jr. (1987) and Winkler (2020) have posited that both races meet at a zone of secondary contact in Montana and possibly other sites east of the N Rockies, where intermediate phenotypes are encountered. In-hand measurements have been derived to estimate subspecies, but so far no genetic comparisons have been published. Some individuals are comfortably assigned depending on a combination of characteristics, but using wingtip pattern alone is not reliable, especially due to age-related differences and normal variation. Strays found on the Great Lakes have ranged from Ring-billed size with slightly darker upperparts to Herring size with noticeably darker upperparts, which is curious (Ayyash, pers. obs.). Therefore, designating a subspecies for such birds is not possible. Both subspecies suspected in Ontario, with *albertaensis* types being more common (Ron Pittaway, pers. comm.). Many are intermediate in size, between Ring-billed and Herring, with noticeably darker upperparts.

No known differences in 1st-cycle plumages. Some so-called cinnamon types show intense bleaching and may come from alkali lakes where a photochemical reaction degrades juvenile feathers. These types are seen primarily in California for a short period in late summer, after local birds—which display typical plumages—have already arrived (Alvaro Jaramillo, pers. comm.). More study needed from known-origin birds.

19 CALIFORNIA GULL

RANGE
Breeding: Breeds on islands in freshwater lakes, rivers, and reservoirs, as well as on saline lakes. *L.c. albertaensis* nests from south-central Northwest Territories southeastward to SW Manitoba and NE South Dakota. Nominate *L.c. californicus* nests from south-central British Columbia on Okanagan Lake and Osoyoos, and possibly Lillooet and Harrison Lakes, south to east-central California (Mono Lake) and east to W Montana, W Nevada, and S Colorado; has recently also nested in N New Mexico. Isolated colonies far from the core range have been established in the San Francisco Bay Area since the 1980s, nesting on levees at salt-evaporation ponds. Since 1996, a small number of pairs have nested at the Salton Sea in SE California, the southernmost breeding site for the species.

Nonbreeding: Common to abundant. Both subspecies occur throughout much of the principal nonbreeding range. In late summer and fall, most birds begin moving westward to the coast, where the first arrivals, including juveniles, are found already in mid- to late July. Some birds also move northwestward in moderate numbers as far as SE Alaska, and in small numbers to south-central Alaska. These northern birds then move south by late fall. In winter, common along the coast and in interior valleys west of the Cascades and Sierra Nevada. Frequents a wide variety of habitats, from inshore marine waters to beaches, mudflats, harbors, lakes, parks, parking lots, and landfills. Also found commonly in offshore waters, where it may sometimes be the most numerous *Larus* gull, as much as 50 miles from land, with some exceptional records from beyond 100 miles. Numerous as far south as NW Mexico, less common to Nayarit, rare to central Mexico (e.g., Guerrero), and very rare to casual to S Mexico and Central America south to Costa Rica. Farther east, relatively numerous in winter only very locally, such as at Salton Sea, in W Nevada, and along Front Range in Colorado; uncommon along lower Colorado River and at larger lakes in S New Mexico; and rare to W Great Plains and W Texas. Northbound and eastbound birds begin moving in February and are numerous in March and April. A few are found east to central and E Great Plains (e.g., E Nebraska and Kansas, W Oklahoma).

A very rare to casual visitor farther north (e.g., central Yukon, N Northwest Territories, and Nunavut). Rare to very rare to the eastern third of North America, where it is mostly found in fall and winter, but where there are a few records also from spring and summer. Records of adult types greatly outnumber reports of younger ages, which are likely overlooked. Most often recorded in Midwest and Great Lakes. But records are widespread elsewhere, from S New England to mid-Atlantic region, Florida, and Gulf Coast. Accidental to N and W Alaska, N Ontario, Maritime Provinces, Bermuda, Ecuador, and Japan.

---

1 Juvenile. Extensive white edging to upperparts. This white is also found throughout the uppertail coverts and tail tip. Dark eye mask and whitish forehead common at this age. ALEX A. ABELA. CALIFORNIA. AUG.

2 1st cycle. Black bill less common at this date, but not very rare. Unorganized pattern to lesser and median coverts. Mealy texture to breast and underparts. Long wing projection. PHIL PICKERING. OREGON. OCT.

3 1st cycle. Several postjuvenile scapulars. Shallower forehead and skinnier bill than Herring, with paler vent and sleek, long wing appearance. Regularly shows extensive pale bar running across tips of greater coverts tract. AMAR AYYASH. ALASKA. AUG.

4 1st cycle. Cinnamon types, such as this individual, are not uncommon on the Pacific Coast in late summer. It's believed this coloration is due to exposure to alkali lakes, where a photochemical reaction degrades juvenile feathers. ALEX A. ABELA. CALIFORNIA. AUG.

5 1st cycles. Variable bill patterns and variable molt at this age. A bicolored bill and moderate postjuvenile molt is more typical for this date, as seen on the individual on the left. CONOR SCOTLAND. OREGON. NOV.

6 1st cycle with similar-aged Glaucous-winged Gull (right). Note overall slim body with long wing projection, small head, and straight bill. MATT GRUBE. CALIFORNIA. NOV.

7 1st cycle. Rather typical. 1st prealternate molt often replaces most scapulars and, at times, a number of wing coverts. Tubular, bicolored bill, with noticeable frown formed by a gape patch. Shallow forehead and dark postocular line distinctive. ALEX A. ABELA. CALIFORNIA. SEPT.

8 1st cycle. Darker wing panel (juvenile feathers). Several advanced 1st alternate scapulars (gray), not rare, especially late in the season. Note obvious gape patch and bluish tibia. ALEX A. ABELA. CALIFORNIA. FEB.

IDENTIFICATION

Adult: Dark eyed, yellow legged, with a fairly long, straight, tubular bill. The upperparts are between pale and medium gray (see race descriptions in this account's "Taxonomy" section). An elongated look to the body behind the legs, highlighted by the longer wing projection. In alternate plumage, the head is a clean snow white with contrasting dark eyes. Red orbital and red gape. The gape-line is a noticeable downward frown in this species and is consistently seen. Head streaking is variable in basic plumage but is densest around the lower hindneck and back of the head, less dense on the front of the neck and breast. A short, dark postocular line is fairly common. In low breeding condition, the legs may be a dull greenish yellow, and the bill is also an unsaturated greenish gray yellow. Others maintain relatively bright bills. The orbital, gape-line, and gonys spot become duller with an orange hue. A supporting field mark is the black-to-red pattern around the gonys, which is seen for a considerable period throughout the year. Some adults lose this black at the height of breeding condition, but others can be seen with thin black distal mark on the bill even when nesting (Ayyash, pers. obs.). The black pattern on the bill is a semi- to complete subterminal ring with proximal red on the lower mandible. On the open wing, the wingtip often appears to be a black triangle in those individuals with well-marked primaries (more so in nominate types, in which the black is more likely to extend toward base of p8 / p7). Mirror on p10 relatively large, and not uncommon to see an all-white tip (more common in *albertaensis* types). Mirror on p9 variable, at times broken at the shaft, and on rare occasions completely absent (likely age related in such cases). Full subterminal band on p5 is expected, and p4 is typically unmarked. Note the relatively long, narrow wings in flight. On the underwing, p6 / p7–p9 often show a distinctive black pattern extending deep along the edges of the inner webs, forming a series of consecutive spurs. The underside of the remiges has a contrasting gray "shadow band" against the white wing linings and trailing edge. *Similar Species:* Compare to other yellow-legged species—namely, Short-billed, Ring-billed, Lesser Black-backed, and Yellow-footed Gulls. Short-billed Gull is decidedly smaller and more delicate and has a thinner, plain yellow bill with a fine tip. The hindneck is more smudgy with clouded markings compared to the denser and blotchier streaking on California. Lesser Black-backed Gull is pale eyed and decidedly darker on the upperparts and averages a larger red gonys spot, finer head streaking, and smaller p9–p10 mirrors. Yellow-footed is larger and bulkier, has a considerably larger bill, is noticeably darker, has pale eyes, and typically doesn't show a p9 mirror. See also Ring-billed Gull account. Other large four-year gulls, such as Herring and Western, have pink legs, but beware aberrants with yellowish legs. Also, a very small percentage of adult California Gulls may show a paler iris. When lighter eyes are found, they're typically not pale yellow, as in Herring, but a so-called snowflake obsidian pattern. A combination of other field marks should help separate these individuals from lighter-eyed four-year gulls. Be aware that some individuals may appear to have a dark pink orbital as it transitions from its dullest gray color to red, and the reverse. This has the potential to invite thoughts of Thayer's Gull, but in these California Gulls, the gape usually remains orange red. Inspect leg color when possible. Smallest individuals can be Ring-billed size, with others matching Herring

9 1st cycle. Particularly pale and retarded appearance to upperparts, which are largely 1st alternate. Note bill pattern and shape, and dark postocular line. ALEX A. ABELA. CALIFORNIA. MARCH.

10 1st cycle. Visible primaries juvenile. Heavily worn and bleached wing panel, with a mix of advanced 1st alternate scapulars and brown feathers produced by the same molt. A thick bill on this individual. ALEX A. ABELA. CALIFORNIA. APRIL.

11 1st cycle. Dark primaries and primary coverts. The tail is often tipped white (before wear). A densely marked neck is common, with a shallow forehead and straight bill. AMAR AYYASH. CALIFORNIA. SEPT.

12 1st cycle. Commonly shows dark bases to the greater coverts. This individual has a frosted appearance to the upperparts, with lesser and median coverts appearing spotted. White rump, pale uppertail coverts, and dark tail. AMAR AYYASH. WASHINGTON. AUG.

13 1st cycle with adult Ring-billed Gull. Long, narrow wings. Sooty-gray 1st alternate scapulars and white tips to the tail feathers. Inner greater coverts can be heavily barred, and some may have paler inner webs to the inner primaries. AMAR AYYASH. MICHIGAN. OCT.

14 Juvenile. Dark inner primaries, dark greater coverts, and paler lesser/median coverts (far wing). Smooth brown head and underparts not unlike American Herring, but note smaller head and bill, white-tipped tail feathers, and overall white base color to tail coverts and vent. AMAR AYYASH. WASHINGTON. AUG.

15 1st cycle. Long, narrow wings with little indication of a pale window. Whitish vent and pale undertail coverts with variable barring. White tips to tail feathers apparent. WAYNE SLADEK. WASHINGTON. SEPT.

16 2nd cycle. Retarded bill pattern resembles 1st cycle, but note muted pattern to upperparts and rounded primary tips. Shallow forehead, dark postocular line, and bluish tibia are noteworthy. AMAR AYYASH. CALIFORNIA. SEPT.

17 2nd cycle. More advanced adult-like gray across the upperparts (2nd alternate). Note dark greater coverts with fine vermiculations. A bulky individual (*albertaensis*?) with outermost primaries not fully grown. AMAR AYYASH. MICHIGAN. NOV.

18 2nd cycle with adults. A smaller individual (possibly female and/or nominate *californicus*) with darker underparts and head. Note greenish-blue cast to legs. BRIAN SULLIVAN. CALIFORNIA. JAN.

19 2nd cycle with adult Lesser Black-backed (right). These two are often similar in size, but extremes may be explained by sex (female?) and/or subspecies (nominate?). DARREN CLARK. IDAHO. JAN.

20 2nd cycle with similar-aged Western Gull (back). California Gull is paler gray with a decidedly smaller head, bill, and body. Note the prominent gape-line on this individual, seen frequently in this species. ALVARO JARAMILLO. CALIFORNIA. MARCH.

21 2nd cycle. Outer tail feather and outermost primary still growing. Otherwise, a fairly typical 2nd basic individual with dark, grayish-brown wing, dark tail, and heavily marked neck. AMAR AYYASH. ILLINOIS. FEB.

22 2nd cycle. More solid gray lesser and median coverts (2nd alternate) compared to 19.21. This age group can show a few advanced inner primaries with grayish centers and thin white tips. Compare to 19.27. AMAR AYYASH. ILLINOIS. FEB.

23 2nd cycle. Variable brown on the wing linings, but the greater coverts and axillaries are often whitish. Dark tail and inner primaries may be pale at this age. Straight bicolored bill and small head. CONOR SCOTLAND. OREGON. SEPT.

24 2nd cycle. Variable underwing. Note small p10 mirror, which is not unusual at this age. The smudged texture to the head and neck creates a superficial resemblance to Black-tailed Gull, which often shows a longer bill, paler lores, and bold white eye crescents. TED KEYEL. CALIFORNIA. NOV.

25 3rd cycle. Medium gray upperparts blemished by light brown. Frequently has black on tertials. Black primaries with adult-like apicals. Leg color variable, but especially dull on this individual. SEAN MCCANDLES. CALIFORNIA. JAN.

26 2nd cycle. Alternate head pattern and bright bill in high breeding condition. Maintains dark eye and, very often, black-and-red pattern on bill tip. PHIL PICKERING. OREGON. APRIL.

27 3rd cycle. Extensive black on the wingtip and, less commonly, no p9 mirror. Long, narrow wings, tubular bill, and medium gray upperparts distinctive, ruling out American Herring. WAYNE SLADEK. WASHINGTON. OCT.

Gull in size, but generally fall squarely between these two species. In flight, the underwing shows a contrasting darker gray "shadow band" across the flight feathers. This isn't found in paler-backed species such as Herring and Ring-billed. On the upperwing, the white trailing edge and gray upperparts show a deeper contrast than in those paler-backed species.

1st Cycle: Juvenile plumage is variable, but the upperparts are predominated by brown centers with broad pale edging. Others have upperparts with paler centers showing a dilute tricolored pattern which may be notched or have irregular edging. A warm to dark brown head, marked with dark streaking and mottled down the neck. The black bill does not persist, rapidly taking on a pale base throughout the summer.

Pinkish legs. The greater coverts tract commonly shows some dark bases, with weak checkering along the middle of the tract, becoming paler and washed out distally. The tertials are dark based and almost always show some notching or patterning along the distal edges and tips. Blackish-brown primaries with a noticeably long-winged look. The underparts are paler than the upperparts, with the vent region pale based and barred. Some individuals show accelerated bleaching and wear early in the season, with paler forehead, breast, and underparts. The combination of this and the commencement of the 1st prealternate molt make for a very mottled and variegated appearance in many birds. A variable number of scapulars are replaced early in the fall, often with a brownish-gray anchor pattern and dusky centers, some rather pale. Late alternate scapulars are apparently more solidly gray, approaching an adult-like aspect, but variable. The bill takes on a sharply demarcated black tip with a pink base. By late September through early October, most 1st cycles show this pattern. A dark postocular line becomes apparent in many hatch-year birds. Legs can become pinkish gray later in the season, with noticeable blue gray along the tibia. On the open wing, the inner primaries are fairly dark and don't typically show a pale window, but some individuals may show this, exaggerated by a fully extended hand. The dark bases to the greater coverts (particularly middle and outer) contrast with their paler tips and fairly regularly display a "double secondary bar." On the upperwing, the median and lesser coverts are usually distinctly paler. Note the whitish edging to the greater coverts, primary coverts, and secondaries, as well as the tail feathers. The uppertail coverts are variably marked but can be surprisingly pale with sparse dark barring. Others have densely barred uppertail coverts. The eye-catching dark tail typically has a dark base, which may show some sparse barring on the outer edges. The underwing linings have a dark smoky-brown aspect, typically with dark axillaries. The wing linings can be barred or even appear spotted; others are more uniformly dark with no pattern. **Similar Species:** Short-billed Gull is noticeably smaller. Bill size and shape are often sufficient to separate these two. Short-billed has a more domed head and appears bug-eyed, with more-prominent white tongue tips on primaries. See also Ring-billed Gull account. Herring Gull averages larger proportions and a stouter bill, which typically isn't as sharply bicolored; lacks hints of blue on the tibia; and shows a contrasting pale window on the inner primaries. At rest, California appears slimmer and longer winged and has a more tubular bill. However, there is some overlap in size, and fresh juveniles can be difficult to pin sometimes, when the bill is still largely black on California. California averages a more pear-shaped head, with a pinched appearance to bill base and shallower loral region. Note also the whitish forehead seen early in some juvenile California Gulls. The greater coverts on California tend to have a weaker pattern and often show lesser and median coverts with an uneven and messy look, being nonuniform in pattern. Herring averages neater patterns to these feather tracts in early juvenile plumage. On fresh feathers, note the more prominent white edging along the tips of the primary coverts, secondaries, and rectrices on California. Thayer's Gull retains much black in the bill and, month for month, has a neater appearance, with much of its juvenile plumage retained. Thayer's does not appear as long winged at rest and doesn't show bluish-gray tones to legs. Some fresh juvenile Heermann's can do a good job at mimicking juvenile California Gulls (see that species account). Lesser Black-backed Gull, which continues to increase throughout the range of California Gull, is overall more solidly dark on the upperparts with less patterned tertials. Lesser Black-backed retains a mostly black bill in 1st cycle and averages a narrower tailband and coarser-patterned underparts. On rare occasions, some California Gulls may have a relatively narrow tailband with bold white uppertail coverts. Compare greater coverts,

28 3rd-cycle type. A more advanced individual with clean underwing and body. A p9 mirror is fairly common on 3rd cycles. Dark eye, dull grayish yellow legs, and straight tubular bill. AMAR AYYASH. WISCONSIN. FEB.
29 Adult. Alternate head pattern. Dark eye and red orbital. Most black on bill now diminished, but many continue to show a small black mark through the breeding season. Yellow legs. Medium gray upperparts. FRED LELIÈVRE. BRITISH COLUMBIA. APRIL.
30 Adult types. Smaller head, straight tubular bill, dark eye with red orbital, yellow legs, and medium gray upperparts. Note variation in gray upperparts and leg color. LIAM SINGH. BRITISH COLUMBIA. JULY.
31 Adults. Notable variation in size and gray upperpart coloration. Left individual is darker and smaller, presumed nominate *californicus*. Front right paler and larger, possibly *albertatensis* (American Herring directly above *californicus* type). PHIL PICKERING. OREGON. OCT.
32 Adult with similar-aged American Herring Gull (right). Large brutes with paler gray upperparts like this individual are almost assuredly northern *albertatensis*. AMAR AYYASH. NEW YORK. DEC.

head streaking, and bill patterns. Western Gull is larger bodied, has a larger, bulbous-tipped bill that is not sharply bicolored, and shows much broader wings in flight. A somewhat rare occurrence is 1st-cycle California Gulls retaining a diffuse black pattern along the bill base and cutting edge well beyond the expected time. Birds showing black on the bill base into late November and December may be extreme examples of variation, but such individuals should be carefully inspected for other "off" characteristics.

2nd Cycle: Generally more white headed with a paler neck and underparts than 1st cycle, but variable. Head streaking and neck mottling heavy in some individuals, extending down the breast and underparts. Second basic greater coverts are largely dark and plain with some stippling. Some have rather retarded median and lesser coverts showing brown barring similar to that of 1st cycle, while others have a muted, gray-brown aspect. As the season progresses, the scapulars take on a more adult-like appearance (2nd alternate). Dark-centered tertials showing some marbling along the distal edges and tips. Dark eye with short postocular line. Pinkish-gray bill base with black tip, similar to 1st cycle, but at times with a pale nail. Others have cream-colored bill base. Some have bill pattern approaching that of the adult, especially near the end of 2nd cycle. Legs dull, pinkish gray, many with a characteristic sickly blue-gray coloration. Blackish primaries sometimes with thin white crescent tips. By the end of 2nd alternate, some individuals

---

33 Adults in high breeding condition. Presumed *albertaensis*, arriving at the breeding grounds fairly early before much open water can be found. Note long gray tongues on p7–p8. ALAN KNOWLES. ALBERTA. MARCH.

34 Adults. Show variation in gray upperparts, as well as head and bill size. Front individual has a paler iris, which is very rare but not unheard of, and also lacks a mirror on p9. Combined oddities can evoke thoughts of a hybrid or may be explained by unusual variation. PHIL PICKERING. OREGON. OCT.

35 Adult. Fairly typical wingtip showing extensive black triangle, a complete band on p5, and a mirror on both p9 and p10. AMAR AYYASH. OREGON. JAN.

36 Adult. Unmarked p5 somewhat rare. The long gray tongue on p7 and white tip to p10 are expected more in *albertaensis*, but not safely identified to subspecies on wingtip alone. AMAR AYYASH. COLORADO. NOV.

are largely white bodied with little streaking on the head. In these individuals, a variable number of lesser and median coverts and tertials are adult-like. It's not unusual to see a red gape, red orbital, and straw-yellow legs in the most advanced birds. On the open wing, note the dark secondaries with pale tips and the dark greater coverts. Many show a somewhat contrasting pale window on the inner primaries. Noteworthy in this age group is the tendency for some to show one to four advanced inner primaries, approaching those of a 3rd cycle. A small p10 mirror is not rare. A broad black tailband is typical, with largely white uppertail coverts. The underwing has a light brown wash throughout, with paler wing linings—especially the greater coverts—often showing light brownish edging. The axillaries are commonly white. The underside to the inner primaries is relatively pale, but dark brown on the outermost primaries.

Similar Species: Ring-billed Gull is smaller overall, with a shorter bill. First-cycle Ring-billeds with postjuvenile scapulars are paler gray and often with silvery greater coverts not shown by California. Ring-billed shows a brighter pink bill base, whereas in many 2nd-cycle California Gulls the bill base is a dull pinkish yellow to greenish gray. Second-cycle Ring-billeds also have paler gray upperparts, generally show a pale iris, and never show red on the bill. On the open wing, note the adult-like aspect to the remiges and lack of a broad black tailband on Ring-billed. Second-cycle Herring Gull has lighter gray upperparts and

37 Adult. 8th cycle banded as nestling at Mono Lake in 2010 (defaults to nominate). Extensive black on base of p7–p8. Faint subterminal band on p4 rare. NANCY CHRISTENSEN. CALIFORNIA. JAN.
38 Adult. Rather rare is no mirror on p9. Extensive black triangle on wingtip is fairly distinctive. The subterminal band on p5 can sometimes have pigment limited to the inner web. Note dark eye, obvious red gape-line, and bill shape. LIAM SINGH. BRITISH COLUMBIA. MARCH.
39 Adult. Extensive black on the underside of p9–p10, with relatively small p9 mirror. Dark shadow bar on underside of secondaries suggests a darker gray to the upperside. NANCY CHRISTENSEN. CALIFORNIA. JAN.
40 Adult. An individual with reduced black on underwing and all-white tip to p10. Gray shadow bar on underside of secondaries. Greenish tinge to legs often less vivid than bill. AMAR AYYASH. BRITISH COLUMBIA. MARCH.

is pink legged. Many show hints of a paling iris. Overall, Herring appears larger, with bulkier proportions, and not as long winged at rest. Second-cycle Lesser Black-backed has decidedly darker gray upperparts, will many times show hints of a paling iris, and lacks the grayish-blue bill and leg color seen in many California Gulls at this age. The bill on Lesser Black-backed will typically have black smudging along the base and cutting edge and is usually not sharply bicolored, as in California.

3rd Cycle: Rather adult-like, with solid medium gray upperparts, although a noticeable brown wash across various tracts. Mottled head and neck with variable streaking in basic plumage. Dark eyes. Bill is a dull greenish gray yellow, with subterminal black around the gonys. Often a broad black ring with proximal red around the gonys. Leg color variable from grayish blue to straw yellow, becoming adult-like late in the season, as spring approaches. The tertials and some secondaries may show brownish-black blots on their centers. Smaller apicals and smaller p9–p10 mirrors compared to adult's, some without p9 mirror entirely. Very rarely has no mirrors on both p9 and p10. Subterminal marks on p4 seen in some individuals. Primary coverts and alula show variable black. Underwing is largely white, but random dusky blemishes are not uncommon. The tail pattern ranges from a complete all white to broken piano pattern to complete black band (although decidedly narrower than in most 2nd cycles). By late winter, 3rd alternate birds take on an adult-like aspect, but note what were already smaller apicals sometimes significantly reduced or completely worn down. A combination of bare-part features and primary covert and primary patterns help place even the most advanced birds as 3rd cycles. **Similar Species:** Second-cycle Ring-billed is paler gray, generally shows a pale iris, and lacks red on the bill. On the open wing, note the weaker contrast between the gray upperparts and trailing edge on Ring-billed. The primary tips typically don't have the larger apicals shown by California at this age. Ring-billed averages a smaller p10 mirror, with no accompanying p9 mirror. Third-cycle Herring Gull is pink legged and should typically have a paling iris, with lighter gray upperparts. This, coupled with average size differences, should readily separate these two at this age.

MOLT The partial inserted molt in 1st cycle is attributed to prealternate, which commonly includes many to most scapulars, some upperwing coverts, to a lesser extent several tertials, and a variable number of tail feathers (Pyle 2008). Behle and Selander (1953) described in detail a partial "post-juvenal" molt in September–January and a "first-nuptial" molt in February–May. These molts are best interpreted as a 1st prealternate molt which suspends, is relaxed to some extent, or is continuous. Beck (1943) reared young California Gulls beginning at the downy stage and described their postjuvenile feather replacement as a gradual, month-to-month, protracted molt throughout the first winter; but using this as a model for wild birds would be incongruent. As with many species exhibiting SAS, determining whether certain feathers are replaced twice—postjuvenile—needs close examination and will likely only continue to be discovered from known-age individuals that are monitored for months on end. Also, precisely determining which upperparts are replaced at the terminus of the 1st prealternate molt or start of the 2nd prebasic molt may be arbitrary and futile. With this in mind, more study is desired regarding intraspecific variation in extent and timing of molt between both races. Banding programs employing the use of field-readable bands on these two subspecies would be worthwhile.

HYBRIDS Heermann's Gull (see that species account for details). With Herring Gull in Park County, Colorado (1982 and 1983; Chase 1984). Male Herring and female California Gull; chicks fledged. Hybridization likely occurred there beforehand, as an apparent five-year-old male hybrid, banded in 1978, was afterward found breeding at this same colony and later collected. Also suspected in Wyoming (Winkler 2020). Putative hybrid Herring × California Gull individuals seen sporadically from the Colorado Front Range (Steve Mlodinow, pers. comm.) to Michigan (Ayyash, pers. obs.), although identifications uncertain (see plate H.3).

Presumably with Ring-billed Gull and Short-billed Gull, although unconfirmed (Howell & Dunn 2007). Also, an incubating individual—likely a female—was found on eggs in a Ring-billed Gull colony two seasons in a row in Toronto in 1981–1982 (Blokpoel 1987). No mate was found in attendance, but it is said the eggs were larger than Ring-billed eggs. The eggs were not successfully incubated, and no chicks hatched.

HERRING GULL COMPLEX

The Herring Gull complex has long produced some of the thorniest questions in large white-headed gull systematics. A number of taxonomies throughout the world, including the International Ornithologists' Union, the Association of European Rarities Committee, and the Dutch Birding Association, recognize three species: American Herring Gull (*Larus smithsonianus*), European Herring Gull (*L. argentatus*), and Vega Gull (*L. vegae*). This is how they are presented in this guide, with each taxon given its own account and treated separately. Others, including the British Ornithologists' Union, maintain that American Herring and Vega Gulls are conspecific but regard European Herring Gull as a distinct species (Sangster et al. 2007 Collinson et al. 2008a; Dickinson & Remsen 2013). Currently, the American Ornithological Society is one of the only bodies that continues to treat all of these Holarctic taxa as a single species (*L. argentatus*). A proposal to split the three was presented to the North American Classification Committee in September 2023, and we await further discussion and a vote in this regard (Billerman 2023).

There is widely accepted genetic evidence indicating that American and European Herring Gulls are not each other's closest relatives and in fact belong to distinct clades with notably different evolutionary histories (Crochet et al. 2002; Liebers et al. 2004; Sternkopf et al. 2010). Consequently, Herring Gull taxonomy in North America remains woefully stagnant, despite sufficient data to support full species rank for *Larus smithsonianus* (de Knijff et al. 2005; Pons et al. 2005).

Note that the name Herring Gull is used throughout the world, despite all of this unsettled taxonomy. Observers in North America often use Herring Gull to refer to *Larus smithsonianus*, while observers in Europe use it to refer to *Larus argentatus*. Therefore, the usage of this English name should be understood in context.

There is no doubt Herring Gulls can present great identification challenges, if not the greatest, in large-gull identification. They possess a spectacular degree of variation, especially in their earlier plumages. The closer we look, the more we learn about the range of variability in any taxon, which for better or worse can complicate the search for constants. This is perhaps part of the appeal for some: finding unordinary features in an unextraordinary species. Vagrant individuals displaying a full suite of textbook field marks are rare. Therefore, it must be accepted that a number of Herring Gulls might not be safely identified in a vagrant context. Observers in North America are best served by becoming genuinely familiar with American Herring Gull (*L. smithsonianus*)—the default Herring Gull, which predominates throughout the continent.

Mixed flock of American Herring Gulls and Laughing Gulls (back). AMAR AYYASH. FLORIDA. JAN.

20 AMERICAN HERRING GULL *Larus smithsonianus*

L: 22"–27" (56–69 cm) | W: 53.5"–60.0" (136–152 cm) | Four-cycle | SAS | KGS: 4–5

OVERVIEW American Herring Gull is a ubiquitous four-cycle species that embodies the quintessential seagull. Studies using mitochondrial DNA suggest this taxon isn't a Herring Gull at all (see following section, "Taxonomy"). As a number of ornithologists have noted, *Larus smithsonianus* is the most successful North American gull in terms of numbers, geographic distribution, and genetic differentiation (Sangster et al. 2007; Weseloh et al. 2020). This widespread species is a benchmark gull that should be learned well by anyone with a serious interest in large-gull identification. First- and 2nd-cycle birds are quite variable, which can delight and humble both the novice and the aficionado.

The loud, piercing long call of the American Herring Gull is a permanent feature of our landscape throughout the Great Lakes and especially along the Atlantic Coast, where it is the most common large gull. As a generalist predator and opportunistic scavenger, this species forages extensively on marine invertebrates, human refuse, bird chicks and eggs, and insects and has also been documented picking berries, such as crowberries and blueberries. On the Great Lakes, it feeds on smaller fish and washed-up carcasses and is regularly recorded taking migrant passerines on the wing, even targeting Chimney Swifts at their roosts during the breeding season (Ayyash, pers. obs.). Nests singly and in small groups on the tundra and smaller water bodies west through Alaska and less commonly on steep sea cliffs farther north in Canada. On the Great Lakes east through the Atlantic, more colonial, sometimes in the thousands, on secluded rocky islands, barrier beaches, and rooftops. Small numbers have also been found nesting in trees. Longevity record comes from a live adult from Door County, Wisconsin, 29 years and 9 months old, found a couple of hundred miles south of its natal colony on southern Lake Michigan (Ayyash 2015).

TAXONOMY First described as a species by Elliott Coues in 1862, but he later demoted it to a "variety" of *Larus argentatus*. The American Ornithological Society maintains this taxonomy and treats it as a subspecies, *L. argentatus smithsonianus*, with old-world Herring Gulls. The current classification by the American Ornithological Society is obsolete, however, and most taxonomies around the world treat *Larus smithsonianus* as a species. Molecular data suggests *L. smithsonianus* is derived from an E Siberian group which colonized North America before the last glacial maximum, and not a very close relative of the European Herring Gull (Crochet et al. 2002; Liebers et al. 2004; de Knijff et al. 2005; Harrison et al. 2021).

No type specimen or locality was ever provided. In 2007, Olson and Banks designated USNM 18216 as the lectotype, with the restricted type locality becoming Henley Harbour, Strait of Belle Isle, Newfoundland and Labrador. These workers, in accordance with European ornithological bodies, recommended *L. smithsonianus* be given species rank and proposed the name Smithsonian Gull (dedicated to the institution; Coues 1862), and this in fact may be helpful in removing the misleading association with *the* Herring Gull. Other proposed English names include Hudsonian Gull and American Gull. In this guide, American Herring Gull is the designated English name for *L. smithsonianus*.

American Herring Gull exhibits appreciable geographic variation in terms of size and overall wingtip patterns of adults. Migration trends, breeding distributions, and molt timings also appear to differ across the continent. There is some preliminary data, based strictly on field observations, that suggests there may be up to four populations in North America: Alaskan, arctic-subarctic, Great Lakes, and Atlantic. However, long-term studies on breeding populations from Alaskan and arctic populations are sorely lacking, and to suggest particular subspecies at this time is premature. Breeding sites in these regions are largely unsurveyed and understudied. Much more information is available on the more colonial populations of the S Great Lakes and Atlantic populations, but this data needs synthesis.

Great Lakes and Western populations average smaller, and adults generally show a single p10 mirror with more pigment on the wingtip. Birds in the Northeast and along the Atlantic average larger, generally have two mirrors and less black on the wingtip, and sometimes show a *thayeri* pattern on p9. Vocalization differences have not been analyzed in any of these populations. There appears to be an arbitrary line around Quebec City, where highly variable wingtip patterns are on display in winter, and it is possible some gene flow exists between Great Lakes and Atlantic birds in this region. However, Great Lakes birds have traditionally been regarded as an essentially isolated and closed system in terms of immigration and recruitment (Chen et al. 2001). Differences in mitochondrial DNA between birds from the Great Lakes and those in the Maritimes are concordant with long-term band recoveries that suggest the Great Lakes population is mostly resident, particularly in its adult population.

RANGE

Breeding: In Alaska, mainly breeds in the interior (away from the coasts, Aleutian and Bering Sea islands, and the southeast), although exact breeding strongholds in the state need survey. Hybridizes extensively with Glaucous-winged Gull in south-central and SE Alaska. Pure birds there are relatively rare or difficult to discern. Locally in parts of central British Columbia eastward across Canada (north to S Nunavut) to Newfoundland, and east of the Great Plains south to the Great Lakes region, east to Pennsylvania, through the St. Lawrence River and interior New England, and then southward along the Atlantic Coast to South Carolina, where slowly spreading south. Overall, breeding range and breeding population saw great increases during the past century. From the late 1980s through the mid-2000s, several pairs nested on the Chandeleur Islands in Louisiana, where they were also found hybridizing with Kelp Gull (see that hybrid account). Several nest records south of the core breeding range exist, from Wyoming, South Dakota, and Laguna Madre in S Texas. Breeding population in Canada shows a moderate decrease, but with birds spread thinly over such a large expanse of land, numbers are difficult to confidently estimate. Rare breeder in Greenland; first confirmed in 1986, although believed to be European Herring. *L. smithsonianus* confirmed breeding in Nuuk in 2021, nesting only 50 meters from a pair of European Herrings. Identification is based largely on fledglings (Lars Witting, pers. comm.). One pair of *smithsonianus* suspected in S Melville Bay (2012). All others are assigned to European Herring, which is presumably more common. Of three band recoveries in Greenland, two were from Canada and one from Britain (Boertmann & Frederiksen 2016).

Nonbreeding: Dispersal away from nesting areas starts as early as mid-July but isn't widespread until late August. Begins arriving well south of breeding range by mid-September, in large numbers by late October. Largely withdraws from northernmost regions by late October and from other ice-locked regions where it does not regularly overwinter by mid- to late November. In Alaska, for instance, very rare to casual in the N Bering Sea south to the Aleutians. Locally common to fairly common at some sites on the Pacific Coast. In California, less coastal and much more common inland, such as in Davis, San Joaquin, and Stanislaus Counties, where several thousand can be found at landfill sites throughout the winter. Preliminary data suggests American Herring on the Pacific Coast may be more pelagic than once suspected. They are commonly found over the continental shelf and also noted in large numbers on the shores of California during coastal storms with westerly winds. Found in a wide variety of habitats, both coastal and inland, along the Atlantic. Particularly abundant at seafood-processing sites and landfills (where some high counts—e.g., in Bucks County, Pennsylvania—have exceeded 40,000 individuals) and on beaches. Also found at sea routinely, 60 or more miles but drops off dramatically after this, although regularly recorded several hundred miles offshore.

In the interior, most birds are locally common at larger lakes and rivers, dams, and landfills, rare elsewhere. Many adults breeding in the lower 48 states remain relatively close to the breeding grounds throughout the year, and younger individuals, particularly from the Great Lakes, move greater distances. East arctic populations believed to have migrations up to four times longer than those on the Atlantic, generally traveling farther south than populations from Newfoundland and NE Atlantic, which remain along the coast (Anderson et al. 2019). Band recoveries and individuals with satellite transmitters from east Arctic have been found to consistently migrate to the Gulf of Mexico (Allard et al. 2006; Anderson et al. 2020). In winter, can be fairly numerous south to Florida and the Gulf Coast, uncommon in the N West Indies. Along the Pacific Coast from SW Alaska, fairly common to common, only very locally south to extreme NW Mexico. Uncommon to rare south to central Mexico and mostly uncommon to rare in much of southwest desert away from Salton Sea, where in some years it is common. Northbound birds are on the move mostly between late February and early April. Nonbreeders are found regularly in numbers well south of the breeding range in the eastern half of North America throughout late spring and summer, more rarely in the West. At least 14 records in Iceland. Summer visitor to Greenland in very small numbers but unknown whether *L. smithsonianus* or *L.a. argenteus/argentatus*. Casual to extreme in N South America. Very rare to E Asia, including South Korea and E China, and sightings in Japan increasing. Extralimital record from Sea of Okhotsk in Russia, where it is likely overlooked (Artukhin 2022). In Europe, most records are from Ireland (approximately 100), but numbers have tapered off or slightly decreased in recent years. Also found on Azores, NW Europe, and to the Iberian Peninsula, with the southeasternmost record being from Malaga, Spain, where an adult was confirmed via genetic analysis (Garcia-Barcelona et al. 2021).

IDENTIFICATION

Adult: An evenly proportioned large white-headed gull. Pale, silvery-gray upperparts, yellow bill, with little to moderate expansion to the gonys. Bill longer and stronger in some individuals, usually males, and particularly so in those in the Atlantic and Northeast. Orange-red gonys spot. In nonbreeding condition,

1 Juvenile. Dark head, smooth brown underparts, dark tail and vent region, and blackish primaries. Scapular pattern highly variable. DAVE BROWN. NEWFOUNDLAND. AUG.
2 Juvenile. Buffy fringes to the upperparts (white) and commonly found holly-leaf pattern to scapulars. Very pale and heavily barred greater coverts on this individual. AMAR AYYASH. ILLINOIS. AUG.
3 Juvenile. Banded in Erie County, Ohio. Dark sooty brown aspect not unusual in fresh juveniles. Darker bases to greater coverts compared to 20.2. KURT WRAY. OHIO. JULY.
4 Juvenile. Overall uniform appearance not unlike some Slaty-backed Gulls, which generally show less-than-black primaries at this age, with noticeable pale edging. AMAR AYYASH. NEW JERSEY. SEPT.

5 Juvenile. Extensive pale edging to upperparts, with plain, dark-centered scapulars. AMAR AYYASH. MICHIGAN. OCT.
6 Juvenile. Dark aspect may elicit thoughts of Lesser Black-backed, which generally shows a darker bill base and paler underparts and lacks white fringes on the primaries. AMAR AYYASH. ILLINOIS. NOV.
7 Juvenile. Paler individual with white-tipped primaries. The deeply notched tertials and inner greater coverts are not common, but not unheard of, especially in the interior. Bill possibly still developing. AMAR AYYASH. MICHIGAN. OCT.
8 Juvenile. Strong bill and blocky head. Much white fringing on the upperparts. Note dark tail and blackish primaries. LIAM SINGH. BRITISH COLUMBIA. NOV.
9 1st cycle. Strong bill, large head, and blackish primaries. Moderate wear for this date. Typical chocolate-brown *smithsonianus*. AMAR AYYASH. ILLINOIS. SEPT.
10 Juvenile. Remarkable plumage for this date, with mostly 1st basic scapulars. Paling bill base. AMAR AYYASH. CALIFORNIA. JAN.
11 Juvenile. 1st basic plumage persists. Although Glaucous influence may be suspected, the black primaries are well within range for a pale *smithsonianus*. AMAR AYYASH. FLORIDA. JAN.
12 1st cycle. Some have brownish primaries with pale edging (rarely as extensive as typical Thayer's) and paler underside to p10 (far wing). Note strong bill paling at base, head structure, and many replaced scapulars. AMAR AYYASH. ILLINOIS. NOV.
13 1st cycle. Most scapulars postjuvenile (1st alternate). Wing coverts and tertials show moderate wear. Large bill. Dilute appearance to wing coverts and tertials, with paler head. Note densely marked vent/undertail. AMAR AYYASH. MASSACHUSETTS. DEC.
14 1st cycle. Pale, weakly patterned postjuvenile scapulars with thin transverse barring and barred greater coverts recall some *argenteus* European Herring Gulls, which have a whiter base color to the belly and a plainer vent region, and also lack the smooth-washed appearance of the lower neck. See 20.35 for tail pattern. AMAR AYYASH. MICHIGAN. NOV.
15 1st cycle. Whitish head and paler breast found in a small percentage of hatch-year birds. Postjuvenile scapulars show sooty, transverse barring with paler edging. AMAR AYYASH. ILLINOIS. DEC.
16 1st cycle banded as a chick in Maine. Large, powerful bill suggests male. Wing coverts and tertials similar to 20.15, but with considerable wear. Compare variable scapular pattern. AMAR AYYASH. FLORIDA. JAN.

17 1st cycle. Superficial resemblance to Azores Gull (see 23b.5), but American Herring has smoother and denser underparts, and paler regions on wing coverts (compare open wing and tail). AMAR AYYASH. NEWFOUNDLAND. JAN.
18 1st cycle. Banded in Door County, Wisconsin. Very pale postjuvenile scapulars highlight the immense variation found in this age group. Bicolored bill fairly common at this date. AMAR AYYASH. MICHIGAN. DEC.
19 1st cycle. Banded in Erie County, Ohio, just four months prior. Rather extensive 1st prealternate molt has replaced most lesser and median coverts, several inner greater coverts, and upper tertials. JANICE FARRAL. OHIO. OCT.
20 1st cycle with Royal Tern. Pale plumage aspect with contrasting black primaries and tertials. Pale bill base not uncommon, but pale nail appears to be an aberration. COLLIN STEMPIEN. ALABAMA. FEB.
21 1st cycle. Paler aspect likely due to some fading. Much black retained on bill on this individual. A paling iris isn't too unusual at this date. AMAR AYYASH. MASSACHUSETTS. APRIL.
22 1st cycle. Pale appearance recalls Glaucous Gull, especially given the bill pattern. The dark primaries and tertial centers suggest nothing more than a heavily bleached *smithsonianus*. Some individuals show uniformly bleached primaries by late spring, in which case identification is moot. AMAR AYYASH. FLORIDA. JAN.
23 1st cycle with adult. Sharply bicolored bill and pear-shaped head resemble California Gull, which typically has a messier and less uniform wing panel (open wing helpful with such birds). Bicolored bills found in a few as early as Sept. AMAR AYYASH. ILLINOIS. NOV.
24 1st cycle with similar-aged Great Black-backed Gull (left). Bill and head size differences readily apparent, but also note warmer brown underparts on American Herring and dark tail. AMAR AYYASH. FLORIDA. JAN.
25 1st cycles. Immense variation on display. The individual in the front is in nearly complete juvenile plumage, with a sharply bicolored bill. In the back is a highly worn and faded 1st cycle with all scapulars replaced. AMAR AYYASH. FLORIDA. JAN.
26 Juvenile. Fairly typical uniform brown upperwing; all-dark tail with densely marked tail coverts. Holly-leaf pattern to lower scapulars. The inner primary window is subdued on this individual. AMAR AYYASH. ILLINOIS. NOV.
27 1st cycle. Dark tail with light patterning along outer base. Paler inner-primary window than 20.26. Barred tips on greater coverts. Pale bill base, and most scapulars already replaced. AMAR AYYASH. ILLINOIS. OCT.
28 1st cycle. Contrasting white ground color to uppertail coverts found in a minority. Replaced lower scapulars are solidly dark as expected in some black-backeds. White head likely molted and not bleached. AMAR AYYASH. INDIANA. FEB.

the bill tip can show variable black smudging on the gonys, and in some birds a quasi black ring similar to that of Ring-billed. Bright lemon-yellow iris with variable orbital color—orange in some birds, yellow in others, and at times a fiery orange yellow. Legs typically faint pink, but a grayish-blue tinge can be found in a number of Atlantic birds; at times, some with yellow cast to leg color. Head streaking variable in basic plumage. Some birds have thin, faint streaking restricted to the lower neck and around the eyes; others show extensive brown wash down to the breast. At rest, black on primaries fairly extensive, with variable-sized apicals (smaller in younger adults). Tertial and scapular crescents are relatively thin. Wing projection is neither short nor long, which contributes to the even, standard shape of this gull. On the open wing, variable black on wingtip. Northeast and Atlantic birds average less black on inner webs of outer primaries, with moderate to large mirrors on p9–p10, occasionally with all-white p10 tip or broken subterminal band, and some with *thayeri* pattern on p9. White tongue tips variable in size from p6 to p8, although light gray feather centers make for weak contrast. P5 typically shows a complete W-shaped subterminal band, but this is absent in palest-winged birds. This band may average broader and less symmetric in western birds, with black on the outer web being noticeably broad (Ayyash, pers. obs.). Birds in the East show a more pointy W-pattern with a thinner band that is more concave along the proximal edges. A fair number of birds show so-called bayonet pattern on the outer edge of the outer webs of p7 and/or p8, but less common in birds on the Great Lakes and in the West. Subterminal band on p6 is an elongated W-pattern with sharp black on outer edge. On the Great Lakes, through the interior and West, birds average more black on inner webs of outer primaries, more black up the base of p7–p8, complete subterminal band on p5, and, at times, black on the outer web of p4 (more common on the West Coast). A complete p4 is very rare, although a broken band is sometimes found and is likely age related (check primary coverts and tail for dark streaks). Pigment on underside of p10 varies from much black approaching the base of the inner web to a limited blackish-gray medial band (less than the length of p10 mirror). Black on underside of remaining primaries (p5–p9) consistent with pigment on upperside (i.e., limited black on underside corresponds to limited black on upperside, etc.).

Similar Species: American Herring Gull is the only expected large four-cycle gull with black wingtips, pale gray upperparts, and pink legs in the East. Kumlien's Gull generally shows less-than-black (i.e., slate black) coloration to wingtips, limited pigment on its primary pattern, and a shorter and more refined bill. Thayer's Gull is generally the greatest confusion taxon. Thayer's has variable eye color, with some dark eyed and some paler eyed, but does not typically have the piercing, lemon-yellow iris of Herring. Up to 1%–2% of adult American Herrings exhibit heterochromia, with one eye completely dark and the other completely normal. This can obviously lead to confusion, but note these individuals are typical American Herrings in every respect. Thayer's has a pinkish orbital ring, which American Herring never shows.

---

29 1st cycles. An interesting duo showing paler vents and relatively muted inner-primary windows. Note variation in flight-feather coloration (brown versus black). Ribbing on outer edge of outer tail feather common. AMAR AYYASH. FLORIDA. JAN.

30 Juvenile. Banded in Erie County, Ohio. Brown head, uniform chocolate-brown underparts, and plain sooty underwing. Densely barred undertail coverts can be whitish. KURT WRAY. OHIO. JULY.

31 1st cycle. An overall paler individual, with paler underside to primaries, whitish vent, and evenly barred undertail coverts. AMAR AYYASH. FLORIDA. JAN.

32 1st cycle. Even brown underparts and dark tail. Flight feathers and wing coverts show moderate pale edging on this individual. LIAM SINGH. BRITISH COLUMBIA. NOV.

33 1st cycle. Same individual as 20.16. Light barring on the axillaries and stippling throughout the wing coverts. Uppertail coverts appear spotted rather than barred. AMAR AYYASH. FLORIDA. JAN.

34 1st cycle showing consistent barring across tail base. Dark marginal coverts, as in some Slaty-backeds, but contrast here due to bleached median and greater coverts. Whitish post-juvenile scapulars. AMAR AYYASH. INDIANA. FEB.

35 1st cycle. Same individual as 20.14. Barring on tail base limited to outermost tail feather, unlike 20.34. White ground color to uppertail coverts. AMAR AYYASH. MICHIGAN. NOV.

36 1st cycle. Somewhat defined tailband with whitish uppertail coverts. Outermost tail feathers barred to their bases, a pattern seen fairly regularly with American Herring. The sooty postjuvenile scapulars, smooth brown neck, and dark greater coverts are rather typical of *smithsonianus*. AMAR AYYASH. FLORIDA. JAN.

37 1st cycle. A somewhat perplexing individual recalling old-world Herrings, given the paler body, barred greater coverts, and tail pattern. European Herring averages more white on the tail base, with less fine patterning. Vega Gull averages paler regions on the outer greater coverts, although appreciable overlap exists. Such tail patterns are not common in *smithsonianus*, but are found with some regularity. AMAR AYYASH. INDIANA. OCT.

Structurally, Thayer's averages more refined proportions (skinnier bill, shorter legs, and more rounded head), but large male brutes are to be expected. Head streaking on Thayer's not as fine and thin as on American Herring, and typically more dense and smudged. Above all, wingtip pattern is usually sufficient for identification, especially in the West, where American Herring shows more extensive black than Atlantic birds. In Thayer's, pale pattern on underside of wingtip usually sees black restricted along tips to the primaries and fainter blackish gray on underside of p10. On the upper wingtip, Thayer's typically shows more-prominent white tongue tips on p6–p8. California Gull averages darker gray upperparts and has a dark eye and yellow legs. Ring-billed is smaller, with yellow legs. Western Gull is noticeably darker, with bulbous-tipped bill. Glaucous-winged does not show black wingtips and is typically dark eyed. Much less expected, compare to Vega and European Herring Gulls (see those accounts). Also, some American Herrings are difficult to separate from extreme, pale-end Glaucous-winged × Herring hybrids. Hybrids usually show some pink on the orbital, an iris that is clouded or with brown flecking, denser head markings, and more diffuse, paler black inner webs on the outer primaries; in particular, they are more expected to show a paler underside to p10 (see plate H2.33 for details).

1st Cycle: Juvenile plumage uniformly dark brown with pale fringes to upperparts. Head, neck, breast, and belly typically with smooth, solid brown appearance, although some birds with more coarse appearance, especially late in 1st cycle. The greater coverts are often largely dark, especially the outer tract, but some show neat, checkered pattern along the tips of these feathers and throughout the inner tract. Tertials have blackish-brown centers, showing variable pale tips, sometimes notched. Primaries all blackish brown, some with faint, white crescents on the tips, others with moderately pale chevrons approaching Thayer's (most common in Western populations).

Dark eyes. Legs dingy pink, sometimes with dark markings on shins, but typically pale pink by midwinter. Dark black bill becomes variably dull colored around base. In some extreme birds, a sharply demarcated black tip may develop; in others, an all-black bill may persist through 1st cycle and beyond. By midwinter, some birds show a contrasting white head and neck, which can be due to bleaching or molt. In birds undergoing 1st prealternate molt, note dark, sooty-gray flanks, sides, and hindneck, and a combination of new and old scapulars. Second-generation scapulars highly variable in color: some silvery white with thin shaft streaks, others with variable brownish-gray barring and anchor pattern. The undertail coverts and vent region are typically extensively pigmented, although in some extreme birds, large areas of white can be present with lighter markings. The open wing on 1st-cycle American Herring should be learned well. Although somewhat variable, typical birds show a light inner-primary window, darker greater coverts, densely marked uppertail coverts, and a dark tail. Note that tail pattern can be appreciably variable, with some birds showing moderate to extensive white uppertail coverts, paler bases to the outer tail feathers, and variable barring on the outer edges of the tail, which in effect makes for a relatively narrow tailband (such birds recall old-world Herring Gulls). The outer primaries are typically dark brownish black with slightly paler inner webs. The underparts and wing linings show a mostly uniform brown wash with fairly smooth texture. *Similar Species:* Month for month, American Herring usually shows a messier plumage with more contrast than Thayer's. American Herring overall has more robust head and larger bill with deeper gonys. Herring often lacks the frosted appearance to its upperparts, as well as the extensive pale chevrons found on the primary tips of Thayer's (although some Herrings may show pale edging to the primary tips, but usually reduced). At rest, Thayer's shows less contrast in

38 1st cycle. Noticeable tailband and whitish uppertail coverts. The barred outermost rectrices are similar to 20.36. The plain brown greater coverts also suggest *smithsonianus*, although such birds may defy proper identification. AMAR AYYASH. FLORIDA. JAN.
39 2nd cycle. Banded in Door County, Wisconsin, June 2013. Basic plumage more muted and marbled compared to 1st cycle. Faint gray on scapulars with thin barring. Paling iris and bicolored bill typical. AMAR AYYASH. MICHIGAN. SEPT.
40 2nd cycle. Slightly underexposed. Some adult-like gray on the upperparts (2nd alternate). Dark greater coverts and dark tail base. AMAR AYYASH. OREGON. JAN.
41 2nd cycle. Overall delayed and muted plumage at this age, generally expected in the East (see 20.17). Solid dark greater coverts and overall plainer wing panel. Averages blacker outer primaries than 1st cycle. AMAR AYYASH. NEWFOUNDLAND. JAN.
42 2nd cycle. Fine patterning on lower tertials and greater coverts resembling some European Herrings, but sharply bicolored bill with pink base more typical of American Herring. See 20.50 for open wing. AMAR AYYASH. FLORIDA. JAN.

43 2nd cycle. Somewhat advanced aspect with nearly all scapulars replaced (2nd alternate), along with an upper tertial and several wing coverts. Pale iris. Banded in Door County, Wisconsin, June 2013. AMAR AYYASH. MICHIGAN. DEC.

44 2nd cycle. Transverse barring on upperparts, particularly the tertial tips, recalls some European Herrings. The dense and smooth wash on the neck, sides, and belly is typical of *smithsonianus*. Also note sharply bicolored bill. AMAR AYYASH. FLORIDA. JAN.

45 2nd cycle. Paler individual with strong bill and blocky head. Dark tertial centers with paler tips. Blackish primaries and apparent dark tail visible. Note smooth wash on lower neck and sides. AMAR AYYASH. NEWFOUNDLAND. JAN.

46 2nd cycle. Average appearance. Distinguished from 1st cycle by muted upperparts, rounded primary tips, and pale iris. Plain pattern to scapulars (a mix of 2nd basic and 2nd alternate feathers). AMAR AYYASH. ILLINOIS. NOV.

47 2nd cycle. A darker individual overall, with delayed appearance. Icy-blue on mantle and inner scapulars. Relatively dark inner primaries and marked uppertail coverts. Diffuse black on bill base. AMAR AYYASH. WISCONSIN. DEC.

48 2nd cycle. All-white uppertail coverts not uncommon, but a white tail base with such fine patterning is somewhat rare in 2nd cycles. AMAR AYYASH. ILLINOIS. MARCH.

49 2nd cycle. Smooth brown on underparts and moderately marked uppertail coverts. White outermost tail feather not unusual (asymmetric). AMAR AYYASH. FLORIDA. JAN.

50 2nd cycle. Some 2nd cycles have diffuse pigment on the outer tail feathers, with a finer pattern on greater coverts and secondaries. AMAR AYYASH. FLORIDA. JAN.

51 2nd cycle. Same individual as 20.39. Relatively dark wing linings and axillaries. Well-patterned underparts, including vent, and densely barred undertail coverts. AMAR AYYASH. MICHIGAN. SEPT.

52 2nd cycle. Paler underwing with white axillaries. Note smooth wash on underparts and densely marked undertail coverts. AMAR AYYASH. ILLINOIS. JAN.

53 2nd cycle. Solid gray back, all-white uppertail coverts, and broad black tailband are fairly common in 2nd cycles on the Great Lakes. Very small mirror on p10 is rare at this age, and it's quite unusual to see one this large. AMAR AYYASH. ILLINOIS. DEC.

54 2nd-/3rd-cycle type. Overall appearance is typical of 2nd-cycle Great Lakes type, although the adult-like inner primaries suggest a delayed 3rd cycle. Whether p1–p5 are 2nd alternate or 3rd basic feathers is in need of further study. Currently, it is assumed such birds are 3rd cycles. AMAR AYYASH. ILLINOIS. NOV.

coloration when comparing the primaries and tertials, and when comparing the tertials and upperparts. In Herring, the contrast is dramatic from the tertials to the upperparts, and the impression on Herring is a less neatly pigmented bird. The outer primaries on Herring typically more blotchy brown with darker inner webs than the milky, pale coloration on Thayer's, which typically has paler inner webs and paler primary coverts (although some 1st-cycle Thayer's can be just as dark as Herrings; compare bill structure and scapular pattern on these individuals). In flight, Herring shows greater contrast between scapulars and wing coverts. Juvenile scapulars on Herring more often show riveted notches on fringes with so-called holly-leaf patterns. Thayer's very uncommonly shows some notching on the lowest scapulars and inner greater coverts, but not as deep or as extensive as on Herring. On Thayer's, the bases to the scapulars are sometimes dark, with paler central region, followed by dark, inverted, three-point crown. Underside of wingtip paler on Thayer's. See also Great Black-backed, Lesser Black-backed, California, Slaty-backed, Vega, and European Herring Gulls and Glaucous-winged × American Herring hybrids accounts.

2nd Cycle: Highly variable in every respect. Underparts with grayish-brown wash and uniform in appearance. Birds on the Atlantic and in the Northeast generally with less advanced scapular pattern, closer to 1st alternate feathers with barring and anchor patterns. These individuals typically have correspondingly retarded pattern to wing coverts. Birds from the Great Lakes often develop more advanced adult-like gray scapulars, with paler heads and bodies. On California coast, scapulars variable, but wing panel generally uniform, especially the greater covert tract, unlike some birds on the Atlantic, which can show pale, icy-white greater covert panel with faint, dark markings. The tertials are typically dark in all populations, with some light peppering near the tips, less commonly with thin white notching or subterminal barring. Eye color variable, from all dark to adult-like yellow. Bill pattern ranges from mostly black to sharply bicolored with black tip. Some advanced birds can show adult-like yellow bill with black tip, particularly in late spring slightly before start of 3rd prebasic molt. Legs dull pink. Primaries generally black, at times with thin, pale crescents on tips, which is rather standard on Western populations. On the open wing, the inner primaries are more gray than in 1st cycle, showing a contrasting window against darker outer primaries and blacker secondaries. The tail is typically all dark with variably marked uppertail coverts (moderately to heavily marked in eastern birds; all white in many individuals on the Great Lakes). Very rarely shows a small p10 mirror. The underwing is quite similar to that in 1st cycle, although with paler regions throughout wing linings and paler underside to remiges. Vent can be entirely pale in some birds, and the undertail coverts are typically with dense markings, although less so than in most 1st cycles.

Similar Species: Compared to 2nd-cycle Thayer's, American Herring typically has darker inner webs to outer primaries, more contrasting appearance to wing coverts, more solidly dark centers to tertials, and less extensive pale edging to primaries. Note some 2nd-cycle Thayer's lack the classic milky-brown, two-toned pattern and sometimes show blacker wingtips and a dark tail, but greater coverts usually more marbled and peppered than those on Herring. Thayer's upperwing coverts and uppertail coverts usually with more internal patterning. Compare bill, head, and body proportions, which average larger on Herring. Second-cycle Thayer's more commonly shows (ghost) mirror on p10 and paler underside

---

55 3rd cycle. Average individual for interior and western populations. Adult-like primaries with white apicals are useful for aging, as well as white tips to secondaries. Black tertial centers are commonly found at this age. AMAR AYYASH. BRITISH COLUMBIA. MARCH.
56 3rd cycle. Average Northeast American Herring. Median coverts replaced (3rd alternate) but otherwise dark greater coverts and tertials. Delayed bill pattern. P10 mirror and adult-like apicals rule out 2nd cycle. AMAR AYYASH. NEWFOUNDLAND. JAN.
57 3rd cycle. Broad white tips to trailing edge and p1–p3 used for aging. Most primaries and s1–s3 delayed for this age, but may help clarify 20.54. AMAR AYYASH. FLORIDA. JAN.
58 3rd cycle (alternate). Fairly typical. Commonly found with dark black patches on secondaries, considerable black on tail, and a small to medium p10 mirror. AMAR AYYASH. MASSACHUSETTS. APRIL.
59 3rd-/4th-cycle type. Overall adult-like, but with extensive black band on pinkish bill, dark eye, dirty primary coverts, and smaller apicals. Aging such birds is guesswork. AMAR AYYASH. ILLINOIS. NOV.
60 4th-cycle type. Larger apicals and p10 mirror (underside of far wing) suggest adult, but the black ink spot on the tertials and bill pattern indicate subadult. AMAR AYYASH. ILLINOIS. MARCH.
61 4th cycle. Known-age individual banded in Door County, Wisconsin (June 2013). Overall adult, but small apicals and small p10 mirror, more extensive brown wash down breast, and weak tertial crescent. AMAR AYYASH. ILLINOIS. DEC.
62 4th cycle. Same individual as 20.61. Dark streaks on two outer-primary coverts, but otherwise typical adult upperparts and tail. Small p10 mirror, much black on p8, and mark on p4 also suggest a young adult. Banded June 2013. AMAR AYYASH. ILLINOIS. DEC.

63 3rd cycle and presumed 4th-cycle type (right). 4th-cycle types are not safely aged without life history, but in general have an overall adult appearance, and at times have light streaks on the primary coverts. AMAR AYYASH. WISCONSIN. DEC.
64 Adult. Extensively marked head not very rare in adults. Relatively large apicals and broader tertial crescent, but with smaller p10 mirror. AMAR AYYASH. ILLINOIS. FEB.
65 Adult. A fairly small-billed individual with nicely rounded head. Wingtip shown in figures 12A and 12B in Introduction. Moderately streaked head and neck, pale yellow iris, and orange-yellow orbital. AMAR AYYASH. ILLINOIS. FEB.
66 Adult. Moderate head streaking can be finely patterned. Rather bright legs on this individual, straight bill with limited red on gonys, and smaller p10 mirror. AMAR AYYASH. OREGON. JAN.
67 Adult. Although birds in the Northeast average larger and pudgier bodies, some have surprisingly thin, parallel-edged bills with a pencil-like shape. Short-legged. Large p10 mirror. AMAR AYYASH. NEWFOUNDLAND. JAN.
68 Adult. Much black on the bill tip not rare, especially on the Atlantic. Apicals are unusually large on this bird. Noticeably short-legged. AMAR AYYASH. FLORIDA. JAN.
69 Adult. Small fine streaks on head, with bright, unmarked bill in midwinter. P10 not fully grown (see 20.85 for open wing). AMAR AYYASH. FLORIDA. JAN.
70 Adults with Great Black-backed Gull (left). Body size of some larger male types can match Great Black-backed (female), although bill size is consistently thinner. AMAR AYYASH. NEWFOUNDLAND. JAN.
71 Adult. Alternate plumage. White head, bright yellow bill, and orange-yellow orbital. Small mark on outer web of p4. HARTMUT WALTER. CALIFORNIA. APRIL.
72 Adult. Alternate plumage. Immaculately white-headed, with some new upper scapulars and perhaps some tertials. AMAR AYYASH. WISCONSIN. FEB.
73 Adults. The individual on the left has grayish-yellow leg coloration, which is found to a variable degree, particularly on the Atlantic. AMAR AYYASH. FLORIDA. JAN.
74 Adults. Note differences in black-and-white wingtip pattern, which may be, in part, age-related and/or geographic variation. AMAR AYYASH. FLORIDA. JAN.

to wingtip. See also California, Western, Vega, and European Herring Gulls and Glaucous-winged × American Herring hybrids accounts.

3rd Cycle: Slightly less variable than 2nd cycle, and now with mostly adult-like-gray scapulars. Head can show considerable streaking, with mottled lower neck, breast, and sides. Generally has cleaner white head later in the season (3rd alternate). Wing coverts variable gray, with some birds still showing extensive brown pattern to upperwing coverts and some on the Atlantic with largely pale wing panel. Tertials may have brown centers with peppering; others have adult-like-gray centers with white tertial crescent, often with some black "ink spots" on centers. Eyes typically pale by now, but some with dusky wash or dark flecking. Bare parts average duller than adult's. Bill may be dull pink or adult yellow, and in some birds a complete black ring is found on the bill tip, resembling that of Ring-billed. Smaller white apicals than adult's. On the open wing, may show extensive black on inner webs of outermost primaries, with variable black markings on primary coverts. Birds on the Great Lakes and in the interior commonly show a small mirror, sometimes limited to the inner web of p10, or no mirror at all. Others with larger p10 mirror. Mirror on p9 rare, although appears to be more common in Newfoundland winter population (Ayyash, pers. obs.). Dark brown wash across upperwing expected. Dark black spots on secondaries fairly common. Uppertail coverts usually all white with variable black tailband. Some birds may have mostly white tail by now, but more

75 Adult with similar-aged Iceland Gull (back). In addition to size and structure differences, note extent of pigment on outer primaries. AMAR AYYASH. ILLINOIS. DEC.
76 Adult with similar-aged Ring-billeds and Laughing Gull. Pinkish legs and red gonys spot. Underside of p10 all-white on this adult. AMAR AYYASH. FLORIDA. JAN.
77 Adult. Well-marked wingtip with no p9 mirror, not unusual for interior types. Extensive black on underside of p10 also found with some regularity, especially on the Great Lakes. Distinct W-pattern on p5. AMAR AYYASH. FLORIDA. JAN.
78 Adult. Extensive head streaking. Small mirror on p10 and no mirror on p9 not typical pattern in Northeast. But note gray base to p9, long tongue on p8, bayonet on outer web of p7, and W-band on p5. AMAR AYYASH. NEWFOUNDLAND. JAN.

20 AMERICAN HERRING GULL

typical to show a jagged black quasi tailband. ***Similar Species:*** Compared to Thayer's, Herring often shows blacker wingtips, especially on the inner webs. The underside to the wingtip is also blacker on Herring, and similarly the primary coverts and wing linings are messier on Herring. Thayer's averages a more polished look to the wingtip, shows bolder white tongue tips and a larger p10 mirror, and can more commonly show a p9 mirror. Thayer's also averages deeper pink legs and can sometimes show a pinkish orbital ring at this age, although both of these features should be used with caution. Compare size, structure, and eye color. An all-dark-eyed Herring Gull in 3rd cycle is not unheard of, but it is uncommon. See also California, Vega, and European Herring Gulls and Glaucous-winged × American Herring hybrids accounts.

MOLT The partial inserted molt in 1st cycle is attributed to a prealternate molt. There is much geographic and individual molt variation in this age group. Local and resident populations—for instance, those on the Great Lakes—begin their 1st prealternate molt in earnest, replacing many of their scapulars from September through November, and also show considerable signs of wear and fading on their wing coverts before winter. In some extreme birds, a variable number of tertials and upperwing coverts may be replaced in this partial molt (Ayyash, pers. obs.). Others, presumably arctic/subarctic populations,

79 Adult. Same individual as 20.66. Similar to 20.78 but with more black on outer webs of p8–p9. Some adults, particularly in the West, average a broader and less symmetrical p5 band, lacking the defined and pointy W-shape. AMAR AYYASH. OREGON. JAN.
80 Adult. Small p9 mirror with black not reaching the primary coverts on that feather. Bayonet on outer edge of p8 and W-band on p5 are *smithsonianus* features. AMAR AYYASH. FLORIDA. JAN.
81 Adult. Black reaching primary coverts on p9 with small mirror. Long tongue on p8, bayonet on outer edge of p7, and black on p5 limited to outer web on this individual. AMAR AYYASH. FLORIDA. JAN.
82 Adult. Extensive black on p9 with long tongue on p8. It is quite rare to find p4 marked on both edges, especially with a long p8 tongue. AMAR AYYASH. ILLINOIS. JAN.
83 Adult. Considerable gray on p7–p9, small p9 mirror, and faint W-band on p5. Note pointed wedges on p6–p8. AMAR AYYASH. FLORIDA. JAN.

maintain rather pristine juvenile plumages into late winter, similar to some 1st-cycle Glaucous and Iceland Gulls. A number of these types are found throughout the Gulf Coast and peninsular Florida in late winter, which is consistent with satellite-tracking studies which have found many arctic / subarctic American Herrings move well south of Great Lakes and Atlantic populations (Allard et al. 2006; Anderson et al. 2020). Winter populations in California also appear to have less extensive and later 1st prealternate molts than their New England counterparts (Howell et al. 1999).

Completion of prebasic molt in adults also variable in timing. Great Lakes and Atlantic populations appear to complete flight feather molt earlier (late Oct. to early Nov.) than the waves of arriving birds in late fall, which are presumably of northern origins (Ayyash, pers. obs.). In E Newfoundland, the majority of adults do not complete primary molt until late November (B. Mactavish, pers. comm.). In California, it's not rare to find adults still growing primaries into mid- to late December (A. Jaramillo, pers. comm.).

A small percentage of 2nd- / 3rd-cycle types acquire advanced, adult-like inner primaries (p1–p3 / p4) in what appear to be otherwise typical 2nd-cycle individuals. Whether these are advanced 2nd basic, 2nd

alternate, or typical 3rd basic feathers needs further study on known-age birds (Pyle et al. 2018; see plate 20.54).

HYBRIDS Well-known hybrid zone shared with Glaucous-winged Gull in the Cook Inlet region of Alaska (Patten 1980). Presumed hybrids with Great Black-backed Gull found annually on Great Lakes (10–15 per winter), where some are misidentified as Slaty-backed Gull. Interestingly, this hybrid is still quite rare in St. John's, Newfoundland; the Maritimes; and New England, where Great Black-backed and Herring Gulls winter together in relatively large numbers in these regions. This suggests hybrids may be originating on the Great Lakes—an example being a suspected breeding adult summering in the upper peninsula of Michigan in 2016–2021 (Ayyash, pers. obs.). Small hybrid zone shared with Glaucous Gull in the Mackenzie Delta of NW Canada (Spear 1987). Population size uncertain and largely unsurveyed. Hybridization with Glaucous Gull also strongly suspected in NE Canada, but breeding sites unknown and no data from known-origin birds available. Glaucous × Herring recorded annually in St. John's, Newfoundland, and scattered records throughout Eastern Seaboard, Great Lakes, and Pacific Coast. Also with Lesser Black-backed Gull: nesting records in Maine and Alaska (Ellis et al. 2014; van Vliet et al. 1993). Hybrids are increasing along Atlantic Coast, where some are suspected of being Yellow-legged Gulls. Also with California Gull (see that species account for details). With Kelp Gull on the Chandeleur Islands (Dittman & Cardiff 2005). The islands were ravaged by Hurricane Katrina in 2005, which displaced this hybrid for several years. Small numbers returned from 2015, with an estimated 10–15 pairs nesting here (Oscar Johnson, pers. comm.).

84 Adult. All-white tip to p10 with long gray tongue (gray visible next to outer web of p9). *Thayeri* pattern on p9 and relatively large white tongue tips on p7–p8. AMAR AYYASH. FLORIDA. JAN.
85 Adult. Same individual as 20.69. Small spot on p5 not rare, especially in birds with white tip to p10. Long tongue on p10 (underside of far wing), gray base to p9 with longer gray tongue, and p9 mirror confined to inner web found in *smithsonianus* in the East. AMAR AYYASH. FLORIDA. JAN.
86 Adult. So-called white-winged type due to limited black on underwing, recalling Thayer's. All-white tip to p10 and *thayeri* pattern on p9. Such wingtips are unexpected in breeding Great Lakes birds, but are found routinely in the nonbreeding season. AMAR AYYASH. WISCONSIN. DEC.
87 Adult. Extensive black on underside of p10 and only a single mirror are not rare in Great Lakes adults. Note that heavily marked neck has smooth pattern. AMAR AYYASH. ILLINOIS. NOV.
88 Adult. Same individual as 20.80. Short gray tongue on underside of p10 somewhat variable. P10 tip nearly all white. AMAR AYYASH. FLORIDA. JAN.
89 Adult. Underside of p10 has extensive gray on inner web, merging with mirror, yet still shows a complete subterminal band. On the far wing, *thayeri* pattern also on p9, with mirror mostly restricted to inner web; long tongue on p8 with bayonet on outer edge. AMAR AYYASH. FLORIDA. JAN.

21 EUROPEAN HERRING GULL *Larus argentatus*

L: 22"–26" (55–67 cm) | W: 49"–61" (125–155 cm) | Four-cycle | SAS | KGS: 3.5–7.0

OVERVIEW Simply known as Herring Gull in the Old World. European Herring Gull consists of two subspecies, *Larus argentatus argentatus* and *Larus argentatus argenteus*. It breeds throughout N and W Europe, where it is one of the most commonly encountered large white-headed gulls. Its behavior and ecology have long been studied by naturalists and ornithologists alike. Foraging habits are quite similar to American Herring's, although European Herring seems to more regularly frequent open, wet fields, where it feeds on earthworms and other insects. It nests colonially on islands, coastal cliffs, salt marshes, rooftops and near bodies of water. In North America, regularly recorded only in St. John's, Newfoundland, where it is considered a rarity (one to three annually). When found, it's typically with flocks of American Herring, Kumlien's, and Great Black-backed Gulls. Likely overlooked in other regions throughout the Eastern Seaboard; therefore, its true footprint in North America may be underestimated.

TAXONOMY
L.a. argentatus (KGS: 5.5–7.0): Nominate *argentatus*, sometimes referred to as Scandinavian Herring Gull. Averages larger body and bill, with stronger build. Darker gray upperparts, with more extensive white in wingtip pattern. Considerable geographic variation. Some adults from farther north and E Baltic population regularly show dull to bright yellow legs.

L.a. argenteus (KGS: 3.5–5.0): Also known as British Herring Gull. Generally smaller and more refined than nominate. Averages more black in wingtip pattern, with paler gray upperparts (more similar to American Herring).

RANGE
Breeding
L.a. argentatus: Northern, nominate race. Breeds throughout Scandinavia, Poland, the Baltic region, and NW Russia.

L.a. argenteus: Breeds in N Atlantic from Iceland east to Ireland, Britain, Belgium, and N France. Clinal variation exists with an intergrade population spanning from the Netherlands to N Germany and SW Denmark. Rare breeder in Greenland. First confirmed in 1986, believed to be European Herring (likely *argenteus*; more study needed). Identification is based largely on fledglings (Lars Witting, pers. comm.).
Nonbreeding: Nominate *argentatus* largely migratory, moving south and west, where it winters with southern populations. British *argenteus* more resident, with shorter movements.

In North America, a rare to very rare but somewhat regular visitor only to E Newfoundland, mostly in winter. Extralimital to most of North America, but reported casually, with several scattered records from MA, NJ, MD, FL, NS, and ON. Reports from Maritimes need review. A handful of European Herring Gull candidates are found annually throughout the Great Lakes and Atlantic Coast, particularly in Florida, although identification is confounding, and almost all of these are tabled or left unidentified. Both Texas and Florida have had several convincing candidates that are without question of old-world ilk and need closer review.

IDENTIFICATION
Adult: A compact and evenly proportioned large white-headed gull with moderate wing projection. Legs pale pink, with pale yellow eyes and a yellow-orange orbital ring. The features detailed below are based on averages. Just as with any large white-headed gull, it is important to consider individual variation and extremes, which sometimes belie expectations.

21 EUROPEAN HERRING GULL

L.a. argenteus: British *argenteus* has pale, silvery-gray upperparts and averages less streaking on the head in winter. The wingtip has more black and fewer white regions when compared to nominate *argentatus*. Often with a complete or broken subterminal band on p10, which has a moderate-sized to large mirror. Generally shows considerable black extending up the base of the inner web of p10 with a short gray tongue showing an oblique angle. The mirror on p9 averages small, sometimes restricted to the inner web or absent, with considerable black extending up the base of the inner web. White tongue tips on outer primaries, if present, are usually indistinct. A small black mark on the outer web of p5 is typical but variable.

L.a. argentatus: Darker gray upperparts and less black in the wingtip when compared to *argenteus*. Larger p9–p10 mirrors, often with all-white tip to p10. Unmarked p5 is typical but sometimes with insignificant diffuse marks. Pigment on p6 variable, commonly with more prominent white tongue tips extending across p6–p8. Averages larger proportions, particularly with deeper and stronger bill, more sloped forehead, and slightly longer winged than *argenteus*. Head streaking generally heavier, persisting longer through the winter months. Birds with most extensive white in wingtip recorded from the northernmost

1. 1st cycle. Largely juvenile. A grayish-brown individual recalling Vega Gull. Unlike American Herring, texture to breast and underparts average coarser with white base color. Consider notched edges to tertials, apparent pale tip to tail, and whitish undertail coverts with widely spaced bars. JULIAN HOUGH. ENGLAND. NOV.
2. 1st cycle. A heavyset individual with strong bill, retained juvenile plumage, and dark brown aspect suggesting nominate. Dense brown lower hindneck recalls American Herring, but note whiter greater coverts with barred, sawtooth pattern, and notched tertial edges. MARS MUUSSE. BELGIUM. DEC.
3. 1st cycle. Fairly typical *argenteus* type. Whitish underparts, unlike American Herring at this date. Note extensive white on outer tail feathers and vent. Replaced scapulars show thin subterminal bars with extensive pale background. HARRY HUSSEY. IRELAND. OCT.
4. 1st cycle. Note heavily notched edge to lowest tertial, pale undertail coverts with weak markings, and predominantly pale 2nd-generation scapulars showing thin transverse barring. White head and neck show finer streaking, unlike heavy smudged pattern in many *smithsonianus*. AMAR AYYASH. ENGLAND. DEC.

5   1st cycle with similar-aged Common Gull. Darker wing panel and tertials on this individual, with duskier head. Note tail pattern. AMAR AYYASH. ENGLAND. DEC.
6   1st cycle. Some dark nominates can very much resemble American Herring, including smudged lower hindneck and overall chocolate aspect. Consider widely spaced barring on undertail coverts with whitish background, paler vent, and overall longer and more defined streaking on face and chin. Uppertail pattern necessary with such birds, which may defy ID. Banded in hatch year from Norway. MARS MUUSSE. THE NETHERLANDS. NOV.
7   1st cycle. A very small percentage of 1st cycles combine pale upperparts and paler primaries, inviting thoughts of Herring × Glaucous hybrids. Such birds are often just pale-end Scandinavian *argentatus*. MARS MUUSSE. THE NETHERLANDS. DEC.
8   Juvenile. *Argenteus* type showing warm-brown upperparts. Strong inner-primary window appears grayish. Whitish tail base with speckled patterning on uppertail coverts, and defined tailband shows fine patterning on proximal edge. RICHARD BONSER. ENGLAND. AUG.
9   1st cycle. Most scapulars replaced with paler feathers showing thin barring. Finely patterned upperwing with extensive barring on greater coverts. AMAR AYYASH. ENGLAND. DEC.
10 1st cycle. Darker upperwing with bolder tailband. Note white ground color to uppertail coverts and distinct brown spots on tips of outer webs of inner primaries. Whiter head than *smithsonianus* and extensive white on tail base. AMAR AYYASH. ENGLAND. DEC.
11 1st cycle. Barring on uppertail coverts widely spaced with white background. Barred greater coverts, and postjuvenile scapulars shown thin anchor pattern. AMAR AYYASH. ENGLAND. DEC.
12 1st cycle. A paler individual with paler head and neck. Weak patterning on bases of tail feathers, and overall finely marked against a white background. The inner primaries form an obvious pale window. AMAR AYYASH. ENGLAND. DEC.
13 1st cycle. Darker inner primaries showing less contrast. Much zigzag patterning on tail base, not as uniformly barred on the outermost tail feathers as in *smithsonianus*. Banded in hatch year from Germany. MARS MUUSSE. THE NETHERLANDS. NOV.
14 1st cycle. Barred underwing coverts and axillaries different than American Herring's plain, smoky-brown underwing. When American Herring shows barred axillaries, they are quite faint and difficult to discern from a distance. Paler body has gritty texture. White tail base and undertail coverts showing widely spaced barring. AMAR AYYASH. ENGLAND. DEC.
15 1st cycle. More heavily patterned tail base, but generally lacks organized and more uniform rows of barring seen on some *smithsonianus*. MARS MUUSSE. THE NETHERLANDS. NOV.
16 1st cycle. Nominate type retaining most of juvenile scapulars. Some have a weak, broken tail pattern similar to Great Black-backed Gull. That species shows a more massive bill with a deeper gonys and a larger, blockier head. MARS MUUSSE. THE NETHERLANDS. MARCH.
17 2nd cycle. Paling eye and bill base. Rounded primary tips. Also note more finely patterned greater coverts showing strong vermiculations (more crisply barred in 1st cycle). AMAR AYYASH. ENGLAND. DEC.

part of the range and routinely show *thayeri* pattern. Many of these are with maximum darkness to upperparts, often with a distinct bluish quality. Some birds from eastern part of range have darker orbital ring, at times coral red. Nominate *argentatus* also comprises subpopulations with yellow legs.
Similar Species: Challenges surrounding identification in North America are two-fold: first, eliminating American Herring, and second, satisfying subspecific criteria (i.e., *argentatus* versus *argenteus*). Field-identification criteria for out-of-range European Herrings is fraught with doubt, especially in the case of adults. This is mostly due to the lack of field studies on geographic differences in American Herring Gull populations. From a North American vantage point, the conversation regarding European Herring Gull candidates usually revolves around what one should *not* look like, rather than what one *should* look like. American Herring Gulls in the Northeast are structurally closer to nominate *argentatus*, in that they are large and heavy-billed and show less black on the wingtip. Ideally, nominate *argentatus* adults would be visually detectable in a flock of American Herrings, since the former are darker, and this is often the first tip-off when they're found in Newfoundland. Elimination of Herring × Lesser Black-backed hybrids, or even Yellow-legged Gull, should be carefully addressed, however, especially in the case of candidates showing yellowish legs.

As far as gray upperparts are concerned, American Herring overlaps in color with British *argenteus*, and this identification question is more difficult. Identifying adult *argenteus* in North America has not been met with success, due to the wide range of variation found in American Herring. In theory, *argenteus* found in Newfoundland should appear slight and lanky, with more black on the inner webs of p9–p10. But such field marks fall apart, for example, on the Great Lakes and in the interior, where many American Herrings are smaller, with slighter bills, and average more black on the wingtip. American Herring overall shows heavier and denser head markings, although variable. Some features that should dissolve the potential of a pale European Herring Gull candidate in North America include an all-black W-band with sharp upturned edges on p5, a long gray tongue on p10 showing a square terminal edge, and a distinct bayonet pattern on p7–p8. Supporting features to help eliminate adult *argenteus* would be an extensively marked head, neck, and breast or gray pseudo-mirrors on the underside of p9/p10 (Adriaens 2013). Gray pseudo-mirrors on the underside of p9/p10 in European Herrings are rare. Note that an individual lacking these features does not default to European Herring. Assuming the above features are absent, and remaining conservatively grounded, the following characteristics combined may support an *argenteus* candidate: the proximal edges to the black borders on p7–p8 should be distinctly flat, not sharply wedged; the subterminal band to p6 should show an oblique, rounded, or straight (not W-shaped) border along the outer edge; and markings on p5 should be limited to a black spot on the outer edge of the outer web (see Adriaens & Mactavish 2004; Adriaens et al. 2022 for critical details). Additionally, such a candidate should show other standard features, such as a complete subterminal band on p10, a smaller mirror on p9, and short gray tongues on the inner web of both p9 and p10 (best inspected from underside).

---

18 2nd cycle. Dilute brown barring and vermiculations across upperparts, including tertials. Banded as a chick in the Netherlands. MARS MUUSSE. THE NETHERLANDS. OCT.
19 2nd cycle. Finer pattern on upperwing coverts, including greater coverts that show extensive vermiculations. American Herring averages overall darker bases to greater coverts, with more uniform brown wash on neck and breast. MARS MUUSSE. THE NETHERLANDS. JAN.
20 2nd cycle. Small dark bill and overall neat and uniform frosty appearance to plumage similar to 2nd-cycle Thayer's Gull in some respects. See also Vega Gull. AMAR AYYASH. ENGLAND. DEC.
21 2nd cycle. Finely patterned upperwing and largely white uppertail coverts. Solid, adult-like scapulars and pale eye suggest *argenteus*. MARS MUUSSE. THE NETHERLANDS. APRIL.
22 2nd cycle. A somewhat delayed individual. Rounded primary tips difficult to appreciate in flight, but note vermiculations on secondaries and greater coverts (more neatly barred in 1st cycle) and incoming gray on scapulars. Pale body and whitish neck. AMAR AYYASH. ENGLAND. DEC.
23 2nd cycle. Paler wing linings with barring, unlike *smithsonianus*. Paler vent and undertail coverts with white base to tail. Advanced bill pattern showing red gonys spot, but dark eye. GARY THOBURNS. ENGLAND. MARCH.
24 3rd cycle. Pale, adult-like gray scapulars, pale eye, and blackish primaries with smaller apicals. Greater coverts and tertials show vermiculated pattern with fine barring. MARS MUUSSE. THE NETHERLANDS. OCT.
25 3rd cycle. Likely nominate *argentatus*, which can often show a delayed plumage aspect, even at this age. White primary tips and broad white tips to secondaries useful for aging. Incoming gray on scapulars appears dark. MARS MUUSSE. THE NETHERLANDS. DEC.

1st Cycle: Overall, averages dilute-brown plumage with whitish base color to body, some with darker brown tones. Underparts appear streaky with coarse texture. Wing coverts largely barred with considerable pale regions. Juvenile scapulars brown with pale fringing. Postjuvenile scapulars show thin anchors, commonly with transverse barring, but variable. Paler head shows distinct longitudinal streaking. Pinkish legs and dark eye. Black bill overall with some gradual paling around the base, but typically not sharply bicolored or boldly pink. Tertials often with notched edges and tips. Dark, blackish-brown primaries may show some pale edging. At rest, the undertail coverts and vent region are largely pale with reduced markings.

The open wing reveals a highly barred and checkered upperwing, with a prominent pale window to the inner primaries. The inner primaries are generally pale on both webs, and, upon close inspection, some splotched marbling can often be seen. The tips to the inner primaries also commonly show dark segments on the outer webs with proximal white capsules. The tailband is relatively narrow and defined, with lightly barred uppertail coverts against a notably pale base color. The bases to the outer tail feathers range from being boldly white to lightly speckled or with moderate vermiculations in some individuals. The underwing shows a gritty and somewhat spotted appearance with a light base color throughout. *Similar Species:* Compared to American Herring, 1st-cycle European Herring typically has a paler base color to the body with a streakier and grittier appearance. American Herring is more uniformly colored, with a solid, smooth brown appearance to its underparts, and also shows a darker lower hindneck that can appear smudged (more finely streaked in European Herring). The greater coverts on American Herring often show dark bases, especially on the outer tract, which is quite conspicuous in flight. In European Herring, it's more typical to see these feathers largely barred or with more patterning, and the overall appearance is increased regions of white. The postjuvenile scapulars on American Herring often combine more variation in pattern between the upper and lower scapulars, whereas in European Herring, 2nd-generation scapulars are somewhat consistently patterned, often with thin, transverse barring, but appreciably variable. The tertials on European Herring also average more patterning, sometimes with deep notches and subterminal barring. In many American Herring, the tertials average much less pattern, without bold pale regions found along the tertial tips and edges, although some do have extensive notching and patterned tips (particularly Great Lakes and interior birds). The tertial edges and tips often become worn by midwinter, making these patterns less noticeable. The undertail coverts and vent region are densely marked on American Herring, whereas in European Herring, the barring is widely spaced, with a predominantly pale background. Analyzing an open wing is critical. Note the white base color to the uppertail coverts on European Herring, complemented by paler and lightly marked outer tail feathers and

---

26 3rd cycle. Aged by adult-like inner primaries. A well-patterned individual with much markings on tail and secondaries. Markings average finer vermiculations, especially on the secondaries, which lack the bold, solid-black patches found on American Herring. AMAR AYYASH. ENGLAND. DEC.

27 3rd cycle. Tail pattern averages weaker pigment than similar-aged American Herring, and overall pigment on upperwing more dilute. Importantly, note fine barring on innermost wing coverts. AMAR AYYASH. ENGLAND. DEC.

28 3rd cycle. Adult-like primaries but with rather brown greater coverts, showing fine vermiculated pattern. Markings on secondaries are faint and not blotted dark black. Diffuse tail markings. Underwing variably marked. MARS MUUSSE. THE NETHERLANDS. MARCH.

29 Adults. Example of *argenteus* type (front left) and nominate *argentatus*. Center bird banded as 1st cycle in England (not as nestling; Feb.). Note typical darker upperparts, streaked head at this date, larger proportions, and extensive white on p9–p10 and unmarked p5. Paler *argenteus* often shows white head at this date and averages delicate proportions. JEAN MICHEL SAUVAGE. FRANCE. JAN.

30 Adult. Presumed nominate *argentatus*. Darker upperparts, strong bill, and sloping forehead. Streaked head on this individual similar to some American Herrings. Orange-yellow orbital, pale eye, and pink legs are typical. All-white tip to p10 and large p9 mirror common. PETER ADRIAENS. BELGIUM. MARCH.

31 Adult. Presumed *argenteus*, banded as breeding adult in the Netherlands (May 2008). A compact individual. Relatively small bill with saturated yellow coloration. Red gonys spot averages larger than on American Herring, but variable. Fine head streaking. PETER ADRIAENS. BELGIUM. DEC.

32 Adult. Some nominate types can show yellow legs (which are, at times, quite vivid). This individual also has a reddish orbital and slightly darker upperparts, recalling Yellow-legged Gull. However, the all-white tip to p10, largely unmarked p5, and moderate head streaking at this time of year all comfortably point away from *michahellis*. See American Herring × Lesser Black-backed hybrids also. DIRK VAN GANSBERGHE. SWEDEN. DEC.

33 Adult. Smaller p9 mirror restricted to inner web. Compared to American Herring, consider the flat proximal edges to the black on the outer web of p7–p8, and lack of distinct W-shape bands on p5 and p6. MARS MUUSSE. THE NETHERLANDS. JAN.

more of a defined tailband. American Herring can show paler outer tail feathers with a somewhat defined tailband, but usually with considerable vermiculations on the bases of the outer tail feathers, and not predominated by a white background. Finally, the underwing on American Herring has more of a uniform appearance, especially the axillaries, unlike the paler underwing of European Herring, which can be finely barred and speckled with pale zigzagging. When American Herring does show barring on the axillaries, it is very faint and difficult to detect with the naked eye.

First-cycle European Herring Gull is generally not safely identified to subspecies. On the whole, 1st-cycle *argentatus* is more likely to approach American Herring, in that some can have relatively dark tails, with more densely marked undertail coverts, darker bodies, and, of course, larger bill and body sizes. In Europe, there are known "*smithsonianus* lookalikes" that are disturbingly similar to American Herring but on the whole appear more finely patterned on the head and neck, with scalier underparts and paler vents. By virtue of its being a more southern resident, *argenteus* sometimes averages greater wear and fading on the wing coverts early in the season, similar to many Great Lakes American Herrings, whereas some northern *argentatus* retain crisper juvenile plumage, sometimes resembling white-winged taxa from arctic North America. Some of these individuals can be superficially similar to Thayer's Gull and Glaucous × Herring hybrids. By mid- to late winter, some faded 1st-cycle American Herring produce paler bodies that may appear somewhat streaked or coarse, and identification of European Herring vagrants only becomes trickier. Conversely, a number of 1st-cycle "Herring Gulls" wintering in east-central Florida retain rather pristine juvenile plumages with variably patterned underparts, narrower tailbands, and exaggerated pale tail bases. These individuals display intermediate, and generally confusing, characteristics not known anywhere else in North America (Ayyash, pers. obs.; Michael Brothers, pers. comm.). They're suspected of having their origins in the Arctic, quite possibly from a mixed-stock population that has recently taken shape (more study needed on these types). Compare also to Lesser Black-backed, Great Black-backed, and Vega Gulls.

2nd Cycle: Highly variable. Head and body now paler than in 1st cycle, usually with some streaking throughout. On the whole, *argenteus* is more advanced in plumage aspect, with more adult-like gray scapulars, and averages a paler iris. Nominate *argentatus* has more subdued, ashy-brown upperparts and averages a darker eye but is more likely to show a brighter bicolored bill pattern. Upperparts still somewhat barred with considerable pale regions, but the greater coverts can now have a solid brown wash, unlike in 1st cycles. Tertials often have brown centers, displaying broad white tips and dark distal bars. Whitish vent and undertail coverts. Blackish-brown primaries now round tipped, sometimes with pale fringes. In flight, inner-primary window more muted than in 1st cycle, with secondaries appearing more dilute brown. Brownish

---

34 Adult. Limited black on the wingtip with an all-white tip to p10 and *thayeri* pattern on p9. Faint spot on p5. Fine head streaking and slightly darker gray upperparts. AMAR AYYASH. ENGLAND. DEC.

35 Adult. Extensive black on wingtip, including sharp W-band on p5. Not safely distinguished from American Herring, especially those on the Great Lakes, which also show short tongue on p10 (underside). European Herring averages an oblique edge to the tongue on p10, whereas American Herring tends to have a flatter edge. MARS MUUSSE. THE NETHERLANDS. MARCH.

outer primaries, sometimes with a small mirror on p10 (particularly in nominate *argentatus*). Tailband often broader and more solidly filled than in 1st cycle, with spangled pattern to outer tail feathers. At times, rectrices show irregular white tips. Uppertail coverts range from nearly all white to moderately barred. The underwing somewhat paler and less patterned than 1st cycle's. ***Similar Species:*** In general, separation from American Herring becomes more difficult with age, due to the greater degree of overlap in various features. Just as with some 2nd-cycle American Herrings, individuals range in aspect from advanced, resembling

36 Adult. Presumed *argentatus* with limited black on the wingtip and moderately streaked head at this date. Large white tip to p10, extensive gray on p8–p9 and unmarked p5 expected in nominate. MARS MUUSSE. THE NETHERLANDS. FEB.
37 Adult type. Complete subterminal band on p10 and absence of p9 mirror (small pinhole mirror on left wing) suggest *argenteus*. Compared to American Herring, consider flat proximal edge to black on p7 and much black on inner web of p9, although may defy identification out of range. AMAR AYYASH. ENGLAND. DEC.
38 Adults. Both individuals show an oblique edge to their p10 tongues (undersides). Upper individual has an extensive block of black on p7–p9, with flat proximal edges on p7–p8. This would be unusual for an American Herring with all-white tip to p10 and unmarked p5. The thin and flat, faded band on p5 on the lower individual is also unexpected in American Herrings, which show extensive black on p9–p10. MARS MUUSSE. THE NETHERLANDS. JAN.

3rd cycle, to extremely delayed, resembling 1st cycle. As in 1st cycles, there are some parallels that can be drawn from population to population. Second-cycle Great Lakes American Herrings often develop advanced, adult-like gray mantles, as found in a fair number of *argenteus*. Many American Herrings from the Northeast are more likely to have a less advanced appearance, mirroring the subdued patterns found in *argentatus*. On average, 2nd-cycle American Herring shows darker and denser markings on the hindneck, which often extend down to the breast and sides, whereas in European Herring, the hindneck and sides tend to be paler and more streaked. Overall, European Herring has body markings that appear more patchily mottled and spotted, whereas many American Herring still have an even, uniform look to the underparts. European Herring averages more barring and vermiculations between considerable regions of white on the tertials and coverts, especially on the inner greater and median coverts. American Herring can regularly show a sharply bicolored bill with bright base, which generally isn't found in old-world Herrings, even at this age, as they often retain some black markings along the cutting edge. American Herring averages a darker vent region, although highly variable. The tailband on American Herring averages denser too, and more solidly filled, particularly on the outermost tail feathers, whereas European Herring commonly shows white on the tail base with white outer edges and a finer pattern to the proximal border of the tailband. Compare also to Vega and Yellow-legged Gulls and Lesser Black-backed × Herring hybrids.

3rd Cycle: Variable, but with obvious adult-like gray found throughout, some averaging more advanced aspect (*argenteus*) than others (*argentatus*). Some nominate *argentatus* are now appreciably darker above, with bold, bluish-gray coloration, compared to paler *argenteus*. The head is largely pale with finer streaks, coarser on the lower neck. Pale eye or with dark clouding. Bill pattern variable. Legs pale pink. Some individuals have delayed wing coverts still showing much fine barring, especially on the inner greater coverts and tertials. Advanced birds can show nearly solid, adult-like wing coverts but with faded brown wash. On the open wing, this faint brown wash is found across the greater coverts and secondaries, which are often finely patterned and peppered. Mirror on p10 not uncommon. The tail can be similar to that of some 2nd cycles, but generally a rather weak and choppy pattern of diffuse markings is found. Some with all-white tails. *Similar Species:* Separation from 3rd-cycle American Herring largely inhibited by broad subspecific variation in both species. With that said, there are particular features that are commonly found in each taxon that can be helpful. American Herring very commonly shows well-defined, solid black marks on the secondary centers and notable black spotting, or ink spots, on the tertials. In European Herring, the tendency is much more of a washed-out, brown appearance with fine vermiculations. Ink spots have been recorded in some adult-type European Herrings, but appearing paler, and they certainly are not a regular or expected feature (Maarten van Kleinwee, pers. comm.). American Herring averages a fuller black pattern on the tailband, unlike the commonly seen diffuse markings on European Herring. On the whole, American Herring can appear dingier, especially some individuals with extensive lower neck and side markings. Differences in primary patterns ambiguous and largely unreliable. One feature that may be helpful in eliminating European Herring is an individual already showing extensive gray tongues on the inner webs of p9–p10, with a full W-band on p5, and marked p4, while with the dark, dingy brown body of American Herring (Lonergan & Mullarney 2004). Vega Gull is a serious contender; compare especially to darker *argentatus*, which averages less black in the wingtip. Also compare to Lesser Black-backed × Herring hybrids.

MOLT First prealternate molt generally limited to head and body feathers. This molt is most evident in the scapulars, especially in more southern *argenteus*, on which many upper scapulars are replaced by September. Longest lower scapulars often remain juvenile into late winter. Does not typically molt wing coverts or tertials, but this may be found in a few random individuals. Not uncommon for some nominate *argentatus* to remain entirely juvenile into late winter (a pattern echoed in other taxa of the far north). Adult *argenteus* often complete primary molt by late November, later in Scandinavian populations.

HYBRIDS Extensive hybridization with Glaucous Gull in Iceland (Ingolfsson et al. 2008). Well-documented and widespread hybridization with Caspian Gull (*Larus cachinnans*) in Poland, Belarus, and Germany, and smaller occurrences to W Europe (Litwiniak et al. 2021). Hybrids with Lesser Black-backed, Yellow-legged, and Great Black-backed Gulls reported regularly.

22 VEGA GULL *Larus vegae*

L: 22"–26" (55–67 cm) | W: 49"–61" (125–155 cm) | Four-cycle | SAS | KGS: see "Taxonomy" section

OVERVIEW A dark-mantled "Herring Gull." Preliminary genetic studies have found that Vega Gull is not a true Herring Gull (de Knijff et al. 2005). The name East Siberian Gull is sometimes used for this large Asian gull, which breeds primarily in NE arctic Russia and occupies only a small niche in North America, in a narrow region where a void is left by American Herring (to which it is very closely related). Vega Gull commonly nests singly and in smaller colonies of 25–30 pairs. It nests at the top of cliffs, above seabird colonies, and in coastal lowlands, where it can often be found with Glaucous Gull (Kondratyev et al. 2000). Primarily coastal throughout the nonbreeding season, favoring fishing harbors, piers, and beaches. Vagrants in North America are usually found in the same settings as out-of-range Slaty-backed Gulls, which involve feeding habits associated with landfills. Adults are rather distinctive and, interestingly, possess a combination of features from various large gulls across the Holarctic. They often show a broad white trailing edge against medium to dark gray upperparts, often appearing dark eyed, with a red orbital ring and rich pink legs. The initial impression is that of a dark-backed Thayer's Gull with a Herring Gull's body.

TAXONOMY
L.v. vegae (KGS: 5.5–7.0 / 8.0): Vega Gull. The American Ornithological Society treats Vega and American Herring Gulls as subspecies of *Larus argentatus*, but most authorities have moved away from this antiquated classification. Vega Gull is the easternmost of an allied group of Asian taxa whose taxonomic positions are unresolved. It is now generally treated as a species, often with *L.v. mongolicus* as a subspecies (Yésou 2002). A controversial and more disputed form, *taimyrensis / birulai*, is sometimes grouped with Vega (Yésou 2002; Harrison et al. 2021), although others treat *taimyrensis* as a hybrid of Heuglin's Gull and Vega Gull (Ujihara & Ujihara 2019). See Appendix for details on Heuglin's and Taimyr Gulls.

L.v. mongolicus (KGS: 5–6): Mongolian Gull. Formerly treated as a subspecies of Caspian Gull (*L. cachinnans*), but an increasing number of taxonomies now treat Mongolian Gull as a subspecies of Vega (Yésou 2002). Some maintain it is a good species (Moores 2011; Moores et al. 2014), but here it is treated as a subspecies of Vega Gull, with the caveat that a future split is quite likely. First cycles are fairly distinctive once they've molted their juvenile scapulars, and it is this plumage (1st alternate) that is most likely to capture one's attention in North America.

Adult *mongolicus* averages paler gray upperparts than *vegae*, appearing slightly longer winged and stately. Leg color ranges from pink to yellowish to orange pink, but generally duller than Vega's (Yésou 2001). Shows more extensive black on the wingtip, typically with p9–p10 mirrors, and very often with black down to p4, sometimes to p3. Head streaking is overall limited to finer streaks on the lower nape and hindneck; may be white headed throughout winter. Primary molt completed considerably earlier than Vega's, usually by mid- to late November. First cycles average earlier and more extensive scapular molt and show paler postjuvenile scapulars with contrasting dark chevrons and anchor markings. Overall, paler body and head with white aspect. Paler underwing, also with barring. The uppertail coverts become largely white with narrow tailband showing intricate proximal markings, with more contrast than in Vega. A fresh juvenile in North America would prove difficult to identify but should show paler underparts and underwing.

Breeds in central Asia beginning in April, mainly in Mongolia, parts of NE China, in smaller numbers to Lake Khanka, and along the Yellow Sea coast of South Korea. Winters primarily around the coasts of Korea, China, and south to Taiwan and in smaller numbers east to Japan (more commonly in W Japan). It occupies an ecological niche remarkably different from that of Vega Gull in the nonbreeding season, preferring inland lakes and rivers, as well as floodplains and wetlands. No confirmed reports in North America (although possibly overlooked); consequently, *mongolicus* is not fully detailed here.

RANGE

Breeding (L.v. vegae only): In North America, breeds only on St. Lawrence Island, where it is locally common in summer (Lehman 2019). High count of 200 at Gambell (May 2012). Core breeding range is NE Siberia, from E Taimyr to Chukotka, including larger islands in the East Siberian Sea, as far east as Wrangel. Southern coastal limit in the W Bering Sea extends to northern boundary of Kamchatka.

Nonbreeding: Reports in Alaska are positively correlated with observer coverage, and the status of this species in the N Bering Sea is likely underestimated. It is fairly common on St. Lawrence Island in spring and fall, mostly at Gambell, where day counts are under 100 during sea watches. Locally common to rare in late summer and early fall on mainland W Alaska, mostly on Seward Peninsula around Nome. Rare to uncommon south to Pribilof Islands, with most reports from St. Paul. Uncommon to rare in the Aleutians, with most reports from Shemya, Attu, and Adak. Rare to very rare in North Slope, primarily in June from Utqiagvik (formerly Barrow). In winter, casual in N Bering Sea and very rare to casual in S Bering Sea and Aleutians. Again, sightings are directly related to observer presence.

Sightings outside St. Lawrence are rarely sufficiently documented. At least a dozen sight records from British Columbia need official review. Most states currently do not vote on Herring Gull subspecies; therefore, data is fragmentary. Noteworthy, at the time of publication, are approximately 30 convincing reports from California, split mostly between adults and 1st cycles (more than half from San Mateo County). Washington has 4 accepted records, with a 1st cycle furnishing the first state record in 2006, and in 2023, a convincing 2nd cycle from Oregon (see plate 22.18). Likely overlooked in other places in the Pacific Northwest and West Coast. May be more common than currently estimated in the Pacific waters around Hawaii, although some may involve Mongolian. For example, around 10 specimens held at the American Museum of Natural History and labeled *vegae* were collected in and around Hawaii during the Pacific Project in the 1960s, many younger birds from March (Ayyash, pers. obs.). Bona fide reports away from the West Coast, some of which have not been voted on, include those from ON (Zufelt 2012), TX (five), FL, PA, MN, MI, MD, MA, and VA. Vagrants to Europe include an adult to Ireland (Jan. 2016; Barton 2017), Italy (Dec. 2021), and Spain (Feb. 2021) and a subadult to France (Nov. 2016).

In Asia, arrives on the coasts of Japan, Korea, and SE China mostly in early to mid-October, wintering mostly through April. Some also remain in the W Bering Sea in winter, although exact status unknown due to shortage of observations (Kondratyev et al. 2000). A recent study which tracked 28 individuals with GPS loggers from the breeding grounds in N Russia and the wintering grounds in Korea and Japan found that adults had prolonged autumn migrations from late August to early December and a rather rapid spring migration in May (Gilg et al. 2023). A preference for coastal rather than inland or offshore routes was noted, as well as strong site fidelity to wintering and breeding sites.

IDENTIFICATION

Adult (L.v. vegae only): Adults are medium to dark gray with a distinctive matte-blue finish to the upperparts. Head streaking is variable, from finer streaking on the face to commonly coarse markings densely concentrated around the lowest part of the neck and upper breast. The head is pear shaped with an elongated forehead. Somewhat long necked in appearance, especially powerful male types, which can show a typical Pacific large-gull shape akin to Glaucous-winged Gull's shape, with others more refined and Thayer's-like. The bill is fairly straight with a relatively small gonys angle, but the culmen varies in shape, from noticeably sloped to blunt tipped. In nonbreeding condition, the bill can be rather dull, often appearing greenish yellow. Oval-shaped, orange-red gonys spot, limited to the lower mandible. Black markings on the bill tip are not regularly found, although not rare. Eye color is variable, ranging from golden yellow (rare) to completely dark. From a distance, many birds appear dark eyed, but upon

1. Juvenile. Primaries (and bill) still growing. Cold brownish-gray neck and underparts show gritty texture. White base color to vent and undertail coverts with widely spaced barring. Greater coverts predominantly pale with irregular dark barring. Deeply notched lower scapulars and patterned tertial tips. JULIO MULERO. ALASKA. SEPT.
2. Juvenile. Underparts may approach American Herring's in darker birds, but note sparsely barred undertail coverts with white base color, weakly barred greater coverts with extensive regions of white, and deeply notched lower scapulars with large arrowhead pattern. Blotted tertial tips unlike American Herring. MIKE DANZENBAKER. SOUTH KOREA. JAN.
3. 1st cycle. Average individual. Greater coverts (except for outers) predominated by pale regions with dark-spotted pattern. Deep notching on lowest scapulars with arrowhead tips. Averages paler head and body than American Herring overall. Unmarked vent on this individual. AMAR AYYASH. JAPAN. DEC.
4. 1st cycle. Overall paler body with whitish head and dilute brown upperparts. Pale ghosting on bill base not uncommon. Postjuvenile scapulars show thin anchor bars. Much white on tertial tips with deep notches. Exposed tail base largely white. JOSH JONES. JAPAN. JAN.
5. 1st cycle. An individual with warmer brown aspect, but note overall frosted appearance to upperparts. Dark eye mask. Some have solid dark outermost greater covert bases. Note pale, patterned tertial tips and arrowhead centers on lower scapulars. White outer edge to tail feather exposed. JOSH JONES. JAPAN. JAN.
6. 1st cycle. Another pale-bodied individual. Not unlike some European Herring Gulls, which can show similarly thin-barred postjuvenile scapulars and pale greater coverts. Pattern on tertial tips unknown due to wear, suggesting paler pigment. Paling bill not uncommon for this date. AKIMICHI ARIGA. JAPAN. APRIL.

7   1st cycle. White head and neck not unexpected, especially late in the season. Postjuvenile scapulars with obvious thin anchors; others more solid grayish-brown. Black tail with well-marked flanks and belly point away from Mongolian Gull. AKIMICHI ARIGA. JAPAN. APRIL.
8   1st cycle. Wing panel appears almost spotted due to intricate barring. Some show bright subterminal spots on the inner webs of the inner primaries and pale capsules on the outer webs. White tail base with narrow tailband. AMAR AYYASH. JAPAN. JAN.
9   1st cycle. Pale individual with uppertail recalling Mongolian Gull, which averages paler uppertail coverts with consistent rows of proximal spotting along tailband and distinctive postjuvenile scapulars by this date. Compare also to European Herring and Taimyr Gull. JOSH JONES. JAPAN. JAN.
10 1st cycle. A relatively dark tail base, but with fine patterning and pale uppertail coverts. Note the deeply notched lower scapulars (juvenile). Some show plainer greater coverts, especially on the outer tract. AMAR AYYASH. JAPAN. DEC.
11 1st cycle. A browner individual with heavily marked head and darker inner primaries. Weak contrast to uppertail not unlike some European Herrings. Pale bases to lesser and median coverts recall some *smithsonianus* look-alikes from Iceland and the Faroe Islands. MARTEN MULLER. SOUTH KOREA. FEB.
12 1st cycles (left and center) with similar-aged Slaty-backed (right). Vega often has darker flight feathers than Slaty-backed, with more consistently barred greater coverts. In flight, Slaty-backed averages a broader wing with relatively plain greater coverts. AMAR AYYASH. JAPAN. JAN.
13 1st cycle. Flanks and belly have gritty pattern with pale base color (not solid, smooth texture of American Herring). Wing linings are generally dark but variable, at times with noticeable barring. Note white pearls on tips of inner webs of inner primaries. AMAR AYYASH. JAPAN. JAN.
14 1st cycle. Presumed *mongolicus*. White aspect with faint neck boa. Narrow tailband shows high contrast with white uppertail coverts. Noticeably pale underwing on many, and variable inner-primary window. Note contrasting brown chevrons on otherwise whitish scapulars. CHRIS GIBBINS. SOUTH KOREA. DEC.
15 1st-cycle female Vega Gull (left). Offshore Japan. Nov. 1st-cycle female American Herring Gull (right). South Carolina. Dec. When seen closely, the base of the outer tail is more intricately patterned on Vega, showing quasi leopard pattern. Note also widely spaced and thin barring on uppertail coverts. *Smithsonianus* averages dense barring on uppertail coverts with largely dark tail base. Also compare tertial patterns. AMAR AYYASH. BURKE MUSEUM. WASHINGTON.
16 2nd cycle. Dark eyes and pale bill base expected. Wing coverts are commonly pale at this age with predominantly white background. Dark tertial centers with prominent white on tips. Pale scapulars on this individual are delayed with thin subterminal bars, but upper scapulars and mantle reveal darker gray. CHRIS GIBBINS. SOUTH KOREA. DEC.
17 2nd cycle. A bulkier individual with much black on bill. Adult-like scapulars (2nd alternate), which are darker than American Herring. Streaking on ear coverts and face average finer and longer than American Herring. Greater covert and tertial pattern may overlap with European Herring and Thayer's. JOSH JONES. JAPAN. JAN.
18 2nd cycle. Plain pattern with much white to median and lesser coverts, with finely patterned greater coverts. Contrasting dark tertials with prominent white tips. Blackish primaries showed faint p10 mirror in flight. DAVE IRONS. OREGON. FEB.
19 2nd-cycle type. Similar to 22.18, but open wing needed for accurate molt/age assessment at this time of year. Compare overall body size and gray scapulars to Slaty-backed Gull (upper right) and apparent *pallidissimus* Glaucous Gull (upper left). MICHAEL D. STUBBLEFIELD. RUSSIA. JUNE.

closer inspection, the iris is usually dull with dark flecking. Importantly, the orbital is dark crimson red, sometimes appearing brownish orange when dull. Leg color variable, from standard pink to often bright purple pink. Some variants presumably show yellowish legs (although consider other taxa; see Taimyr Gull account and summary of Mongolian Gull under "Taxonomy" in this account). At rest, the scapular and tertial crescents, especially the latter, are noticeably broad. Some individuals show a moderate secondary skirt. Can project a long-winged look as a result of posture, but just as commonly appears truncated, due to late primary molt.

On the open wing, the broad white trailing edge to the secondaries shows a bold contrast with the medium gray upperparts. The tips to the inner primaries are sometimes noticeably broad but variable. Wingtip pattern generally combines a large rectangular p10 mirror with full subterminal band, a smaller p9 mirror, and extensive black up the bases of the inner webs, especially on p10. But p10 sometimes with all-white tip, and p9 can be without a mirror (or show a typical *thayeri* pattern in extreme birds). P5 variable but often with broad W-band. P4 sometimes with markings, typically a smaller mark restricted to the outer web, particularly on birds with small or absent p9 mirror, and therefore may be age related. White tongue tips are commonly found from p5 to p7, at times to p8, and much rarer to p9, forming a noticeable string-of-pearls in the boldest birds. More study needed on average wingtip patterns and extremes in this taxon. The black pattern to the underside of the wingtip tends overall to have a clean and well-defined L- or V-shape. *Similar Species:* In North America, adults should be compared to Thayer's and American Herring Gulls, Glaucous-winged × Herring hybrids, and Slaty-backed Gull (a formidable challenge in W Alaska). See also European Herring (particularly darker nominate *argentatus*), Taimyr, and Mongolian Gulls accounts. In addition to the darker upperparts, two very helpful field marks that are often utilized with vagrants is the red orbital and the late primary molt. Adults can be found with old p9/p10 in mid-December, and many are commonly found growing these two primaries in late December through early January. Molt pattern should be used cautiously with vagrants, however. Compared to Thayer's, Vega averages darker gray upperparts, but extremes in both taxa can be quite similar and overlooked in the field. Vega averages larger, and its neck appears more bulgy, with a bigger head and higher eye placement. When together, Thayer's is almost always shorter legged. The red orbital on Vega is very useful if properly discerned, although some are so deeply red that they appear purplish, and caution must be exercised in various lighting conditions. Both taxa can have a greenish cast to the bill with similar, small gonys spots, but Vega averages a stronger and longer bill. Green cast on Thayer's bill usually more polished olive green, while Vega's is duller and a sickly greenish gray. On the open wing, Vega has more black on inner webs of p9–p10, which are largely gray on many Thayer's Gull. Also, the white tongue tips on Thayer's are typically more defined and flashy. The pigment on the underside of p10 is unmistakably black in Vega but often grayish and reduced in Thayer's. Head streaking in Vega tends to be coarser and streakier, and it extends farther down the neck and upper breast, although variable.

American Herring Gull almost always has a piercing yellow iris, orange-yellow orbital, and noticeably pale gray upperparts. Particularly in the West, American Herring averages more black on p7–p8, typically without a p9 mirror, and without noticeable white tongue tips. In the East, Vega's darker upperparts would be the first tip-off of a different type of "Herring Gull." Leg color averages richer pink in Vega. Glaucous-winged × American Herring Gull hybrids can be similar in structure, with similar head markings, but they often show a pinkish orbital, paler gray upperparts, and more washed-out black on the wingtip with diffuse proximal edges. Also, the underside to the outer primaries on hybrids is sometimes obviously slate colored or light gray, unlike the jet-black coloration on Vega. Some hybrids also have the tendency to show considerable black around the gonys in nonbreeding condition, which is not expected in Vega.

Slaty-backed Gull typically shows noticeably darker upperparts, which is often sufficient to separate the two. However, individuals at the dark extreme of Vega and pale extreme of Slaty-backed can be surprisingly tricky, and doubly so in unfavorable lighting conditions. Structurally, Slaty-backed is more potbellied, with shorter wing projection. Slaty-backed has a broader wing with a short, blunt hand in flight and almost always shows continuous broad white tips from the outer secondaries to the inner primaries, whereas Vega averages thinner tips to the inner primaries and less contrast between the upperparts

22 VEGA GULL

20 2nd cycle. Dark tail common with contrasting paler inner primaries. Moderately marked uppertail coverts delayed on this individual. Distinctive wing panel washed white with fine patterning. Neck often densely marked. AMAR AYYASH. JAPAN. JAN.
21 2nd cycle. More advanced than 22.20, with white uppertail coverts and adult-like scapulars. Fine pattern on whitish wing coverts, here with plainer outer greater coverts. MARTEN MULLER. SOUTH KOREA. FEB.
22 2nd cycle. Pale-bodied individual with contrasting upperwing. Undertail coverts and vent largely white. Underwing coverts variable, often remaining dark. CHRIS GIBBINS. SOUTH KOREA. DEC.
23 3rd cycle. Aged by adult-like primaries (surprisingly large and fresh for this date). Bill pattern and wing panel delayed. Note broad white tertial tips. Some have dark tertial bases, recalling American Herring, although typically with reduced and paler pigment. MICHIAKI UJIHARA. JAPAN. MARCH.
24 3rd cycle. Medium-gray upperparts with light brown throughout. Pale eye and delayed bill pattern on this individual. Note broad white tips to tertials and extensive lower neck mottling, with finer streaking on face. Potbelly and overall silhouette strangely recall Slaty-backed. JEFF HIGGOTT. SOUTH KOREA. JAN.
25 3rd cycle. Typical pear-shaped head and dark eye, with fine streaks on face and upper neck. Interesting combination of advanced gray upperparts and white tail, paired with primaries lacking apicals, which oddly are truncated (p9–p10 growing?). AKIMICHI ARIGA. JAPAN. APRIL.

and trailing edge. The white tongue tips on p5–p7 on Slaty-backed are often bolder and much more noticeable on the underwing, which shows darker remiges. Compare Vega's reddish orbital to Slaty-backed's purple-pink orbital. Primary molt completion averages a month earlier in Slaty-backed. Despite all of this, there are some birds not safely separated, which may represent Vega × Slaty-backed Gull hybrids (see this account's "Hybrids" section).

1st Cycle: Overall, grayish-brown body with darker tertials and blackish primaries. The underparts have a scaly texture with paler base color, some strikingly white in late winter. Head shows thinner streaks, at times with a moderately mottled shawl on the lower neck. The bill is largely black through the winter, but dull pinkish undertones on the bill base are noticeable in many birds. Pink legs. The juvenile scapulars show relatively broad pale fringes with variably shaped centers: some notched, others more plain and spade shaped. The lower scapulars often have deeply notched edges, forming distinct brown arrowheads on the centers. The postjuvenile scapulars are generally cold brownish gray, showing thin anchors and barring, or mostly plain with thin shaft streaks. The tertials range from solidly dark with pale fringing to often notched with noticeably pale, broad tips. Whitish-buff base color to the wing panel, which appears somewhat frosted and checkered. The greater covert tract is striking, showing parallel rows of isolated dark notches against what is typically predominated by paler regions. The lesser and median coverts are evenly patterned with smaller dark centers, showing equal regions of pale and dark checkering. The vent is washed out, averaging paler than the belly, sometimes white, and this color extends onto the undertail coverts, which typically show widely spaced bars. In flight, an obvious inner-primary window is found, although subdued in some birds. The inner webs on the inner primaries are pale, at times with brighter tips, and the darker outer webs show distinctive pale subterminal capsules. White rump with whitish uppertail coverts showing weak brown barring. Although 1st-cycle Vega Gull has a somewhat restricted and narrow tailband, the pattern on the outermost tail feathers is appreciably variable, ranging from nearly all white to relatively dense rows of speckling and barring, which may appear dark at a distance. In most cases, the tailband appears defined and contrasts with the whiter uppertail coverts. Relatively dark underwing coverts showing faint barring, often uniform in color with flanks, which commonly appear coarsely spotted. *Similar Species:* Compare to American Herring, European Herring, Slaty-backed, and Thayer's Gulls. See also Mongolian and Taimyr Gulls accounts. A small percentage of Vega Gulls are surprisingly dark bodied, vaguely recalling American Herring, but seldom showing the solid, smooth textures to the underparts and lower neck found in American Herring. The overall coloration is a cold brown gray. The appearance of the breast and underparts on Vega is scaly and almost spotted, with a paler base color to the body. Individual features show overlap, but a combination of field marks makes for two different-looking gulls, which can usually be separated with reasonable confidence. Vega has largely checkered and spotted greater coverts with considerable pale regions and typically lacks dark solid brown bases to this tract. The lower juvenile scapulars often show distinct arrowheads formed from deeply notched edges. On American Herring, the lower tertials show a riveted or weakly notched pattern, or holly-leaf pattern, with largely dark, plain centers. Also consider Vega's notched tertials with broader pale tips. On the whole, Vega almost always appears more patterned, with a paler aspect, and underparts appear more gritty. On the open wing, the wing panel of Vega looks uniformly barred, whereas in American Herring there is more contrast from one tract to the next, due to a greater variety in patterns. The outer greater coverts on American Herring often show dark bases when the wing is open,

26 3rd cycle. Broad white trailing edge, white tongue tips to p5–p7 and distinctive, pale-washed wing panel. Small mirror on p10 variable, at times absent. MICHIAKI UJIHARA. JAPAN. MARCH.
27 3rd cycle. Extensive black on wingtip, including outer-primary coverts. Thin white crescents on p5–p6. Variable black on tail. Note contrasting pale-washed greater coverts. Advanced bill pattern. MICHIAKI UJIHARA. JAPAN. FEB.
28 Subadult. Composite of the same individual. Black on alula and primary coverts as well as extensive, diffuse pigment on base of p6–p8. Bill tip also with extensive black ring. Otherwise adult-like upperparts, tail, and wing linings. MARTEN MULLER. SOUTH KOREA. FEB.
29 Adult. Gonys spot often small with orange hue as in Thayer's, with greenish cast to bill. Dark eye with heavy head mottling readily recalls that taxon. Vega has darker gray upperparts, more black in the wingtip, bulkier proportions, and a dark red orbital ring. OSAO UJIHARA. JAPAN. JAN.
30 Adult. Honey-colored iris with much speckling appears all-dark from a distance. Moderate head streaking and greenish cast to bill (compare to Thayer's 32a.31, which averages paler underside to p10). JOSH JONES. JAPAN. JAN.

31 Adult. Pale-eyed individual with noticeable dark red orbital (can appear burgundy). Fully grown secondaries form a moderate skirt that merges with broad tertial crescent. JOSH JONES. JAPAN. JAN.
32 Adult. Somewhat typical head pattern with long, coarse streaks, becoming heavier and densely concentrated on lower neck. Pear-shaped head. Darker iris and long, dull bill on this individual. AMAR AYYASH. JAPAN. DEC.
33 Adult. Medium-dark gray upperparts and prominent tertial crescent. Darkish eye (variable) and dense lower-neck markings extending onto breast. MICHIAKI UJIHARA. JAPAN. FEB.
34 Adult type with American Herring Gulls. Upperparts noticeably darker than *smithsonianus*, although some may be very similar in shade and overlooked in North America. Molt gap in primaries not uncommon at this date (outermost primaries old with worn tips, p5–p6 new with fresh apicals). TOM JOHNSON. PENNSYLVANIA. DEC.
35 Adults. Variable bill patterns and head streaking. Can show dull greenish bill base (left) and dense lower-neck markings extending onto the breast. Broad tertial crescents. Compare to Thayer's and Glaucous-winged × Herring hybrids. MICHIAKI UJIHARA. JAPAN. DEC.

but largely barred in Vega. Tail patterns also show appreciable average differences, but some American Herrings can show a relatively defined tailband with less pigment on the outer tail. Vega almost always averages whiter outer tail feathers with finer vermiculations. On the outermost tail feathers of the "tailband" American Herrings, the barring appears somewhat more orderly and consistent, whereas in Vega the vermiculations appear more random and asymmetric (see plate 22.15). American Herring also averages darker uppertail coverts that are more densely and boldly barred. Overlap is expected, and these differences are somewhat subjective, especially without detailed images of a fully spread tail, so other field marks must be used in combination, such as scapular and tertial patterns, greater coverts pattern, and the overall scaly versus smooth texture to the body. American Herring shows smoother brown flanks and underwing coverts. In Vega, the flanks are paler, with weak brown spotting, and the underwing coverts show faint barring and pale spotting. Also, as supporting features, the bill on American Herring may have a bright pink base earlier in the season, whereas in Vega, the tendency is to show more black with dull pink undertones. Note also the tendency for American Herring to show a more smudged and heavily marked lower hindneck and its likelihood of replacing all of its scapulars, whereas Vega often molts scapulars slowly and retains lower juvenile scapulars for a longer part of the winter. Again, molt patterns with vagrants should be used cautiously.

The more formidable identification challenge at this age is separation from European Herring. Average differences may have limited use with vagrants, especially juveniles. There is much overlap in several helpful features among Vega and European Herring Gulls that are used for separation from American Herring. For instance, both Vega and European Herring show reduced pigment on the tail base with paler uppertail coverts and reduced barring. Vega tends to show denser barring on the proximal edge of the tailband and outermost tail feathers. Both can show notched tertials with broader pale tips. The lowest juvenile scapulars on Vega often have a pale-colored deep notch toward the base, with defined dark tips in the shape of arrowheads. The pattern is neater than in some European Herrings, which have broader brown centers with weaker pale notches. Although there is great variation and definite overlap in the lower scapular pattern, a large arrowhead may be considered a plus for juvenile Vegas when these feathers are juvenile. Both species have postjuvenile scapulars that can show thin terminal barring, although some European Herrings average plainer centers and unmarked shafts. Both can show highly patterned greater coverts with neat barring. Vega tends to have paler regions on the inner greater coverts, at times appearing faintly and randomly speckled. Both have reduced lower hindneck mottling, unlike the smooth, dense lower neck of American Herring. European Herring may average a whiter head at this age. Both Vega and European Herring have similar underpart textures (i.e., scaly, with pale base color to the body). Vega may average warmer tones to the underparts, although highly variable. Some European Herrings (*L.a. argenteus*) average an overall more dilute brown aspect to the upperparts and paler wing linings with more defined barring. None of the features mentioned above is diagnostic, however, and some out-of-range birds will defy identification.

Slaty-backed Gull has, among other features, a more rotund body with fuller vent region; a shorter wing projection; plainer greater coverts; a distinct diamond-tipped pattern to the lesser and median coverts; browner primaries, which may show more extensive pale edging; inner primaries showing less of a contrasting window; and a darker tail base (see that species account). Thayer's Gull overall is more refined in structure; shows frostier upperparts with fine, vermiculated markings on the wing panel; has more obvious and consistent pale edging to the outer primaries and, on the open wing, a two-toned pattern to the mid-outer primaries; and has a more pigmented tail base and tail coverts. Note also that Thayer's remains largely juvenile on the upper scapulars and averages a more refined bill.

2nd Cycle: Can show extensive longitudinal head streaks becoming coarse on the mid-neck and rather dense brown markings on the lower neck. Bill pattern variable, ranging from bone colored and a dullish green to pink, commonly with black smudging against pale base or with black tip. Others more advanced, often in the breeding season, with brighter yellow bill and dark tip. Dark eyes and pink legs. Scapulars vary from some appearing like those of 1st cycles, with buffy-gray feathers with thin anchors, to more advanced individuals showing solid adult-like gray mantle and scapular feathers. The wing coverts also can take on adult-like gray, but patchily, depending on extent of 2nd prealternate molt. Second basic median and lesser coverts are often all white with thin brown shaft streak and thin brown terminal bars. Others with more prominent barring or dark, spade-shaped tips with pale base. Greater coverts vary from largely white to intricately patterned with fine, wavy bars, but generally not solid dark. The tertials often have darker brown centers with limited markings bordered by relatively broad white tips. In flight, the upper surface to the wing is rather lackluster with little patterning. A pale inner-primary window is visible, contrasting with the secondaries, which are generally paler than the outer primaries. Second basic secondaries can show relatively broad white tips, although lacking the adult-like gray centers. A small percentage can show a small to medium p10 mirror. The uppertail coverts are largely white, but the tailband is variable. Some have full-black tails; others have tail patterns resembling those of 1st cycles. The underwing pattern is generally dark, but some individuals show surprisingly pale wing linings at this age, recalling 3rd cycles. *Similar Species:* Thayer's, American Herring, and Slaty-backed Gulls. See also details of 2nd and 3rd cycles in European Herring, Taimyr, and Mongolian Gulls accounts. Second-cycle Vega Gull commonly shows some gray on the mantle and upper scapulars, revealing a darker blue-gray coloration not found in American Herring. This, combined with a rather pale wing panel with finer markings, is distinctive. Thayer's Gull tends to look closer to Vega at this age (unlike in 1st cycle).

---

36 Adult. Saturated bare parts in breeding condition. Vibrant red orbital and gape help eliminate pale-end Slaty-backed, which would average darker, charcoal tones, although some can be surprisingly similar to Vega. IAN DAVIES. ALASKA. MAY.

37 Adult with Slaty-backed Gull (right). Alternate plumage with worn primary tips. Vega has a more athletic body, but averages narrower white tips to the tertials and duller leg color. MICHAEL D. STUBBLEFIELD. RUSSIA. JUNE.

38 Adult. Individual with extensive black on the wingtip, no p9 mirror, black on the outer web of p4, and no noticeable white tongue tips. MICHIAKI UJIHARA. JAPAN. MARCH.

39 Adult. Unmarked p5 and strikingly broad white trailing edge, continuing onto the primary tips, recalling Slaty-backed Gull. Prominent white tongue tips on p6–p7 and p10 with large white tip. Individual showing bright orange on maxilla. Pale eyed. AMAR AYYASH. JAPAN. JAN.

40 Adult. Not unusual for p9–p10 to still be growing at this date, as these two primaries are often retained into Dec. (p9 is 50% grown with no mirror, p10 is ~25% grown with large mirror). Subterminal W-band on p5 averages less symmetrical than American Herring. AMAR AYYASH. JAPAN. JAN.

41 Adult. Darker upperparts with much black on base of p8–p9 and white tongue tips to p5–p7. Fine lower-neck streaks recall Mongolian or Taimyr Gull, although this pattern can be expected late in the season, especially when head is molting (prealternate here). MICHIAKI UJIHARA. JAPAN. MARCH.

42 Adult. Wingtip represents a pale extreme. All-white tip to p10, gray base to outer web of p9, and black limited to outer web of p6. Bold white tongue tips to p7–p8. MICHIAKI UJIHARA. JAPAN. FEB.

43 Adult. Short-billed and pale-eyed individual with Taimyr Gull expression, but striking string-of-pearls on p5–p8 and *thayeri* pattern on p9 commonly found in Vega. Prominent white tips to inner primaries. MICHIAKI UJIHARA. JAPAN. FEB.

44 Adult. Reduced black on underside of wingtip sometimes forms characteristic L- or V-shape. Dense lower-neck shawl not unlike American Herring. Compare wingtip pattern, broad white trailing edge, and darker gray underside to primaries. JOSH JONES. JAPAN. JAN.

45 Adult. Fairly common appearance to underside of wingtip. Others have more gray on the underside of p10 and reduced black along the inner edge of p9, but variable. Thin white tongue tips on p5–p7 washed out in part due to lighting. MARTEN MULLER. SOUTH KOREA. FEB.

Those Vegas with delayed scapular patterns should show a combination of anchors on their scapulars, whereas in Thayer's, isolated thin bars are expected with muted, grayish-brown tones. Note also the more pronounced two-toned pattern on the outer primaries of Thayer's. Compared to both American Herring and Thayer's, on Vega the tertials average broader white tips and whiter greater coverts. Slaty-backed, again, shows different body structure, with 2nd alternate scapulars that are decidedly darker.

3rd Cycle: Overall adult-like aspect with lead-gray upperparts, showing brown wash throughout. Commonly shows extensive head markings, becoming denser on the lower neck and upper breast. Generally dark eyed, but some dirty yellow becoming more apparent at closer range. Bill pattern often dull pinkish yellow with sizable black tip, and legs average paler pink than most adults. White scapular and tertial crescent noticeably broad even at this age, but some individuals show considerable brown on tertial centers. Some can show a relatively dense brown belly, which contrasts sharply with a whitish breast and neck (usually late-spring birds at the start of 3rd prebasic molt). The open wing reveals a relatively broad white trailing edge, and the secondaries in general are largely clean. The primary coverts have well-defined black markings, as does the tail, but typically in a fragmented pattern without a full tailband. Typically with medium-sized p10 mirror, and not very rare for p9 to have a small mirror. Considerable black is found on p6–p8, sometimes with variable white tongue tips, and regularly shows mark on p4. *Similar Species:* Third-cycle Thayer's overall has paler gray upperparts, but identification can be surprisingly tricky in some individuals. On average, Thayer's shows a smaller head and bill, with eye evenly centered, although some males can belie expectations (some which may be Glaucous-winged × Herring hybrids). Thayer's has paler inner webs to the outermost primaries, producing more of a venetian-blind pattern. Also, the underside to the wingtip is noticeably paler on Thayer's. Orbital color can be helpful if accurately interpreted, but largely unreliable at this age. American Herring is paler gray and often shows distinctive black centers on secondaries and tertials. The secondaries on Vega are generally cleaner, with tertial centers showing diffuse brown markings, not triple-black like American Herring . Glaucous-winged × Herring hybrids should be paler gray and can show more smudged head markings and less-than-black wingtips. A suite of field marks, including structure, can usually be employed to eliminate one or the other. See also details of 3rd cycles in Slaty-backed, European Herring, Taimyr, and Mongolian Gulls accounts.

MOLT Partial 1st prealternate molt variable in timing and extent. Most evident in scapulars by November. Others remain juvenile into late winter. Birds replacing scapulars earliest often show subtle brown patterning resembling juvenile feathers. Those molting later in the season average more grayish-white coloration, still showing considerable anchor pattern and barring. Not uncommon for the longest lower scapulars to remain juvenile into late winter, while the uppers are replaced early. Seldom replaces (a few) coverts and / or tertials. A very small percentage of 2nd cycles have been recorded with several 3rd-generation-like inner secondaries (possibly 2nd alternate).

Adults typically complete primary molt through December and into early January (seldom into Feb.). From a sample of 65 adults in Choshi, Japan, only 7 had completed primary molt in mid-December. In a similar sample, 32 of 63 had fully grown primaries on 1 January (Ayyash, pers. obs.). Those still molting in early January typically had p9 more than half grown and p10 about one-third grown: final stages of primary molt. Most adults still show substantial head markings into mid-March, averaging later than other winter congeners.

HYBRIDS With Glaucous, Glaucous-winged, and Slaty-backed Gulls. In Bluff, Alaska, a mixed Vega / Glaucous pair produced young that were identified as *vegae × hyperboreus* (Kessel 1989). More recently, an apparent Vega / Glaucous-winged pair produced at least one nestling at Bull Seal Point on St. Matthew Island: the first documentation of this hybrid pairing in North America (Robinson et al. 2020). Kishchinsky (1980) found Vega and Slaty-backed Gulls hybridizing in the Koryak Upland and described coloration in primaries and upperparts of hybrids in this region. Portenko (cited in Vaurie 1965) also provided a collection of intermediate specimens and went so far as to assign Slaty-backed Gull a subspecies of Herring, *L. argentatus schistisagus*. This hybrid also occurs throughout the N Sea of Okhotsk, where both taxa overlap and reportedly interbreed. More study needed on extent of contact here.

23A YELLOW-LEGGED GULL *Larus michahellis michahellis*

L: 21"–26" (53–67 cm) | W: 47"–55" (120–140 cm) | Four-cycle | SAS | KGS: 5–7

OVERVIEW The predominate large white-headed gull of S Europe, the Mediterranean Sea region, N Africa, and Macaronesia. Yellow-legged Gull is a large, four-cycle species with three recognized races. Nominate *Larus michahellis michahellis* is discussed in detail here, along with some rudimentary notes on *L.m. lusitanius* (from N Spain and Portugal). Account 23b is devoted to *L.m. atlantis*, the population breeding on the Azores. This subspecies is most distinct and is presented separately to allow for more conciseness and ease of accessibility (with no implied taxonomic separation). Nominate, also referred to as Mediterranean *michahellis*, is a stocky gull with upperparts generally darker than those of American Herring Gull and closest to those of California Gull (although some variation is known). A rapid range expansion took place in the 1980–1990s, but this has apparently ceased since the turn of the century (Adriaens et al. 2022). Interestingly, this race is also the most migratory, particularly young birds, undergoing a south-to-north postbreeding dispersal starting in July before migrating back south in the fall (Garner & Quinn 1997).

In general, the status of Yellow-legged Gull in North America is without question the most obscure of all our rare gulls (and one of the most interesting). Its identification is gripping, with multifaceted challenges. One of these is the question of separating adult Yellow-legged Gull from the more expected Lesser Black-backed × American Herring hybrid. This hybrid is spread thinly but has rapidly increased along the entire Atlantic seaboard, E Great Lakes, and particularly in the Northeast. Another challenge is separating 1st-cycle Lesser Black-backed Gull and Yellow-legged Gull. This appears to be most relevant in the southern part of the continent, where larger concentrations of young Lesser Black-backeds are found, some of which are worn and faded, while showing advanced wing covert molt (e.g., in Florida and Texas). Ultimately, the question which overshadows this entire topic is: Which subspecies of Yellow-legged Gull should be used as an anchor? It is a seemingly simple question, but answering it is not easy.

Some of the earliest records of "Yellow-legged Gull" in North America come from a time when *michahellis* was a subspecies in the Herring Gull complex and subsequently conspecific with what is today known as Caspian Gull (*L. cachinnans*). Much has been learned since. In this account and in the account for Azores Gull (*L.m. atlantis*), I attempt to dissect some historic records and explicate the status of Yellow-legged Gull in North America. It would appear that many of the historic records I reviewed are in error or need further clarification (see this account's "Nonbreeding" section).

TAXONOMY Modern genetic studies position Yellow-legged Gull in a unique Atlantic-Mediterranean clade (Crochet et al. 2002; Yésou 2002). Until recently conspecific with Caspian Gull (*Larus cachinnans*), and before that both were part of larger Herring Gull complex (Collinson et al. 2008b). Polytypic. Two subspecies described, *michahellis* and *atlantis*, and a third "type," *lusitanius*, informally introduced by Olsen & Larsson (2004) and described in more detail by Adriaens et al. (2020). Within the complex, three broad groups are identified via preliminary genetic work: Mediterranean (nominate *michahellis*), Atlantic (*lusitanius*), and Macaronesian (Arizaga 2018). There are several points of interest that await clarification. The first concerns where to draw the western limits of nominate *michahellis* and the southern limits of *lusitanius*. Another is where to place populations from the Canary Islands, Madeira, and Morocco. Genetically, it appears the island populations from the Canary Islands and Madeira may be allied to Azores Gull *atlantis*. Phenotypically, these island populations have intermediate characteristics also shared with nominate *michahellis* and *lusitanius*, and overall they appear to be a population of intergrades (Peter Adriaens, pers. comm.). For this reason, they are not assigned a specific trinomial here. For more detail, see the study published in 2020 by Adriaens et al., which includes discussion on taxonomy.

L.m. michahellis (KGS: 5–7): Yellow-legged Gull. The largest subspecies and averages the palest. Most ubiquitous large white-headed gull of the Mediterranean region, extending east to the Black Sea region and west to parts of the Iberian Peninsula. Adults in the eastern part of the range average less black on the wingtip. A recent northern expansion of *michahellis* through interior Europe has resulted in well-documented hybridization events with Herring, Caspian (*Larus cachinnans*), and Lesser Black-backed Gulls (Adriaens et al. 2022).

L.m. atlantis (KGS: 7–9): Azores Gull. See that account.

L.m. "lusitanius" (KGS: 5–8): Sometimes referred to as Cantabrian Gull. A brief description was first provided by Joiris (1978) near Peniche, Portugal, and later endorsed as a valid taxon based on genetic work by Pons et al. (2004). No type specimen, however. Breeds along the Atlantic Coast of Iberia. Southern limits not entirely clear, although some authorities suggest as far south as the Algarve, in S Portugal, with perhaps a zone of intergradation with nominate *michahellis* east of Faro (Baggott 2022). Mostly sedentary, with approximately 70% of banded birds from one study found within about 30 miles of natal colony (Arizaga et al. 2011). However, banded birds have been sighted in England, France, the Netherlands, and Morocco, suggesting more mobility than expected (Adriaens et al. 2020).

Geographically, *lusitanius* is situated between the Azores Gull population to the west and nominate *michahellis* to the east. In theory, its more sedentary habits suggest it is the least likely to cross the Atlantic and wander to North America, but more long-term data is needed on its movements. As for identification, some overlap with nominate *michahellis* is expected in all ages, and some 1st cycles are not safely separated from Lesser Black-backed Gull. Therefore, vagrants to North America will almost certainly lose in a tie with Lesser Black-backed based on location or will defy subspecific identification based on current knowledge. Some workers in Europe will only label known-origin *lusitanius* (i.e., banded as chicks) away from their core range.

Adult: Lusitanius averages smaller and shorter legged than nominate *michahellis*, appears potbellied and shorter winged, and shows a slighter bill. Differences in races most pronounced between female *lusitanius* and male nominate *michahellis*. Variable gray upperparts. Some, from Cantabria and east, are paler, and gray tone overlaps with that of nominate *michahellis* (some virtually as pale as American Herring). Others, west of there and south into Portugal, are darker. Therefore, KGS values are of little use when used in isolation. Compared to nominate *michahellis*, head streaking more prominent in winter, with fine streaking on crown and around eyes, face, and nape. At times, a pseudo-hood is formed, recalling a weakly marked Azores Gull. The most-marked individuals have blotchy markings down to the lower neck forming a denser shawl (similar to American Herring's). Averages more black on bill tip in nonbreeding condition. Also has variable orbital color, some unmistakably orange. Mirror on p10, sometimes with all-white tip, and mirror on p9 is not rare but averages small. Commonly shows gray on base of outer web of p8, and p4 is typically unmarked. Differences in long calls documented, with *lusitanius* said to have higher pitch than nominate *michahellis* (Adriaens et al. 2020).

1st Cycle: Lusitanius has less extensive and later postjuvenile molt than nominate *michahellis* and averages less wear in late summer into the fall. Earlier in the season, the head is darker and more streaked, and the underparts as well may average darker than nominate's. The tailband averages broader with more proximal markings than that of nominate. The postjuvenile scapulars are variably patterned, commonly with a large, dark diamond on the feather center; are anchor patterned; or show a double-anchor with dark transverse bar along the shaft. Some *lusitanius* cannot be safely separated from similar-aged Lesser Black-backed Gull based on current knowledge, especially early in the season. Lesser Black-backed averages a more pear-shaped head with longer wing appearance, but variable. Similarities with Lesser Black-backed, in addition to size, include moderate head streaking, dark outer greater coverts, and overall plain upperparts and tertials with neat, pale edging. Average differences should be applied to perplexing birds, including full assessment of the upperwing and tail, as well as patterns on postjuvenile scapulars when present (see plates 23a.4 and 23a.5). The inner-primary window averages paler on *lusitanius*, but some overlap exists. Lesser Black-backed averages more earth-brown plumage aspect, while some *lusitanius* show subtle reddish cast to upperparts. Other 1st-cycle *lusitanius* can have a paler plumage aspect, resembling European Herring, with notched tertials, notched inner greater coverts, and prominent pale checkering on the wing panel. European Herring shows a noticeable inner-primary window at this age and often lacks the large, heart-shaped, brown spots that dot the uppertail coverts.

2nd Cycle: Lusitanius averages less adult-like gray in scapulars and coverts when compared to nominate *michahellis*, with an admixture of brown patterning throughout. The greater coverts show moderate to extensive barring. The lower neck and breast average more dark wash than those of nominate, sometimes with markings continuing onto the underparts. The inner primaries and tail pattern are similar to nominate's.

3rd Cycle: Difficult to ascertain at this age and may not be safely identified to subspecies, but size and structure may offer some indication of race. Pale-eyed; retains considerable head streaking, which can extend down the neck, breast, and sides; and shows an increased amount of black on the bill and duller yellow legs, which may be pinkish. Regularly has p10 mirror and more rarely may show small p9 mirror.

RANGE

Breeding (L.m. michahellis only): Breeds throughout the Mediterranean Sea, extending east to S Black Sea, west to at least Faro along the southern Iberian Peninsula, and north along the French Atlantic Coast and into parts of Central Europe. Birds breeding on Mediterranean coasts of N Africa are assigned to *michahellis*, but those on the Atlantic Coast of Morocco appear to be intergrades and need clarification. Nesting starts fairly early, typically at the beginning of April, in Mediterranean population.

Nonbreeding (North America only): Extralimital throughout the continent, with the exception of Newfoundland, where it is recorded almost annually. Exact status outside Newfoundland confused by similar-looking hybrids involving Lesser Black-backed × Herring Gulls. Interestingly, Yellow-legged Gull remains unrecorded in Greenland. Found almost annually in St. John's since the mid-1990s (first record in 1985). Often arrives in mid- to late August, with most being in heavy molt (Bruce Mactavish, pers. comm.). In "good" years, two to three are present, mainly September–October, with four being an apparent high count for the region (Oct. 2009). Found in fields and on small lakes. Some spend the winter around St. John's but can go missing for weeks and months on end. Less regular here from 2015 onward, perhaps suggesting some were returning birds that have perished or identification is now eluding observers. The overwhelming majority of sightings are of adults; these are followed in number of sightings by 3rd-cycle types. There are no confirmed sightings of 1st-cycle Yellow-legged Gulls in Newfoundland. Birds in Newfoundland traditionally believed to belong to the Azores race, *L.m. atlantis*. Although some do appear to be *atlantis*, a more rigorous approach is needed to objectively identify this gull in a vagrant context (see Azores Gull account).

Two accepted records are from Quebec, including the first for North America, from the Madeleine Islands (Aug. 1973). In error, Howell et al. (2014) state both individuals were from the Madeleine Islands, but the second record comes from La Malbaie, on the St. Lawrence River (June 2003). Both were assigned to *atlantis*, although after reviewing these records, I've found they need reevaluation. The Madeleine Islands bird was an adult, and the specimen has been well documented and studied by several workers, including myself (National Museum of Natural Sciences, 60750). Gosselin et al. (1986) initially surmised that it may represent a hybrid between Lesser Black-backed and Herring Gulls, but it was later reassigned to *atlantis* by Wilds & Czaplak (1994), who make a convincing case, especially given the head pattern and early primary molt in mid-August, and the bird does appear to be of Macaronesian ilk. However, p5 has black limited to the outer web, which should exclude Azores Gull (*atlantis*). Any vagrant Yellow-legged Gull (*sensu lato*) to North America should involve birds with classic features, including a complete subterminal band on p5. The second record from Quebec is of an adult with an all-white head and yellow legs. It was accepted without images of an open wing, which is problematic. Although it may be a Yellow-legged Gull, ruling out a hybrid Lesser Black-backed × Herring is not possible. Therefore, the "trend" of summering birds in Quebec suggested by Howell et al. (2014) cannot be considered part of a pattern—certainly not based on two contestable records documented 30 years apart.

An adult that returned to Washington, DC, for several winters in the early to mid-1990s is widely cited in the literature as a bona fide Yellow-legged Gull (Wilds et al. 1994). No subspecies was ever assigned. However, the original finder leaned toward *atlantis* in his eBird notes, although the photos do not positively support this claim (Czaplak 1990). Texas has two accepted records of 1st cycles (Nueces County, 2004),

with other potential candidates, although separation from Lesser Black-backed Gull is a formidable challenge here, and these records should be reevaluated as field-identification criteria develop (Martin Reid, pers. comm.). An adult with white head and yellow legs found in Hyannis, Massachusetts, in April 2011 was expected to be a state first record. It returned to the same beach the following December with surprising head and neck markings and pinkish-yellow legs. These features, along with a review of vocal recordings from April, supported a putative Lesser Black-backed × Herring Gull (Garvey & Iliff 2012). This individual is an important example of the challenges involved in eliminating this hybrid. Virginia's only accepted Yellow-legged Gull, reported as a 1st cycle, appears not to be that species (Brinkley & Patteson 2001). My review of photos revealed a 2nd-cycle bird, which in all likelihood is a Lesser Black-backed Gull. Therefore, the identification criteria used to support this record are moot. North Carolina's record of an adult shows a pale gray bird with upperparts similar to those of American Herring. This is troubling, especially considering the same site harbored several presumed Lesser Black-backed × Herring hybrids at the time (Lewis 1996). Although the record was endorsed by experienced observers from Europe, the photos do not lend themselves to objective analysis, and the paler upperparts alone make this a nonideal record. Several 1st-cycle Yellow-legged Gull candidates have been discovered in Florida in recent years, but the substantial variation shown by 1st-cycle Lesser Black-backed Gulls in this region, as well as a perpetual deluge of apparent hybrids, blocks reliable identification of younger birds given current knowledge. See Azores Gull account for more information on the status of Yellow-legged Gull in North America.

IDENTIFICATION (*L.M. MICHAHELLIS* ONLY)

Adult: Adult upperparts are generally darker than those of American Herring and decidedly paler than those of Lesser Black-backed Gull (approaching California Gull darkness). Body size is similar to Herring Gull's, averaging a deeper breast and more svelte rear. Somewhat blocky headed and longer legged, although variation among sexes quite evident. In breeding condition, the legs are a deep, saturated yellow, at times orange yellow; duller in nonbreeding condition, appearing straw yellow at their palest. Importantly, the head is commonly white in midwinter, but some show light streaking, mostly around the eyes and crown. Maximum head streaking is found in late summer through fall (Aug.–Nov.), although geographic variation exists, and some may show moderate head streaking in winter (i.e., *lusitanius*). Pale iris with red, at times orange-red, orbital. Stout yellow bill with something of a wider base. The bill is described as having an upside-down "butter-knife" shape, due to its even edges and noticeably blunt tip. The red gonys spot can be considerably large and sometimes intrudes onto the upper mandible. Some black markings, often reduced, are found around the gonys in nonbreeding condition.

The wingtip is predominated by black, with minimal white. A rather ordinary adult wingtip has much black coming up the bases of p8–p10, with little to no white on the tongue tips on p6–p7, and a broad subterminal band on p5. A thin white tongue tip can sometimes be found on p5, and very thin and subtle

---

1 Juvenile. Recalls Lesser Black-backed with rust-brown tones. Whiter underparts and head with distinct eye mask. Strong bill and often blocky-headed. Tertials and outer greater coverts have dark centers and plain pale fringes. Notched inner greater coverts. MARS MUUSSE. THE NETHERLANDS. AUG.

2 1st cycle. Superficially recalls something intermediate between Great Black-backed and Lesser Black-backed. Most scapulars replaced and, as expected, some upperwing coverts (dropped upper tertial). GERALD SEGELBACHER. ENGLAND. SEPT.

3 1st cycle with 2nd-cycle Lesser Black-backed (right). Mediterranean *michahellis* type with heavier proportions, considerable wear and fading, and many replaced coverts (1st alternate). MIGUEL RODRIGUEZ. SPAIN. FEB.

4 1st cycle. *Lusitanius*. Largely juvenile (not unusual at this date). Refined proportions, including shorter leg appearance. Approaches Lesser Black-backed, some with more extensive notching on (inner) greater coverts. Pale head with contrasting eye mask. Banded as chick in Basque Country. ASIER ALDALUR. SPAIN. SEPT.

5 1st cycle. *Lusitanius*. A bulkier individual with moderate wear. Pale head with eye mask. Postjuvenile scapulars often show double-anchor pattern with dark shaft bar. Banded on Berlengas. AMAR AYYASH. PORTUGAL. DEC.

6 1st cycles with Lesser Black-backed (right). Note overall larger head and deep breast. Similarly barred postjuvenile scapulars on Yellow-legged Gulls. AMAR AYYASH. PORTUGAL. DEC.

7 1st cycle with Lesser Black-backed (left) and Caspian Gull (center). Powerful bill, thickset legs, and large, blocky head suggest male. AMAR AYYASH. PORTUGAL. DEC.

8 1st-cycle nominate type (center) with Great Black-backed Gull (left). Rightmost bird problematic, likely Yellow-legged, given the large, strong bill and several replaced inner coverts, but ruling out Lesser Black-backed Gull difficult. AMAR AYYASH. PORTUGAL. DEC.

9   1st cycles. Yellow A4U banded as a chick in Mogan, Canary Islands (June 2010). Such birds resemble nominate *michahellis*, with large bodies and white underparts, but with more solidly dark upperparts and short-legged appearance. Tentatively designated as an intermediate population. XABIER REMIERZ. CANARY ISLANDS. MARCH.
10 1st cycles. Large, paler *michahellis* on the right with extensive postjuvenile molt. Smaller and darker individual perhaps not safely separated from Lesser Black-backed, but rufous tone to wing panel in line with Yellow-legged (assess open wing and tail). AMAR AYYASH. PORTUGAL. DEC.
11 Juvenile. Strikingly white tail base with reduced uppertail covert markings. Inner primaries form a subtle window due to pale, grayish inner webs. RICHARD BONSER. ENGLAND. AUG.
12 1st cycle. Very narrow tailband tapers on the edges, with much white on outermost tail feathers. A tinge of rufous to the upperwing. Notched inner greater coverts with plainer outer tract. Some postjuvenile scapulars. AMAR AYYASH. PORTUGAL. DEC.
13 1st cycle. All scapulars already replaced with many molted and dropped lesser and median coverts. White tail base, white head, and subtly pale inner webs to inner primaries. JOSH JONES. ENGLAND. SEPT.
14 1st cycle. Tawny-brown wings. Darker greater coverts and inner primaries compared to European Herring. The combination of white tail base, pale postjuvenile scapulars, white head, and paling bill point away from Lesser Black-backed. AMAR AYYASH. PORTUGAL. DEC.
15 1st cycle. Pale nominate type with a fair number of replaced lesser and median coverts. Multiple stress bars across primaries. AMAR AYYASH. PORTUGAL. DEC.
16 1st cycle. Same individual as 23a.5 (Yellow PNG). Approaching Lesser Black-backed, but reduced tailband and whiter tail base averaging a different look. AMAR AYYASH. PORTUGAL. DEC.
17 1st cycle. Presumed *lusitanius* type, which averages more compact wings and tail, and can show a relatively broad band. Note white face and lores, as well as pale fringes on wing coverts. Subtle inner-primary window shows grayish inner webs. AMAR AYYASH. PORTUGAL. DEC.
18 1st cycle. *Lusitanius*. Strikingly similar to Lesser Black-backed Gull, with much dark on the tail base. Although appreciably variable, average differences include paler regions to inner greater coverts and larger leopard spots on the perimeter of the uppertail coverts (sometimes distinctly heart-shaped). Banded on Berlengas (Yellow YPK). CARL BAGGOTT. PORTUGAL. OCT.
19 1st cycle. Quite difficult to rule out Lesser Black-backed, but tawny coloration to greater coverts, much pale coloration to postjuvenile scapulars, and tapered tailband favor Yellow-legged Gull. Some individuals will defy identification. AMAR AYYASH. PORTUGAL. DEC.
20 1st cycle. Same individual as 23a.5 and 23a.16. Relatively pale body, sides, and undertail with minimal markings. Variably brown wing linings and axillaries barred. AMAR AYYASH. PORTUGAL. DEC.
21 Juvenile. White tail base, brownish underwing coverts, and slightly paler inner webs to inner primaries. CHRIS GIBBINS. SPAIN. JUNE.

22 2nd cycle. Adult-like medium gray scapulars. White tail base, patterned inner greater coverts, and relatively clean head and underparts. Pinkish legs beginning to show yellow tinge. Bill pattern variable. PETER ADRIAENS. GEORGIA. DEC.
23 2nd cycle. *Lusitanius*. Short legs with delicate head and bill proportions, not uncommon in this subspecies, which can also look more potbellied. Densely streaked on neck and breast extends down to the belly. Banded as chick in Basque Country. ASIER ALDALUS. PORTUGAL. JAN.
24 2nd cycle with 1st-cycle Lesser Black-backed (left). Yellow-legged averages larger and blockier head with thicker legs. High contrast greater coverts. Paling bill base on this individual with dusky iris. AMAR AYYASH. PORTUGAL. DEC.
25 2nd cycle. Relatively pale inner primaries and moderately barred inner greater coverts. Largely white tail base. Medium gray scapulars show noticeable bluish hue. AMAR AYYASH. PORTUGAL. DEC.
26 2nd cycle. Darker inner primaries than 23a.25. Lesser Black-backed averages darker scapulars, broader tailband, with narrower and longer-winged look. AMAR AYYASH. ENGLAND. DEC.
27 3rd cycle. Paler than Lesser Black-backed and darker than American and European Herring (although may overlap with some nominates). Extensive black on bill not uncommon at this date. Dull yellow legs. MARS MUUSSE. THE NETHERLANDS. OCT.
28 3rd cycle. Lusitanius. Rather advanced upperparts. Finishing primary molt. Short legged; much duskiness around eye and face. Fleshy legs not rare. Banded as nestling on Berlengas (July 2017). CARL BAGGOTT. PORTUGAL. OCT.
29 3rd cycle. Much black on bill, dusky eye, and less advanced greater coverts. Consider slightly darker and "bluer" gray upperparts than European Herrings in background. MARS MUUSSE. THE NETHERLANDS. NOV.
30 3rd cycle. Entirely black outer primaries all the way to the primary coverts. Small p10 mirror and black on secondaries not uncommon. Clean white head and tail notable. AMAR AYYASH. ENGLAND. DEC.
31 3rd cycle. Much black on wingtip, variable pigment on tail, and dense lower-neck markings. AMAR AYYASH. PORTUGAL. DEC.
32 3rd-cycle type. Rather advanced individual with dirty primary coverts and alula. Marked p4. AMAR AYYASH. PORTUGAL. ENGLAND. DEC.
33 3rd cycle. Extensive black on wingtip and markings to p4. Dirty wing linings and streaked face. Dull yellow legs with black on bill tip. AMAR AYYASH. PORTUGAL. DEC.

white crescents out to p7. A large, rectangular-shaped mirror on p10 with complete black subterminal band is standard. Frequency of mirror on p9 is variable, found in about half of adults, although typically not large. It is not very rare for the subterminal band on p5 to be interrupted, and sometimes p4 shows a smaller mark on the outer web. The most marked birds form a plain black triangle pattern across the wingtip. Others can have black on the wingtip, appearing more concave with longer gray tongues on p7–p9, as some regional variation is known. Nominate *michahellis* in the eastern part of the range average less black, sometimes showing an all-white tip to p10 with a longer gray tongue, and more commonly a p9 mirror (Adriaens et al. 2020; see plate 23a.44). This population is geographically farthest from North America, with less likelihood of occurring in the Nearctic. In favorable lighting, the gray underside to the remiges can show appreciable contrast between the white wing linings and black wingtip, but this is quickly diminished in brighter lighting. *Similar Species:* In North America, the primary identification challenge is separation from Lesser Black-backed × American Herring hybrid. This hybrid is regularly reported throughout the Atlantic Coast and patchily on the Great Lakes, primarily from late fall through early spring. It is much more probable than Yellow-legged Gull, and it is essential to first and foremost eliminate this hybrid. Comparison of gray upperpart coloration is of limited use, since hybrids presumably present a range of grays that can approach that of either parent species. The three most helpful features appear to be head streaking, bare-part coloration, and wingtip pattern. Yellow-legged Gull is generally white headed throughout winter, whereas many presumed hybrids commonly show considerable dark streaking around the face, nape, and forneck at this time. A moderately to heavily streaked head and breast after November should favor a hybrid, provided other features align. Hybrids become white headed by late March and into April, rendering this feature inoperative. Leg color in the hybrid is likely more variable than presently known, but birds comfortably identified as Lesser Black-backed × American Herrings often show yellowish legs blemished with pink around the joints and webbing or pinkish legs with a yellow tinge throughout. In the most obvious cases, the result is an unevenly distributed sherbet-orange coloration. Adult Yellow-legged, even at its dullest, shows more uniform yellow legs. With that said, it should be expected that some hybrids will have all-pink or all-yellow legs, particularly in high breeding condition. If seen well, an obviously yellow-orange or orange (not orange-red) orbital may support a hybrid. Also, notable red bleeding onto the upper mandible is a Yellow-legged feature. The primary pattern on hybrids generally *averages* more-conspicuous white tongue tips on p6–p7 (and at times on p8), along with a thinner, or more diffuse, subterminal band on p5 (sometimes broken or limited to the outer web). Yellow-legged typically has a broad band on p5 and at times a spot on the outer web of p4. These features are variable and are not necessarily shown in all presumed hybrids, nor do they particularly exclude Yellow-legged Gull singly. Nominate *michahellis* generally completes primary molt in early October, whereas presumed hybrids are commonly found growing a few secondaries and p9–p10 into late November. Some putative hybrids have even been found growing p10 in late January (Ayyash, pers. obs.). As usual, the limited use of molt with vagrants should be taken into account. See account for Azores Gull, which has different identification criteria. See also account for European Herring Gull, which sometimes has yellow legs and can show red orbital rings (specifically nominate *argentatus*). Wingtip pattern differs, as well as average head streaking and hue to upperparts.

In a comparison of museum skins, American Herring and nominate *michahellis* showed surprisingly similar gray values, although the latter is almost always noticeably darker, possessing a matte gray coloration without the silvery sheen of American Herring. With that said, some adult American Herrings display noticeably yellow-tinged legs (primarily in the East), with a sickly blue-gray coloration around the joints (not as intense as that found in some subadult California Gulls). The leg color on these birds, even at its maximum, does not compare to the genuinely yellow legs of Yellow-legged Gull. American Herring has denser and more extensive lower neck markings (although see plate 23a.36 of subspecies *lusitanius*), has sharp black wedges on the outer webs of p6–p7, and does not have a red orbital.

34 Adults. Medium-gray upperparts, square-headed at times, with thick, blunt-tipped bill. Relatively large red gonys spot (bleeding onto upper mandible on right individual). By this date, very little head streaking. CARL BAGGOTT. PORTUGAL. NOV.
35 Adult. Clean white head and vivid yellow bare parts in winter. Red orbital. AMAR AYYASH. PORTUGAL. DEC.
36 Adult. Some show head streaking in winter, particularly in *lusitanius* types (may be even denser). Faint pinkish hue may also be found on legs in this subspecies. Separation from American Herring × Lesser Black-backed hybrids not trivial, although hybrids average more pink on legs and less black on wingtip. AMAR AYYASH. PORTUGAL. DEC.
37 Adult with 3rd-cycle-type Lesser Black-backed (left). Decidedly paler than Lesser Black-backed, often with a thicker bill, although there is some overlap. AMAR AYYASH. PORTUGAL. DEC.
38 Adult. Large p10 mirror and broad subterminal band on p5 are commonly found. Rather extensive black on wingtip. AMAR AYYASH. PORTUGAL. DEC.

1st Cycle: Juvenile upperparts are a variable mid-brown with rufous tones. The scapulars and coverts are neatly fringed with white appearing scalloped when fresh. The head and neck show dusky streaking, with lightly mottled brown underparts. Powerful, all-black bill may give way to some light paling around the base, but usually remains dark. Pale pink legs. The greater coverts, especially the outers, can show extensive dark bases and centers but become more barred along the inner tract. The tertials are typically plain with thin, pale edging.

Juvenile plumage is retained only for a short period, as a somewhat rapid postjuvenile molt soon replaces most, if not all, scapulars in many individuals. A variable number of wing coverts are commonly replaced, and even tertials in most-advanced birds. The head, neck, and underparts become largely pale with an obvious white ground color. Fine, light streaking remains on the back of the head and neck, and it can be boldly marked on the lower neck. A contrasting, dark eye mask persists in many. Appreciable variation is found in the postjuvenile scapulars, with some showing broad brown diamonds and others more buffy, predominated by white with thin anchors (somewhat recalling Great Black-backed Gull). On the open wing, a faint inner-primary window is sometimes found, in part due to the inner webs being buffy gray with darker brown outer webs. The blackish secondary bar contrasts against paler greater coverts. The focal points in flight are a white rump, largely white uppertail coverts with light patterning, and a narrow black tailband. The outer edges of the tailband are tapered, and the tail base is very often solid white. The underwing coverts are relatively dark, with noticeable grayish-brown barring throughout. ***Similar Species:*** The question of which population of Yellow-legged Gull to expect in North America is paramount when discussing this age group. Currently, there is no data on 1st cycles to suggest a pattern for subspecies. Although birds from Newfoundland are assigned to Azores Gull, no 1st cycles have been recorded there to bolster this hypothesis. This is a curious phenomenon, but it appears to hold in parts of Europe as well. Therefore, observers should not limit their search image of this vagrant to Azores Gull (discussed at length in that account). Lesser Black-backed Gull is the primary confusion species at this age. Compared to Yellow-legged Gull, Lesser Black-backed Gull averages smaller and lankier, appears more svelte throughout its hind parts, and has a shallower bill tip and a more pear-shaped head. These size and structural features are highly variable, and overlap should be expected. On the whole, 1st-cycle Lesser Black-backed Gull retains more streaking on the head, face, and neck with darker underparts compared to Yellow-legged. It has a deeper earth-brown plumage aspect, with more uniformly filled upperparts than Yellow-legged. Yellow-legged can appear more dilute, sometimes creamier, other times tawnier with rust-colored tones, and commonly shows larger pale regions on its postjuvenile scapulars and coverts. Yellow-legged is expected to show a narrower tailband that is wedge shaped, with a clean white tail base. In Lesser Black-backed, more patterning is found on the proximal edge of the tail, with more-marked outermost tail feathers and tail base, although variable. Molt is helpful, especially in late summer and fall. Advanced scapular molt and / or replaced coverts and tertials should be pro-Yellow-legged Gull features, although use them cautiously, as a few 1st-cycle Lesser Black-backeds may show advanced molt by mid- to late October. Furthermore, the reliability of molt in vagrants limits this to a supporting feature at best. Identification in the south (Florida and Texas in particular) is most troublesome, as 1st-cycle Lesser Black-backeds wintering there can look considerably different from those in the Northeast and Great Lakes, inviting confusion with Yellow-legged. This is mostly the case in mid- to late winter, when the seasonal effects of wear and bleaching have settled in, with many actively molting upperparts (Ayyash, pers. obs.). The samples of Lesser Black-backed Gulls found wintering in Florida (100 or more 1st cycles on a good day) reveal some of the scope of variation found in this taxon, much of which has yet to be popularized in North American birding literature. First-cycle Lesser Black-backeds with subdued, dark gray postjuvenile scapulars are rather straightforward, but some also show paler feathers which overlap with Yellow-legged and furthermore can have somewhat extensive postjuvenile molts. Yellow-legged should not show plain, dark gray postjuvenile scapulars. The other layer at work here is the possibility of some of these 1st cycles being hybrids with American Herring Gull: this age group with this combination remains poorly understood.

Great Black-backed Gull superficially resembles some larger Yellow-legged Gulls from the Mediterranean, in that both share paler plumage aspects, thick black bills, and whitish heads. Furthermore, some Yellow-legged Gulls with extensive postjuvenile molt (i.e., 1st prealternate) can acquire predominantly white upperparts with thin brown barring that further resembles Great Black-backed. Great Black-backed does not show such a postjuvenile molt in the coverts and tertials. The juvenile greater coverts are more purposely barred in Great Black-backed, and the tertials average more notching with paler tips. Compare also the weaker tailband of Great Black-backed, which averages more patterning on both distal and proximal edges. The bill is stouter on Great Black-backed, which also has a larger and flatter head, is more thickset behind the legs, and has a shorter wing projection. See also Azores and European Herring Gulls accounts.

2nd Cycle: Largely white headed, although with visible streaking on head and nape. The upperparts are barred brown and white in 2nd basic plumage. The greater coverts are usually darker and contrast with the rest of the wing panel, showing internal barring, especially on the inner tract. Lesser and median coverts more boldly marked, almost appearing spotted. Second alternate plumage can have entirely adult-like gray scapulars, as well as a variable number of advanced gray coverts (and upper tertials), although it is more typical to find a mixture of brown-pattern scapulars combined with gray feathers. Bill pattern variable, from entirely smudged black to adult-like yellow base with black tip, although typically with obvious black persisting. Iris variable, from all dark to some paling and others strikingly pale. On average, legs remain a dull, pink color in fall and winter, often becoming yellowish by late winter into spring. On the open wing, the darker secondaries and darker outer primaries are interrupted by a small,

39 Adult. Similar to 23a.38 with much black on wingtip, although black on outer web of p8 not reaching primary coverts. Note extensive black on underside of p10 (far wing). AMAR AYYASH. ENGLAND. DEC.
40 Subadult. Small p9 mirror. Aged by obvious brown wash on primary coverts, with diffuse pigment on proximal edges of black on wingtip. Black on bill and "dirty" head likely age-related too. AMAR AYYASH. PORTUGAL. DEC.
41 Adult. Reduced pigment on p7–p9, thinner black subterminal band on p10, and broken p5 band. Note some red bleeding onto the upper mandible. AMAR AYYASH. PORTUGAL. DEC.
42 Adult. Similar to 23a.41, but with small p9 mirror and thin white tongue tips to p7. Unknown origins, although such wingtips with reduced black, especially on p8, are more common in eastern *michahellis*. PETER ADRIAENS. FRANCE. DEC.

contrasting inner-primary window, subdued in some birds. The uppertail coverts are predominantly white with a narrow black tailband. The underwing shows some dark edging on the wing linings and is patchily marked, although not solidly dark. *Similar Species:* At this age, Lesser Black-backed × American Herring hybrids might be expected to resemble nominate *michahellis* and Yellow-legged populations from Atlantic Portugal and N Spain (*lusitanius*). Green F02, a known-origin hybrid from Appledore Island, offspring of Green F07 (see plates H7.9–H7.10), is our only example of this hybrid in 2nd cycle and showed medium gray upperparts and wing covert patterns that cannot be safely distinguished from those of Yellow-legged Gull. The bill pattern and leg color also matched what is found in many 2nd-cycle Yellow-legged Gulls. However, this hybrid averaged paler inner primaries and more barring on the upper- and undertail coverts and showed a dark tail base. It had a shorter bill, rounder head, and shorter legs, appearing much slighter when compared to typical *michahellis* (although not unlike the gestalt of some *lusitanius*). The breast and sides were heavily washed and marked with brown blotching throughout the winter (mid-Dec. through Feb.). Nominate *michahellis* typically has whiter underparts at this time of year, although such underparts are not unexpected in some *lusitanius*. A larger sample of known-origin 1st- and 2nd-cycle hybrids is required before we can form a reliable impression of how to separate them from Yellow-legged Gull.

Separation from American Herring can usually be accomplished through darkness of gray upperparts. By late fall, most 2nd-cycle Yellow-legged Gulls already show considerable gray on the upperparts, which should assist in direct comparisons. But based on comparisons of museum specimens, some *michahellis* (presumably nominate) can have strikingly similar gray upperparts to American Herring (Ayyash, pers. obs.). Second-cycle American Herrings sometimes have an added tinge of gray brown to their scapulars, and this makes for a darker effect. Also, recall that some American Herrings can show a yellow tinge to the legs (particularly Atlantic Coast populations), which may invite confusion. American Herring shows paler inner primaries and a darker tail base, and some have sharply bicolored bills with a bright pink base, which is unexpected in Yellow-legged. Also compare head and neck patterns: American Herring often has a denser pattern, especially on the lower hindneck. Lesser Black-backed averages darker gray on the upperparts, although some retarded birds at this age are without gray and can cause confusion, especially with large, robust male types. Compare patterns on upperwing coverts and underparts. See also European Herring and Azores Gulls accounts.

3rd Cycle: Head largely white, with fine streaking, often concentrated around the eyes and face. Pale eyes in most, although some can have dusky coloration. Dull yellow legs, brighter in others, with a small percentage still showing dull, pink-colored legs at this age. The bill is typically dull yellow with variable black around the tip, some with extensive black on base. Adult-like gray upperparts predominate in a fair number of individuals, with less advanced birds showing brown wash across wing and bolder, almost spotted, dark markings on the lesser and median coverts. Commonly shows patches or vermiculated markings on tertials. Smaller white apicals on the outer primaries. On the open wing, extensive black is found to the bases of p7–p10, virtually reaching the primary coverts. Mirror on p10 is relatively small, sometimes absent, and a small mirror on the inner web of p9 is rare. Black commonly found down to p4. Blackish-brown markings on the primary coverts and alula are common, as well as on the secondaries in the most delayed individuals (recalling American Herring). The tail feathers regularly show weaker black markings, at times forming a diffuse tailband; entirely white tail in others. The underwing is mostly white, although some individuals sporadically show faint brown markings throughout, at times extending to the edges of the axillaries. *Similar Species:* Separated from American Herring by darker upperparts and leg color. Lesser Black-backed shows darker upperparts and is noticeably more streaked on the head and neck (in basic plumage). The main identification hurdle would be a Lesser Black-backed × American Herring hybrid. A hybrid would presumably average more head and neck streaking in basic plumage, with an admixture of pink and yellow on the legs. More study needed. California Gull may appear similar in gray upperpart coloration and yellowish legs, but this species is completely dark eyed, with a more tubular bill. See also European Herring and Azores Gulls accounts.

MOLT The partial inserted molt in 1st cycle is attributed to a prealternate molt, which includes the scapulars, head, neck, underparts, and very commonly a variable number of upperwing coverts and tertials n nominate *michahellis*. A somewhat rapid postjuvenile molt begins in late July but slows by November. Subspecies *lusitanius* has a less extensive and later postjuvenile molt than nominate. Some *lusitanius*, from the Berlenga Islands, for example, still have young in the nest in late August. It is not rare to find pristine juvenile *lusitanius* in September–October while some nominates already show whitish heads and have molted most scapulars and a variable number of wing coverts. Some 1st-cycle nominates have postjuvenile molt limited to most of the scapulars but average more wear and fading than *lusitanius*, due to earlier hatching. By late November, *lusitanius* "catches up" to nominate with most scapulars replaced, but wing covert and tertial molt much more limited. Molt differences in 1st-cycle nominate and *lusitanius* show some statistically significant differences, but not diagnostic, especially with out-of-range birds. Recently, a known-origin 1st-cycle *lusitanius* was documented with roughly half of its wing coverts and upper tertials replaced in early December (Baggott 2022). Although this appears to be the exception rather than the rule, without a leg band, this individual would have easily been written off as a nominate. Interestingly, some nominates appear to molt (upper) scapulars a second time well before the 2nd prebasic molt—as early as November in some—suggesting a 2nd inserted molt in some 1st cycles (e.g., Martin et al. 2013). In adult nominates from the Mediterranean (i.e., southern birds), the prebasic molt can conclude in October, and in others in November. In *lusitanius*, which breeds on average one month later, adults complete primary molt in late November and into December (Adriaens et al. 2020). However, there is considerable overlap in primary molt timing in adults of both subspecies (Baggott 2022).

HYBRIDS Well-documented and ongoing hybridization throughout various parts of Europe with Caspian (*Larus cachinnans*), European Herring, and Lesser Black-backed Gulls (Adriaens et al. 2012; Olsen 2018; Litwiniak et al. 2021). Confirmed hybridization with Great Black-backed Gull (male) from coastal Portugal. The pair produced three chicks in 2004 but failed in subsequent breeding attempts from 2005 to 2009 (Goncalves 2010). Also reported with Cape Gull (*L.d. vetula*) in S Morocco.

43 Adult. Two mirrors and marked p4. Upperparts similar to what is expected on American Herring × Lesser Black-backed, which averages more head streaking in winter, more gray on inner webs of p8–p9, and more noticeable white tongue tips to p7, although variable. JOSH JONES. ENGLAND. FEB.

44 Adult. Somewhat rare all-white tip to p10, with long tongue on that primary (underside). Small p9 mirror, moderate black on p8–p9, complete band on p5, white head, and red intruding onto upper mandible point away from *argentatus* with yellow legs. SERGEY ELISEEV. TURKEY. DEC.

45 Adult. Extensive black on underside of p10 not uncommon. Rather distinct black triangle on wingtip with moderate black on p8–p9. AMAR AYYASH. PORTUGAL. DEC.

23B AZORES GULL *Larus michahellis atlantis*

L: 20"–23" (52–58 cm) | W: 45"–53" (115–135 cm) | Four-cycle | SAS | KGS: 7–9

OVERVIEW A dark, four-cycle gull with yellow legs, endemic to the Azores, located roughly 1,000 miles west of mainland Portugal. Jonathan Dwight (1922) was the first to designate this taxon and chose the name *atlantis* in reference to the fabled island from the Atlantic Ocean. On the islands, Azores Gulls are often found in open fields and pastures but also frequent landfills, beaches, and harbors. This subspecies averages darkest and smallest of the Yellow-legged Gulls and is the most distinct race in the Yellow-legged Gull complex. Importantly, adults have extensive black on the wingtip, which is a critical feature to assess with vagrants. A number of potential vagrants are first spotted by the neatly streaked "hood" effect on their heads, but this feature is shown only for a short period in late summer through fall, and it may be mimicked by other populations of Yellow-legged Gull, particularly those from S Macaronesia (Dubois 2001).

In North America, one to two Yellow-legged Gulls are recorded almost annually in St. John's, Newfoundland, and these, believed to be from the Azores, have traditionally been assigned to *L.m. atlantis*. Although some do appear to be *atlantis*, a more rigorous approach is needed to objectively identify Azores Gull in a vagrant context.

TAXONOMY Type is from Fayal Island, Azores. Also known as Azorean Yellow-legged Gull. After reviewing a small series from the Azores and Canary Islands, Dwight (1922) referred these birds to *Larus fuscus atlantis*. The treatment of *atlantis* as a pale "Lesser Black-backed Gull" persisted for some 40 years before it was lumped with the Herring Gull complex (Vaurie 1965). Subsequently, it was placed in a southern *cachinnans* group of yellow-legged taxa, until recent genetic data revealed that *atlantis* and *michahellis* were closely allied and part of an Atlantic-Mediterranean clade (Crochet et al. 2002; Liebers et al. 2004). Now treated as a subspecies of Yellow-legged Gull, in which it appears to be the primitive form (Sangster et al. 2005; Harrison et al. 2021).

Which populations make up *atlantis* has seldom been agreed, from the inception of this taxon (Stoddart & McInerny 2017). For varying reasons, some authors have included birds from the S Macaronesian islands (i.e., Canary Islands and Madeira), and preliminary genetic data suggests these populations may in fact make up a larger Macaronesian group with birds from the Azores (Arizaga 2018). As stated in the Yellow-legged Gull account, these southern populations are rather intermediate and appear to be intergrades (Adriaens et al. 2020), and preliminary banding studies suggest they are largely resident to those islands (Remirez et al. 2023). Therefore, this guide restricts *atlantis* to the Azores for the time being, with the admission that there is no simple placement for birds from S Macaronesia (see also this account's "Nonbreeding" section).

RANGE
Breeding: Restricted to the Azores. Nesting begins in mid- to late April.
Nonbreeding: Largely resident (but some slight movement between local islands). Vagrants have been recorded in Spain, Great Britain, Ireland, Iceland, and Newfoundland (Stoddart & McInerny 2017; Adriaens et al. 2020). Notably, Greenland has no records. Ireland and Newfoundland have a disproportionate number of sightings, perhaps pertaining to returning birds that have completely forsaken the Azores. Habits of vagrant Azores Gulls are poorly known, but there is some slight evidence to suggest 1st cycles are inclined to follow ships out at sea. This is not much of a surprise to anyone who has ever been on a deep-water pelagic with gulls in tow. Colm C. Moore (1996) documented five 1st cycles attending to ships that he was on during transect surveys in Macaronesia in August of 1987–1994. There is no way to know if these birds actually originated from the Azores, especially two of them which were initially sighted south of Madeira. Distances traveled range between 100 and 1,120 kilometers (62–695 miles), with a mean of 718 kilometers (446 miles). Moore specifically noted adults and subadults being scarce away from land. Interestingly, records of vagrant 1st cycles are the rarest, with only a single individual, from Iceland, known outside Macaronesia as of this writing.

In North America, records of Azores Gull (*sensu stricto*) and Yellow-legged Gull (*sensu lato*) are obscured for several reasons. When *michahellis* and *atlantis* were first suspected in the New World, they were both smaller units in a larger complex of white-headed gulls. There was very little nuanced literature available to help distinguish these forms or to assist in eliminating hybrids. The sharing of photographs was not by any means easily achieved, and many photos through the 1980s produced grainy images that distorted critical details. Grant's (1986) monumental gull guide devoted no more than 75 words to *atlantis*, with only one black-and-white photo provided. That single photo, interestingly, is from Madeira and not the Azores. Different geographic boundaries drawn for *atlantis* have in turn resulted in varying ideas of what this taxon looks like. For example, Howell and Dunn (2007) and Olsen and Larsson (2004) included birds from the Canary Islands and Madeira, despite the difficulties in safely separating many of them from Portuguese and Mediterranean populations. These populations are better off being cataloged as *michahellis* subspecies. More recent workers have reserved *atlantis* strictly for birds on the Azores (Olsen 2018; Adriaens et al. 2020; Harrison et al. 2021). As stated earlier, this guide also restricts *atlantis* to the Azores and relies on rigid criteria for diagnosing vagrants (outlined by Adriaens et al. 2020). This account's "Identification" section discusses some of these subtleties.

IDENTIFICATION

Adult: Dark gray upperparts with an ashy-gray quality. Bright yellow legs in high breeding condition. Appears slightly smaller than nominate *michahellis*, with relatively short legs and compact stature. Somewhat blocky headed. Piercing pale yellow eyes, sometimes whitish. Red orbital. The yellow bill is stout and evenly deep, with a blunt tip. The red gonys spot is large and oval shaped and sometimes bleeds onto the upper mandible. In nonbreeding condition, black markings are found around gonys, commonly on the proximal

1 Juvenile. Dark, chocolate-brown plumage recalls American Herring or even Western Gull. Dark face and strong black bill; often appears short-legged. RICHARD BONSER. AZORES. AUG.
2 1st cycle. Dark face and underparts with overall sooty aspect. Bulkier and more swarthy than Lesser Black-backed. Pale notching on upperparts soon gives way to wear, but some do show white regions on greater coverts. Banded on Azores. RICHARD BONSER. AZORES. AUG.
3 1st cycle. Scapulars replaced (1st alternate). Dense brownish-gray neck pattern appears gritty. Mottled underparts, often with distinctive "zebra" barring on flanks above the legs. Tawny wing panel with worn fringes. Dark shins. AMAR AYYASH. AZORES. DEC.
4 1st cycle. Not uncommon for postjuvenile molt to include a number of wing coverts and some tertials. Overall appearance lacks uniformity, with variable patterns on the same generation of feathers. White tail base. AMAR AYYASH. AZORES. DEC.

edge of the red spot and back along the cutting edge and onto the nostrils. The legs become a duller yellow, with some showing remnant shin markings on the tarsi. From August through October / November, a strikingly dark head pattern emerges in many. Heavy streaking forms a semi-hooded appearance not extending to the lower neck or breast. In some birds (in part due to posture), the markings are restricted to the face and crown, with a cleaner white nape, imparting a masked look. Others have weaker head streaking and don't show dark hood effect. Many are white headed by mid- to late November, remaining so through July. Examination of museum specimens shows that a small percentage of adults may retain faint streaking around the lores into December, although this is difficult to detect at a distance (Ayyash, pers. obs.).

In flight, the wing appears somewhat broad and relatively short. The impression is that of a compact "large" gull. The typical pattern on p10 is a moderate-sized to large rectangular mirror, with a complete black

5   1st cycle. An average appearance, with wing coverts and tertials showing typical plain, chestnut-brown coloration. Dark face and dense neck pattern persist. Barred flanks above legs. CHRIS GIBBINS. AZORES. FEB.
6   1st cycle. Dark face with obvious streaking on chin. Heavily mottled underparts. Legs not as pink now, but still showing obvious dark shin markings. Whitish undertail coverts with sparse spotting. CHRIS GIBBINS. AZORES. FEB.
7   1st cycle. Rather classic sooty brown individual that was clinched with additional open-wing photos. As of this publication, this is the northernmost-accepted record of a vagrant Azores Gull. INGVAR ATLI SIGURDSSON. SOUDURLAND, ICELAND. MAY.
8   1st-cycle duo in front with extensive postjuvenile molts. Note overall dark brown plumage recalls a swarthy Lesser Black-backed. 2nd-cycle Great Black-backed (back left) and 2nd-cycle Azores Gull (back right). AMAR AYYASH. AZORES. DEC.
9   Juvenile. Dark, chestnut-brown plumage with uniformly dark inner primaries. Broad tailband with contrasting white tail base. RICHARD BONSER. AZORES. AUG.
10 Juvenile. Slightly paler fringes on wing coverts, not rare in fresh juveniles, but soon diminished by wear. Broad tailband distinctive, showing moderate pattern on outer tail feathers. RICHARD BONSER. AZORES. AUG.
11 1st cycle. All scapulars and many wing coverts replaced. Slight paling to base of lower mandible not rare. Bold barring on uppertail coverts with plain white outer tail feathers on this individual. AMAR AYYASH. AZORES. DEC.
12 1st cycle. Dark inner primaries and oiled-brown appearance to upperwing. Postjuvenile scapulars not uniformly patterned. Broad tailband. AMAR AYYASH. AZORES. DEC.
13 1st cycle. Noticeably dark underwing coverts, and at times, faint barring on axillaries. Dark face, mottled underparts, with faint "zebra" barring on rear flanks, and white undertail coverts showing sparse markings. AMAR AYYASH. AZORES. DEC.

subterminal band, and an absent or short gray tongue on the inner web (best viewed from the underside of this primary). Only about 5% of adults have a p9 mirror, usually restricted to the inner web. Markings on p4 are very common: typically a noticeable spot on the outer web but sometimes with a thin band. A small percentage may also show a black mark on p3, which is expected in this subspecies and extremely rare in nominate *michahellis* (Olsen & Larsson 2004; Adriaens et al. 2020; Alex Boldrini, pers. comm.). Overall, the black on the outermost primaries runs deep toward the primary coverts. The base to the outer web of p9 is generally all black. At times, the base to the outer web of p8 is also entirely black, but variable: some with black only on outer edge of outer web with small gray wedge here, and others with black not reaching primary coverts (see discussion in following section, "Similar Species"). Interestingly, p5–p7 can sometimes resemble those of American Herring, with complete W-band on p5 and sharp black wedges on the outer webs of p6–p7. However, no noticeable white tongue tips are seen on p6–p7 in Azores Gull (thin white slivers are sometimes apparent upon close inspection, especially on the inner webs). A thin white tongue tip is often seen on the inner web of p5. On the underwing, the gray remiges show appreciable contrast between the white wing linings and black wingtip. The underside to p10 is largely black with an absent or short gray tongue found on the inner web. ***Similar Species:*** In accordance with Adriaens et al. (2020), only about one-third of adults can be objectively identified in a vagrant context. To summarize, the following features should all be present on the wingtip: absent or short tongue on p10, no p9 mirror, completely black base to the outer web of p8, and some black on p4. This is not to say birds not meeting these strict criteria are not Azores Gulls, but rather, they cannot be positively assigned. Such birds should of course show suitably dark upperparts and in late summer through early fall (basic plumage) should show the characteristic streaked hood pattern. The head should be white or nearly all white by late fall and into winter. Caution must be taken, however, as the streaked hood pattern that has been used to typify Azores Gull is not unique to this taxon. See other Macaronesian populations for comparison (e.g., plate 23b.27). Furthermore, any adult with a broken p5 band or an all-white tip to p10 should not be identified as an Azores Gull in a vagrant context.

Nominate *michahellis* has paler gray upperparts (some only marginally darker than those of American Herring), averages a larger body and longer legs, and shows less black on the outermost primaries. It is much more likely to show a p9 mirror and averages less black on the bases of p8–p9. Azores has a shorter gray tongue on p10 (see underside) and is more likely to have a complete black base to the outer web of p8 and a black mark on p4. When head streaking is present, Azores shows distinctive hood appearance with markings evenly extending onto the lores and throat. Compared to Lesser Black-backed, Azores has paler upperparts, although extremes at both ends can be surprisingly similar and perplexing on lone birds. Azores appears more squat, with a deeper breast, and has a broader wing and stouter bill. On the open wing, Lesser Black-backed will show less contrast between black wingtip and gray upperparts and greater contrast between gray upperparts and white trailing edge. Extent and pattern of head streaking are considerably different.

The most substantial identification pitfall is Lesser Black-backed × American Herring hybrids. Many of the same provisions that apply to nominate *michahellis* apply here (see adult "Similar Species" section

---

14 2nd cycle. Basic plumage overall cryptic-brown, dark face, and pale iris. Greater coverts largely unpatterned. RICHARD BONSER. AZORES. AUG.
15 2nd cycle. Somewhat advanced bill pattern, paling iris, and red orbital. Demarcated "hood" effect and isolated dark belly patch noteworthy. Yellowish legs with faint shin markings on this individual. Dark underwing. CHRIS GIBBINS. AZORES. FEB.
16 2nd cycle. Adult-like scapulars (alternate), white neck and breast, and largely plain wing. The streaked "hood" and solid-brown greater coverts are two key features found in this age group. CHRIS GIBBINS. AZORES. FEB.
17 2nd cycle. Not identified to subspecies. Similar to Azores Gull in many respects; it is for this reason that out-of-range 2nd cycles can be objectively identified only with sharply demarcated streaked hoods with unmarked neck, isolated belly patch, and, most importantly, plain brown greater coverts. MARIN ROOME. CANARY ISLANDS. DEC.
18 2nd cycle. Strongly streaked "hood" includes lores and chin. The finely barred greater coverts are not rare in 2nd cycles, but would lose points in a vagrant context. CHRIS GIBBINS. AZORES. FEB.
19 2nd cycle. Greater coverts adult-like gray (2nd alternate) and thus not possible to interpret 2nd basic pattern. Short legs, streaked "hood," and dark shin markings typical of Azores Gull. JOSH JONES. AZORES. FEB.
20 2nd cycle. Fresh 2nd basic individual with mostly solid brown upperwing, marked uppertail coverts, dark bill, and very heavily marked neck. Note plain greater coverts. RICHARD BONSER. AZORES. AUG.
21 2nd cycle. The combination of sharply demarcated "hood," isolated belly patch, and plain greater coverts make this a diagnosable individual from a vagrancy context. AMAR AYYASH. AZORES. DEC.

in Yellow-legged Gull account). Consider timing and extent of head streaking, timing of primary molt (averages earlier completion in Azores), bare-part coloration, and more importantly wingtip patterns (see Lesser Black-backed × American Herring hybrid account). Hybrids are likely to show noticeable white tongue tips on p6–p7, at times also on p8, which are not shown by Azores. Also, hybrids much more likely to show mirror on p9, less black on p5, and unmarked p4. The tongues on p8–p9 average longer on hybrids (more gray and less black on the inner webs). Structurally, Azores averages more squat, with a squarer head, steeper forehead, and more blunt-tipped bill, although variable, and structural features are not likely to tip the scale in this identification.

1st Cycle: Fresh juveniles have a rich, dark brown plumage aspect, with uniform sooty brown head, neck, and underparts. Some weak notching can be found on the juvenile scapulars, and at times modest barring on the inner greater coverts, with light to moderate pale edging throughout. The eyes are dark, but a small percentage of individuals show faint paling in the iris by midwinter. The legs often appear dingy and commonly show dark bars on the tarsi (so-called shin pads). The bill remains largely black, with little paling around the base of the lower mandible. In short order, a somewhat disorganized look emerges on the back, due to the irregular patterns of the postjuvenile scapulars (often by late Oct.). At this time, 1st-cycle Azores Gulls become most distinct. The juvenile wing coverts and tertials are particularly prone to fading, with few if any discernible pale fringes on these feathers by midwinter. The back of the head and neck becomes slightly paler but overall retains coarse streaking, with a dark, contrasting face mask. A noticeable percentage of 1st cycles show a contrasting, dark earspot. The head and neck have a somewhat gritty and grimy appearance. A variable number of upperwing coverts may be replaced, and tertials in advanced birds (typically limited to a few upper tertials). The underparts and sides are mottled brown, and, importantly, distinct "zebra" barring is commonly found on the flanks just above the legs.

In flight, 1st cycles can appear even more broad winged than adults. The remiges are uniformly dark brown, and there is typically no pale inner-primary window (faint in some). The black tailband is strikingly wide, contrasting with a bright white base color to the uppertail coverts. Variable barring and speckling are found along the proximal edge of the tailband and along the outer edge of the outermost tail feathers. Some are plain white here, others showing a series of short bars, although typically not reaching the central region or base of the tail. The undertail coverts are white and variably marked: some are strongly barred, and others have a sparser pattern. The underwing coverts are overall solid dark brown with faint, intermittent barring. Often, the greater underwing coverts show a distinct sooty-gray coloration, which contrasts with the browner lesser and median wing linings. **Similar Species:** Only about half of vagrant 1st-cycle Azores Gulls are likely to show the full suite of features to be diagnosed (Adriaens et al. 2020). In North America, compare to Lesser Black-backed Gull. Azores appears more squat, with a deeper breast, and averages a stouter bill and chunkier head. On the perched bird, the wing projection is noticeably shorter, and the open wing appears broader on Azores. Despite all of these structural differences, overlap exists, and structure should only be used to support a bird that has correct plumage features. The plumage aspect on Azores at this age is made up of uneven burnt-brown tones. Azores has darker lores with a grimy head pattern and a chestnut-brown coloration to the wing coverts, which average less pale edging than those of Lesser Black-backed. The postjuvenile scapulars are less uniform on Azores, lacking the neater anchor pattern found on Lesser Black-backed. When found, the characteristic zigzag, zebra barring on the flanks and the dark shin pads should support Azores, although dark shin markings may also be found on some Lesser Black-backeds.

Separation from nominate *michahellis* types is rather straightforward in most cases. Azores averages darker overall, with browner head and neck, darker underparts, darker underwing coverts, plainer wing coverts, and a broader tailband. Nominate *michahellis* averages a neater and more regular pattern to its postjuvenile scapulars, with obvious barring, anchors, and diamonds set on pale feathers. The appearance of Azores is a variety of disorganized, dark grayish-brown patterns lacking uniformity (similar to some American Herrings). Shin pads are often found on the legs of Azores Gull, whereas the norm in nominate *michahellis* is plain dull-colored legs. Again, use this feature judiciously. See other Macaronesian populations for comparison (plate 23a.9). Interestingly, juvenile Azores Gulls can sometimes have a very American Herring feel to them, much of which is superficial. In juvenile plumage, both share a rather rich,

2nd cycle. Dark and plain greater coverts and heavily streaked head. The "hood" effect is lost, as some individuals have variably marked neck and sides, making messy underparts. AMAR AYYASH. AZORES. DEC.

2nd cycle. Slightly paler inner primaries on this individual, with pale edging on outer greater coverts. Intermediate type with demarcated "hood," but without a dark belly patch. CHRIS GIBBINS. AZORES. FEB.

2nd cycle. Considerable barring on the inner greater coverts and very dense lower-neck markings. Such birds overlap with those from other Macaronesian Islands and cannot be objectively assigned to Azores Gull in a vagrant context. AMAR AYYASH. AZORES. DEC.

3rd cycle. Active primary molt. "Hood" effect, with piercing pale iris typical. Much black on bill at this age, as is found in some *michahellis*. Remnant shin markings visible. Dark gray upperparts approaching pale Lesser Black-backed. RICHARD BONSER. AUG.

3rd cycle. Noticeable brown wash throughout wing coverts and tertials. Duller legs than adult. Extent of head streaking variable at this time of year. CHRIS GIBBINS. AZORES. FEB.

---

ocolate-brown aspect. An open wing readily settles all uncertainty, as American Herring generally has a rk tail base with densely barred upper and undertail coverts and a prominent primary window.

d Cycle: Quite distinctive at this age, developing a demarcated hood effect with dark streaking restricted ostly to the face and head. Hood effect can be found throughout the winter (unlike in adults), although variable blotched markings can be strewn down the neck and breast in some individuals. These neck and breast markings are heaviest and most expected in fresh 2nd basic birds but are also found to a lesser extent later in the season (inviting confusion with other Macaronesian island populations). In general, a streaked "hood," largely white neck, and white underparts with a contrasting dark belly patch is the ideal search image. Bill pattern ranges from darker in the fall and early winter to yellow based in late winter through spring. Eye shows obvious paling at this age, some approaching pale eye of adult, and may even show red orbital. Second basic coverts are rather plain, solid dark brown in classic birds; others with variable buffy patterning, appearing barred in fresh birds, becoming a faint stippled pattern on worn feathers. Second alternate scapulars mostly solid dark gray; this can include some upperwing coverts and occasionally several tertials. Dull, pink-colored legs, some with strong yellow tinge. Dark shin pads still present on many individuals. On the open wing, the inner primaries can be marginally paler than the mid-outer primaries, but overall, the dorsal side to the wing appears uniformly dark. The uppertail coverts are mostly white with faint markings at times, showing a relatively broad black tailband with white tail base. The underwing coverts average dark but are somewhat variable. *Similar Species:* An estimated 30%–40% of 2nd-cycle Azores Gulls are likely to show the full suite of features that would lead to satisfactory diagnosis (Adriaens et al. 2020). In North America, Lesser Black-backed is the main confusion species at this age. Azores averages paler gray scapulars, although it is quite possible to find overlap with 2nd-cycle Lesser Black-backed. Azores Gull with the full package of demarcated hood effect, dark belly patch, shin pads, and solid brown greater coverts should be distinctive. Of course, it's unlikely that all of these features will be combined in a single vagrant, in which case overall plumage aspect and structural features should be weighed carefully. Consider average differences of shorter legs, broader wings in flight, shorter wing projection at rest, and deeper, more blunt-tipped bill on Azores. Compared to nominate *michahellis*, Azores shows darker ashy-gray scapulars (more bluish in nominate) and plainer, dark brown coverts. Compare greater coverts in particular, which are more patterned and fairly barred in nominate *michahellis*. Although somewhat variable with some overlap, Azores averages darker inner primaries and darker underwing coverts and is more likely to show bold shin marks (although these may be found on the legs of other populations of Yellow-legged Gulls at this age). Also consider hybrid Lesser Black-backed × Herring (see plates H7.9–H7.10), which theoretically would have intermediate gray scapular color that can approach or match that of Azores. Known-origin 2nd-cycle hybrid showed busier pattern on greater coverts, paler inner-primary window, and extensive brown wash on breast and sides (Ayyash, pers. obs.). Overall, this individual looked more like Yellow-legged Gulls from the Portuguese coast (*lusitanius*). An identification pitfall that bears mentioning is overlap with other Macaronesian island populations. Adriaens et al. (2020) describe four broad types of 2nd-cycle Azores Gulls, and only those showing the following features can safely be labeled Azores when dealing with vagrants: demarcated streaked hood covering the chin, face, and head; dark, isolated belly

27 3rd cycle. Aged with open wing, but note exposed black primary coverts. Overall appearance is of Macaronesian ilk. Dark upperparts, "hood" effect, and short legs invite thoughts of Azores Gull, but assigning such birds to *atlantis* difficult due to nonfunctional wingtip criteria. Increased black on the outer primaries is often age-related and not telling. Currently, 3rd-cycle types cannot be safely identified in a vagrant context. ANTONIO GUTIERREZ. SPAIN. JAN.
28 3rd cycle. Advanced individual from the Canary Islands with fully grown primaries (banded as chick in Agaete, May 2010). Short legs, darker upperparts, and stocky bill recall Azores Gull, although note markings on lower nape. Such birds are assigned to *michahellis* ssp. XABIER REMIREZ. CANARY ISLANDS. NOV.
29 3rd cycle. Primary pattern overall similar to adult's. Extensive swaths of brown are often found on the primary coverts, and less often across the greater coverts. RICHARD SMITH. AZORES. NOV.
30 3rd cycle. Somewhat advanced wing coverts. Note streaked "hood" effect, variable. Often with clean tail at this age. MARTIN GRIMM. AZORES. NOV.
31 Adult. Darker upperparts approach pale Lesser Black-backed, but decidedly paler. Red bleeding onto the upper mandible. Red orbital and pale eye. Extensive black on outer-primary bases (below tertials). AMAR AYYASH. AZORES. DEC.
32 Adults. Obvious variation in head and bill size and shape (presumed male on right). Adults have clean white heads from late fall through early summer. Bright bare parts are expected at this time of year. AMAR AYYASH. AZORES. DEC.
33 Adults. Canary Island birds. Very much approaching Azores Gull with short-legged appearance and strikingly dark upperparts. Thus wingtip criteria are crucial to objectively identify vagrant Azores Gulls. XABIER REMIREZ. CANARY ISLANDS. MAY.
34 Adult. The "hood" effect often coincides with primary molt from late summer through fall. Black bill smudges expected. Dull legs at times with remnant shin markings. RICHARD BONSER. AZORES. AUG.

patch surrounded by white; and mostly solid brown greater covert tract. The last feature is critical, as other populations, from the Canary Islands and Madeira, may combine a streaked hood and belly patch.

3rd Cycle: Mostly dark gray, adult-like upperparts. Some less advanced birds can show brownish wash across wing, with faint barring and dark across the entire greater covert tract, as well as on the tertial centers. The underparts are largely white. The demarcated hood effect is quite obvious in many, contrasting with a clean white neck and breast. Winter head pattern is predominantly white, presumably via 3rd prealternate molt, but some keep head markings through mid- to late winter (Ayyash, pers. obs.). Piercing pale eye in most. Bill pattern variable, from tricolored black, yellow, and red to bright, adult-like yellow with red gonys spot. Legs are a duller yellow in most, with very few still showing delayed pink-colored legs. Some show dark shin pads on tarsi even at this age. Smaller white apicals on the outer primaries. On the open wing, much black is found on the outer primaries, with black commonly covering the bases of p7–p10. A small mirror may be found on p10, often indistinct or absent altogether. The primary coverts and alula are often heavily marked, especially the outer greater primary coverts, which are commonly solid black across their entire length. Black commonly found down to p4, and p3 in some individuals. The secondaries and greater coverts can be washed black in the most retarded birds. The tail feathers regularly show small black patches and streaks, although usually fragmented, and in some birds

---

35 Adult with all visible primaries new, but not fully grown. The border to the back of the "hood" appears vertical, in part due to posture. DOMINIC MITCHELL. AZORES. OCT.
36 Adult. Primaries now almost fully grown. Bare parts brighter and head streaking diminished, soon to be lost. RICHARD BONSER. AZORES. OCT.
37 Adult type with Lesser Black-backed Gull (left), which incidentally has very vivid bare parts here. Compare blockier head, shorter legs, and paler upperparts of Azores Gull. ANTONIO A. GONCALVES. AZORES. SEPT.
38 Adult. Presumed Azores Gull (left) with Lesser Black-backed (center) and American Herring (right). Azores is intermediate in gray coloration. Primary molt almost complete and head streaking diminished. DAVE BROWN. NEWFOUNDLAND. OCT.

the tail is entirely white. The underwing is largely white, but faint traces of brown can be expected on the linings. *Similar Species:* At present, safely pinning down a 3rd-cycle Azores Gull in a vagrant context is likely not possible. The challenge at this age is separation from other Macaronesian populations which share several characteristics with Azores Gull. With that said, it may be possible to assign a 3rd cycle to the Atlantic islands group *sensu lato*. In North America, the primary identification challenge is Lesser Black-backed × Herrings hybrids. Hybrids at this age are more likely to show blemished yellow-pink legs and more extensive longitudinal streaking down the nape and lower neck. Identification is more complicated in breeding condition, when leg color in hybrids may become uniform, or when white alternate head pattern emerges in late winter and early spring. Compared to Azores Gull, 3rd-cycle hybrids average a larger p10 mirror and less black on base of p7–p8 and have cleaner greater primary coverts, with fewer swaths of solid black (Ayyash, pers. obs.; more study needed). Compared to Lesser Black-backed Gull, Azores averages paler upperparts, showing greater contrast between gray and black on hand. Compared to nominate *michahellis*, Azores has darker gray upperparts and may show a combination of any of the following: characteristic streaked hood pattern, dark shin markings, and more extensive black on outer primaries.

39 Adult. Near-perfect wingtip pattern with overall impression forming a black triangle. No p9 mirror, base of outer web of p8 completely black, full band on p5, and some black on p4. AMAR AYYASH. AZORES. DEC.
40 Adult. Similar to 23b.38. Interestingly, p3 has a small black mark, which is expected in adult Azores Gulls and is extremely rare in other subspecies in this complex. AMAR AYYASH. AZORES. DEC.
41 Adult. Presumed Azores Gull. Wingtip pattern approaches desired criteria, including very short gray tongue on p10 (underside of left wing). Compare to hybrid American Herring × Lesser Black-backed and other Yellow-legged Gull taxa. BRUCE MACTAVISH. NEWFOUNDLAND. FEB.
42 Adult. Some individuals have a near-complete subterminal band on p4. Note all-black bases of p8–p10 and single squarish mirror. AMAR AYYASH. AZORES. DEC.

MOLT The partial inserted molt in 1st cycle is attributed to a prealternate molt, which includes the scapulars, head, neck, and underparts. Commonly includes a variable number of upperwing coverts (in all tracts) and less commonly a few (upper) tertials. Head and scapular molt is detected through August, and many have entirely replaced scapulars by the end of November. Active molt in lower scapulars, some upperwing coverts, and tertials observed through December (Ayyash, pers. obs.). A sample of 418 1st cycles in mid-December had no individuals entirely in juvenile plumage. Second prealternate molt produces obvious adult-like gray scapulars and regularly includes a variable number of upperwing coverts (often innermost greater coverts) and, less commonly, a few tertials. An important identification note here is that the 2nd basic brown inner greater coverts—which are most likely to show barring—are sometimes replaced with new, adult-like 2nd alternate feathers by early to midwinter. These new feathers obscure any previous patterns used to home in on 2nd-cycle Azorean birds. Inspect the adjacent, mid-greater coverts in such birds for hints of white barring. Adults begin prebasic molt in late May and complete primary molt through November. Basic head pattern is usually found August–November, with maximum streaking in October. Prealternate molt readily produces white head by mid- to late November in many. Note that the streaked hood on Azores Gull often coincides with the mid- to outer-primary molt, which can hinder interpretation of wingtip patterns on birds of interest. Asynchronous molt patterns should also be considered with vagrants.

HYBRIDS None recorded.

---

43 Adult. Clean black triangle on wingtip. Underside of p10 shows extensive black, covering nearly the entire feather. The first gray feather base seen on the far wing is that of p7. AMAR AYYASH. AZORES. DEC.

44 Adult type with lingering hood pattern, getting late. Small black streaks on outer-primary coverts should be taken into consideration when assessing wingtip. Interestingly, base on outer web of p8 is gray here, but otherwise this is a typical Azores Gull primary pattern. RICHARD BONSER. AZORES. NOV.

45 Adult. Another individual with gray base of outer web of p8, but with otherwise normal Azores Gull wingtip. AMAR AYYASH. AZORES. DEC.

46 Adult. Combination of gray base to p8 and unmarked p4 is found in some adults. However, such a wingtip could not be positively attributed to Azores Gull in a vagrant context. AMAR AYYASH. AZORES. DEC.

24 LESSER BLACK-BACKED GULL *Larus fuscus*

L: 19"–24" (49–60 cm) | W: 46"–59" (118–150 cm) | Four-cycle | SAS | KGS: 9–17

OVERVIEW Primarily an old-world black-backed gull. Lesser Black-backed is a relatively recent addition to the avifauna of the New World. The first US record came in 1934 from Ocean County, New Jersey (Edwards 1935), and Canada's first record was in 1968 from Churchill, Manitoba (Godfrey 1986). A notable incursion in North America began in the early 1980s, with numbers soaring in the mid- to late 1990s, mostly in the northeast. Its colonization of North America is compelling in two respects. First, no other gull species is known to have undergone so rapid a transatlantic expansion. Second, there is no record of a Lesser Black-backed pair that has nested in the United States or Canada as of this writing. Its presence in the New World is truly an enigma. This is our only gull species whose breeding range in the ABA Area—assuming there is one in the ABA Area—remains completely unknown. It is steadily increasing throughout the interior and the western half of the continent, as well as in the Caribbean, Middle America, and the northern rim of South America. Feeding preferences are similar to those of other large gulls, from offshore to coastlines, lakes and reservoirs, large rivers, dams, open fields, food-processing plants, and landfills. Averages smaller and sleeker than Herring Gull, with a decidedly longer wing projection.

TAXONOMY Traditionally, three subspecies recognized in Europe: *Larus fuscus graellsii*, *Larus fuscus intermedius*, and *Larus fuscus fuscus*. However, this grouping is an oversimplification, as several eastern taxa—*heuglini*, *barabensis*, and *taimyrensis*—are affiliated with this complex. The Association of European Records and Rarities Committees and the International Ornithologists' Union treat *graellsii*, *intermedius*, *fuscus*, *heuglini*, and *barabensis* as valid subspecies of Lesser Black-backed Gull, but *taimyrensis* remains unresolved (see Appendix for more on *heuglini* and *taimyrensis*). Various authorities endorse a division of two species: Lesser Black-backed Gull (*L. fuscus*), comprising nominate *fuscus*, *intermedius*, and *graellsii*; and Heuglin's Gull (*L. heuglini*), comprising nominate *heuglini* and *barabensis*; with *taimyrensis* disputed. Although these proposals have gained some traction in recent years, more work is needed to better understand the evolutionary histories of these taxa, particularly the relationship of the adjacent populations *fuscus* and *heuglini* (Liebers & Helbig 2002).

Lesser Black-backed Gulls in North America are presumed to be derivatives of the largest and palest European race, *L.f. graellsii*. Therefore, in this book, Lesser Black-backed Gull generally refers to *graellsii*. This subspecies saw a major boom in its traditional range in the 20th century, with a rapid expansion throughout W Europe and the upper N Atlantic, particularly to Iceland and the Faroe Islands, and then to Greenland. Its increasing presence in Canada and the United States coincided with this range extension (Ayyash 2013). The discussion that follows is centered around *fuscus*, *intermedius*, and *graellsii*, with much emphasis on the last of these.

L.f. fuscus (KGS: 13–17): Nominate race, commonly referred to as Baltic Gull. Found throughout the northern and eastern parts of the species' range, in N Norway, Finland, Sweden, and the Baltic Sea region. A true long-distance migrant, wintering to E and S Africa and the Middle East. Rather elegant, with refined proportions and noticeably long, narrow wings. Smallest and darkest subspecies. Adults typically show little contrast between their jet-black upperparts and black wingtip. Averages a small to medium-sized p10 mirror, and only a small percentage acquire a p9 mirror. Regarded as a three-cycle gull, unlike *L.f. intermedius / graellsii*. Declining throughout its range. No substantiated records in North America (although see plate 24.43).

L.f. intermedius (KGS: 11–13): Upperparts and size "intermediate," with clinal characteristics that overlap nominate *fuscus* on the dark end and *graellsii* on the pale end. Breeding range more south and west in Scandinavia, from W Norway, SW Sweden, and Denmark to Germany and the Netherlands. Also established in mixed colonies with nominate in N Norway, where historically not known. The overall trend is an increased expansion at the expense of nominate. Some of the earliest reports of Lesser Black-backed Gull in North America were attributed to this race, but detailed documentation and descriptions are lacking (Post & Lewis 1995b). Therefore, its status in North America is hypothetical, if not impossible to verify.

L.f. graellsii (KGS: 9–11): Palest and largest of the European subspecies, situated at the westernmost and southernmost parts of the species' range. Breeds along the coasts of W Europe from the Iberian Peninsula, north through the Netherlands, Britain, Ireland, Faroe Islands, Iceland, and Greenland. Increasing throughout its range and has likely established a self-sustaining population in the New World. Commonly shows a slate-black to bluish hue to the upperparts, with discernible contrast to the black wingtip. Averages robust proportions, appearing bulkier and shorter winged than typical nominate *fuscus*. Mirror on p10 relatively large, and mirror on p9 commonly present.

There is appreciable overlap in appearance between some *graellsii* and *intermedius*, and between *intermedius* and *fuscus*. An approximate 15% overlap in upperpart coloration exists between nominate from the E Swedish coast and *intermedius* from W Sweden (Jonsson 1998). Furthermore, from the early 2000s until the time of writing, a mixture of *fuscus* and *intermedius* types has been documented in 9 of 12 colonies along the coast of N Norway, where only nominate was historically expected (Helberg et al. 2009). Although there are general distinctions that can be made when combining size, structure, upperpart coloration, migration corridors, and extent of molt, various gull experts and rarities committees agree that subspecific identification for out-of-range birds should be reserved for individuals that are banded (i.e., of known origin), especially beyond 1st cycle, when molt patterns become ambiguous (Altenburg et al. 2011). Observers in North America are advised to follow this practice.

The presence of so-called Dutch intergrades in North America is worth contemplating. This term refers to the intergrade breeding populations centered around Belgium and the Netherlands, which is geographically situated between the westernmost populations of the British Isles (*graellsii*) and those from parts of W Scandinavia (*intermedius*). Dutch intergrades average darker than classic *graellsii* but paler than classic *intermedius*, with much overlap on both ends, and identification can be rather arbitrary.

Dark-backed adults are not uncommon in the United States and Canada, such as Green F05, an individual found hybridizing with Herring Gulls off the coast of Maine in 2007–2015 (Ellis et al. 2014; see this account's "Hybrids" section). Whether this is simply an example of dark-end *graellsii* or in fact influenced by *intermedius* cannot be known given current knowledge. Another case in point is a banded adult discovered on Long Island, New York, in October 1997, which was traced back to the Netherlands (Hallgrimsson et al. 2011). This record is curious, but it seems somewhat negligible, given that more than 20,000 Lesser black-backed Gulls were banded in the Netherlands between 1990 and 2008, with no others reported on this side of the Atlantic since.

At present, North American Lesser Black-backeds are attributed to *graellsii*, despite the increasing number of dark birds resembling Dutch intergrades and even *intermedius*. The question is not so much whether *intermedius* types are found in the New World as whether we can prove a bird is of *intermedius* stock.

RANGE

Breeding (L.f. graellsii only): Breeds colonially along coastlines, on islands, and on buildings in W Europe, north through the Faroe Islands, Iceland, and Greenland, since at least 1990 (Boertmann 1994, 2008). No known nesting of a pure pair in Canada or the United States. Geographically, the closest known breeding colonies are in S Greenland. Most breeding is around the capital city of Nuuk, but it extends as far north as Melville Bay in the west (Boertmann & Frederiksen 2016) and to several sites around Tasiilaq and Kuummiut in the southeast (Boertmann & Rosing-Asvid 2017). Two hybridization records with American Herring from the United States (see this account's "Hybrids" section).

Nonbreeding: After breeding, migrants begin to appear beginning in late August, more regularly during September. Mostly a migrant in the Maritime Provinces, N New England, and N Great Lakes, where rare to uncommon in winter. Larger waves begin moving through in October, particularly in the mid-Atlantic and south to Florida. Noteworthy is a fall interior high count of 438 individuals in October 2023 from Hancock County, Ohio.

Winters in small numbers in S Great Lakes region and interior Midwest, Atlantic coastal region (including the coastal plain) from Massachusetts south to S Florida, Bermuda, the Bahamas, and along the entire Gulf of Mexico coast to NE Mexico and the Yucatán Peninsula. Rare to Panama (primarily Panama City), although increasing. Increasing through the West Indies, N Colombia, N Venezuela, and Brazil. Accidental to Ecuador but likely overlooked. Its presence in this region has outpaced most print publications and range maps.

Locally up to several hundred winter in the mid-Atlantic and Northeast region, found at mass staging sites in late winter and early spring, such as in Bucks County, Pennsylvania, where counts of 1,000 and more individuals have been recorded beginning in 2023. Regular counts of 200–300 in Virginia and Florida not uncommon, with some exceeding 500, and now more scattered with a southward expansion into the Caribbean. These highs are associated with landfills and nearby loafing sites on the coast in mid- to late winter. The species appeared to be increasing as a winter resident on the Great Lakes in the early 2000s, although these numbers have tapered off with a notable shift southward, possibly due to a preference for ice-free feeding grounds and less competition with Herring Gull.

Uncommon to rare in much of the remaining interior, mostly at large lakes, dams, and landfills. Increasing at several sites in S Manitoba, Kentucky, Texas, Nebraska, and especially Colorado. Farther north and west, rare but regular to even locally uncommon almost everywhere except Alaska, where casual. Several records from the Mackenzie Delta and North Slope, one of which involves an adult "billing and bonding" with a hybrid Glaucous × Herring in the breeding season of 2001 (Karl Bardon, pers. comm.). Casual visitor to north-central Canada, which may be due to a lack of observers. Has become routine in small numbers in the West, where it was once considered very rare—both coastal and inland. Expected annually at the Salton Sea. Also increasing throughout Gulf Coast of Mexico, particularly Yucatán and Tamaulipas, and as a scarce winter visitor to NW Mexico, where high counts of up to eight individuals have been found in Puerto Peñasco, Sonora (2023).

Generally begins to push north in mid- to late March, signaled by the departure of Florida's large winter flocks, followed by those of the mid-Atlantic, with a corresponding surge in numbers in the Maritime Provinces. However, uncommon to locally fairly common well into April at some sites. Concurrent with the overall increase in the population in North America, counts of oversummering nonbreeders have increased substantially in recent years. Formerly very rare during this season, now fairly routine, and some recent totals have reached as many as 70 birds on Long Island, New York, and much larger numbers in SE Virginia and Maryland in July through early August. On Lake Michigan, 69 individuals from Wisconsin in mid-July is a summer high count for the Great Lakes. The majority of summer clubs are of one-year-olds in their 2nd molt cycle, although 3rd-cycle types increased in recent years (Ayyash, pers. obs.). At the time of this writing, definitive adults are rare from late May through early August in Canada and the United States. This further supports the "exploratory migration model" practiced by this species (Zawadzki et al. 2021). During the first year and a half of life, Lesser Black-backeds overwinter at suitable sites that eventually influence migratory and wintering behavior as they age. It may be expected, then, that sites with an increase in young birds summering later see a spike in adult birds during migration and winter.

In late March 2018, a small cohort of 9 wintering adults in Bucks County, Pennsylvania, was fitted with GPS trackers by the Pennsylvania Game Commission (2018), revealing some much-anticipated data. The majority departed on a north-northeast trajectory in mid-spring, with 5 individuals traversing the Labrador Sea by May and settling in S Greenland for the breeding season. In February 2023, 15 individuals were fitted with trackers in Nantucket, Massachusetts. By mid-May, 2 of these had already touched down in SW Greenland (Richard R. Veit, pers. comm.).

IDENTIFICATION (*L.F. GRAELLSII* ONLY)
Adult: Dark slaty-gray upperparts, less than true black, at times with a bluish hue. Variation in upperparts appreciable, with some similar to Laughing Gull, others approaching pale Great Black-backed Gull.

Contrast between black primaries and dark upperparts is noticeable at rest. Yellow legs and pale eyes. The legs and bill can become dull yellow in nonbreeding condition, with black markings on the bill found along the cutting edge and nostril, but generally less prominent than in subadult. Yellow bill generally straight with a weak gonydeal expansion. The red gonys spot can be noticeably large, with red sometimes extending to the upper mandible. Orbital ring red to orange red. Wing projection is long compared to other large gulls in North America, giving the bird an attenuated feel to the rear, with a shallower chest. Body size variable, with some female types slightly larger than Ring-billed on the small end, and male types matching and exceeding the size of American Herring at the large end. In basic plumage, particularly September–December, the head can be well marked with much fine streaking covering the face and down the neck or can show faint streaking concentrated around the eye. The pattern may become blotted or dense on the lower neck. Generally becomes white headed by late February through March, but variable. On the open wing, p10 mirror is relatively large, uncommonly with an all-white tip. Mirror on p9 is generally small or absent, but not uncommon. Complete subterminal band to p5 is a standard feature, sometimes with markings to p4 and less commonly to p3. Generally lacks white tongue tips, but thin slivers sometimes found on p5–p6 / p7, mostly on the inner web. Variable black markings on the primary coverts of otherwise perfect-looking adults not rare, although such birds relegated to "adult type" or "subadult" label. On the underside of the wingtip, prominent black is often found up the base of p10 with a step-down appearance in black on underside of p7–p9. Smoky-gray underside to remiges contrasting with white wing coverts. ***Similar Species:*** See accounts for Yellow-footed and Kelp Gulls, which are the only other confusion black-backeds with yellow legs. Adult California Gull has noticeably paler upperparts, although various lighting conditions can diminish this difference. California Gull is dark eyed, with more white in wingtip. Compare to Great Black-backed, Slaty-backed, and Western Gulls, which are pink legged. Beware Great Black-backed and Western Gulls with yellowish legs, in which body size and bill structure are radically different. See Yellow-legged Gull account and Lesser Black-backed × Herring and Kelp × Herring hybrids accounts. See also account for Heuglin's Gull, for which separation in North America is generally problematic given current knowledge.

1st Cycle: Juveniles largely dark above with pale edging to upperparts. Head densely streaked with dark concentrated around eye. Foreneck, breast, and belly with coarse markings, generally with pale base color to body becoming more white later in the season. Straight, black bill retained for much of the season, although not terribly rare for bill base to start paling before end of winter. Scapulars have solid dark centers and on the whole can look scalloped or, less commonly, lightly notched. Dull, pinkish legs, some with dark shin marks, although often diminished and absent later in the season. Pale edging to upperparts gives way to wear. Postjuvenile molt (e.g., 1st prealternate) is mostly limited to the scapulars, which can be an ashy grayish brown with dark shaft streaks and pale fringes or retarded in appearance with light bases, barred, or with anchor pattern. Juvenile wing coverts with solid centers and pale fringes.

---

1 Juvenile. Long wings with svelte appearance to hind parts. Dark tertials and outer greater coverts with thin, pale edging. MARS MUUSSE. THE NETHERLANDS. AUG.

2 Juvenile. Much less commonly, tertial tips may be patterned with slight notching across the upperparts. Outermost greater coverts concealed by flank feathers. AMAR AYYASH. MICHIGAN. OCT.

3 1st cycle. Chocolate-brown tones. Dark face with heavily marked lower hindneck found in some individuals, even late in the season. Straight bill. Noticeable white base color to underparts. More patterned mid- and inner greater coverts than 24.1. AMAR AYYASH. INDIANA. NOV.

4 1st cycle. Female type of unknown origin. Banded in east-central Florida, Jan. 2015 (first Lesser Black-backed banded in the state). Bears some resemblance to Azores Gull, which shows a less uniform pattern on the postjuvenile scapulars, and a grimy head and neck pattern. American Herring averages darker vent, with barring and smoother brown breast and belly. AMAR AYYASH. FLORIDA. JAN.

5 1st cycle. A large-billed individual with faded brown tones. Note plain and dark outer greater coverts and tertials. Pale head and base color to underparts. AMAR AYYASH. FLORIDA. JAN.

6 1st cycle. Dark sooty-gray postjuvenile scapulars and sides, with white head and underparts, make identification rather straightforward. AMAR AYYASH. PORTUGAL. DEC.

7 1st cycle. A number of wing coverts and upper tertials replaced (1st alternate). The blocky white head and reddish juvenile coverts recall Yellow-legged Gull, which often shows more wear at this date, with paler and more barred upperparts, and a whiter tail base. AMAR AYYASH. FLORIDA. JAN.

8 1st cycle. A small, short-legged individual with very dark and plain upperparts (see open wing in 24.15). AMAR AYYASH. MEXICO. DEC.

24 LESSER BLACK-BACKED GULL

9   1st cycle with similar-aged American Herring (left). Lesser averages darker-brown wing coverts, slimmer body, and blacker bill base, and generally lacks smooth underparts of *smithsonianus*. AMAR AYYASH. FLORIDA. JAN.
10 1st cycle. Paling bill base and many fresh 1st alternate wing coverts and upper tertials, not rare at this date. Flight feathers juvenile (1st basic). MARS MUUSSE. THE NETHERLANDS. APRIL.
11 1st-cycle *intermedius*. Blue V.JY4 from Denmark. A few darker postjuvenile scapulars visible. White head with dark eye mask. Subspecies not safely assigned without life history (i.e., leg band). PETER ADRIAENS. PORTUGAL. NOV.
12 Largely juvenile. Presumed nominate *fuscus* based on location and overall appearance. Small body, thin bill, short legs, long wings, and considerable white on underparts. Plain upperparts with bold white edging. Not safely assigned out of range without life history. AMIR BEN DOV. ISRAEL. SEPT.
13 1st cycle. Nominate *fuscus* based on location and molt. Complete 1st prealternate molt occurs on the wintering grounds, expected in this subspecies. All flight feathers replaced (2nd-generation remiges on a one-year-old). Plain, brown upperparts, white body, and pale iris resemble 2nd-cycle large gull. Slim and long-winged. HANNU KOSKINEN. FINLAND. JUNE.
14 1st cycle. Same individual as 24.3. Much pale edging on coverts. Plain outer greater coverts, sometimes with noticeably barred inner tract. Pale uppertail coverts with spotted tail base. AMAR AYYASH. INDIANA. NOV.
15 1st cycle. Same individual as 24.8. Long, narrow wings. Whitish head with fine streaking on crown and nape. Broader tailband. Some show paler tones to inner webs of inner primaries. AMAR AYYASH. MEXICO. DEC.
16 1st cycle. Tail pattern is somewhat variable. This individual has extensive white tail base and narrow tailband, recalling Yellow-legged, but note streaked head and plain sooty-gray postjuvenile scapulars. AMAR AYYASH. FLORIDA. FEB.
17 1st cycle. Same as 24.5. Narrow tailband and pale rump. Dilute appearance to plumage likely due to fading. Bill often remains black into early spring. Lower scapulars juvenile. AMAR AYYASH. FLORIDA. MARCH.
18 1st cycle with similar-aged American Herring (left). Lesser averages smaller, with narrower wings and darker brown upperparts. Note pale inner-primary window on American Herring. AMAR AYYASH. FLORIDA. JAN.
19 1st cycle. Pale-bodied individual with barred axillaries. Wing linings medium brown. Narrow tailband and white undertail coverts sparsely marked. MARS MUUSSE. THE NETHERLANDS. OCT.
20 1st cycle. Darker underparts than 24.19, but still coarsely marked with pale base color to body. Plain axillaries here with darker tail base. Note dark inner primaries and greater coverts (far wing). AMAR AYYASH. FLORIDA. JAN.

Greater coverts, especially the outer tract, show dark bases, while the central and inner greater coverts can often be checkered, but with pale fringes. Tertials generally have narrow, pale fringes with plain, solid dark centers, but a small percentage may show notched edging. At rest, note the long-winged look, with blackish primaries showing large contrast with the upperparts. The undertail coverts have variable barring or well-spaced chevron markings with white base color. Vent is generally poorly marked and at times largely white and unmarked. On the open wing, note lack of prominent pale window, with a more uniform look to the hand, which overall has dark primary coverts adjacent to dark outer greater coverts. Some show lighter inner primaries but lacking an impressive contrast with the outer primaries, and a small percentage may show pale "lozenges" on the outer webs of the outer primaries. The uppertail pattern is noteworthy. A dark, relatively narrow tailband often contrasts with whitish uppertail coverts and rump that show variable brown barring. Tailband somewhat demarcated but can have intricate patterning along the proximal edge. Outer web of outer rectrix more often with whitish base and some light barring; however, some birds can show darker base to outer tail feathers, which contrasts sharply with white uppertail coverts. The flanks are mottled and coarsely streaked, sometimes with patchy brown spotting, but belly overall paler. The underwing coverts and axillaries can show indistinct brown barring, although greater coverts on the underwing more grayish and uniform with secondaries. *Similar Species:* In North America, most likely to be overlooked as American Herring Gull at this age. Herring averages a deeper and more rotund body, with a larger bill that commonly shows some paling, whereas Lesser Black-backed more likely to retain a mostly jet-black bill at this age. Note that body size overlap exists with female-type Herrings and male-type Lesser Black-backeds, but body structures differ, with Lesser Black-backed quite often appearing thinner in the vent region; also averages a smaller and "cuter" head. Lesser Black-backed longer winged at rest and shows a more pointed hand in flight. The upperparts on Lesser Black-backed are more solidly dark, which produces a marked contrast with the white base color to the head, neck, and underparts. American Herring Gull is more uniformly brown on the underparts. The open wing is quite different, with American Herring generally showing a more prominent primary window and dark uppertail, with darker base color to uppertail coverts. See also Great Black-backed, Yellow-legged, and European Herring Gulls accounts. American Herring × Lesser Black-backed hybrids are poorly described at this age, with data only available from two known-origin birds (Ayyash et al. 2015).

2nd Cycle: Both plumage and bare parts are highly variable at this age—a common theme with 2nd-cycle large gulls. Variation in appearances is due to some individuals retaining a retarded 2nd basic plumage while others take on advanced 2nd alternate plumage (see this account's "Molt" section). The result is some birds resembling 1st cycle and others appearing like 3rd cycle. Early in the season, the bill can still be largely black, often with a yellow nail. Some paling throughout the bill base is expected, but a sharply bicolored bill is not typical at this age. Advanced individuals, particularly in early spring, can have an adult-like yellow bill with orange-red gonys spot surrounded by black subterminal markings. Legs range from pink colored to dull yellow, at times with peach tones as leg color transitions from pink to yellow. Eye color can be dark, but generally some bluish-gray paling is detected, with advanced birds approaching a clear iris. Head, breast, and sides variably streaked, although a pale white body predominates. Second basic wing coverts can be plain and solid brown with little patterning or more boldly patterned with pale edging. Greater coverts are now with more pale regions and finer markings, although some with muted and plain, unmarked pattern (e.g., 2nd basic). At times, the entire wing panel can fade to a creamy white,

21   2nd-cycle duo with American Herring Gulls in background. Note refined structure, blacker bills, and dark upperparts. 2nd prebasic molt. AMAR AYYASH. MASSACHUSETTS. JULY.
22   A smaller 2nd cycle (center), with similar-aged American Herring (left) and Great Black-backed Gull. AMAR AYYASH. NEW JERSEY. JULY.
23   2nd cycle. Still much black on the bill and fine head streaking. Outer primaries (2nd generation) not fully grown. Several noticeably dark gray scapulars and coverts. AMAR AYYASH. MICHIGAN. SEPT.
24   2nd cycle. Same individual as 24.4. Superficial resemblance to Azores Gull, but lacks important features of that taxon (e.g., sharply demarcated hood, belly patch, or shin markings). AMAR AYYASH. FLORIDA. JAN.
25   2nd cycle. Fairly advanced blackish scapulars and median coverts (2nd alternate). Dark tail and absence of white apicals point away from 3rd cycle, but inspect open wing. AMAR AYYASH. PORTUGAL. DEC.

with contrasting adult-like scapulars. Tertials with solid brown centers and wider white tips than in 1st cycle. Second alternate scapulars and wing coverts, often beginning with the median coverts, are more adult-like but with brown wash, and similarly the tertials now show broader, adult-like white tips that form a weak tertial crescent. On the open wing, the inner primaries are a lighter brown, showing more contrast with the blackish outer primaries when compared to the uniform look of 1st cycle. Secondaries now with distinct thin white tips, extending to the inner primaries, but not broad or adult-like as in 3rd cycle. Second basic tail may average more black and a wider tailband than 1st cycle, but with plainer uppertail coverts and rump. A variable number of tail feathers can be replaced in 2nd prealternate, producing all-white rectrices and all-white uppertail coverts. Underwing coverts with light brown edging and distinctly paler centers than in 1st cycle. *Similar Species:* American Herring typically shows some gray on scapulars by now, which will be distinctly paler than those of Lesser Black-backed. Compare size and structure and note darker inner primaries and inner-primary coverts on Lesser Black-backed. See also California, Yellow-footed, Great Black-backed, Kelp, and Yellow-legged Gulls accounts.

3rd Cycle: Aging can be tricky, as some resemble 2nd-cycle Lesser Black-backed, and others are more adult-like. Pale eyes now expected, although not as distinctly clear as adult's, and a minority still with dark eyes. Bill can still show pinkish base, often with dirty black markings, coupled with duller orange-red gonys spot. Late in the season, many are seen with vivid yellow bill and brighter gonys spot, complemented by a noticeable red orbital ring. In basic plumage, fine head streaking can be extensive, becoming more concentrated and denser on lower hindneck. Birds that combine a very pale iris with dense markings concentrated around the eye portray a menacing look, superficially recalling Slaty-backed Gull. Generally with white underparts, but some mottling can be found on the breast and sides. Tertials can show brownish-black centers with less demarcation to white tips. Apicals average noticeably smaller than adult's, especially on outer primaries, and may be absent entirely. On the open wing, remiges are generally adult-like with gray centers, although secondaries can have blackish-brown centers with brown freckling. Trailing edge now with relatively broad white tips. Diffuse subterminal bands forming on p3–p5, not as neatly demarcated as the adult's. P10 mirror smaller than adult's and sometimes absent, and mirror typically not found on p9. Some can have retarded, brownish wing coverts, especially greater coverts, and primary coverts commonly show dark, subadult markings. Uppertail sparsely marked with some dark patches or all white. Underside to wing largely white, but with dusky markings mostly on the lesser and median coverts. Underside of wingtip with much pigment up the bases of p7–p10. *Similar Species:* Compare to other 3rd-cycle black-backeds with yellow legs—namely, Yellow-footed and Kelp Gulls, and also to Yellow-legged Gull. Also consider Herring × Kelp and Herring × Lesser Black-backed hybrids, which are poorly described (but see plates H7.13 and H7.14). By now, leg color and upperparts are unmistakably different from those of American Herring, which has pink legs and pale gray

---

26 2nd cycle *intermedius*. Banded as chick in Norway, June 2015. Compact body with plain, dark upperparts. Subspecies not safely assigned without life history (i.e., leg band). JOSE MARQUES. PORTUGAL. NOV. 2016.
27 2nd cycle. Somewhat delayed with tail pattern and coverts resembling 1st cycle. Note rounded primary tips, white tips to inner secondaries, and pale eye. AMAR AYYASH. FLORIDA. JAN.
28 2nd cycle. More advanced than 24.27, with many adult-like gray coverts (2nd alternate). Delayed bill pattern, but note dark inner primaries. AMAR AYYASH. PENNSYLVANIA. DEC.
29 2nd cycle. Same as 24.4 and 24.24. Variable brown on wing linings, with largely white underparts. AMAR AYYASH. FLORIDA. JAN.
30 3rd cycle. Delayed. Note mirror on underside of p10, white apical on p5, and white tips to secondaries. Retarded bill pattern and fleshy legs. AMAR AYYASH. ILLINOIS. DEC.
31 3rd cycle. Advanced, adult-like appearance, but note reduced apicals, small mirror on underside of p10, black on tail, narrow tertial crescent, and brown wash across upperparts. Yellowish legs. AMAR AYYASH. FLORIDA. JAN.
32 3rd cycle. Third-prebasic molt in motion with new p1–p4 and incoming p5 (3rd basic). Secondaries and outer primaries 2nd generation (2nd basic). AMAR AYYASH. NEW JERSEY. AUG.
33 3rd cycle. Largely white tail not uncommon at this age. Delayed remiges, but note adult-like p1–p2 and broad white tips to secondaries. Much black on bill not unexpected. AMAR AYYASH. PORTUGAL. DEC.
34 3rd cycle. Advanced with bright yellow bill, clean white tail, and adult-like gray upperparts, but note brown secondary centers and marked primary coverts, as well as reduced p10 mirror. MARS MUUSSE. THE NETHERLANDS. MARCH.
35 3rd cycle. Some show small p10 mirror at this age. Dirty wing linings and few markings on tail. Note delayed bill pattern and fleshy color to legs. AMAR AYYASH. AZORES. DEC.

scapulars. However, some Lesser Black-backeds retain pinkish legs at this age, inviting confusion with Great Black-backed and Slaty-backed Gulls (see those species accounts).

MOLT Molt strategy is highly variable across subspecies, perhaps more so than in any other gull. In Eurasia, molt in the Lesser Black-backed Gull complex is a topic of much interest, due to implications for identification. North American observers have just recently begun to form an impression of molt patterns in this "founder" population, but data involving known-age and known-origin birds is virtually nonexistent. Therefore, much of the information provided here is based on a database of my own photo collections and personal observations. The overall picture mirrors data collected by the Gull Research Organisation (2017), with a few new observations previously undocumented in North America. General patterns of what is presumably the westernmost race, *graellsii*, are provided in the following section.

L.f. graellsii: The subspecies *graellsii* averages the earliest and most limited 1st prealternate molt, generally restricted to scapulars, and less commonly may include a variable number of wing coverts and tertials, and the odd rectrix here and there. Does not include secondaries or primaries. Extent of 1st prealternate appears more advanced in some southern birds (Florida), with year-to-year variation among these populations (Ayyash, pers. obs., 2011–2022). For example, 64% of 1st cycles in a sample of 79 in E Florida in 2016 showed some replaced upperwing coverts and one to three replaced tertials by the end of January. Northern birds (Great Lakes and NE Atlantic) consistently show juvenile wing coverts and tertials through late winter (through March / April).

Second prebasic molt generally detected in May, but some individuals can already be molting innermost primaries and wing coverts by the end of April. By May, a variable number of 2nd basic scapulars and upperwing coverts are on display, some with advanced, plain, ashy-gray aspect; others with brown barring, appearing closer to 1st basic. Determining which feathers are late 1st alternate and which are early 2nd basic is not always straightforward (Muusse et al. 2005). By June, generally shows a whitish head with limited streaking, with neck and underparts showing less mottling. By July, approximately half of primaries renewed with variable tail molt. Many juvenile secondaries are retained through the better part of July. Second prebasic molt continues into August–September, with upperparts and flight feathers appearing patchy, but with fewer conspicuous molt gaps. Most individuals have renewed tails and secondaries by end of September. By October, primary molt is generally completed, and 2nd basic plumage is maintained for only a short while as the 2nd prealternate molt advances.

The 2nd prealternate molt is more variable than its counterpart in 1st cycle. Some individuals show little to no active molt from fall through winter (Oct.–March), while others may undergo moderate to extensive molt of the scapulars and wing coverts, which generally produces adult-like, 3rd-generation feathers with a brown hue (2nd alternate). Again, individuals at more southern latitudes appear to average more extensive molts when compared to those wintering farther north, but larger sample sizes needed.

---

36 Adult. Dark slate-gray upperparts. Fairly large apicals and p10 mirror. Yellow legs, pale eye, and considerable red on the gonys. This individual has fine pencil streaks on hindneck, recalling Heuglin's Gull. AMAR AYYASH. FLORIDA. JAN.
37 Adult. A dark individual with shorter legs and deep breast showing a quasi "Slaty-backed Gull" silhouette (although note long wing projection and leg color). Blackish upperparts in line with Dutch intergrade (banded as adult type in the Netherlands and assigned to *graellsii*). JOSE MARQUES. PORTUGAL. NOV.
38 Adult. Paler *graellsii* type with sharply demarcated head pattern. Recalls Azores Gull, which averages paler gray and is largely white-headed at this date. Outer edge of p9 mirror exposed and p10 even with tail tip in late Jan. AMAR AYYASH. FLORIDA. JAN.
39 Adult with similar-aged American Herring (back) and Laughing Gulls. Green F05 is a large dark male from Appledore Island that hybridized with American Herrings for several years. Already white-headed. AMAR AYYASH. FLORIDA. JAN.
40 Adult. Fairly long, straight bill with tapered tip. Worn apicals. Red AZ was banded and radio-tagged as an adult on Nantucket Island in Feb. 2023, and is pictured here five months later still on the "nonbreeding" grounds. Interestingly, it's pictured here just a rock's throw from Appledore Island. LEO MCKILLOP. NEW HAMPSHIRE. JULY.
41 Adults showing variable upperpart coloration. The individual on the left is a pale-end *graellsii*. The darker individual matches a Dutch intergrade or paler *intermedius*, although such birds are tentatively assigned to *graellsii* in North America. AMAR AYYASH. FLORIDA. JAN.
42 Adult *intermedius* (center) banded as a chick in Norway. Decidedly blacker upperparts. Subspecies, however, not safely assigned without life history (i.e., leg band). MARS MUUSSE. THE NETHERLANDS. SEPT.

It's not uncommon for a variable number of tail feathers and tertials to be replaced in 2nd prealternate before April. Known-origin birds from Europe have been recorded replacing secondaries and a variable number of primaries, although this appears to be a rare occurrence in *graellsii* (Muusse et al. 2005). Pyle et al. (2018) documented the first known example of a Lesser Black-backed Gull replacing *all* remiges and rectrices in North America. Between January and April a 2nd cycle from Brazoria County, Texas, was regularly monitored in the field as it replaced these tracts in typical, sequential order in 2nd prealternate. Interestingly, the secondary coverts remained largely 2nd basic during this period. Also during this same period, a second individual was found molting secondaries and primaries (s1–s3 / p1–p7) in 2nd prealternate. How frequently this occurs in North American populations remains to be seen. Given that both birds were found in the same season at such a southern latitude is noteworthy (compare Lesser Black-backeds wintering farther south in the Old World, which average more extensive prealternate molts). I have recorded several similar individuals in Florida (Jan.) and know of no such examples from the Great Lakes or Northeast. Finally, very rarely recorded with "jump" molt or Staffelmauser-like pattern. An April specimen from Louisiana had p6–p8 as newer 3rd-generation-like (i.e., 2nd alternate), with older 2nd basic p1–p5 and p9–p10 (Ayyash, pers. obs.; Louisiana State University Museum of Natural Science, #21852). Similarly, a live bird in Florida showed replaced p5–p6 (presumably 2nd alternate), with remaining primaries 2nd basic in late January (Ayyash, pers. obs.).

The 3rd prebasic molt is generally detected in May and June. Note, however, that two-year-old Lesser Black-backed Gulls become less common in the United States and Canada at this time. Good sites to study 3rd cycles from May to July include Assateague Island, Cape May, Long Island, and Cape Cod. Numbers are unpredictable from year to year but likely to increase as the species continues to settle the continent. By the end of October, many have completed primary molt and are in 3rd basic plumage.

The 3rd prealternate molt is slightly more difficult to detect at this age, but incoming adult-like upperparts appear decidedly fresh and more solidly gray, although sometimes with noticeable dark markings and a brown wash throughout. The uppertail may be solidly white or with black swaths, forming a fragmented tailband. The 4th prebasic molt gives way to an adult plumage, generally beginning in May–June. This molt can continue into November–December and sometimes is protracted through late winter. In late January, for example, nearly one in five adult types has still not finished completely growing p9 and p10 in E Florida, and one in eight is growing one to four secondaries (Ayyash, pers. obs.). Adults observed on the Great Lakes typically have fully grown primaries by December. More data needed on birds in the interior and western half of the continent.

It's not rare for the primary coverts to show some blackish markings at this age. Those with large, diffuse, dark marks on the primary coverts average younger than those with smaller, teardrop spots that are neatly demarcated (Muusse et al. 2011). Such birds are commonly referred to as "subadults" by field observers, but this term is better reserved for those showing marked primary coverts and a combination of other subadult features (i.e., markings on the tail, small mirror and apicals, brown wash on the upperparts, and / or duller bare parts). By April, most adults have immaculately clean white heads and can show abraded white tips to the primaries. By the end of the month, adults are virtually absent across much of the United States and Canada, presumably en route to remote breeding grounds.

L.f. intermedius and L.f. fuscus: The 1st prealternate molt in *intermedius* is quite variable but averages more extensive than that in *graellsii* and not as extensive as that in nominate *fuscus*; therefore, it is an "intermediate" molt strategy (Mars Muusse, pers. comm.). Hatch-year birds generally have restricted body molt until arriving on the winter grounds, where a surge in molt takes place in late winter through early spring. The 1st prealternate molt may include a variable number of rectrices (sometimes the entire tail) and secondaries. Less commonly, a variable number of primaries may be replaced before migration north to the breeding grounds.

The majority of nominate *fuscus* appear to undergo an extensive to complete 1st prealternate molt. This molt takes place on the wintering grounds, averaging more advanced than that of *intermedius* and considerably more advanced than that of *graellsii*. First-cycle birds returning to Europe in the spring with all flight feathers and tail feathers replaced are typically relegated to nominate, provided they're

43 Adult type with Great Black-backed Gull. Head markings and bill pattern suggest 3rd cycle. Perhaps the best-known nominate *fuscus* candidate recorded in North America. Triple black upperparts, sleek long wings, and noticeably thin bill. A leg band is needed to confirm provenance. KIRK ZUFELT. NEWFOUNDLAND. JAN.

44 Adult nominate *fuscus* type (right). Although unbanded, subspecific identification supported by limited primary molt at this date (only two inner primaries replaced). Also note truly black upperparts, shorter legs, slimmer body, longer wings, and thin bill. VILHELM FAGERSTROM. PORTUGAL. SEPT.

45 Adult *graellsii* type, on the darker end. Typically, with a complete subterminal band on p5, sometimes with no p9 mirror and little to no traces of white tongue tips on the outer primaries. Considerable black on both webs of p8–p9. Relatively narrow white trailing edge. AMAR AYYASH. PENNSYLVANIA. DEC.

46 Adult. Pale-end *graellsii* showing considerable contrast between black wingtip and upperparts. An all-white tip to p10 is somewhat rare, especially coupled with a thin subterminal band on p4. Small p9 mirror. AMAR AYYASH. ENGLAND. DEC.

47 Adult. Presumed nominate *fuscus*, identified by overall appearance and location. Long, narrow wings with little to no contrast with black wingtip. Small p10 mirror typical. Largely white-headed on nonbreeding grounds. AMIR BEN DOV. ISRAEL. APRIL.

supported by plain, dark upperparts. Furthermore, those showing three generations of primaries while undergoing the 2nd prebasic molt default to *fuscus*. The 2nd prebasic molt is generally restricted to a few inner primaries on the breeding grounds and commences once they've reached the nonbreeding grounds (E Africa and Middle East). Identifying individuals of unknown origin (without leg band) cannot be done with certainty once the 2nd prebasic molt has concluded (Altenburg et al. 2011). An adult-like plumage is typically acquired in three molt cycles.

HYBRIDS Two records of hybridization with American Herring Gull, both from the United States. The first of these, quite surprisingly, was outside Juneau, Alaska (June 1993). An adult Lesser Black-backed Gull was found consorting with a Herring Gull in a small mixed colony of American Herring and Glaucous-winged Gulls, observed on a nest with two eggs. No young were seen during a subsequent visit in late July (van Vliet et al. 1993). The second record spans from 2007–2011 and from 2013–2015 on Appledore Island, off the coast of Maine and New Hampshire. This event was well documented, as the island receives detailed coverage by biologists and ornithologists conducting summer fieldwork in a mixed colony of Herring and Great Black-backed Gulls. At least six hybrid chicks, as well as the adult Lesser Black-backed (male) and Herring were banded with green field-readable bands over several seasons. The Lesser Black-backed Gull, Green F05, and several hybrids were spotted up and down the Atlantic Coast during the nonbreeding season during this period (Ellis et al. 2014). There were no reports of Green F05 or any of the hybrid offspring after 2015. Hybridization in other American Herring colonies is very likely, as this hybrid has steadily increased as a winter visitor on the Atlantic Coast in recent years. Whether these are predominantly American Herring × Lesser Black-backed hybrids or include a mixture of European Herring × Lesser Black-backeds remains to be learned.

An adult individual found in St. John's, Newfoundland, in the winter of 2018 had intermediate characteristics of what was suspected of being a Lesser Black-backed × Kumlien's Gull (Alvan Buckley, pers. comm.; see plate 5, p. 437). It is worth noting that nominate Iceland and Lesser Black-backed Gulls have hybridized in captivity, producing two clutches of three eggs in which the young were reared to adulthood (Lonnberg 1919). Hybridization with European Herring and Yellow-legged Gulls reported in Europe (Olsen & Larsson 2004; Adriaens et al. 2012). Apparent hybrids with Ring-billed Gull suspected in Europe (Lowe 2012; see plates 6, p. 438, H8.20).

---

48 Adult *graellsii* type. Moderate black on underside of wingtip with smaller p9 mirror. Obvious slate-gray remiges suggest darker dorsal surface. Large red gonys spot and yellow legs. GARY THOBURNS. ENGLAND. APRIL.
49 Adult Great black-backed (left) and Lesser Black-backed Gull. The Lesser is still a way from completing its prebasic molt with tail feathers, most secondaries and p9–p10 old. It's not unusual to find some adult Lessers still growing p9–p10 well into Jan., especially throughout the southern parts of the winter range. BRIAN PATTERSON. NORTH CAROLINA. NOV.

25 KELP GULL *Larus dominicanus*

L: 21"–26" (54–65 cm) | W: 50"–56" (128–142 cm) | Four-cycle | SAS | KGS: 12.5–16.0

OVERVIEW The only gull with a circumpolar distribution throughout the southern hemisphere. It is *the* counterpart to the large white-headed gulls of the north. This gregarious black-backed species is among the darkest of gulls as an adult and is difficult to misidentify where expected. Highly coastal. Breeds primarily on small offshore islands and prefers this over mainland nesting. Its distribution has only increased since the mid-1970s, with surprising range expansions. An aggressive forager with a versatile diet. Kelp Gull's migration is of relatively short distance, with many populations being semisedentary.

Its presence in North America is very rare. However, there has been a recent uptick in sporadic records, spanning from California to Ohio, and from Ontario to Newfoundland. The species may be expected to turn up just about anywhere throughout the continent. Hybridization with American Herring Gull on the Chandeleur Islands in Louisiana's Gulf Coast, as well as the widespread increase in Lesser Black-backed Gull, has certainly changed the dynamics for field identification of this species. Still, adults are distinctive enough and may be identified without much trouble, and, indeed, almost all records in North America pertain to adult and adult-type birds.

TAXONOMY Tentatively, six subspecies recognized, mostly using biometrics (Jiguet 2002; Jiguet et al. 2012) and the first molecular-based phylogeny of the species (Sternkopf 2011). Jiguet et al. (2012) conducted a sex-separated analysis of morphometrics—a most welcome practice that should serve as a model for future field studies on gull populations. The conclusions reached by Sternkopf (2011) have not been reassessed by other workers and should be considered tentative. Confounding mismatches between morphological and molecular data remain. For instance, birds in Chile appear to be more closely genetically related to the African subspecies (*L.d. vetula*) than they are to the adjacent nominate (*L.d. dominicanus*) from mainland South America. It is unclear where in S South America populations shift from nominate *dominicanus* to *L.d. austrinus*, particularly along the coasts of Argentina and S Patagonia, and how birds from the Pacific (via Chile) connect to those on the Atlantic (via Argentina). Jiguet et al. (2012) proposed recognizing subspecies where genetic and phenotypic data was congruent—a difficult proposition.

L.d. dominicanus (KGS: 14–16): Dominican Gull. Brazil, N Argentina, Ecuador, Peru, and south to Chile. Not unusual for adults to show small p10 mirror and sometimes no mirror at all.

L.d. austrinus (KGS: 12.5–14.0): Antarctic Gull. Antarctic Peninsula, South Shetland Islands, and Falklands. Averages shorter wings, and shorter billed. Adult upperparts average more slaty and slightly paler than nominate's, showing an observable contrast between the black primaries and back. Broader white trailing edge, and consistently shows white tongue tips to p7 and a few to p8. Birds from Tierra del Fuego and other points in S Patagonia may be intermediate between mainland types and *austrinus* (more study needed).

L.d. vetula (KGS: 13.5–15.0): Cape Gull. South Africa. Characteristic darker iris and orange-yellow orbital (matched by a few birds in South America). An appreciable grayish-green cast to the legs that is almost always duller than bill.

L.d. judithae: Kerguelen Gull. Kerguelen Islands (subantarctic islands in S Indian Ocean). Adults almost always show two mirrors, and white tip to p10 not uncommon.

L.d. antipodus: Southern Black-backed Gull. New Zealand and nearby Chatham Islands. About 30% of birds show two mirrors as adults.

L.d. melisandae: Malagasy Gull. Madagascar. Phenotypic differences documented by Jiguet et al. (2012), but genetic relationship needs clarification.

RANGE (*L.D. DOMINICANUS* ONLY)
Breeding: Circumpolar in the southern hemisphere. Generally increasing, with a recent presence northward in the Americas. Nominate is discussed here, as multiple North American records suspected to be *L.d. dominicanus*. However, vagrants not safely assignable to subspecies as of this writing. On the Pacific Coast, breeds from S Ecuador, where it is locally common, south through Peru and Chile. On the Atlantic Coast, breeds from S Brazil and Uruguay south to Argentina. The breeding range has expanded north since the mid-1980s but is patchy, presumably where coastal habitat is suitable. In S Patagonia, population is believed to be nominate, although some *austrinus* may cross the Drake Passage from Antarctic Peninsula and winter here, as indicated by a banded hatch-year bird from Palmer Station Area found in Tierra del Fuego (Higgins & Davies 1996; Jiguet et al. 2001) and by sightings of adults with *austrinus* features (Alvaro Jaramillo, pers. comm.). The extent of oversummering *austrinus* in S South America and whether there is gene flow here between the two subspecies are unknown. Note that breeding times vary latitudinally in South America, with southern populations breeding later. Harrison et al. (2021) state breeding occurs year-round in the Peruvian Humboldt, but there is no evidence for this (Adriaens et al. 2023).

Small numbers reached the Gulf of Mexico during the late 1980s and 1990s, with up to several pairs breeding on Chandeleur Islands, Louisiana, and possibly on or near the Yucatán Peninsula, Mexico. But these birds fairly quickly began hybridizing with American Herring Gulls, and pure Kelp Gull pairings ceased. These few nesting birds were then displaced by Hurricane Katrina in 2005, but some had returned by 2015 (Oscar Johnson, pers. comm.).

First record for United States in July 1989 (Curlew Island, Louisiana), believed to be a nesting pair. During the following breeding season, another Kelp Gull was found here paired with a Herring Gull, leading to the discovery of an extraordinary hybrid event on the Chandeleur Islands (see "Other Hybrids" account).

Nonbreeding (North America only): Outside the Chandeleur Islands, casual mostly in fall and winter from coast to coast. North American birds have been found along shorelines and at landfills and nearby loafing sites in CA, CO (Semo 2007), DE, FL, IN, MD, OH, PA, TX, WV, ON, NS, and NL. States with multiple records, and Canada's records, are as follows.

Maryland: Record of a recurring wintering adult dubbed Shrimpy, first found in January 1999 and returning every year until March 2005, St. Mary's County and Calvert County. Small, square p10 mirror.

California: Records are believed to be of the same individual. First found in San Mateo County (April–May 2015) and subsequently in San Francisco County (May 2015), Los Angeles County (Feb. 2016), Southeast Farallon Island (April 2016), and San Mateo County (May 2016). Adult bird with relatively slim proportions, showing no p10 mirror. No active flight feather molt, but outer primaries frayed and worn. Presumed to be nominate from Pacific population.

Texas: Six records, which may involve repeat individuals.

Galveston County, 15 Jan.–5 April 1996: Adult following southern-hemisphere molt regimen.

Kleberg County, 4 May 1996: Adult type, no p10 mirror.

Galveston County, 30 Nov.–21 April 1997: Presumed to be same individual from 1996.

Brazoria County, 8 Nov.–24 Dec. 2008: Adult type in active flight feather molt with new p1–p2/p3, with remaining primaries and all secondaries old. Small, broken p10 mirror.

Brazoria County, 19 Dec. 2008: First cycle. To date, this is the most convincing 1st-cycle record in the ABA Area. A hybrid/backcross Chandeleur Gull is likely impossible to rule out without DNA analysis. No obvious features point away from a pure Kelp Gull. Also noteworthy is that an adult Kelp Gull was at this site on this date.

Cameron County, 20 April 2022: Adult with medium-sized p10 mirror. Fully intact flight feathers but with abraded tips to outer primaries.

Canada: Four records.

Essex County, Ontario, 7–9 Sept. 2012: Adult type in active flight feather molt, p8 growing, p9–p10 old; with small, square mirror on p10.

Mohawk Island, Haldimand County, Ontario, 12 July 2013: Adult in alternate plumage. Possibly the same individual from 2012.

St. John's County, Newfoundland, 23–25 Dec. 2016: Northernmost occurrence in the ABA Area. Adult with p10 growing with medium-sized mirror. Inner secondaries molting.

Halifax County, Nova Scotia, 21–27 Jan. 2018: Adult with all flight feathers fully grown. Medium-sized p10 mirror.

South of the ABA Area, Kelp Gull is rare on the northern coast of the Yucatán Peninsula, with around 12 reports that I know of, beginning perhaps as early as 1989 (Howell et al. 1993). At least 16 reports for all of Mexico, but not officially voted on by any ornithological body. At least 3 unpublished reports from Tamaulipas that are convincing and should be added to the local record. Recent reports in Colima (Jan. 2013) and Sinaloa (Feb. 2019) suggest the species may also make it to Baja, where there are no current records. Central America and the Caribbean Sea have also seen a marked increase in reports, mostly from Panama City. In June 2021, 10 individuals were discovered together in Falcon, Venezuela, furnishing a first record for the country (Freddy Antonio Velasquez, pers. comm.). A similar situation has unfolded in N Colombia, which, incidentally, coincides with the rapid range expansion of Lesser Black-backed here from 2001 to 2023.

IDENTIFICATION (*L.D. DOMINICANUS* ONLY)

Vagrants to North America tentatively assigned to the nominate subspecies, *L.d. dominicanus*, although origins unknown.

Adult: Black upperparts show little to no contrast with black wingtip. Similar in size to Herring Gull, but stronger looking, with some structural differences that recall Great Black-backed. Powerful bill with a bulge to the gonys, less so in female types. Many birds show a noticeable downward gape. Generally long legged (but see subspecies *austrinus*, plate 25.30). White tertial crescent is often broad and merges with a drooping white secondary skirt when feathers are fully grown. Leg color varies in intensity from yellow to greenish yellow. Leg and bill color seldom concolorous, with the bill typically brighter yellow. The gonys spot can be boldly dark red, large, and circular, at times with red spilling over to the upper mandible. Eye color is variable, ranging from light to a honey brown and to all dark. Orbital rings commonly reddish or a

---

1 Juvenile. Powerful bill with deep gonys, flat head, and dark face. Bulky body. Largely plain brown greater coverts. ETIENNE ARTIGAU. CHILE. MARCH.

2 Largely juvenile. A rusty-brown aspect with darker underparts. Plain inner greater coverts with pale edging. Already showing wear on upperparts and bleaching/molt on neck, reflecting earlier hatch date of this region. AMED HERNANDEZ. URUGUAY. JAN.

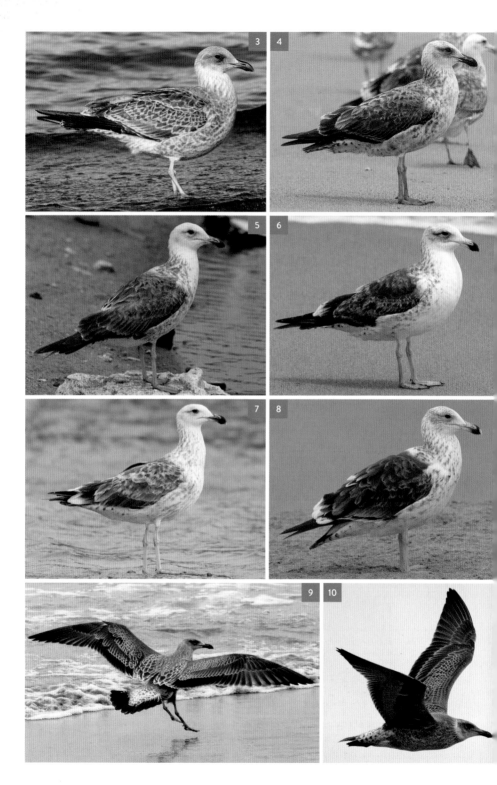

3   Juvenile. Presumed nominate. A colder, grayish-brown aspect with more patterned inner greater coverts. Note larger spotting on flanks and fresh white edging on tertials. Lesser Black-backed not in juvenile plumage at this date. RICHARD BONSER. PUNTA ARENAS, CHILE. APRIL.
4   1st cycle. Bulky structure with large, blocky head, powerful bill, and longer legs. Very commonly shows exposed secondaries beneath plain greater coverts. Most scapulars replaced. Whitish vent with sparse barring. Dark tail. AMAR AYYASH. PERU. NOV.
5   1st cycle. A long-winged individual, with slimmer appearance recalling Lesser Black-backed, which does not typically show such uniform inner greater coverts. AMAR AYYASH. PERU. NOV.
6   1st cycle. All scapulars, upper tertials, and several wing coverts replaced (1st alternate) already. Note thick legs, longer bill with slight droop, and dark postocular line. AMAR AYYASH. PERU. OCT.
7   1st cycle. Recalls Great Black-backed with white underparts and vent, and blocky head. Long legs and bluish-green tibia not rare. Extensive 1st prealternate molt with most upperparts renewed and new incoming 2nd-generation primaries (old p10 down to shaft). AMAR AYYASH. PERU. OCT.
8   1st cycle. Extensive 1st prealternate molt in some Peruvian birds renders them almost inseparable from 2nd cycle (as found in some Yellow-footed Gulls), but note pointed brown 1st-generation primaries at this date. AMAR AYYASH. PERU. OCT.
9   Juvenile. Plumage similar to Lesser Black-backed, but averages plainer inner greater coverts, more extensive dark tail, and less patterned uppertail coverts. Compare bill size and shape. DAVE CZAPLAK. BRAZIL. NOV.
10 Juvenile. Dark brown wing linings, dark inner primaries seldom show contrasting pale inner webs. Plain greater coverts. JOSE LUIS BLAZQUEZ. BUENOS AIRES, ARGENTINA. DEC.
11 1st cycle. Head and underparts now with obvious white base color, but note dark wing linings, long droopy bill, and dark inner primaries. Widely spaced, even barring on undertail coverts somewhat variable. ROGER AHLMAN. ECUADOR. JULY.
12 1st cycle. Presumed one-year-old. Plain postjuvenile scapulars from limited 1st prealternate molt (often highly barred/anchored on Lesser Black-backed). Dark tail with largely white coverts and plain inner greater coverts uniform in appearance. AMAR AYYASH. PERU. NOV.
13 1st cycle. Plain and dark postjuvenile scapulars. Some individuals have a relatively narrow tailband with mostly white uppertail coverts. Bulbous bill tip apparent. AMAR AYYASH. PERU. OCT.
14 1st cycle. Same individual as 25.6. Limited 1st prealternate molt on a presumed one-year-old. Strong bill, dark tail, and plain greater coverts. Whitish head and bicolored bill not rare at this date. AMAR AYYASH. PERU. OCT.

15 1st cycle. All scapulars and tail feathers, as well as many wing coverts and inner secondaries, replaced in 1st prealternate molt. Still, note plain pattern on coverts and scapulars and massive bill. AMAR AYYASH. PERU. OCT.

16 1st cycle. Extensive 1st prealternate molt with all secondaries and rectrices replaced. A few Peruvian birds exhibit eccentric primary molt, replacing a variable number of outer primaries while retaining juvenile inners. AMAR AYYASH. PERU. OCT.

17 1st cycles. Extremes in molt patterns and plumage at this age. The individual on the right has undergone a complete 1st prealternate molt with all flight feathers replaced, while, on the left, all flight feathers and most wing coverts are juvenile. Bulky bodies and massive bills distinctive. AMAR AYYASH. PERU. NOV.

18 2nd cycle. Completing 2nd prebasic molt with outer primaries not fully grown. Overall plain upperparts, large bill, and thick legs. RICHARD BONSER. CHILE. APRIL.

19 2nd cycle. Presumed nominate. Adult-like scapulars, but wing coverts largely plain with pale edging. A shorter-legged and shorter-billed individual with compact body and small rounded head (female?). JOHN CHARDINE. TIERRA DEL FUEGO, ARGENTINA. DEC.

20 2nd cycle with 1st cycle. Blocky head with strong bill, advanced and adult-like. The 2nd prealternate molt in Peruvian populations sometimes produces a plumage that very much resembles 3rd cycle. Visible primaries 2nd generation. AMAR AYYASH. PERU. OCT.

21 2nd cycle. A somewhat delayed individual with a mix of brown and slate-colored scapulars and much pale edging on the lesser and median coverts. BERNARDO ALPS. MAGALLANES, CHILE. NOV.

22 2nd cycle. Variable brown on wing linings. Advanced, adult-like body and bill pattern. Dark eye and greenish-yellow legs. All-dark tail fairly standard in 2nd basic. PETER KENNERLEY. VALPARAISO, CHILE. OCT.

23 2nd cycle. Similar to 25.20, this individual appears much more like a 3rd-cycle large gull, but note 2nd-generation primaries. All secondaries and their respective coverts, most tail feathers, scapulars, and head feathers replaced in advanced 2nd prealternate molt. AMAR AYYASH. PERU. NOV.

24 2nd-/3rd-cycle type. Adult-like upperparts, but note extensive head streaking, bill pattern and dull legs. New p5–p7 (white tips), old p8–p10 (duller brown). Given time of year and location, this molt limit suggests an advanced 2nd alternate bird rather than 3rd cycle. In other parts of South America, such an appearance would default to 3rd cycle. DAVID BEADLE. PERU. OCT.

25 3rd-cycle type with 1st alternate individuals. Long legs, broad secondary skirt, and powerful bill. Small p10 mirror and abraded primary tips inspected on open wing, although secondaries appear new (4th alternate). Aging such birds can prove problematic. AMAR AYYASH. PERU. NOV.

26 3rd-cycle types. Similar to 25.25, an open wing revealed older primaries with abraded tips (not fresh and recently replaced), although individual on left replacing secondaries. Note variation in head shape. Thick, yellow legs duller than bill. AMAR AYYASH. PERU. NOV.

rich orange red, but a few orange yellow or simply yellow. On the open wing, note the broad white trailing edge to the secondaries and innermost primaries. The leading edge can show a lot of white along the wrist and arm, similar to some Slaty-backed adults. Primary pattern is variable, but most individuals recorded in North America have shown a medium-sized, squarish p10 mirror. However, some with no p10 mirror at all, such as the individual sighted in California in 2015–2016. Pale tongue tips variable, from having thin crescents on p4–p5/p6 to completely absent. White headed in alternate plumage. In basic plumage, the head is also largely white with light, faint streaks, mostly on the mid- to lower hindneck. May show a brownish hue on upperparts that become worn and faded. White apicals can also appear nonexistent due to heavy wear. *Similar Species:* Other black-backed gulls with yellow legs are Lesser Black-backed and Yellow-footed. Regional overlap with Yellow-footed is not an immediate concern. Yellow-footed is paler, and many show a bluer hue to the upperparts. Also note the mustard-yellow orbital on Yellow-footed, as opposed to the more common orange-red orbital of Kelp (although some may be yellowish). Lesser Black-backed is the primary confusion species in the ABA Area and points south. Bill structure is a key field mark in this identification. Kelp typically shows a larger and thicker bill and often has a bulbous tip not typical of Lesser Black-backed. At the very least, Kelp will show a consistently stouter bill, and the gonys spot is generally restricted to the lower mandible and averages smaller than that of Lesser Black-backed, in which species the red can appear as a large blob and is sometimes found above the cutting edge. In addition, most Lesser Black-backeds in North America average paler black upperparts than Kelp Gull. Overall, Lesser Black-backed is slimmer, with an attenuated long-wing appearance, but some female-type Kelps may be similar in size and structure. Kelp averages a more chesty appearance, with a squarer head and longer legs. Kelp is broader winged, and this is quite noticeable upon comparison. Leg color on Kelp, especially in low breeding condition, is duller and more greenish gray than the yellow on Lesser Black-backed. On the open wing, the white trailing edge in Lesser Black-backed is not as wide and bold as in Kelp. At rest, Kelp Gulls show a larger white tertial crescent, but this should be used with caution, especially on molting or worn birds. Also of importance is head streaking in basic plumage: any adult with substantial head and neck streaking is not a Kelp Gull.

In recent years, separation from American Herring × Kelp hybrids (Chandeleur Gull) has become an identification obstacle. It should be noted that in the Gulf of Mexico, hybrids are more likely than pure Kelps, but data is evolving. For example, birds found on the N Yucatán Peninsula appear to be pure Kelps, but birds found throughout Louisiana to Alabama are almost always suspected to have Herring influence. Adult hybrids average paler upperparts and long, bulbous-tipped bills and have moderate to heavy head streaking in basic plumage. Primary patterns of hybrids need more study, due to small samples, but black found on more hybrid primaries (down to p2–p3 on some) than on those of pure Kelps, showing irregular subterminal bands.

1st Cycle: Juveniles are a warm brown with pale-edged upperparts. Black bill. Dusky pink legs. Some dark flecking on the chins and around the joints is observed for a considerable period in 1st cycle. The greater coverts are typically plain with dark bases and paler tips, sometimes notched, especially on the

---

27 3rd cycle. Typical prebasic molt sequence beginning at inner primaries (broad white tips to p1–p3). All other flight feathers 2nd basic. Note extensively dark tail, broad wing, and massive bill. JIM WILSON. USHUAIA, ARGENTINA. JAN.

28 3rd-cycle type. Note brown coverts. Adult-like white trailing edge, all-white tail, and yellow bill with red gonys spot. No p10 mirror and dark eyed. JON IRVINE. PERU. JULY.

29 Adult. Black upperparts show little contrast with primaries. Broad white tips to secondaries form a skirt. Bulbous bill tip, deep red gonys spot, red orbital, and pale eye. AMAR AYYASH. PERU. OCT.

30 Adult. Presumed *austrinus* ssp. Short legs, short bill, and black upperparts regularly show a paler, slaty-like appearance. JOHN CHARDINE. DECEPTION ISLAND, ANTARCTICA. FEB.

31 Adult. African ssp. *vetula* (Cape Gull). Dark eye, orange-yellow orbital, and massive bill. Long, greenish-gray legs. Small p10 mirror. RICHARD BONSER. NAMIBIA. AUG.

32 Adult with Laughing Gulls. This individual was the sixth state record for Texas. Bulky gull with heavy chest and longer legs (typically not concolorous with bill). Darkish eye and jet-black upperparts. Worn apicals. ALEX LAMOREAUX. TEXAS. APRIL.

33 Adult with similar-aged Belcher's Gull (left). Note grayish wash on neck and underparts of Belcher's, which has a longer and slimmer bill with black/red pattern. AMAR AYYASH. PERU. OCT.

34 Adult. Broad white trailing edge to secondaries often extends to inner primaries. Black wingtip shows little to no contrast with upperparts. Small p10 mirror not uncommon. AMAR AYYASH. PERU. NOV.

inner greater coverts. The head, neck, and breast have dense streaking with a white base color. A dark postocular line is often visible, as well as dark shadowing surrounding the eye and face. On the open wing, the uniform brown upperparts are maintained with little to no trace of a pale window. The underwing is a smoky brown with some barring on the axillaries. The tail is largely dark with white tips, and occasional barring and spotting are seen along the outermost tail feathers, particularly along the bases. The uppertail coverts have variable thin brown barring, but the base color of the tail coverts is a flashy white, boldly contrasting with the entire dorsal side of this plumage. The undertail coverts are also white with well-defined segments and barring that is evenly spaced. Postjuvenile scapulars are variable but are plain with dark centers, or show thin shaft streaks, or have dark anchor markings. Those 1st cycles replacing tertials will often show noticeably broad white tips. Later in 1st cycle, the bill begins to show a pale base with variable black along the cutting edge, accentuating a bulbous tip. This gives the bill a semi-drooped look. At times, the bill may show a demarcated black tip with an all-pale base, but this is exceptional. Also late in 1st cycle, the dull pink legs may take on a grayish-blue color, especially on the upper tibia.

Similar Species: The identification of 1st-cycle Kelp Gulls in the ABA Area remains tricky. In North America, the primary confusion species is Lesser Black-backed. Structure is indispensable here, as Kelp appears more heavyset and stockier throughout, but some overlap may exist among female Kelps and male Lesser Black-backeds. Lesser Black-backed is rangier, with a longer and thinner impression to the body behind the legs. The bill on Lesser Black-backed is thinner and not as stout through the tip. Head streaking is finer on Lesser Black-backed and not as densely spotted and mottled, especially on the lower neck. The 2nd-generation upperwing coverts on most Lesser Black-backeds will show more intricate

---

35 Adults. Subspecies *austrinus* is distinctively more compact than nominate, averaging shorter legs, bill, and shorter wing projection. OTTO PLANTEMA. FALKLAND ISLANDS. DEC.
36 Adults. Broad white trailing edge and contrasting leading edge. Nominate often shows a small, squarish p10 mirror and overall plain wingtip. DOUGLAS FAULDER. PERU. JAN.
37 Adult type. This individual shows a complete subterminal band on p4. The absence of a p10 mirror is not rare in nominate types from western South America. FABRICE SCHMITT. PUNTA ARENAS, CHILE. NOV.

patterning and overall with paler centers. Finally, the eye on Kelp appears to be higher and more forward than the evenly centered eye of Lesser Black-backed. On the open wing, Lesser Black-backed may show more of a window, display longer and thinner wings, and average a narrower tailband. Compared to larger species in W North America, Kelp averages less of a bulbous-tipped bill and a shallower bill base than Western and Yellow-footed, but some overlap exists. The outer edges of the outer tail feathers on the two northern species are typically entirely dark, and those species average slightly paler inner webs to the inner primaries. Finally, some paler Kelps can take on a white base color to the upperparts, especially those with extensive postjuvenile molts, recalling Great Black-backed Gull. But Great Black-backed is noticeably larger, will inevitably show a more peppered and checkered tailband, and has a thicker bill and blockier head. Also consider Kelp × Herring hybrids, which, admittedly, are poorly known at this age. **2nd Cycle:** Mostly white head with finer, diffuse streaking on crown and hindneck. Lower neck streaking becomes distinctly spotted and coarse. Faint postocular line often seen. Eyes variable, typically dark, but some becoming honey colored. Bill generally pink based with a mixture of black smudging and dark along cutting edge. Some advanced birds may show yellow coloration to bill similar to adult's, with dark tip and red on gonys; others remain almost entirely black billed, as in 1st cycle. Feathers on 2nd cycles show variable dark brown centers, with grayish-black scapulars approaching adult's. The wing coverts are generally plain when compared to those of other large gulls of the north. The tertials have dark centers which are dominated by broad white edges and tips. On the open wing, the outer primaries are largely black, while the inner primaries are a dark brown with slightly paler inner webs, at times with noticeable white subterminal pearls on inner webs. Individuals with these markings will often show broader white edging on the tips of the inner primaries, but not as broad as in 3rd cycle. Thin white trailing edge to the secondaries. Uppertail coverts predominantly white with mostly black tail, but with less black on outermost tail feather when compared to 1st cycle. Body and vent region largely white with intermittent dusky markings. *Similar Species:* Lesser Black-backed averages a thinner bill, which is not as stout out to the tip. Lesser Black-backed is longer winged, with thinner tailband, and averages a richer leg color at

38 Adult. Small p10 mirror and extensive black on underside of wingtip. Orange-red orbital, orange gape, and heavy bill. AMAR AYYASH. PERU. NOV.

39 Adult, ssp. *austrinus*. Regularly shows noticeable white tongue tips to p5–p7, and some to p8. Note restricted black on underside of wingtip. Undergoing prebasic molt (p1–p2 growing, p3 dropped, p4–p10 old). STEVE KELLING. ANTARCTICA. JAN.

40 Adult, ssp. *austrinus*. Commonly shows white tongue tips to p7, and some to p8, as seen here. Note the broad white trailing edge extends to inner primaries. Yellow legs typically don't match bill color. JINING HAN. ANTARCTICA. NOV.

this age. Markings on upper flanks and lower hindneck are more spotted and blotchier on Kelp compared to finer streaking found in Lesser Black-backed. Second-generation tertials are typically not so broadly tipped with white on Lesser Black-backed. Body structure and timing of molt should be useful in separating the two.

3rd Cycle: Similar to adult but with variable brown on upperparts. Averages duller bare parts and more black markings on bill, some with complete black subterminal tip. May show light to moderate head markings that lightly extend onto the lower neck and breast. Generally has a thinner tertial crescent than adult, with more black intruding onto the feather centers. Faded adults or those with worn-down apicals may be confused with this age. On the open wing, black on secondaries extends deeper into the feather centers and is more extensive than adult's, resulting in a weaker trailing edge. Often with no mirrors, and may show variable black on uppertail.

MOLT Molt timings variable, based on geography and breeding schedules. In South America, complete molt in southernmost populations appears to commence earlier in the calendar year, while northern populations average slightly later molts (Jiguet et al. 2001; Howell & Dunn 2007). First prealternate molt variable, from limited (S and E South America) to extensive, especially in nominate population along the Pacific Coast of Peru. Peruvian birds appear to undergo an accelerated and extensive prealternate molt, whereby many replace rectrices and a variable number of secondaries by November, and a small percentage replace a variable number of primaries (rarely all). Primaries are typically replaced last in these birds, at times in an irregular and nonsequential pattern (Ayyash, pers. obs.), and it is not unusual to see several primaries dropped at once, resulting in large molt gaps in what are presumed to be largely resident birds. Similarly, many 2nd-cycle Peruvian birds are found with new secondaries and rectrices (2nd alternate) by November, and about one-third will have molted some primaries as well. Those that undergo an extensive or complete 2nd prealternate molt very much resemble adults with retarded patterns and may prove difficult, if not impossible, to age out of context (Adriaens et al. 2023).

HYBRIDS Small-scale hybridization with American Herring Gull on the Chandeleur Islands. First suspected in 1989, and confirmed in 1990. The Chandeleur Islands make up an important bird area, found some 50 miles off the coast of Louisiana in the Gulf of Mexico. The islands have the misfortune of being pummeled by one hurricane after the next, with Hurricanes George and Katrina washing away much of the land mass here in 1998 and 2005, respectively. In the late 1990s, a small colony of roughly 75 individuals of mixed pairings and backcrosses was reported on Curlew and Gosier Islands. In 2004, 18 pairs of hybrids were recorded, with both first- and second-generation (F1 and F2) hybrids present. No large gulls were found breeding on the islands immediately after Hurricane Katrina, but, as they did after Hurricane George, the gulls eventually returned; they were found breeding here in small numbers in 2015. In 2019, at least 30 hybrids were found with active nests (Oscar Johnson, pers. comm.). A medley of upperpart coloration was documented, with some birds slightly paler than pure Kelp and others slightly darker than pure Herring.

The recent hybrid phenomenon in the Gulf of Mexico has produced some puzzling birds and further complicates identification of Kelp Gulls in North America. For instance, the last accepted Kelp in Louisiana was in 2000 at the time of this writing, but photos taken since then suggest some pure birds may still be present in this region, although most are suspected of being hybrids. Lighting conditions, molt, and the time of year are critical factors when deciding which labels these individuals receive. Adult hybrids have a long-legged appearance and tend to show more head streaking than expected in pure Kelps; streaking is reminiscent of Herring Gull's head pattern put patchy and less consistent throughout. Identification of young Kelp Gulls in the ABA Area has for the most part been unbroached and has become trickier with the presence of hybrids and young Lesser Black-backed Gulls. The strongest candidate I know of for a pure 1st-cycle Kelp Gull comes from Brazoria County, Texas (Martin Reid, pers. comm.). An article on field identification of this hybrid combination, including younger birds, was published by Dittman and Cardiff (2005) using museum specimens that they collected.

A report exists of a Cape Gull (*L.d. vetula*) hybrid suspected with Yellow-legged Gull (*L. michahellis*) in S Morocco (Jonsson 2011).

26 WESTERN GULL *Larus occidentalis*

L: 21"–26" (53–66 cm) | W: 53"–55" (135–140 cm) | Four-cycle | SAS | KGS: 8–11

OVERVIEW Suitably named, Western Gull is restricted to parts of the Pacific shores of the United States, south to the Baja California peninsula of Mexico, and north to SW British Columbia. It is the darkest large white-headed gull throughout its range, except where it overlaps with Yellow-footed Gull in Baja California. Western Gull is a regular fixture of beaches, marinas, river mouths, and urban parking lots along the Pacific Coast and generally isn't found very far inland. A generalist with some rather interesting feeding behaviors, it catches live fish and marine invertebrates, scavenges refuse, plunders eggs, and has been recorded eating dead seal pups and sometimes stealing milk from lactating females. Can be seen feeding 1st-cycle conspecifics—presumably their own young—year-round, into the start of the next breeding season (Byron & Joanna Chin, pers. comm.). Regularly feeds offshore. Nests on small rocky islands along the coastline. Approximately one-third of the world's population nests on Southeast Farallon Island in California—the largest known colony for the species. In the 1970s, female–female pairings of Western Gulls were discovered nesting on Santa Barbara Island after an increasing number of supernormal clutches of four to six eggs were investigated. This was the first documentation of homosexual pairing in gulls, and likely the first for all wild birds (Hunt & Hunt 1977). The longevity record for this species is from a known-age adult in California found alive at the time just one month shy of 34 years old (Kennard 1975).

TAXONOMY Among the large white-headed gulls of North America, Western Gull is a primitive taxon with a highly distinct mitochondrial DNA sequence (de Knijff et al. 2005). Surprisingly, it shows no close genetic relation to Glaucous-winged Gull, with which it frequently hybridizes. Its closest relative is likely Yellow-footed Gull, which was once treated as a subspecies of Western.

Two subspecies. Not always assignable, especially in parts of central California, where clinal step pervades. Average differences in body size, upperpart coloration, and eye color exist, but marginally so in many individuals.

L.o. occidentalis (KGS: 8.0–9.5): Nominate *occidentalis* is found throughout the northern portion of the range. Has paler upperparts and averages slightly larger proportions. Gray tone has a bluish hue. Populations farthest north are palest, may show more dusky head markings, and average darker eyes. Some of this is likely due to introgression with Glaucous-winged and should be assessed on a case-by-case basis (see this account's "Hybrids" section and the Glaucous-winged × Western hybrid account).

L.o. wymani (KGS: 9.5–11.0): Found throughout the southern portion of range, *wymani* averages slightly smaller (Bell 1992). Darkest adults have a slaty-black quality to their upperparts and are in line with true black-backed species. Black wingtip shows little contrast with upperwing, and white tongue tips on p5–p6 are often minimal or absent.

RANGE

Breeding: Subspecies *L.o. occidentalis* nests from central California northward, and subspecies *L.o. wymani* to the south. Common resident along Pacific Coast from S Washington south along the coasts of the Baja California peninsula. Nests mostly in small groups or colonies on islands, mainland cliffs and ledges, piers, and buildings. Reports of breeding farther north in Washington and in SW British Columbia are problematic, due to the difficulty in distinguishing pure Western Gulls from hybrid Glaucous-winged × Western Gulls, but Bell (1997) reported apparently pure Westerns in Juan de Fuca Strait. Since 2001, a few pairs have also bred in the Gulf of California in N Sinaloa, Mexico (Howell & Dunn 2007).

Nonbreeding: Mostly sedentary, with short-distance movements. Some populations nonmigratory. Frequents a wide variety of marine and other coastal environments, including offshore out to 40 miles, sandy and rocky coastlines, mudflats, harbors, lakes, parks, parking lots, fish-processing plants, and landfills and nearby loafing areas. Fairly common visitor north to the W Strait of Juan de Fuca in Washington / British Columbia, mostly between early fall and early winter; uncommon farther east to Vancouver area and Puget Sound, north to N British Columbia, and south to the southern tip of Baja California. Pure birds are very rare to casual north to SE Alaska (including one bird banded on the Farallon Islands, California). Small numbers are found inland up the Columbia River as far as SE Washington and southward in central Oregon to the central Willamette Valley. They are rare but have increased in recent decades and are now regular in small numbers inland to the Salton Sea in SE California and in the N Gulf of California, Mexico. Casual south along the Pacific Coast to S Mexico. In N California, Western Gulls occur regularly in the Delta region of inland San Francisco Bay / Sacramento Valley, and in S California they have increased as year-round visitors to lakes and landfills on the coastal plain up to 20 miles from the coast, where they are fairly common to locally common. Rare elsewhere in inland Oregon and California. At the time of this writing, there are 14 records in Utah, where it is recorded more than in any other state in the interior West. Very rare to accidental visitor farther east, such as NM, CO, TX (six records), NY, WY, AB, and SK; and westward to south-central Alaska, Hawaii, and Korea. Several reports of individuals outside the normal range are problematic, due to difficulty distinguishing pure Western Gulls from hybrid Glaucous-winged × Western Gulls, which also have been recorded north to Alaska and well inland, such as in Alabama and Florida. For example, a long-standing record from Chicago, believed for many years to be the only "good" Western Gull for E North America, was recently delisted and found to be a hybrid (Ayyash 2014). A 2006 record from offshore Long Island, New York, is extraordinary, serving as the first and only known occurrence of the species on the Atlantic (New York State Avian Records Committee 2009, 25–27).

IDENTIFICATION

Adult: Dark backed, with dull-pink legs. Adults largely white headed throughout the year. Some head markings not rare, however, especially during complete molt in late summer through early fall, more so in northern population. Yellow to orange-yellow orbital ring, at times mustard yellow. Eye color variable, from pale and dusky to dark olive and exceptionally dark in some. Thick, banana-yellow bill generally shows a bulbous tip. At rest, moderate tertial crescent merges with fully grown secondaries, often showing a secondary skirt. *Broad wings:* On the open wing, the broad white trailing edge to the secondaries boldly contrasts with the dark upperwing. White tips to inner primaries tend to taper compared to those on secondaries. Wingtip with moderate black, typically with complete band on p5, and some with broken subterminal band on p4. Black on outer webs of p7/8–p10 regularly reaches primary coverts. Mirror on p10 averages intermediate size, with narrow black subterminal band. All-white tip to p10 very rare; more likely to have broken black subterminal band with white outer vane and black

---

1 Juvenile. Fresh birds have dark brown head and underparts. Eyes placed relatively high on the face. Pale edging throughout upperparts, often with dark bases to outer greater coverts. ALISON DAVIES. CALIFORNIA. AUG.
2 1st cycle. Dark face mask and well-marked underparts, including vent region and undertail coverts. A stocky bird with dilute appearance to wing coverts. ALEX A. ABELA. CALIFORNIA. OCT.
3 1st cycle. Large head and relatively small eye, at times appearing very beady. Many juvenile scapulars replaced, but all wing coverts juvenile. Note largely dark tail. Bill typically becomes pale along base of lower mandible at this age. BARBARA SWANSON. CALIFORNIA. DEC.
4 1st cycle with similar-aged Glaucous-winged Gull (left). Size and shape can be very similar in these two. Western has darker remiges, a less uniform appearance, often due to replaced scapulars, and a pale bar running across the tips of the greater coverts tract. MARIO BALTIBIT. CALIFORNIA. OCT.
5 1st cycle with rather dilute look to wing panel. Dark grayish-brown 1st alternate scapulars are fairly common. The underparts often become pale by early winter. ALEX A. ABELA. CALIFORNIA. JAN.
6 1st cycle. Hatched on Farallon Islands June 2020. Now with evident wear and fading throughout, with head and neck becoming noticeably pale. Diffuse black on bill base with yellowish cast similar to some Yellow-footed Gulls. PHIL PICKERING. OREGON. MAY.
7 Juvenile. Overall broad dark wings with little to no sign of a contrasting inner-primary window. Dark bases to outer greater coverts. ALEX A. ABELA. CALIFORNIA. AUG.
8 Juvenile. Contrasting white uppertail coverts typically show dense patterning (spotting and barring). Dark tail. Chocolate brown plumage, generally without pale inner-primary window. STEVE ROTTENBORN. CALIFORNIA. AUG.

26 WESTERN GULL

9. 1st cycle. Dark tail base and dense brown head and neck recall American Herring, but note thicker and more bulbous-tipped bill, larger head, and beady eye. Washed-out upperparts this early in the season and hints of a primary window invite thoughts of Glaucous-winged influence. BRIAN SULLIVAN. CALIFORNIA. NOV.

10. 1st cycle. Broad wing with fairly uniform primaries, dark tail, and contrasting white uppertail coverts with barring. Sooty-brown head and neck. Deeply notched inner greater coverts on this individual. GREG GILLSON. CALIFORNIA. OCT.

11. 1st cycle. Late-season individual with variable 1st alternate scapulars. Many individuals have a fairly noticeable white trailing edge to the secondaries. JOACHIM BERTRANDS. CALIFORNIA. FEB.

12. Juvenile. Sooty-brown face and underparts have smooth texture continuing to vent region. Patterned undertail coverts and dark tail base. Plain, brown wing linings. PAUL FENWICK. CALIFORNIA. AUG.

13. 1st cycle. Some juveniles can have a surprisingly pale belly, at times whiter than this, inviting confusion with Yellow-footed. That species typically has little to no noticeable barring on the uppertail coverts and, month for month, shows a whiter head, more wear, and/or accelerated molt. ALEX LAMOREAUX. CALIFORNIA. AUG.

14. 1st cycle. Smooth texture to underparts and densely patterned tail coverts recall American Herring, but note very broad wing, bulbous bill tip, and distinct white tips to secondaries. AMAR AYYASH. CALIFORNIA. JAN.

15. 1st cycle. Late-season individual with paler underparts now and pale, pink bill base. Long, broad wings and fairly uniform inner primaries distinctive in range. SASHA CAHILL. CALIFORNIA. APRIL.

16. 2nd cycle. Completing 2nd prebasic molt with outer primaries still not fully grown. Primaries and tail now more blackish. Plain upperparts with marbled appearance. Regressive bill pattern largely black, but variable. AMAR AYYASH. CALIFORNIA. SEPT.

17. 2nd cycle with juvenile (left). Plain 2nd basic upperparts showing paler plumage. Large head with proportionally small eye. Bill with diffuse pattern, not typically sharply bicolored at this age. ALEX A. ABELA. CALIFORNIA. AUG.

18. 2nd cycle hatched on Farallon Islands (2021). Now with white underparts and adult-like, gray scapulars (2nd alternate). Smudged head pattern not unusual in this species, which is also found in hybrids with Glaucous-winged. STEVE HOVEY. CALIFORNIA. FEB. 2023.

19. 2nd cycle. Rather retarded and dilute plumage aspect, demonstrating the wide range of variation in this species. Diffuse black on pinkish bill base. ALEX A. ABELA. CALIFORNIA. JAN.

20. 2nd cycle. Similar to 26.18, but with very plain 2nd basic wing panel. Yellow bill base with large black tip. White head (2nd alternate), dark eyes, and pink legs. MIKE SEWELL. CALIFORNIA. FEB.

21 2nd cycle with fairly worn and bleached wing panel. Bright, adult-like, yellow bill is found in some younger birds, particularly in spring. Olive-gray iris with yellow orbital and dull pinkish legs. KENT LELAND. CALIFORNIA. MAY.

22 2nd cycle. Fairly plain 2nd basic individual with broad wing, blackish primaries, and large blob-tipped bill. BRIAN SULLIVAN. CALIFORNIA. DEC.

23 2nd cycle. Fairly typical individual with dark tail, white uppertail coverts, grayish-brown upperwing, and dark remiges. Regressive bill pattern. AMAR AYYASH. CALIFORNIA. JAN.

24 2nd cycle, same as 26.21. Largely bleached wing panel with few gray (2nd alternate) inner median coverts. Some 2nd cycles can have surprisingly broad white tips to secondaries and inner primaries, but note brown (not gray) centers. KENT LELAND. CALIFORNIA. MAY.

25 3rd cycle. P10 still growing, with small mirror (underside of far wing). Some individuals retain much smudging on the head and neck, with regressive bill pattern at this age. Black on tail and brown on coverts expected. ALEX A. ABELA. CALIFORNIA. OCT.

26 3rd cycle. More advanced than 26.25, but note later date. 3rd prealternate molt has presumably resulted in white head and newer (fresher) upperparts. Note lack of adult-like apicals on primaries. Presumed *wymani* ssp. ALISON DAVIES. CALIFORNIA. JAN.

27 3rd cycle (center) with adult and 1st-cycle Western Gulls. A fairly typical appearance, with bulky body, large head, and powerful bill. PHIL PICKERING. OREGON. FEB.

28 Subadult features include smaller white apicals, brown wash to greater coverts, and bill pattern. ALEX A. ABELA. CALIFORNIA. NOV.

29 3rd cycle. Complete "tailband" is found in some individuals, but averages narrower now. Much brown and black marking across upperparts and small p10 mirror, fairly common at this age. JIM TIETZ. CALIFORNIA. NOV.

30 3rd cycle. Late-season individuals are often white-headed, with brighter bills. This individual shows a rather delayed upperwing. ASHER PERLA. CALIFORNIA. MAY.

31 3rd-cycle type with advanced features. Black streaking on primary coverts and tail excludes an adult label. Superficial resemblance to Lesser Black-backed, which differs in size and shape. JOACHIM BERTRANDS. CALIFORNIA. FEB.

32 3rd cycle. Massive bill, smudged head pattern, and broad white trailing edge are classic. Some residual black and brown markings remain on the wing linings at this age. Note asymmetric p10 mirror. BILL BOUTON. CALIFORNIA. OCT.

inner vane, but this pattern too is rare. Up to 5% reported to have mirror on p9 (Howell & Dunn 2007). Often with white tongue tip to p5, sometimes to p6–p7, very rarely a thin sliver to inner web of p8. The tongue on p6 can sometimes be quite noticeable, but best viewed on the underwing. Underwing coverts are white, with contrasting slaty-gray remiges, becoming black on underside of outer wingtip. By early spring, white apicals on adults commonly suffer from wear, becoming similar in appearance to those of 3rd cycles. *Similar Species:* Early-spring adults in high breeding condition will often show vibrant orange-yellow bills, and in the brightest individuals, legs may have strong yellow tinge, inviting confusion with Yellow-footed and other "yellow-legged" black-backed gulls, such as Kelp and Lesser Black-backed. This yellow tinge may also be present in nonbreeding condition but is more faint and typically limited to the shins. Slaty-backed Gull has much broader white tips on the inner primaries, moderate to extensive white tongue tips on the outer primaries, pink orbital, and thinner bill. See also Yellow-footed and Kelp Gull accounts and accounts for hybrids with Glaucous-winged Gull.

1st Cycle: Juveniles are a uniform sooty brown with a dark face, especially around the eyes and cheeks. A heavyset black bill with a swollen gonys is evident on many young, soon developing a pale base to the lower mandible. Dull pink legs, some surprisingly dirty looking throughout 1st cycle. Upperparts variable, but fairly plain and dark, often with a distinct row of paler, broad covert tips. Underparts also dark brown on youngest birds, but a paler scaly pattern is also fairly common. Others have a paler belly patch and vent region, and some with almost entirely whitish lower breast and belly (compare to Yellow-footed). Outer greater coverts often show dark bases with a paler distal region along the edges and tips. Dark tertial centers with variable pale edging. Dark primaries. On the open wing, the inner primaries show little contrast with the outer primaries. The secondaries show noticeable pale tips on fresh birds. Dark uppertail typically shows outer tail feathers with completely dark bases, but some show barring along outer edge of outermost tail feather. On the freshest birds, a thin white border to the tail is seen, soon giving way to wear. Uppertail coverts have a white base color with variable spotted and barred pattern. Undertail coverts show variable dark barring, often dense. Dark underwing coverts and axillaries. Many have a sooty, silvery-gray aspect to underside of remiges. First alternate birds in winter and spring show variable postjuvenile scapulars, some largely plain and dark brownish-gray, others pale with dark shaft streaks. At this time, many birds are untidy, with worn and faded upperwing coverts and with head and body feathers becoming increasingly pale. *Similar Species:* Fairly distinct in its range, but compare to Yellow-footed, Herring, and California Gulls; also to hybrids with Glaucous-winged Gull. Southernmost populations of Western Gull can look much like Yellow-footed Gull, especially those individuals that become pale headed and pale bodied by midwinter. Yellow-footed averages more thickset bill with more exaggerated bulbous tip, but overlap exists. Yellow-footed more advanced in both wear and molt, having many more juvenile wing coverts replaced in fall. Yellow-footed at this age is pale on the belly and vent region, while Western averages more markings and overall sootier base color (although some 1st-cycle Westerns can show pale underparts). Uppertail coverts less patterned on Yellow-footed, which also

33 Adult. Large, bright-yellow bill with bulbous tip, classic. Dark eyes, variable. Pink legs. Upperpart coloration highly variable in this species, with this individual showing mid-pigment. Note jet-black primaries. AMAR AYYASH. WASHINGTON. JAN.
34 Adult. Southern *wymani* type based on blackish upperparts. This race averages smaller overall. A drooping white secondary skirt is fairly common. Saturated yellow orbital (among other features) eliminates Slaty-backed Gull. MICHAEL STUBBLEFIELD. CALIFORNIA. APRIL.
35 Adult type. Head markings invite suspicion of Glaucous-winged influence, although apparently within range, based on known-origin and known-age adults (see "Hybrids" section). ALEX A. ABELA. CALIFORNIA. NOV.
36 Adults displaying the wide range of variation found in upperpart coloration. Note large, powerful bills with swollen tips. ANDREW BIRCH. CALIFORNIA. FEB.
37 Adult. It's not rare to see yellowish legs and an overly saturated orange bill in late winter through early spring (primarily March–April). Yellow orbital and overall bill size and shape are typical of Western Gull. ALEX A. ABELA. CALIFORNIA. MARCH.
38 Adults. Presumably paler *occidentalis* subspecies. This trio displays the paler extreme found in Western Gull. GEORGE CHRISMAN. CALIFORNIA. JAN.
39 Adult. Rather dark upperparts, so likely *wymani* ssp. A medium-sized to large p10 mirror is typical, with little to no white in the wingtip. Stray mark on outer web of p4. TED KEYEL. CALIFORNIA. JAN.
40 Adult type. Relatively pale upperparts on this individual, showing more contrast with black wingtip. A mirror on p9 is found in a small percentage of adults. Also here, a rather noticeable white tongue tip on p5. FRODE JACOBSEN. CALIFORNIA. FEB.

shows a moderately pale bill base in winter. Herring Gull is slighter, with narrower wings and thinner bill, and shows a contrasting pale window on the inner primaries. California Gull is noticeably smaller, long winged, with bill mostly thin and straight, becoming sharply bicolored before winter. Hybrids with Glaucous-winged often show pale edging to primaries, a paler plumage aspect, and more boldly and intricately patterned wing coverts.

2nd Cycle: Mottled brown similar to 1st cycle, but appearance plainer and more uniform in 2nd basic. Upperparts have contrasting colors and patterns due to timing of growth as well as overlap in 2nd prebasic and 2nd prealternate molt. Therefore, gray wing coverts are likely 2nd alternate. Howell & Dunn (2007) described two broad types in this age group: gray and brown. The former, apparently much more common in *wymani*, is more advanced in appearance, with noticeable gray scapulars coming in early with paler head and body. Brown types are more typical of the overall standard mottled, 2nd-cycle, large gulls of the north, which many nominate *occidentalis* birds adhere to. Bill color variable but becoming progressively

---

41 Adult. Large, bright-yellow bill with bulbous tip, and extensive black on wingtip with single mirror. Broken subterminal band on p4. First and only Atlantic record of this species, and easternmost known occurrence. ANDY GUTHRIE. NEW YORK. FEB.

42 Adult. Long, broad wings with prominent white trailing edge to secondaries, typically tapering off and becoming thinner on inner primaries. Undergoing prebasic molt (p1–p3 new, p6–p10 old). ALEX A. ABELA. CALIFORNIA. AUG.

43 Adult. Underside to flight feathers noticeably dark. Compared to Slaty-backed Gull, thin white tongue tips are typically restricted to p5 and sometimes p6 (see underside of near wing). STEVEN MLODINOW. MEXICO. OCT.

44 Adult type with similar-aged California Gull (right). In addition to the substantially larger proportions, the dark underside to the remiges on Western indicate a black-backed species. ALVARO JARAMILLO. CALIFORNIA. APRIL.

pale yellow with a darker tip through late winter into spring. Some advanced birds acquire an entirely pale bill with dark tip, with others remaining mostly dark and apparently taking on a regressive, darker bill pattern in fall and winter which resembles that of 1st cycle (Howell & Dunn 2007). Bill pattern also likely tied to "type," but much overlap and variability. Legs dull pink. Eyes largely dark or olive colored but become paler in some. Tertials often with relatively broad pale tips. On the open wing, note relatively broad white trailing edge of 2nd-generation secondaries and paler inner primaries. All-dark tail and all-white uppertail coverts common, but some less advanced—brown—types with considerable patterning on uppertail coverts. Underwing coverts generally dark and plain, but greater underwing coverts more similar in color to underside of remiges. *Similar Species:* First-cycle Yellow-footed Gull in fall and winter can look quite similar to some 2nd-cycle Westerns, due to the former being a three-year gull. Yellow-footed's bill is more evenly thick throughout its length and stouter. Pointed primaries on 1st-cycle Yellow-footed often bleach to paler brown. Juvenile wing coverts on Yellow-footed average older appearance with more wear. Second-cycle Western should be fresher overall. Also, compare to 2nd-cycle Yellow-footed in late summer, which overall has a cleaner white body and paler underwing, and often with more adult-like coverts by late winter. See also similar-aged Slaty-backed Gull in that account and accounts of similar-aged hybrids with Glaucous-winged Gull.

3rd Cycle: Head and body becoming largely white, but some birds show extensive dusky, smudged head, neck, and breast. Paling eye most notable at this age, with some becoming almost clear yellow. Others remain dark eyed. A bright yellow bill is acquired with some dirty flecking and pigment, but, as in 2nd cycle, some known-age birds have shown bill pattern that regresses in nonbreeding season. Small red gonys spot on proximal half of gonys is common, with a black subterminal tip. Others very advanced, with adult-like bill. At rest, upperparts are largely adult-like slaty gray, with a thin white secondary skirt and broad white tertial tips. Outer primaries have few if any white tips. On the open wing, note adult-like inner primaries with slaty-gray centers and broad white tips. Broad white trailing edge to secondaries mostly clean white, with gray bases dominated by dark centers at times. Some show strong brown tinge to upperparts and may have retarded plain brown greater coverts. Therefore, some 3rd cycles often appear darker than adults, due to accentuating brown fused with slaty-gray coloration. Outer primaries black, often with no or relatively small and/or asymmetric p10 mirror. Primary coverts range from mostly adult-like to largely brown and 2nd-cycle-like. Underwing coverts overall white with dusky flecking throughout. Uppertail can show a black-and-white piano pattern with variable black markings and at times extensive black. A bright yellow bill is acquired with some dirty flecking and pigment, but, as in 2nd cycle, some known-age birds have shown bill pattern that regresses in nonbreeding season. Small red gonys spot on proximal half of gonys is common with a black subterminal tip. *Similar Species:* Second-cycle Yellow-footed can look much like 3rd-cycle Western, so careful attention should be given to confusing birds, especially in southernmost portion of Western's range, where overlap with Yellow-footed exists. Again, Yellow-footed averages a thicker bill throughout its length, with more bulbous tip. Some Yellow-footed Gulls are easily identified by molt patterns, as prealternate molt is more advanced, which may include a variable number of remiges and/or rectrices. These individuals will have adult-like black subterminal bands and white tips to inner primaries, with molt contrast against brown, often-worn, 2nd basic outer primaries. At this age, Yellow-footed can show dull yellowing on legs, with some advanced birds showing adult-like bright mustard-yellow legs. See also similar-aged Slaty-backed Gull in that account.

MOLT SAS most likely, but some individuals are thought to replace feathers twice, suggesting a limited preformative molt in 1st cycle (Pyle 2008). Southern birds begin molt slightly earlier than northernmost populations, with southern 1st cycles averaging more advanced 1st prealternate molt than their counterparts in the north. Much overlap in start of prealternate and end of prebasic molts, to the point that it appears one continuous molt is at work at times, and delineating the two isn't always easy.

HYBRIDS Hybridizes extensively with Glaucous-winged Gull. Surprisingly, the two are not closely related, with no sharing of mitochondrial DNA lineages. See account H1.

27 YELLOW-FOOTED GULL *Larus livens*

L: 22.0"–26.5" (56–67 cm) | W: 58"–61" (148–155 cm) | Three-cycle | SAS | KGS: 9.0–10.5

OVERVIEW Yellow-footed Gull is unique among North America's large white-headed gulls, in that it occupies a small niche in the Gulf of California, where it is an endemic breeder. It nests in fairly small, loose colonies on rocky and sandy islands and islets, and sometimes solitarily, away from conspecifics. Nesting requirements differ from Western Gull's. Adults lay eggs in small scrape on ground, lined in formation, parallel to the tideline (Spear & Anderson 1989). Some interannual variation in peak egg-laying period, likely due to food availability. Earliest birds hatch in late April, many in May. By the end of June, most have fledged and are free flying. A small contingent of varying ages carries out a short postbreeding dispersal north to the Salton Sea annually, peaking in August. An obligatory trip here is in order for anyone wanting to see this species in the United States. Adults that arrive in late spring through early summer almost surely did not breed. Still, little is known about the feeding ecology and particular niche requirements of this species. A study found that in some years, likely tied to food shortages, Yellow-footed preys extensively on Least Storm-Petrels and Black Storm-Petrels (Flores-Martinez et al. 2015). In other years, large-scale conspecific predation of chicks occurs on various islands, likely due to nesting-habitat disturbances (Patten 2020). Yellow-footed is highly vulnerable to human disturbance, and egg collecting is still of concern (much more problematic historically), along with uncontrolled tourist visitation to islands where it breeds. Some protections have been put in place, but with limited enforcement. Estimated 20,000 breeding pairs in Gulf of California, making this one of the smallest populations of any North American gull species (Anderson 1983; Wilbur 1987).

TAXONOMY Monotypic.

RANGE
Breeding: Resident in Gulf of California, NW Mexico, where it is a fairly common breeder south to N Sinaloa.
Nonbreeding: Frequents rocky and sandy shorelines, mudflats, harbors, pilings, and various other marine structures. Most birds remain in the Gulf of California; a few disperse farther south along the Sinaloa coast, June–December. An uncommon to fairly common postbreeding visitor, mostly June–September (a few in late May and into Oct.), north to the Salton Sea in SE California, where first recorded in 1965. Formerly more numerous at that site, where maximum counts approached 1,000, but due to increased salinity and a receding shoreline, substantially fewer birds are now found in both the south and north ends of Salton Sea (Guy McCaskie, pers. comm.). A few individuals may remain into late fall, but very rare to absent there in winter and spring. Confusion with increasing numbers of Western Gulls found here since 2005 has clouded the current true off-season status of Yellow-footed there. Casual inland elsewhere in SW United States, mostly in summer and fall, north to E California (Mono County), S Nevada (Lake Mead), and N Arizona / S Utah (Lake Powell); and east to extreme SE California (Senator Wash Reservoir), SW Arizona (Painted Rock Dam), and in December 2023, Texas's first state record (Randall County), serving as the easternmost record for the species at the time of this writing. Uncommon locally to very rare along Pacific Coast of S Baja Peninsula, but increasing from southern tip of Baja north to Magdalena Bay in recent years; multiple erroneous reports of subadults are misidentified Western Gulls. Casual north to S California. Also a casual visitor south along Pacific Coast as far as Oaxaca, Mexico.

IDENTIFICATION
Adult: A large black-backed gull with yellow legs and feet. Large bill, stout throughout its length, with a deep tip and exaggerated gonys. Thickset body with broad wings. Head is generally bright white in alternate plumage and most other times of the year, with minimal light gray clouding and streaks. Eye is usually pale, but some exceptional birds show dark olive to brown eyes at this age (perhaps 3rd cycles

inseparable from older adults). Orbital ring saturated yellow in high breeding condition, sometimes orangey. Broad white tertial crescent merges with drooping secondaries when feathers fully grown. Freshly grown primaries show moderate-sized white apicals (Sept.–Nov.) but are often completely worn down by February/March. Open wing reveals broad white trailing edge and largely black wingtip. Black on p8–p10 reaches primary coverts and frequently along outer web of p7. P10 commonly shows medium-sized to large mirror, often C-shaped with thin subterminal band that may be broken and at times with white outer web. Very rarely a small mirror on p9, likely as a restricted stray mark on outer web (concealed by p10). Often with thin white tongue tip on p5, which regularly has full black subterminal band. Quite regularly shows restricted black mark on outer web of p4. *Similar Species:* Southern race of Western Gull (*wymani*) closely matches gray upperpart coloration of Yellow-footed, and some Westerns may even be darker; these darker Westerns average smaller body size than Yellow-footed, but overlap exists. Western Gull is pink legged, although beware individuals with hypervivid bare-part colors, especially approaching the breeding season (March/April). These individuals show intensely bright yellow bills (bordering on orange), with a pinkish-yellow coloration to the legs. Some Yellow-footed Gulls show pinkish tinge on upper surface of feet, more so on underside, especially in low breeding condition, when leg color may be duller. Voices considerably different and very helpful in separating the two, with Yellow-footed having a deeper and lower croaking vocalization. Western generally has shallower bill tip

1 Juvenile. Large, dark bill, scaly upperparts, with dilute appearance to wing coverts, and already pale belly and vent. SEAN FITZGERALD. MEXICO. JULY.
2 Juvenile. Slimmer individual with attenuated look. White belly and vent and dark greater covert bases. Bulbous-tipped bill distinctive. MARK CHAPPELL. CALIFORNIA. JULY.
3 1st cycle. Grayish postjuvenile upper scapulars and mantle feathers. Dilute appearance to wing coverts. Blocky head, strong bill, and whitish underparts. KENNETH Z. KURLAND. CALIFORNIA. AUG.
4 1st cycle. Swollen bill tip, dark greater coverts, and pale underparts. Obvious wear apparent. OWEN HILCHEY. NEW MEXICO. SEPT.

5 1st cycle. Now with most scapulars replaced, as well as some inner-wing coverts (1st alternate). This extent of molt and yellow-tinged legs are not expected in Western Gull at this date. DAVID VANDER PLUYM. MEXICO. OCT.

6 1st cycle. Dull yellowish bill base and legs. An advanced individual with all visible wing coverts and tertials replaced (1st alternate). The appearance is similar to a 2nd-cycle large gull. Flight feathers 1st basic. ALVARO JARAMILLO. MEXICO. JAN.

7 1st cycle. Red M020 banded as a chick near La Paz in April 2018. Fleshy legs, but with yellow bill base. The calico appearance to the upperparts is due to a mix of juvenile and 1st alternate feathers. FLOYD HAYES. MEXICO. JAN. 2019.

8 An advanced 1st cycle with some adult-like scapulars (some possibly 3rd generation or late 1st alternate). The advanced bare parts, paling iris, and white head look remarkably similar to a 2nd cycle, although this bird is merely a one-year-old. CHRIS GIBBINS. MEXICO. MARCH.

9 1st cycle with 2nd-cycle Lesser Black-backed Gull (right). Yellow-footed has a noticeably larger bill and body, with overall more bulky appearance. BRANDON MILLER. NEVADA. MARCH.

10 Juvenile. Contrasting white uppertail coverts, with indistinct light barring against an all-dark tail. Dark inner primaries and dark bases to outer greater coverts. Massive bill already paling. MIKE DANZENBAKER. MEXICO. AUG.

11 1st cycle. Similar to 21.10, but now with some apparent wear and fading. Little indication of a pale inner-primary window. White vent and uppertail coverts distinctive. JACK PARLAPIANO. NEW MEXICO. SEPT.

12 Juvenile. Largely white belly and vent, carrying over to uppertail coverts. Broad wing with brownish wing linings. ADRIANA H. ALVAREZ. MEXICO. JULY.

13 1st cycle. Massive bill and dark head resemble Western Gull, which at this date has darker juvenile uppertail coverts (molting here) and overall fresher appearance to upperparts. CHRISTOPHER LINDSEY. ARIZONA. SEPT.

14 1st cycle. Same individual as 27.13, now with more median and lesser coverts replaced, newer uppertail coverts, and more sooty gray lower scapular and mantle feathers. Two replaced secondaries likely adventitious. KAVANAGH MCGEOUGH. ARIZONA. DEC.

15 1st cycle. Thick bill with overall whitish body and uppertail coverts. An advanced 1st prealternate molt has included most wing coverts and roughly half of tail feathers (glossy black), which is not expected in our northern 1st-cycle gulls. ALVARO JARAMILLO. MEXICO. JAN.

16 1st cycle. Broad wings, with little indication of a window, and dark wing linings. Whitish head and body expected at this date. PETER ADRIAENS. MEXICO. MARCH.

17 2nd cycle completing primary molt here (2nd prebasic). Plain grayish-brown upperparts and pale body are typical for this date. Some delayed birds remain with pinkish bill base and legs well into 2nd cycle. AMAR AYYASH. CALIFORNIA. AUG.

18 2nd cycle. A somewhat more advanced appearance for this date, now with adult-like upperparts and yellow legs. The appearance recalls a 3rd cycle of other large gulls. AMAR AYYASH. CALIFORNIA. SEPT.

19 2nd cycle. Advanced bare parts, with whitish head and body. Note newer, 2nd alternate p5–p6 (glossy black with white tips) and old p7–p10 (2nd basic). ALEX LAMOREAUX. MEXICO. FEB.

20 2nd cycle. Brilliantly adult-like for a two-year-old. Distinguished from adult by lack of white apicals on primaries (2nd basic) and brown outer greater coverts. Massive bill and deep chest, unlike Lesser Black-backed. ALEX LAMOREAUX. MEXICO. FEB.

21 2nd cycle (front) with 1st cycle. A rather typical 2nd alternate individual, but with most primaries replaced (p9–p10 old, 2nd basic). Thick legs and powerful bill tip distinctive. JOACHIM BERTRANDS. MEXICO. FEB.

22 2nd cycle. Same individual as 27.7, now with largely adult-like aspect and yellow legs. Note brown wash to upperparts and black on tail. STEVE MLODINOW. MEXICO. JAN. 2020.

23 2nd cycle. Less advanced than 27.22 with brownish inner primaries (presumed 2nd basic). Broad wing, bulbous bill tip, and yellow feet apparent. TOM GRAY. MEXICO. DEC.

24 2nd cycle. Some, presumably 2nd basic, retain much black in the tail. Note broad white tips to secondaries (s1 growing). AMAR AYYASH. CALIFORNIA. SEPT.

25 2nd cycle. Large bill becomes brighter yellow late in the season. Molted outermost secondaries (2nd alternate) immediately point away from other species such as Western Gull. JOACHIM BERTRANDS. MEXICO. FEB.

---

with smaller red gonys spot, and bill is not as wide throughout its length. Again, considerable overlap exists. Western adults average less black on outer wingtip, often show thinner outer web to subterminal band on p5, and more likely to show white tongue tip on p6. Leg color is of utmost importance in this identification. A recent observation that may help separate these two is the color of the greater underwing coverts, particularly the outermost underwing primary coverts: Yellow-footed often has noticeable, smoky-gray wash throughout, while Western adult coloration is overall uniformly white (Chris Gibbins & Peter Adriaens, pers. comm.). Although some overlap exists, especially with southernmost population of Western Gull, month for month, primary molt in Western starts six to eight weeks later and

ends later in the season than in Yellow-footed. For instance, an adult Yellow-footed at the Salton Sea may show new p1–p5 in early July, while adult Western has only begun to drop p1/p2. Some Yellow-footeds already with primary molt completed by late September. Therefore, apicals on Yellow-footed often show more wear around the start of the calendar year compared to fresher Westerns.

Lesser Black-backed Gull has become a legitimate confusion species in recent years, especially among novices visiting the Salton Sea. Lesser Black-backed has a more slender body, with longer wings, and a noticeably thinner bill. The red gonys spot on Lesser Black-backed is often larger and more oblong, and the orbital ring is typically red—not yellow. Also, Lesser Black-backed often shows denser and darker head streaking, which is not found in Yellow-footed. Body size and structure should be the first things that grab one's attention when a Lesser Black-backed is among a group of Yellow-footed Gulls. Kelp Gull is the least expected confusion species but quite similar, with yellow legs and large, robust body. Kelp is noticeably darker and generally has duller legs that often cast a greenish hue. On the open wing, Kelp averages a broader white trailing edge, especially so on the inner primaries.

1st Cycle: As with other large white-headed gulls, juvenile upperparts are various shades of brown. Upperwing coverts, tertials, and scapulars relatively plain with pale fringes, often notched. Tertial tips can sometimes be boldly pale. Head and neck show coarse markings with uneven white coloration to neck and chest. Some juveniles have a distinct dark face mask. Dirty grayish-pink legs becoming dull, pale color. Black bill often begins to show some paling around the base fairly soon after fledging. The most distinctive features on 1st cycles are the large, stocky bill and noticeably pale belly and vent regions. Importantly, on the open wing, note the plain white uppertail coverts with light barring (appearing largely white in flight). Inner primaries can show light contrast with outer primaries, and secondaries have thin, pale trailing edge. Tail feathers are all dark with pale edges giving way to wear. In juveniles, upper flanks are spotted with blotchy brown markings, very often with prominent patch along rear flanks above the legs. Many hatch-year birds are prone to accelerated wear, likely related to latitude and weathering. By late fall, upperparts become heavily frayed and disheveled, and a limited to extensive postjuvenile molt ensues (1st prealternate molt). Here, 2nd-generation scapulars can range from dark centered with plain edges to mature, adult-like slaty gray. Wing coverts, tertials, and some rectrices may also be replaced (see this account's "Molt" section). Head and neck may become considerably white with dense streaking throughout. Not uncommon for bill to show a largely pink base with black smudges and black tip. Legs remain dull pink for most of the winter, with yellow tones developing in late winter to early spring. Around this time, some individuals may take on a paling iris, and in advanced birds, overall appearance is reminiscent of 2nd cycle of other large white-headed gull species. *Similar Species:* Yellow-footed Gull's bill tip is generally deeper, and bill base seems to pale more evenly on both mandibles, when compared to similar-aged Western, which often has pale region restricted to base of lower mandible. Yellow-footed averages paler underparts, particularly throughout the belly and vent. Uppertail coverts are noticeably different, with Western typically showing darker and more boldly barred feathers, whereas on Yellow-

26 2nd cycle with most flight feathers presumably 2nd alternate. P9–p10 are old 2nd basic. Again, molt pattern alone should rule out similar large taxa. JOACHIM BERTRANDS. MEXICO. FEB.
27 2nd cycle. Yellow legs apparent with large, powerful bill. Extensive 2nd prealternate molt included all secondaries, p1–p6, and some tail feathers. p7–p10 old 2nd basic. Note small mirror on outer web of p10, apparently not very rare at this age. PETER ADRIAENS. MEXICO. MARCH.
28 Adult (4th cycle). Red M171 banded as a chick in May 2019 (Santo Island). Yellow legs and bill, with moderate red gonys spot. Pale eye with orange-yellow orbital. JUSTIN PETER. LA PAZ. MEXICO. DEC.
29 Adult type. Distinctive yellow legs, dark upperparts, and deep bill tip. Besides the dark eye, all features point to a definitive adult. Outermost primaries growing (large p10 mirror emerging below primary stack). AMAR AYYASH. CALIFORNIA. SEPT.
30 A somewhat refined adult (female?) with similar-aged California Gull (right). Yellow-footed is darker, with a noticeably thicker bill, and adults typically have a pale iris. MARK CHAPPELL. CALIFORNIA. FEB.
31 Adult types. Dark upperparts, large head, and massive bill. Disheveled look to molting upperparts and flight feathers. New p6–p7 with retained outermost primaries. AMAR AYYASH. CALIFORNIA. AUG.
32 Adult type with similar-aged Western Gull (right). Southern *wymani* can be just as dark, or darker, than Yellow-footed, although Western averages a shallower bill tip. AMAR AYYASH. CALIFORNIA. SEPT.
33 Adult type. Another individual with darkish iris and small, dark smudge near bill tip, suggesting a 3rd cycle. Aging such birds is not always straightforward. AMAR AYYASH. CALIFORNIA. SEPT.

footed, uppertail coverts are typically pale with faint barring or all white. Extensive upperwing covert, tertial, and / or tail feather replacement in midwinter is a safe indication of a Yellow-footed.

2nd Cycle: Head markings generally light with dusky clouding and diffuse streaking on back of neck, not typically heavy. Upperparts variable, some retarded grayish brown (2nd basic), others more advanced and adult-like slaty gray (2nd alternate). On the open wing, not uncommon to see brownish-black primary coverts and black centers to a variable number of secondaries. Primary tips generally without prominent white apicals, except in those individuals with 2nd alternate primaries, which are often more advanced in appearance. Sometimes with small p10 mirror (typically restricted to outer web on 2nd-generation primary). Tail ranges from solid black band to black-and-white piano pattern to mostly white in the most advanced birds (2nd alternate). Bill pattern varies from dull pink to dull yellow with large black tip, sometimes showing small red on gonys and becoming brighter yellow. Leg color variable, mostly by season, from ambiguous dull pink color with light yellow tinge to dull yellow, and in late winter to early spring, some become vivid and adult-like. Eyes are often pale, dusky gray olive, and in early spring, orbital ring can be bright yellow or with orange tinge. Wing linings white overall, with random marked feathers; some birds are less advanced, with variable brown wash throughout, but axillaries largely white. Outer greater primary coverts on the underwing can be surprisingly dark, similar in coloration to underside of outer primaries. Note that from below, the trailing edge to the 2nd basic secondaries can

34 Adult. Broad wing with extensive black on wingtip. Medium-sized p10 mirror, often with thin subterminal band that is sometimes broken. MICHAEL O'BRIEN. MEXICO. JAN.
35 Adult. Bulbous-tipped bill. Note smoky-gray greater primary coverts on underside of far wing. Small mark on p4 fairly common. MORGAN EDWARDS. MEXICO. FEB.
36 Adult. Extensive black on underside of wingtip. Triangular p10, small mark on p4, and virtually no white tongue tips on this individual. TIM MELLING. MEXICO. MARCH.
37 Adult type completing prebasic molt. Perhaps a 3rd cycle, but not safely aged. AMAR AYYASH. CALIFORNIA. SEPT.

appear very advanced, giving the appearance of what is typically found in 3rd cycles of other large gulls.
Similar Species: Third-cycle Western, Kelp, and Lesser Black-backed Gulls. Prealternate molt pattern can be very telling in all ages in this species.
3rd Cycle: Usually synonymous with adult, and by 3rd alternate should be indistinguishable. Random tail or primary covert markings and brown hues on portions of the upperparts suggest 3rd-cycle type. These individuals may show smaller p10 mirror, dark markings around bill tip, and darker iris. More study needed from known-age birds.

MOLT Inserted molt in 1st cycle is a variable prealternate molt, often including many to most body feathers. Regularly includes some wing coverts (nearly all or all in advanced birds), some tertials and uppertail coverts, and a variable number of rectrices (Ayyash, pers. obs.). Secondary and primary molt unrecorded in 1st prealternate, although likely replaces some innermost secondaries (adjacent to tertials). Birds less than a year old often have a 2nd-cycle-like appearance of four-cycle gulls. Molt averages earlier and more advanced than in northern gulls, as some chicks have hatched already by late April.

Variable number of primaries, secondaries, and tail feathers may be replaced in subsequent prealternate molts (Howell & Dunn 2007). For instance, in a sample of 46 birds, a combination of 8 adult-type and 3rd-cycle individuals showed two waves of concurrent primary molt in late September at the Salton Sea (Ayyash, pers. obs.). As inner primaries were being replaced via definitive prealternate molt, the outer primaries were also molting at tail end of definitive prebasic molt (Pyle et al. 2018). See plate 27.22, showing Red M020, a known-age 2nd cycle with two generations of primaries. Replacement of p9–p10 unrecorded, with p1–p5/p6 most commonly included in prealternate molts.

HYBRIDS None.

38 Adult type with two waves of concurrent primary molt, with p1–p4 presumably via definitive prealternate molt and p5–p10 a result of definitive prebasic molt. This molt pattern is not expected in any other *Larus* adults in North America. AMAR AYYASH. CALIFORNIA. SEPT.

39 Adult type with prebasic molt near its end. Perhaps a 3rd cycle based on dusky eye, small black streaks on primary coverts, and extent of flight feather molt at this date. Moderate white tongue tip on p5 and a thin white sliver on p6. AMAR AYYASH. CALIFORNIA. SEPT.

40 Adult. Yellow legs and dark remiges hint at Yellow-footed. Note smoky-gray greater coverts, especially outer greater primary coverts. STEVEN MLODINOW. MEXICO. JAN.

28 GREAT BLACK-BACKED GULL *Larus marinus*

L: 25"–31" (64–79 cm) | W: 57"–65" (145–165 cm) | Four-cycle | SAS | KGS: 13–15

OVERVIEW The largest of all the world's gulls; or, as the American ornithologist Arthur Cleveland Bent (1921) put it, "The largest and strongest of its tribe." Great Black-backed Gull is our darkest regularly occurring species. An imposing gull that is difficult to misidentify throughout its range, especially when seen with other birds for scale. Generally more shy and skittish than other gulls, and once disturbed takes longer to resettle. Flight is lumbering but powerful and eagle-like. When taking off, runs a few steps to liftoff. A voracious feeder and predator, preying upon a wide variety of small mammals and birds, exceedingly more than any other *Larus* species in North America. Commonly takes eggs and chicks of nearby seabirds in the breeding season, especially to the north of its range in years when spawning capelin are delayed (Veitch et al. 2016). Its exceptionally large gullet allows it to carry prey items up to one-fourth its weight. This majestic bird was on the verge of being extirpated from the New World, due to its feathers being exploited in the 19th century. Now generally expanding its breeding range south along the Atlantic, perhaps at the expense of Herring Gull, which it freely associates with year-round and nests alongside in mixed colonies. Common name is sometimes given as *Greater* Black-backed Gull, in error.

TAXONOMY Monotypic.

RANGE
Breeding: NE North America across N Europe, and recently to SW Morocco. Breeds in small, loose colonies and as single pairs on flat rocky coasts, sandy and dredge-spoil islands, salt marshes, buildings, and locally on lakeshores from S and W Greenland and Hudson Strait, Gulf of St. Lawrence, and Bay of Fundy, south along the Atlantic Coast (where common) to North Carolina. Farther west, also breeds on at least one islet in James Bay, in small numbers on Lake Huron, and in larger numbers on Lake Ontario (where uncommon). Away from salt water, nests at Lake Saint-Jean, Quebec; at Lake Champlain, New York/Vermont; and locally inland in Atlantic Provinces. Declined during 1800s due to hunting and egging. First nested again in Maine in 1928, in New York in 1942, then on Great Lakes in 1954; first times in Virginia and North Carolina in 1970, on Lake Champlain in New York in 1975, and in Vermont in 1983. Attempts to establish itself as a breeder on E Lake Ontario slowed by its susceptibility to type-E botulism (Shutt et al. 2014). On Lake Erie, first breeding records in 1993. Limited expansion here may be due to both susceptibility to botulism and lack of suitable breeding habitat, as the species seldom nests on woody islands on the Great Lakes (Weseloh et al. 2007). Some have suggested the species is poorly adapted to inland freshwater habitats, due to lack of food availability. Breeding range expansion to Lake Huron also not momentous, but annual in very small numbers on rocky islands. In 1990, nested on Goose Island, Michigan, in the Straits of Mackinac (Ewins et al. 1992). One breeding record from Lake Michigan in 1995, on Spider Island, Wisconsin. This is the westernmost breeding record I know of. Two chicks banded by US Fish and Wildlife Service and successfully fledged (Cutright et al. 2006). Farther north and inland, first nest record for inland Nunavut was for 10 adults breeding in Akimiski Strait, in lower Hudson Bay (June 2007).
Nonbreeding: Found in the same habitats as other large *Larus* gulls, including well offshore, all coastlines, harbors, salt marshes, lakes, reservoirs, larger rivers, dams, parks, parking lots, fish and other animal processing plants, and landfills, but uncommonly at many sites, likely due to its being less abundant overall. Occurs regularly well offshore out to 80 miles. Northerly breeders mostly or partly migratory, whereas southern breeders mostly resident. Southbound movements begin by late August or early September. Most begin to appear far from breeding sites by late October or November. Winters commonly in St. John's, Newfoundland, south along Atlantic Coast to NE Florida, with small numbers to S Florida and N Bahamas. Casual as far south as Cuba, Puerto Rico, Yucatán Peninsula, Mexico, Belize, Aruba, and Venezuela. Accidental to Costa Rica. Nonbreeding wanderers to S Baffin Island and west to

Hudson Bay. Also found regularly in small to moderate numbers at larger lakes, rivers, and landfills well inland on the coastal plain. Fairly common to common on the E Great Lakes, with a patchy distribution, uncommon on Lake Michigan, and most uncommon on Lake Superior. Moves northbound along Atlantic Coast late February–April. Small numbers of nonbreeders summer nearly throughout the winter range, including south to Florida and recently singletons on the Gulf Coast. European breeders winter as far south as N Africa. West of the Appalachians and away from the Great Lakes, mostly rare to very rare migrant and winter visitor in Midwest, found most regularly at lakes and rivers, as well as landfill sites. Rare from Kentucky south to Gulf Coast. Rare on SW Hudson Bay and west to W Great Plains from central Alberta southward to E Colorado and N Texas, but annual. Casual or accidental farther west to north and south-central Alaska, Prudhoe Bay, British Columbia, Washington, Idaho, and S California (Salton Sea). Noteworthy is an individual presumed to be a returning bird (nicknamed Murray) in Pueblo County, Colorado, for at least 30 consecutive winters (1993–2024)—a rather remarkable record on its own, but unfortunately, this individual remained unbanded at the time of this writing.

IDENTIFICATION

Adult: A large and bulky gull with thickset bill, often swollen at tip. Blocky head structure, especially in males. Black upperparts may show obvious fuscous-like quality or fade to dark blackish gray. Remains largely white headed throughout the year, with some fine but faint streaking on crown and hindneck in basic plumage. Eye color varies from pale to all dark, but generally with some pigment. Reddish orbital, orange red at times. Gonys spot also red to orange red. In nonbreeding condition, bill can appear dull pink at the base, and often with some black marks around the tip. Leg color overall weak pink, some with noticeable yellow tinge, creating confusion for some observers. Note the relative girth and length to the legs compared to most other large gulls. Of all our black-backed species, adult Great Black-backed averages the most white on the two outermost primaries, and hints of this can be detected at rest. On

Juvenile. Heavyset bird with thick legs, noticeably checkered wing coverts, and overall pale base color to body. AMAR AYYASH. MARYLAND. AUG.

2. Juvenile with a relatively large 2nd-cycle Lesser Black-backed Gull (right). The Great Black-backed is likely female, but some lens compression may be at work. AMAR AYYASH. NEW JERSEY. AUG.
3. Juvenile with adult Herring Gulls. Still showing immaculate juvenile scapulars with much duskiness on the face and neck. Powerful bill and white on tail base. AMAR AYYASH. PENNSYLVANIA. DEC.
4. 1st cycle. An individual with darker wing coverts. Most upper scapulars replaced. Short wing projection and blocky head. AMAR AYYASH. NEW YORK. DEC.
5. 1st cycle. A remarkably pale and fresh individual with much white on the tertials and uppertail. Chunky bill and short wing projection. AMAR AYYASH. FLORIDA. JAN.
6. 1st cycle. Square-headed and stout bill; inflated appearance to the body with short wing projection. Anchor pattern on postjuvenile scapulars is rather common. AMAR AYYASH. FLORIDA. JAN.
7. 1st cycle with similar-aged Lesser Black-backed (center) and American Herring (left). A true-to-life comparison of average size differences with these three species. AMAR AYYASH. FLORIDA. JAN.
8. Juvenile. Peppered black-and-white appearance, often with subtle inner-primary window, and much white on the uppertail coverts. The tailband is narrow and patterned on the proximal edges. AMAR AYYASH. ILLINOIS. OCT.
9. 1st cycle. Bolder inner-primary window than 28.8, but note distinctive long, broad wings, large head with massive bill, and narrow tailband with white rump. AMAR AYYASH. PENNSYLVANIA. DEC.
10. 1st cycle. Pale body and mostly white head. Broad wings. Wing linings often show thin "zebra" barring. AMAR AYYASH. INDIANA. MARCH.
11. 1st cycle. Still with moderate markings on head, neck, and flanks. Barred axillaries common. Some show white pearls on inner primaries. AMAR AYYASH. INDIANA. JAN.
12. 1st cycle. Blocky head and bull neck protrude well past leading edge. Darker upperwings on this individual, with distinct white capsules on inner primaries. Typical tail pattern. AMAR AYYASH. WISCONSIN. MARCH.
13. 1st cycle (same individual as 28.8). Impressive wingspan, large white body, and distinctive underside to tail pattern. AMAR AYYASH. ILLINOIS. OCT.

14 1st cycle (lower right) with American Herring Gulls. A warmer brown individual, but compare overall body girth and much broader and longer wing. CHUCK SLUSARCZYK JR. OHIO. DEC.

15 2nd cycle. Overall resembles 1st cycle at this time of year, but note darker greater coverts and new (black) incoming primaries (p10 old). AMAR AYYASH. WISCONSIN. JUNE.

16 2nd cycle. Same individual as 28.15. p1–p8 new (2nd basic) with old p10 retained. Secondaries in heavy molt. A mostly black bill at this age is fairly common. AMAR AYYASH. WISCONSIN. JUNE.

17 2nd cycle with 1st cycle (right). Aging points on 2nd cycle include rounded primary tips, fine pattern on greater coverts, and a number of dark, slate-colored scapulars. MARS MUUSSE. THE NETHERLANDS. MARCH.

18 2nd cycle. Darker greater coverts with finer pattern. White body with stout black bill rather common. Small white crescents on the primary tips aren't rare. AMAR AYYASH. MICHIGAN. NOV.

19 2nd cycle. Dark median coverts (2nd alternate) showing "black-backed" tones. Rounded primary tips. Blocky head and bulgy body. AMAR AYYASH. FLORIDA. JAN.

20 2nd cycle. Note dark wing panel. A number of adult-like median coverts and scapulars (2nd alternate). AMAR AYYASH. ILLINOIS. FEB.

21 2nd cycle. A rather advanced individual with much jet black on the upperparts (2nd alternate feathers). The outer primaries lack white tips, which helps with aging, but such individuals are best aged with an open wing. AMAR AYYASH. FLORIDA. JAN.

22 2nd cycles. Variation in this age group is striking, ranging from aspects that resemble 1st cycle (left) and others that approach 3rd cycle (right) and presumably reflect hormonal levels at the time of feather growth. MARIO ESTEVENS. PORTUGAL. NOV.

23 2nd cycle. Aging points include brownish-gray inner primaries with rounded tips, broader white trailing edge, marbled greater coverts, slaty-black scapulars, and paling bill base. AMAR AYYASH. INDIANA. MARCH.

24 2nd and 1st cycle (right). 2nd cycle overall with paler underparts and bill base, with finely patterned greater coverts (noticeably barred and checkered on 1st cycle). Faint mirror on p10 (left wing). MARS MUUSSE. THE NETHERLANDS. OCT.

25 2nd cycle. It's rather common to find a variably sized p10 mirror in this age group. The mirror here is asymmetric. AMAR AYYASH. WISCONSIN. MARCH.

the open wing, commonly shows an all-white tip to p10, a large p9 mirror (not uncommon to see a broken subterminal band on p9), and sometimes a small mirror on p8. Relatively thin white tongue tips on p6–p7, and less commonly on p8. P8 may have a prominent white tongue tip restricted to inner web and thereby resemble a "mirror" when inner web is partially covered by p7. P5 pattern ranges from full subterminal band with white tongue tip to completely unmarked. Inner primaries have relatively thin white trailing edge. Long, broad wings in flight with a hulking body and thick neck. The underwing shows an obvious contrast of white wing linings with dark gray to black remiges. Although thin and indistinct, the white tongue tips on p6–p7 are routinely seen from below. *Similar Species:* Distinctive throughout its range, but compare to other pink-legged black-backed gulls—namely, Slaty-backed Gull and, incidentally, Great Black-backed × Herring hybrids (see those accounts for details). Beware the regularly occurring phenomenon of adult Great Black-backed Gulls with variable yellow coloration to their legs. This invites confusion mostly with Lesser Black-backed Gull. That species is substantially smaller, thinner, and longer winged. Lesser Black-backed averages paler upperparts and a larger red gonys spot and often has more extensive and darker head streaking in basic plumage. Yellow-footed Gull is genuinely yellow legged, with a yellowish orbital, and is typically without a p9 mirror. Kelp Gull is much rarer and has a broader white trailing edge to the wing, including the inner primaries. Adult Kelp Gulls sighted in North America also average a medium to small p10 mirror, as opposed to the large mirrors found on p9–p10 in Great Black-backed. Western Gull averages paler upperparts and has a yellow orbital, a broader white trailing edge to the inner primaries, and smaller mirrors on p9/p10. The bill on Western is typically brighter yellow without much black blemish in nonbreeding condition.

1st Cycle: Fresh juveniles have warm brown streaking on the head and neck, with markings down the breast and sides, but typically not solidly dark. The base color to the body is pale, and by mid- to late winter many have a contrastingly white head and neck with light streaking. Thick, all-black bill can show diffuse paling, especially late in 1st cycle. Dull pinkish legs. Upperparts variable, from mostly solid brown with golden fringes to predominantly pale with small diamond-like centers in lighter birds. Overall, the wing coverts have a checkered pattern. Light marbling on the bases of the greater coverts not uncommon, and sometimes with much white across that tract. Postjuvenile scapulars can appear buff colored with brown barring or dark anchor pattern. Tertials typically dark centered with light patterning on the fringes. Dark brown to black primaries with thin, pale edges. Vent region predominantly white, while undertail coverts generally have sparse barring or chevron markings. On the open wing, a subtle primary window is common, sometimes with some marbling around the tips. Individuals with darker inner primaries show lower contrast with outer wing for a more uniform appearance. Indistinct pale fringes and tips to trailing edge of secondaries. Rump is boldly white, as are uppertail coverts, which have variable dark flecking. Tailband is decidedly narrow, weakening distally. Proximal edge to tailband commonly shows a speckled black-and-white pattern, while bases to outermost tail feathers are white. Wing linings and

---

26 2nd cycle. A dark-winged bird with much black on the scapulars and median and lesser coverts. The tail is well-marked on this individual. Broad wing with large head and impressive bill. MAX EPSTEIN. NEW YORK. NOV.

27 3rd cycle with similar-aged Lesser Black-backed (left), both undergoing 3rd prebasic molt. The Great Black-backed has retained p9–p10 (2nd basic), with new 3rd basic primaries emerging (noticeable white tips). Note thick pinkish legs, thick neck, blocky head, and massive bill. AMAR AYYASH. MASSACHUSETTS. JULY.

28 3rd cycle. Aged by incoming, adult-like primaries and retained 2nd-generation outer primaries. A relatively slim body, but note blocky head, thick legs, and deep breast. Head streaking does not generally exceed this example. AMAR AYYASH. NEW JERSEY. AUG.

29 3rd cycle. Noticeable white tips on primaries average smaller than adults. Broad white tips to secondaries (below greater coverts). Delayed bill pattern not unusual at this age, often regressing in appearance in the nonbreeding season. JEAN-MICHEL SAUVAGE. FRANCE. OCT.

30 3rd cycle. Adult-like secondaries and inner primaries with broad white tips. Relatively large p10 mirror. Brownish wing coverts presumably 3rd basic. Moderate black on tail quite common. MARS MUUSSE. THE NETHERLANDS. NOV.

31 3rd cycle. Variable tail markings and obvious brown tones to upperwing. Advanced yellow bill pattern averages more black on gonys than adult. Mirror on p10 expected, and mirror on p9 fairly regular. MARS MUUSSE. THE NETHERLANDS. MARCH.

32 3rd cycle. Large bill with swollen tip, deep belly and noticeably dark underside to remiges. Asymmetric p9 mirrors. Pinkish feet. DEBBIE PARKER. OHIO. FEB.

33 Adult. Generally white-headed year-round. Jet-black upperparts, large blocky head, and stout bill. Dark eyes (variable). Dull, pinkish legs. Extensive white on underside of far wing (white tip to p10 and large p9 mirror). AMAR AYYASH. FLORIDA. JAN.

axillaries are light brown to brown, with distinct (zebra) barring. ***Similar Species:*** Again, size often settles any uncertainties with this species, but sometimes difficult to judge without comparisons. Lesser Black-backed Gull is smaller, sleek looking, and noticeably long winged at rest. The bill is fairly evenly shaped on Lesser Black-backed, as opposed to the noticeably thick bill with swollen tip of Great Black-backed. Overall, most 1st-cycle Lesser Black-backed Gulls appear darker with more solidly filled upperparts, especially across bases of greater coverts. Head, neck, and breast are with a darker brown base. Tailband on Lesser Black-backed averages wider and fuller. When in doubt, also consider the thicker and longer tarsi on Great Black-backed. Herring Gull as well is often seen alongside Great Black-backed Gull, the latter trumping the former in size. Uppertail coverts and tailband on Herring are much more pigmented, lacking the bold white base color seen on Great Black-backed. Herring Gull's inner primaries form a distinctly paler window, and upperparts are consistently darker, as are the warmer brown underparts, head, and neck. Compare bill size as well as wing length and structure.

34 Adult with 1st-cycle Lesser Black-backed Gull. Some adult types show moderate black near the bill tip. Thin secondary skirt doesn't typically "pop." Short wing projection. AMAR AYYASH. FLORIDA. JAN.
35 Adult. Faint head streaking at times, seldom extensive. Dull yellowish legs not very rare. An open wing can readily eliminate Kelp Gull, but note the large white tip to p10 and relatively narrow white tips to the secondaries on perched birds. MARS MUUSSE. THE NETHERLANDS. NOV.
36 Adult male (front) and female pair near their nest, standing guard while their chicks are banded just a few feet away. Note differences in head and bill shape. Worn primaries and active molt. Variable eye color. AMAR AYYASH. MAINE. JULY.

28 GREAT BLACK-BACKED GULL

2nd Cycle: Eyes range from all dark to light brown; rarely pale. Bill pattern can be largely black in some birds with diffuse pale regions. Others with mostly pale and dull-colored bill base with developing black tip and yellow nail, typically with dark along the cutting edge. By late winter to early spring, a dull yellow bill with red on the gonys is found in some individuals. Legs dull pink. Head, neck, and underparts average less streaking than in 1st cycle, with some considerably pallid at this age. Less advanced birds can appear very 1st-cycle-like with lightly barred scapulars and a peppered plumage aspect. More advanced individuals show a variable number of adult-like slaty-black mantle and scapular feathers and at times a

---

37 Adult. Low contrast between black upperparts and wingtip. White tip to p10 common. The mirror on p9 is typically larger than this. p5 is often unmarked, but shows a thin band on this individual. LUKE SEITZ. MASSACHUSETTS. JAN.

38 Adult. Reduced black on the wingtip with all-white tip to p10 and large p9 mirror with partial subterminal band. p6 often has complete band, but is nearly unmarked on this bird. Thin white slivers as tongue tips on p7–p8 not common. ANTONIO GUTIERREZ. SPAIN. JAN.

39 Adult. p8 can sometimes show a white pearl (as seen here) or a small mirror. Relatively thin white tips to inner primaries help rule out Slaty-backed. Massive bill and red orbital (pale-eyed bird). FRODE FALKENBERG. NORWAY. APRIL.

40 Adult. Slate-colored underside to flight feathers, with thin white tongue tips to p6–p7. Commonly has an all-white tip on p10 with large p9 mirror. Yellowish legs are not very rare. AMAR AYYASH. WISCONSIN. MARCH.

41 Adult. Moderate black on underside of p10, but otherwise limited on rest of wingtip. Complete subterminal band on p6 and p5 often unmarked. Stout bill with submarine body. AMAR AYYASH. WISCONSIN. MARCH.

variable number of wing coverts (2nd alternate). The wing coverts are more contrasted now with multiple generations of feathers commonly seen at once. More speckled and finely marked compared to 1st cycle, especially the greater coverts, which may show broader white tips. The tertials are more finely marked at this age as well. While perched, the tips to the primaries often have distinct white crescents, boldest on p5–p7 (although not large, defined apicals as in 3rd cycle). On the open wing, not uncommon for p10 to show an indistinct to small mirror (some surprisingly large). Similar to 1st cycle, a pale window is seen, and now the trailing edge to the secondaries and inner primaries averages thicker white than in 1st cycle (some with surprisingly adult-like secondaries, usually with diffuse brown). Uppertail coverts can be entirely white or with light flecking. Tailband remains narrow at this age, with variable speckling and a diffuse pattern distally. Wing linings are generally plainer and less marked but still with some barring. Underside of the remiges is grayish brown with light marbling. *Similar Species:* Again, compare to Lesser Black-backed and Herring Gulls. At this age, many Herrings show some paler gray plumes across the scapulars or wing coverts, which readily points away from Great Black-backed. Compare tail patterns and overall size. Some Lesser Black-backed Gulls may look superficially similar with a black-and-white peppered aspect, but overall, most 2nd-cycle Lesser Black-backeds are darker brown and more muted on the upperparts. Consider size and structure, as well as the attenuated look to the rear of the body on Lesser Black-backed. Compare also to 1st-cycle Great Black-backed, which, among other contrasting features, has more-barred greater coverts, darker wing linings, and darker bill.

3rd Cycle: With the acquisition of 3rd basic plumage, many slaty-black adult-like feathers are found on the upperparts. Noticeable brown hues to these feathers, especially the upperwing coverts. Some less advanced individuals show 2nd-cycle-like patterned coverts, but note the broader white tertial tips and blacker primaries with white apicals. The white apicals may be approaching adult size in advanced birds, or merely thin slivers or tips approaching 2nd cycle in least advanced birds. The head, neck, and underparts may be mostly white or with some moderate, fine streaking. Legs pink, but commonly show a sickly blue-gray coloration around joints and tibia. Dull pale-yellow bill with some red and black on gonys is common. Eye color is variable, from dark to light brown, but not commonly pale. On the open wing, note the adult-like trailing edge to the secondaries and inner primaries. The secondaries may be dark centered, some with a grayish-black coloration or diffuse brown. Mirror on p10 relatively large, typically with subterminal black and not an entirely white tip as in adult. Mirror on p9 medium sized, although variable and at times absent. The uppertail pattern ranges from nearly all white to some having considerable black. However, the tailband is typically with a diffuse marbled pattern spangled across the rectrices, rarely solid black. The wing linings become largely white with "dirty" edges and tips. Note the contrasting slate-gray underside to the remiges with adult-like black on the outer primaries. *Similar Species:* Third-cycle Lesser Black-backed Gull will generally have obvious yellow tones to the legs by now, but beware retarded individuals with dull pinkish legs, as well as color distortion at greater distances. Eye color is often wickedly pale on Lesser Black-backed Gull, which also shows lighter upperparts when compared to the blackish value of Great Black-backed. The importance of size and structure cannot be overstated here.

MOLT Appears to have one inserted molt in 1st cycle, with fairly typical SAS. This molt is generally limited to the scapulars and some head and body feathers. Northernmost populations—perhaps those with latest hatch dates—may have limited 1st prealternate molt. Some known-age individuals still predominantly juvenile into February (Mars Muusse, pers. comm.).

HYBRIDS Putative hybrids with Glaucous Gull recorded annually in Newfoundland (fewer than 10 per winter). Rare to very rare outside Newfoundland (B. Mactavish, pers. comm.). Also recorded in Greenland and several European countries on rare occasions. Hybrids with American Herring Gull found annually on Great Lakes (approximately 10–15 per winter), up the St. Lawrence River, and less so along the NE Atlantic. Rare in St. John's, Newfoundland, where the two winter together in relatively large numbers. First reports as early as 1960. Also in Europe, presumably with *L. argentatus / argenteus* (Olsen & Larsson 2004). Single reports with Yellow-legged Gull in Portugal and Morocco (Jonsson 2011). Reportedly with Cape Gull (*L.d. vetula*) from Morocco (Fareh et al. 2020).

29 SLATY-BACKED GULL *Larus schistisagus*

| L: 22.0"–26.5" (56–67 cm) | W: 57"–59" (145–150 cm) | Four-cycle | SAS | KGS: 9.5–13.0 |

OVERVIEW A Beringian species of NE Asia, Slaty-backed Gull is quite distinctive, especially after its 1st plumage cycle. Adults are unique among black-backeds, in that they often show a flashy "string-of-pearls" (a series of prominent white tongue tips on the outer primaries) coupled with a broad white trailing edge to the wing. The specific scientific name, *schistisagus*, translates to "slate-colored cloak" (Jobling 2010). Omnivorous, with similar foraging practices to its northern relatives. However, there is some indication that adult males show preferential feeding behaviors, some that are called "hunting-biased" and others termed "fishing-biased" (Watanuki 1989, 1992; McKee et al. 2014). Nests on rock pinnacles, cliffs, rocky islands, and less so in low vegetation. Nests in segregated colonies, but more commonly among other seabird species (Kondratyev et al. 2000). Majority of population breeds in Russia, with significantly smaller numbers in Japan. Largest colony believed to be on Shelikan Island, in the Russian Far East, with an estimated 11,000 individuals. World population estimated to be 131,000 pairs, but some regions remain largely unsurveyed (Larisa Zelenskaya, pers. comm.; Zelenskaya & Solovyeva 2016). Populations in the Russian Far East increased rapidly from the 1980s onward (Denlinger 2006). In other regions, such as N Japan, the population now appears to be declining, and in 2017 the species was placed on the Hokkaido Red List as near threatened (Ujihara & Ujihara 2019).

The noteworthy pattern of vagrancy to mainland North America is a fairly recent event, and a curious one. No other Asian gull has gained such a foothold throughout the lower continent, although it is far from secure. The species may be on the brink of expansion, but breeding records in W Alaska are stagnant and do not suggest this is so (first breeding record in Alaska was in 1996). Some breeding may be undetected throughout the mammoth landmass from Alaska through N Canada, however. The swelling populations along the W Bering Sea, in addition to shifts in N Pacific fisheries and unpredictable sea-ice levels, may be driving more and more individuals to stray in search of winter feeding sites. The phenomenon of "more people looking, more birds found" must also be at work. The paucity of 1st-cycle individuals reported in North America is almost assuredly due to the challenge in identifying them (Jaramillo 2020).

TAXONOMY Regarded as monotypic by most taxonomies, but Portenko (1963) proposed *ochotensis* for a smaller and darker form of Slaty-backed Gull that he found south of Ayan (type locality, mouth of the Lantar River along the southwest shores of the Sea of Okhotsk). Average measurements suggest *ochotensis* is shorter winged and smaller, but larger samples needed. Portenko also considered *Larus schistisagus* conspecific with *Larus argentatus* [*vegae*], due to a large degree of overlapping and intermediate characteristics that he found on his expeditions to Koryakland along the lower courses of the Apuka and Kultushnaya Rivers. This relationship is now broadly disregarded, as it is undeveloped and likely reflects a small zone of hybridization.

RANGE
Breeding: NE Asia. Found from Cape Navarin, Siberia, south to Kamchatka, Kuril Islands, south to Russia-China border and to N Japan. Reportedly near absent as a breeder on the Commander Islands (Kondratyev et al. 2000). Winters as far south as Taiwan, in small numbers. Some populations nonmigratory (del Hoyo et al. 1996). Nests along rocky and sandy coasts in NE Russia south to N Japan. In North America, first confirmed breeding in 1996 and then again at the same site in 1997, off Cape Romanzof, on Aniktun Island, W Alaska (McCaffery et al. 1997). Behavior and occurrence in W Aleutians suggestive of nesting, and some pairings there believed to involve hybrid Slaty-backed × Glaucous-winged Gulls (unconfirmed; see this account's "Hybrids" section). *Nonbreeding:* Found at a variety of habitats frequented by numbers of large *Larus* gulls, although many of the records away from W Alaska are associated with landfills, fish runs, fish-processing plants, animal-rendering plants, and their associated nearby loafing areas. In Alaska, it is an uncommon visitor in late

summer and early fall to St. Lawrence Island (especially at Gambell) and in the Nome area; rare to uncommon in spring at St. Matthew, the Pribilof, and the Aleutian Islands. Noteworthy is a high count of at least 39 individuals at the Nome River mouth on 9 July 1984 (Kessel 1989). Regular in very small numbers around Utqiagvik in June, mostly around local landfill.

Commonly winters along sea-ice edges in the S Bering Sea. In North America, also coastal, but individuals regularly turn up inland, especially on the Great Lakes. There are numerous records from coast to coast, mostly in winter and mostly involving adult and 3rd-cycle individuals. The number of reported 2nd cycles is steadily increasing, but reports of 1st cycles remain very rare, with few accepted records away from W Alaska. Rare in the Pacific Northwest, but regular. As of 2023, the species was removed from the review list in California, where there are well over 100 records and multiple sightings annually. Washington has around 30 records, mostly in winter. Some winter records are also attributed to returning individuals. It is rare to very rare east to the Great Lakes region but increasing and now annual in S Ontario (over 35 records) and throughout the Great Lakes, where on several occasions 2 individuals have been present at once. Casual to very rare elsewhere, including from S California to Colorado and to S Texas (8), the Carolinas, and Florida, and multiple records northeast through Massachusetts. Louisiana has 2 promising records with images that strongly suggest Slaty-backed Gull (3rd-cycle type, May 2009, and an adult type, Sept. 2011), although unaccepted at the time of this publication. Scattered records from mid-Atlantic north through Maritime Provinces, more commonly inland in this region, with a slight aversion to the coast. Nearly annual to St. John's, Newfoundland, where 1 is usually recorded each winter. Also found in summer on several occasions in arctic Canada from Inuvik, Northwest Territories, to Cambridge Bay, Nunavut. Current status difficult to assess completely. Also recorded in Hawaii (more regularly since 2018), Guam, Thailand, Australia, Philippines, Iceland, and N Europe.

IDENTIFICATION

Adult: A black-backed gull with a noticeably short wing projection. Body appears somewhat rotund and potbellied at rest. Shows a prominent tertial (and scapular) crescent that merges with a well-defined secondary skirt when feathers are fully grown. Pink legs often more vivid than in other accompanying pink-legged species. Even-edged yellow bill, sometimes with a faint greenish cast, with a relatively small red gonys spot. At times, the bill appears to have a slight droop, especially in longer-billed individuals. Some—presumably males—can have wider bill base and show a relatively pronounced gonys. Generally pale eyed or with some light flecking. Small percentage show dark eyes, but uncommon. Purple-pink orbital regularly detected in winter (can sometimes appear crimson-like due to lighting conditions or when transitioning from nonbreeding to breeding condition and the reverse). Dark streaking is usually found around the eye forming an eyeshadow look. The head is variably marked in basic plumage, from

---

1 Juvenile. Would present some challenges out of range, with some aspects resembling Glaucous-winged hybrids and Western Gull. Western averages a shorter look to the wings, with dark bases to the (outer) greater coverts. Note pale underside to p10, with pale edging on primaries. A gape frown is apparent in many. YANN MUZIKA. JAPAN. SEPT.
2 Largely juvenile, with scalloped appearance to upperparts. Dark tertial centers with pale tips. Plain greater coverts important, as well as overall straight bill, potbelly, and short wing projection. ALVARO JARAMILLO. JAPAN. FEB.
3 1st cycle. Chest-heavy, sagging vent, and straight-edged bill. Noticeably frowning gape not uncommon. A longer-winged individual. Plain greater coverts. Pattern on scapulars less common but variable. Brown primaries with pale edging, plain greater coverts, and darker eye-mask are characteristic. JOSH JONES. JAPAN. JAN.
4 1st cycle. A textbook individual showing classic shape and plumage. Plain greater coverts and scapulars, drooping brown secondaries, plain brown tertials with prominent pale tips, dark eye patch, and warm ginger-brown underparts. ALVARO JARAMILLO. JAPAN. FEB.
5 Neck and upper breast are often streaked and spotted with a paler ground color, showing more contrast here than North American hybrids involving Glaucous-winged. Thin barring found in a few scapulars, but not a common or predominating pattern. AMAR AYYASH. JAPAN. DEC.
6 1st cycle. Some noticeable wear. Classic head and bill shape with deep chest and "full diaper." Brown primaries (not black). Muted pattern to postjuvenile scapulars, which should be compared to faded Glaucous-winged × American Herring. AMAR AYYASH. JAPAN. DEC.
7 1st cycle. White base color to plumage, but an otherwise classic rotund body and short wings. Importantly, note the dark and fine shaft streaks on the scapulars, which are found in a fair number of 1st-cycle Slaty-backeds. ALVARO JARAMILLO. JAPAN. FEB.
8 1st cycle. A robust individual with typical "Pacific Northwest" large gull structure, recalling Glaucous-winged. Goose neck, sagging vent, relatively short and straight bill, and short wing projection. Note remnant dark diamond tips on outer median coverts and short, dark shaft streaks on upper scapulars. AMAR AYYASH. JAPAN. DEC.

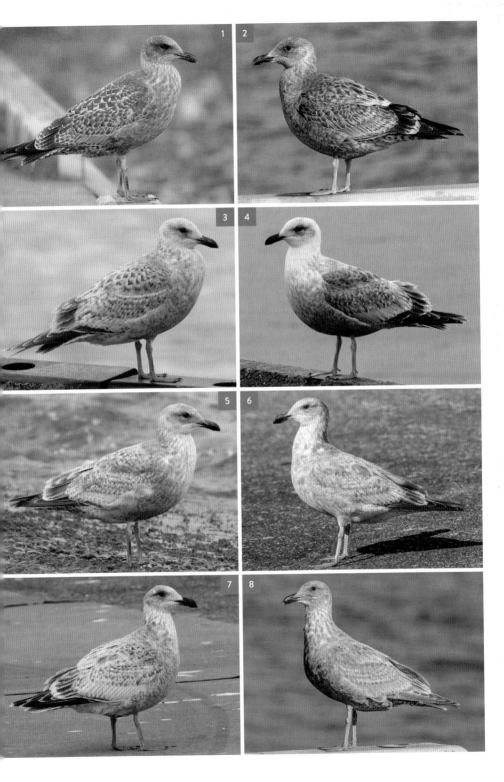

29 SLATY-BACKED GULL

9. 1st cycles. Faded, late-winter individuals, but fairly distinctive. Plain tertial centers show prominent white tips, pale vents, and plain greater coverts lacking the stippling and prominent barring found in some Glaucous-winged hybrids. Dark eye patch and faint earspot most noticeable when the head whitens. JOHN MARTIN. JAPAN. MARCH.
10. Juvenile. Short broad wing, with outer hand appearing noticeably stunted. Inner-primary window is variable, but typically shows gradual, low-contrast pattern (unlike Herring). Uppertail coverts variable, commonly pale with faint barring, and may show a tailband effect. YANN MUZIKA. JAPAN. SEPT.
11. 1st cycle. Broad wing, short hand, and plain greater coverts. Distinct dark diamonds on tips of inner primaries. Dark tail. Sooty and plain postjuvenile scapulars. AMAR AYYASH. JAPAN. JAN.
12. 1st cycle. Cinnamon tones not rare. Note contrasting dark marks on upper scapulars, which are quite characteristic of this age group. AMAR AYYASH. JAPAN. DEC.
13. 1st cycle (same individual as 29.5). Dark tail with average barring on uppertail coverts. Brown remiges, with outer primaries showing two-toned venetian-blind pattern, similar to Thayer's. Note dark diamond tips on median coverts, with paler and plainer greater coverts. AMAR AYYASH. JAPAN. JAN.
14. 1st cycle. A darker individual with uniform appearance and subdued window. Dark postjuvenile scapulars (blackish), broad wing, and short hand. Uppertail coverts can be moderately pale, but note darker tips and brownish rump. CHRIS GIBBINS. SOUTH KOREA. DEC.
15. 1st cycle. Largely faded wing panel, but note how the marginal coverts on the leading edge usually remain darker. Diamond tips on outer median coverts and dark slate-gray postjuvenile scapulars. Pale and unpatterned uppertail coverts are not uncommon. CHRIS GIBBINS. SOUTH KOREA. JAN.
16. 1st cycle. Dark tail and busy uppertail covert pattern recall American Herring, which typically shows more patterned greater coverts and a more contrasting inner-primary window. Shape, leg color, and median-covert pattern should be compared on perched birds. AMAR AYYASH. JAPAN. DEC.
17. 1st cycle. Moderately worn and faded, thus the chances of being identifiable out of range less favorable. Important features include persisting dark marginal coverts, plain whitish greater coverts, and venetian-blind pattern on outer primaries. Compare to Glaucous-winged × American Herring. MARTEN MULLER. SOUTH KOREA. FEB.
18. 1st cycle (right), with two similar-aged Vega Gulls. Slaty-backed averages warmer tones with browner primaries. The greater coverts are often the palest and plainest tract on the upperwing (highly barred on Vega). Note distinct black-and-white barring on tail base of Vegas. AMAR AYYASH. JAPAN. DEC.

29 SLATY-BACKED GULL

19 1st cycle (same as 29.13). Plain brown wing linings and axillaries, typically without barring. The flight feathers have a grayish quality, with dark tips to the outer primaries (similar to Thayer's). AMAR AYYASH. JAPAN. JAN.

20 1st cycle. A paling bill base is sometimes found at this age, particularly on the upper mandible. On the far wing, note the largely pale and plain greater coverts, and darker marginal coverts. Even, smooth texture on the belly with warm tones. Plain brown axillaries. AMAR AYYASH. JAPAN. JAN.

21 2nd cycles. Worn and faded birds in high molt from late May through Aug. Leg brightness, rotund body, bill shape, and plainer scapular patterns point away from more lanky Vega Gull, which averages paler underparts and vent, with more barred upperparts. Bill shape and a paling iris point away from Glaucous-winged. MALTE SEEHAUSEN. KAMCHATKA. JUNE.

22 2nd cycle (left) and 1st-cycle Slaty-backed. Note classic head and potbelly structure, and prominent gape-line. Paling iris and paling bill base, but highly variable at this age. Greater coverts dark and largely unpatterned. MARTEN MULLER. SOUTH KOREA. FEB.

23 2nd cycle. Recalling Vega Gull, which also shows stone-washed greater coverts and broad white tertial tips. Vega has blacker primaries and averages more intricate patterns on the greater coverts. Note slate-colored scapulars, dark eye patch, and "full diaper." JOSH JONES. JAPAN. JAN.

24 2nd cycle. Characteristic earspot, deep gape-line, and "tipped forward" impression are rather classic. Some have slightly paler primaries and a moderate bulge to the bill tip (male?), recalling Glaucous-winged. The short wing projection and slate-colored upper scapulars suggest an acceptable Slaty-backed. JOSH JONES. JAPAN. JAN.

25 2nd cycle. Classic rotund body shape; typical plumage features include dark, slate-colored scapulars (2nd alternate) and less-advanced white feathers mixed in. Plain wing panel with muted patterns. RICHARD BONSER. JAPAN. DEC.

26 2nd cycle. Paling iris with distinctive eye-mascara look. Slightly advanced 2nd alternate scapulars (and few coverts). Western Gull averages blockier head, with higher eye placement, stronger, and more bulbous-tipped bill, smudgier head pattern, and more uniformly black primaries. AKIMICHI ARIGA. JAPAN. FEB.

27 2nd cycle. An individual with uniform upperwing and relatively pale outer primaries (in part due to lighting, and they appear darker at rest). Dark tail with some typical slate-colored scapulars. IAN DAVIES. JAPAN. DEC.

28 2nd cycle. Dark tail and adult-like scapulars distinctive. Not unlike Lesser Black-backed, which averages narrower and longer wings and typically lacks the two-toned pattern on the mid- and outer primaries. MARTEN MULLER. SOUTH KOREA. FEB.

29 2nd cycle (montage). Regularly shows whitish upperwing coverts (2nd basic) with darker underwing. Note two-toned outer primaries with pale edging around tips, eye-mascara, and smooth belly patch. MARTEN MULLER. SOUTH KOREA. FEB.

light streaks on the lower neck to heavy streaking covering the entire head and neck. These markings can be rather coarse and blotched, and, much less commonly, smudged, as in some Glaucous-winged Gulls. Some birds show cinnamon-colored head and neck markings. On the open wing, an eye-catching broad white trailing edge is seen on the secondaries, continuing onto the inner primaries, with white often cutting deep into the black feather centers. There is much to be said about the overall wing shape in this species, which shows a broad arm and "stunted" hand. The flight pattern is somewhat lumbered and seems to lack agility. The wingtip typically shows white tongue tips on p6–p7/p8 (string-of-pearls). Mirror on p10 generally large, and less expected is an entirely white tip. Variable p9 pattern: some are without a mirror with much black coming up the feather base; others show a striking *thayeri* pattern. Subterminal band on p5 may be complete, broken, or entirely absent. Small mark on outer web of p4 uncommon. In flight, note the barrel-shaped body, broad wing, and relatively short, stunted hand. Leading edge of wing also has a relatively broad white border. This is often accentuated by much white around the alula and outer-primary coverts. Sometimes appears hunchbacked in flight. The underside of the remiges is dark gray, becoming black only at the outermost primaries. String-of-pearls commonly seen from below. **Similar Species:** Beware color variation in upperparts. Some pale-end Slaty-backeds have a bluish-gray appearance that approaches pale Lesser Black-backed Gull. These Slaty-backeds most likely to be confused with dark-end Vega Gulls (see that account for details). Others are boldly black, similar to *wymani* Western Gull, and at times approaching Great Black-backed Gull (Gustafon & Peterjohn 1994). Averages larger than southern *wymani* Western Gull, but even on hulking male Slaty-backeds, the bill generally lacks the bulbous appearance and brighter color of Western Gull. Orbital-ring color, bill size and shape, wingtip pattern, and average differences in head streaking and white cutting into inner primaries usually sufficient for separating from Western. Note that on Slaty-backed the outer webs on the wingtip are often darker than the inner webs, and the underside of the wingtip averages noticeable string-of-pearls. Compared to Glaucous-winged × Western hybrids, Slaty-backed averages a smaller and straighter bill, with darker upperparts and blacker wingtips. Head streaking finer in Slaty-backed. Great Black-backed Gull noticeably larger, with much stouter bill, and lacks broad white tips on the inner primaries. Be aware that some large male-type Slaty-backeds can approach Great Black-backed in overall size, but head and body shapes differ. Duller leg color on Great Black-backed, and head markings in basic plumage also sufficiently different. Great Black-backed × Herring hybrids superficially similar, but with thinner inner-primary tips on hybrids (see that account for details). An increasing number of Slaty-backeds reported with dark gray primaries in Alaska suggests hybrid influence (presumably with Glaucous-winged). Primaries on Slaty-backed should appear black, and not dark gray. Those putative hybrids sometimes have darker eyes and some clouded head markings recalling Glaucous-winged, as well as paler pigment on the underside of p9–p10. Leg color easily separates Slaty-backed from yellow-legged black-backed species, such as Lesser Black-backed, Kelp, and Yellow-footed Gulls.

30 3rd cycle. Aged by broad white tips to secondaries, subterminal bands on p4–p5, and larger p10 mirror (underside of far wing). Potbelly, bright legs, slaty-gray scapulars, and broad white tips to tertials. RICHARD BONSER. JAPAN. DEC.
31 3rd cycle. More advanced than 29.30, with blacker primaries showing (worn) white tips, more adult-like wing panel, and yellow bill. Variable black often shown on tail at this age. MARTEN MULLER. SOUTH KOREA. FEB.
32 3rd cycle finishing prebasic molt (p9–p10 and inner secondaries growing). Dirty wing linings. Lightly marked tail on this individual. Recalling Lesser Black-backed, but note prominent white tips to primaries, fairly large p10 mirror, small p9 mirror, and more extensive white on leading edge of hand. YANN MUZIKA. JAPAN. OCT.
33 3rd cycle. Classic. Broad wing, with distinctive broad, white trailing edge to secondaries, extending to the inner primaries. White tongue tips on p6–p8 (string-of-pearls) and moderate-sized p10 mirror. LIAM SINGH. BRITISH COLUMBIA. MARCH.
34 3rd cycle. Moderate white trailing edge, at times not as impressive as adult. Maximum tailband. Head streaking and prominent white tongue tips on p6–p7, unlike Western Gull. IAN DAVIES. JAPAN. DEC.
35 Adult type. Found in Toronto in Sept., continuing for three months. Subadult bill pattern, but in Sept. the bill was bright yellow and adult-like. Molt of outer primaries, rectrices, and about half of secondaries recorded locally. Inner secondary molt slowed, which is a familiar pattern in adult types wintering in North America. AMANDA GUERCIO. ONTARIO. DEC.
36 Adult type. Open wing revealed typical adult plumage, but with smaller mirrors, reduced white tongue tips, and smaller apicals, perhaps age-related. Upperparts on the paler end, dull bill, and pink orbital. Broad white tertial crescent. AMAR AYYASH. JAPAN. DEC.
37 Adult. Classic bird with fierce eye-mascara stare, fairly straight bill, broad white tertial crescent merging with prominent secondary skirt, and slate-colored upperparts. AMAR AYYASH. ILLINOIS. FEB.

1st Cycle: Plumage is highly variable, but there are some classic patterns that are expected, which should be used with body shape and overall structure for identification. Short winged at rest, with deep breast and belly, and even-edged bill. A rotund body is often complemented by a long-goose-neck look, with sagged lower belly and vent region. Drooping tips to the secondaries can often be found, forming a moderate skirt. The bill is often black or mostly black, and by midwinter soft paling can be found on many, especially around the bill base. Pink legs often appear dark and rich in color. A dark eye patch is somewhat common with contrasting paler head and commonly shows a dark earspot. Head markings are not very finely streaked or heavily blotched or smudged. Most have a warm base color to the body and upperparts, but some can be surprisingly pale throughout, even white. Overall, the underparts are mostly plain without much mottling or contrasting blotches, but variable, especially on birds actively molting. A key feature on the wing panel is the pattern to the greater coverts, which commonly appear distinctly plain, and typically paler than the lesser and median coverts. The median coverts often show contrasting dark diamond- or kite-shaped tips with plain, pale bases. The lesser coverts generally mimic the median coverts with a busier and tighter pattern.

Juvenile scapulars can appear scalloped, with dark centers and pale fringes, at other times lightly notched or spade shaped. Patterns found on the postjuvenile scapulars are quite variable, on par with other 1st-cycle gulls of four-year species. Classic and common patterns include pale grayish-brown feathers with thin, dark shaft streaks and solidly gray and muted. Prominently pale (whitish) postjuvenile scapulars are not rare and often show a dark tesla-T or fine shaft streak. A light anchor pattern or thin horizontal barring is less common; if present, it is rarely the predominant pattern and seldom appears bold. Others show solid sooty-brown postjuvenile scapulars. The tertials are generally dark centered and plain, at times with noticeably broad and clean white edging. The primaries are typically brownish and may appear blackish in unfavorable lighting, but seldom truly black. Primaries commonly show thin, pale edging, some approaching that of a white winger. Some variants have considerably pale primaries (similar to those of Glaucous-winged), and others bleach to near white by late winter. Undertail coverts and vent region vary from densely spotted and barred to plain white, but average between the two. Many have a contrasting dark belly that persists throughout much of 1st cycle, and this is often seen up through the flanks.

On the open wing, the outer primaries generally have a tidy two-toned venetian-blind pattern, similar to some Thayer's Gulls, with paler window and darker tips to the inner primaries. Others are darker brown, approaching American Herring, but with a less contrasting window. Commonly, the inner to mid-primaries suggest a string-of-pearls pattern with pale subterminal tips to inner webs of the primaries. Similar to that of adults, the wing is broad with a short-hand appearance. The greater coverts often contrast with the rest of the wing in flight, being less patterned and plainer than surrounding feather tracts. A helpful feature that seems to hold with many individuals is a velvety and contrasting brown band across the leading edge with the marginal coverts being noticeably darker than the lesser and median coverts.

The tail feathers range from largely dark up to the feather bases to having light barring and patterning on the outermost feathers. Others show a relatively narrow band with some light vermiculation on the outer rectrices and pale outer tail feather bases. The uppertail coverts, especially early in the season, can

---

38 Adult. Faint but noticeable cinnamon wash on head and neck; appears cloudy at times. Fairly thin and long bill on this individual, showing a slight greenish cast. Full band on p5 and small mark on outer web of p4. First state record. CHRIS HILL. SOUTH CAROLINA. JAN.

39 Adult. Broad white tertial and scapular crescents and impressively white underside to far wing. Small red gonys spot and deep purple-pink legs common. RICHARD BONSER. JAPAN. DEC.

40 Adult type with American Herring Gull. Some show faint brownish tones to upperparts, perhaps related to fading or age-related (as suggested by bill pattern). Flat crown and short-legged look in part due to posture. Deep gape-line. First state record. KENT MILLER. OHIO. JAN.

41 Adult. Short-bill, gently rounded head, and compact proportions suggest female. Upperparts can appear blacker when photographed on snow. Markings on head variable, although this appearance is more typical in late Feb. RICHARD BONSER. JAPAN. DEC.

42 Adults on nest, with male presumably at left. The combination of bright bill and white head now more closely resembles Western Gull, which averages a heftier bill and duller legs. Compare orbital color and wingtip pattern. PAVEL SHUKOV. RUSSIA. JUNE.

be evenly barred and later in the season may appear sparsely barred or mostly pale, but variable. On the underwing, the primaries are relatively pale, silvery gray, with contrasting dark terminal band on the outer hand (similar to Thayer's). The wing linings and axillaries are rather uniform and plain brown where barring is not expected. **Similar Species:** Several taxa to consider from North America, including Glaucous-winged × American Herring and Glaucous-winged × Western hybrids, as well as Glaucous-winged, American Herring, and Thayer's Gulls. Not always safely identified. In most instances, the greater coverts will be plainer or more evenly marked when compared to these taxa, creating a distinctive contrast with the wing panel. Compared to Glaucous-winged × Herring, note the more patterned scapulars (notched) and tertial tips on hybrids, as well as their lighter gray postjuvenile scapulars. Those hybrids tend to be more athletic in appearance, not as pudgy and short winged, and can show more bulbous-tipped bills. The greater coverts on Glaucous-winged hybrids often have noticeable stippling and/or barring, but these patterns are lost with wear and fading. Slaty-backed often shows a dark eye mask and also averages warmer tones on the plumage with overall plainer scapular patterns. Also, Slaty-backed often shows bolder and more defined white tips on plain brown tertials. Compared to Glaucous-winged × Western, bill size and structure stouter and more bulbous on hybrids, on average, and the greater coverts have more intricate wavy patterning. Also, this hybrid tends to show paler primaries with more extensive pale edging. American Herring Gull is just as wildly variable in 1st cycle as Slaty-backed, but there are obvious differences, including a bolder inner-primary window, the pattern on the greater coverts, and the longer-winged appearance to Herring. But some darker 1st-cycle Slaty-backed Gulls may have barring on the greater coverts with darker bases to the feathers, especially early in the season, when the feathers are fresh. Those individuals will overall be more evenly and lightly patterned with tidier lesser and median coverts when compared to Herring. The median coverts on Slaty-backed average darker tips and paler bases. Slaty-backed doesn't typically show much notching on its juvenile scapulars, as is expected on many Herrings. Structure and open wing are important. On the open wing, Herring has darker inner webs to the outer primaries and lacks the often-seen white winged two-toned pattern on lighter Slaty-backed individuals. Herring also averages darker and more densely marked uppertail coverts. Thayer's averages smaller overall, with a more delicate bill and longer wings. Also, the greater coverts and tertials on Thayer's are decidedly more marked and patterned. Glaucous-winged can be surprisingly similar in structure but generally has a smaller eye appearance against a larger face that is more smudged, muddied greater coverts, paler primaries, and a more uniform grayish-brown aspect to its plumage. Also compare bill size and structure. The uppertail coverts on Glaucous-winged appear more uniform with the rectrices, as opposed to the pale base color on Slaty-backed's coverts, which contrasts with a darker tail. Also compare to Vega Gull, which has a more streaked head and highly barred greater coverts and averages more-patterned outer tail feathers, blacker outer primaries, and a more svelte body.

43 Adult. Eye color somewhat variable, but a dark iris like this is least common. Long, thick goose neck and deep chest (male?) very much resemble the silhouette of Glaucous-winged. PAUL FRENCH. JAPAN. FEB.
44 Adult (left of center), with Great Black-backed Gulls. Very distinctive due to smaller size, brighter leg color, slightly paler upperparts, and, of course, head streaking. A larger white-headed adult might be overlooked, however. BRUCE MACTAVISH. NEWFOUNDLAND. DEC.
45 Adults. Dark gray underside to remiges. The underside to p10 is typically not jet black and varies from blackish to light gray. Note variable eye color and head streaking in these two. CHARLEY HESSE. JAPAN. FEB.
46 Adult (same individual as 29.37). Broad white trailing edge to secondaries typically continues onto primaries. Bold white tongue tips (string-of-pearls) on p6–p8, pseudo *thayeri* pattern on p9 and large p10 mirror. AMAR AYYASH. ILLINOIS. FEB.
47 Adult. Characteristic broad and short-winged look. Cinnamon wash on head and dark eye-mascara. Wingtip pattern variable, with this individual showing small p10 mirror and no p9 mirror (fairly common). AMAR AYYASH. WISCONSIN. DEC.
48 Adult. Noticeably paler upperparts, thus black on wingtip shows more contrast. Extensive black on base of p8 and no pearl (uncommon). Less-than-black underside to p10, somewhat variable. The leading edge, especially around wrist and alula, shows increased white. GRAHAM GERDEMAN. JAPAN. JAN.
49 Adult. Paler upperparts similar to *graellsii* Lesser Black-backed. Inner webs to p9–p10 often dark gray and blackish, paler than outer webs. Lackluster white trailing edge and surprisingly narrow on inner primaries. Such a pattern is far from classic. AMAR AYYASH. JAPAN. JAN.
50 Adult. White-headed adults should be compared closely to Western and Great Black-backed Gull. The fairly thin and straight-edged bill points away from those two. Wingtip pattern similar to some Great Black-backeds, but note broad white tips to inner primaries, with white cutting deep into the feather shafts. MARTEN MULLER. SOUTH KOREA. FEB.

2nd Cycle: Note structure. Plumage rather variable. One of the first giveaways on many 2nd cycles is adult-like slate-gray scapulars that contrast with the rest of the body. Some simply show dark brownish mantle and scapular feathers, with small dark gray shaft streaks or spotted centers. Also seen in many are icy-white scapular fringes that result in a distinctively contrasting appearance against the solid, adult-like 2nd alternate scapulars. Most individuals noticeably pale eyed at this age, with a light olive-brown color to the iris. Fewer completely dark eyed. Bill averages overall dark with diffuse pattern on pink-tinged base. Sharply demarcated pattern with black tip not common. Legs pink to deep pink. Head and body overall with white base color and variably marked throughout. Some individuals take on a very pale aspect, and in these individuals the wing panel is bright white with fine brown markings. Browner birds sometimes mottled and may resemble 1st cycle with more solidly dark bases to the wing coverts. Patchy dark gray coverts seen fairly regularly, mostly limited to the lesser and median coverts. Vent region and undertail coverts mostly pale but variable. Tertials generally dark centered with broad, pale edges, especially along the tips; some with freckling on outer edges, others uncommonly show subterminal barring. The primaries are dark brown to black, paler in some variants. Primary tips generally show indistinct pale edging.

On the open wing, the inner primaries are relatively pale, typically contrasting with darker brown outer primaries. The appearance is usually a gradual step from soft brown to darker brown distally and recalls Thayer's. Infrequently a small p10 mirror, often limited to the inner web. The trailing edge to the secondaries is pale with brown freckling and darker centers. The greater coverts are plain overall with the most patterned tracts on the wing being the lesser and median coverts (somewhat mirroring what we see in 1st cycle). The uppertail can range from all dark to less pigmented on the outer rectrices with a rectangular-shaped tailband. Uppertail coverts are predominantly white, with some showing lighter barring. The wing linings are mostly light brown with plain patterned centers, sometimes showing darker brown fringes, resulting in a scalloped look. A final note on mature 2nd alternate gray upperparts: these feathers often look distinctly paler than in older cycles, which appears to be age related as opposed to hybrid influence. *Similar Species:* Consider bill size and shape, as well as body structure and wing projection. Hints of dark, adult-like scapulars and wing coverts usually sufficient to distinguish this age group from similar taxa, such as Herring, Thayer's, and Glaucous-winged Gulls. The greater challenge with 2nd cycles lies in paler 2nd basic individuals showing few if any adult-like gray scapulars. These birds can be confusing and difficult to place. Glaucous-winged × Herring has decidedly paler gray scapulars and averages plainer, more muted lesser and median coverts. Slaty-backed more often with thicker pale fringes and darker tips to these two tracts or shows distinct contrasting shaft streaks. This hybrid, along with Glaucous-winged × Western, tends to have slightly higher eye placement on a larger face, but not always obvious, especially in supposed female-type hybrids with smaller heads. Broader white tips and edges to lowest scapulars and tertials often seen on Slaty-backed, and head streaking more fine when compared to the clouded and smudgy marked head of Glaucous-winged × Westerns. Pure Westerns tend to show more uneven chestnut browns with larger, blob-tipped bill. The primaries on Westerns are less elegant, with darker pigment on the inner webs and inner primaries.

3rd Cycle: Overall adult-like, with white body and moderate to heavy head and neck streaking. Iris is typically pale, with distinctive dark feathering encircling the eye. Bill mostly dull pink to grayish yellow with dark pigment greatest around the tip. This often creates an illusion of a more blob-tipped bill, recalling Western Gull. Small traces of red may be seen around gonys at this age. Legs bubblegum pink. Wing coverts range from delayed to having a piebald aspect and to almost completely adult-like slaty black. Tertials at this age usually dark centered with broad white tips that merge with secondary tips to form a thin secondary skirt. Black primaries with white apicals smaller than adult's, sometimes limited to very small and indistinct white flecks.

On the open wing, note the broad, adult-like white tips to the inner primaries, and developing tongue tips in some birds. Most have a moderate-sized p10 mirror, but in some, mirror is absent. In these individuals, the outer primaries are often dark brown with delayed aspect and very little hint of subterminal bands on p4–p5. Mirror on p9 not commonly seen. Secondaries with medium gray to slaty-black centers and broad white trailing edge. Some pigment on an otherwise-all-white uppertail is common, but usually not forming a full band. The underwing is much like adult's, but remiges appear more ghosted and slightly paler. Wing linings have a light brown, blemished appearance but overall white. Note that a number of 3rd-cycle individuals average paler gray upperparts when compared to adult types, and this is generally accepted as an age-related attribute. *Similar Species:* At this age, many Slaty-backed Gulls have a largely pale bill with some dark pigment around the tip and show a distinctly pale iris with dark feathering surrounding the eye. This, along with any thinner head streaking, may recall the fierce face of Lesser Black-backed Gull. Note that some 3rd-cycle-type Lesser Black-backeds with delayed leg-color maturation may show pink-colored feet and tarsi, and from a distance, Slaty-backed Gull may come to mind. But Lesser Black-backed is long winged and lanky in the ventral region and averages a thinner scapular crescent. Slaty-backed has thicker trailing edge to the secondaries and inner primaries. Close examination of leg color usually reveals some yellowing on even the most delayed Lesser Black-backeds, whereas in Slaty-backed the legs are decidedly pink or deep pink. Also, when detected, orbital color is red in Lesser Black-backed and pinkish in Slaty-backed. Western shows a yellowish orbital, larger bill, and more clouded head markings and lacks white tongue tips on p6–p7. See also the "Adult" descriptions in similar species accounts and the Vega Gull account.

MOLT Appears to have one inserted molt in 1st cycle, relegated to 1st prealternate. Limited mostly to the head and scapulars, and typically does not include upperwing coverts. More study needed on variability, timing, and extent of 2nd prebasic molt. Molt gaps, molt limits, varied primary lengths, and randomly replaced primaries observed on a number of 2nd cycles (and a few 3rd cycles) in December–February in Japan suggest some interesting patterns (Ayyash, pers. obs.; Jaramillo, pers. comm.). Some appear to be adventitious, but other examples with two to four adult-like inner primaries. Whether these are retarded 3rd basic individuals, advanced 2nd basic primaries, or 2nd alternate primaries (least likely) needs investigation; the possibility of more than one explanation should be explored (e.g., see an early-April individual aged as second winter in Ujihara & Ujihara 2019; Chris Gibbins, pers. comm.). Completion of prebasic molt can be rather late in some individuals in North America. A subadult / 3rd-cycle type in Ontario was still growing p8–p10 in late November, with p10 not fully grown in mid-December and more than half of secondaries still old in mid-December. In the same year, an adult to the Niagara region had p10 only about 75% grown in late December. Also notable are a number of adult and 3rd-cycle types with late inner-secondary molt in mid- to late winter, several with old secondaries into March / April, mostly from Japan (Ayyash, pers. obs.). Others noted with late primary molt in SE Alaska and Washington, close to timing often associated with some Vegas.

HYBRIDS Glaucous-winged and Glaucous Gull. See also Vega Gull account. Unknown extent of hybridization with Glaucous-winged on Kamchatka and Commander Islands. Note that the breeding ranges of these two do not broadly overlap, but on Toporkov Island, for instance, one to five hybrid pairs are readily found annually with little effort (Zelenskaya, pers. comm.). These hybrids are said to be Glaucous-winged Gulls and Slaty-backed Gulls with the "wrong" color primary feathers. Farther west, in Magadan, a male Slaty-backed and female Glaucous-winged nested on a rooftop in the city center annually from 2016 through at least 2020, successfully fledging young (Dorogoi & Zelenskaya 2016). To the south, two pairs on Rishiri Island, Japan, from 2005 to 2010 produced at least eight chicks (Kazama et al. 2011).

Birds that are likely Slaty-backed × Glaucous-winged hybrids are seen somewhat sporadically in W Alaska and on several occasions farther east and south, particularly on the Aleutians. Up to 14 individuals may have been seen here in the breeding season of 2023 (Nathan Dubrow, pers. comm.). In July 2019, an apparent Slaty-backed × Glaucous-winged hybrid was found nesting with a pure Glaucous-winged on mainland Alaska in Bethel County. The nest was occupied with one chick being fed by the adults (Drew Lindow, pers. comm.). Overall, identification criteria not well established, and current identifications in North America are largely speculative. A first for the continent was confirmed hybridizing with Glaucous Gull in 2001 at Prudhoe Bay, Alaska. Both adults were observed alternating at the nest, which produced two chicks that successfully fledged (Karl Bardon, pers. comm.; Declan Troy, pers. comm.). The Slaty-backed Gull in this pair was likely the same individual found here in 2000, when breeding was suspected.

30 GLAUCOUS-WINGED GULL *Larus glaucescens*

L: 22.0"–26.5" (56–67 cm) | W: 54"–60" (137–154 cm) | Four-cycle | SAS | KGS: 5–7

OVERVIEW Glaucous-winged Gull is the most abundant four-year gull across the N Pacific. This is the archetypal seagull, often pictured meddling around bears feeding at salmon runs or working the nets of fishing vessels over the Bering Sea. Omnivorous. Commonly found foraging among mixed flocks of seabirds, more so than other large gulls, and will often exploit prey found by those birds. Mostly a saltwater and fringe brackish-water species but is regularly found inland at landfills and even open fields where reliable food sources exist. The species has clearly increased in recent decades, a direct result of readily accessible human food sources. But the combination of closing landfill sites and increases in Bald Eagle populations may reverse this in local populations. Nesting sites include many human-altered settings, especially in the Puget Sound region. The longevity record for this species is an individual from Vancouver Island that lived 37 years, 2 months (Campbell 2007). The common name, Glaucous-winged, refers to its pearly blue-gray wings, from the Latin *glaucus*, and it is often distinguished from most large gulls by its lighter colored wingtips, which approach the gray color of its upperparts.

Hybridizing with every similar-sized gull with which its breeding range overlaps in North America, Glaucous-winged Gull is a "key" taxon in better understanding the speciation events that have recently unfolded with the large gulls of the North (Patten 1980). With this in mind, some of the most puzzling identification questions surrounding four-year gulls involve this taxon. A consensus on where to draw lines regarding variation in this species has not been reached. As such, the identification of many individuals is difficult, and a true picture of population sizes is clouded by hybridization.

TAXONOMY Monotypic, but much variation within the species. Aleutian Islands birds are largest, averaging paler upperparts and wingtips, believed in part to be due to Glaucous Gull influence, but more study needed. Populations farthest south from Washington and Oregon average darker upperparts and wingtips, surely in part due to extensive interbreeding with Western Gull (see account H1). Siberian breeders, presumably those found in N Japan in winter, show darker upperparts and routinely have darker wingtips, but this is not attributed to hybridization as with individuals in the Pacific Northwest. This presents a dilemma for anyone attempting to label all Glaucous-winged Gulls with confidence. For instance, a darker-winged subadult that showed up in Cork, Ireland, in January 2016—a first for the country—could have originated equally from Asia or from North America (Barton 2017). Although the plumage was acceptable from an Asian perspective, many North American observers simply label such birds as hybrids, with no standard cutoff agreed regarding the degree of pigment on the wingtip. A similar dilemma exists in the interior United States and Canada, where vagrants are put under much scrutiny when displaying contrasting pigment on the primaries, while these same phenotypes get passed off as "okay" in parts of Alaska and British Columbia. Therefore, questions that remain without satisfactory answers with respect to Glaucous-winged Gull phenotypes include: How pale and, more importantly, how dark can the wingtips be on an adult? What range of KGS values represents pure Glaucous-winged Gulls? Are paler eyes due to hybridization or natural variation? Description of variation in 1st cycles also needs synthetization. These questions are beyond the reach of the average binocular-wielding observer and require critical genetic analysis over the species' range coupled with known-origin exemplars.

RANGE

Breeding: Breeds commonly, often colonially, along the coast in a variety of situations, including on both natural and human-made islands and on pilings, piers, and buildings, from SW Alaska (north to St. Matthew Island in central Bering Sea since the 1960s) south to NW Oregon (small numbers). Numbers increased greatly during the 1900s. Also nests on inland lakes in SW Alaska and Alaska Peninsula, in small numbers on fresh water in British Columbia and along the Columbia River to the Columbia River Gorge, Oregon / Washington, and rarely farther east.

Also breeds in Russian Far East, almost entirely on Commander Islands, where it is the most common breeding large gull. Approximately 90% of breeders are found on the smaller islands Toporkov and Ari Kamen (Kondratyev et al. 2000).

Nonbreeding: True abundance in some areas somewhat clouded by the presence of large numbers of Glaucous-winged × Western and Glaucous-winged × Herring Gulls. Arrives well south of breeding range (e.g., S California) as early as mid-September, but most birds don't arrive until late October, November, or December. Very uncommon in spring and early summer north of breeding range in W and N Alaska, becoming uncommon to fairly common in late summer and fall north in Bering Strait. Rare but regular transient (mostly spring) inland in S Yukon.

Winters commonly in a wide variety of habitats frequented by large *Larus* gulls, including well offshore up to several hundred miles, on sandy beaches, rocky shores, mudflats, and harbors, as well as inland at landfills, animal-processing plants, dams, park ponds, and wet fields. Occurs regularly in numbers inland to south-central British Columbia and west of the Cascades in Washington and Oregon, where locally common in Portland area, and in smaller numbers south to Salem and west of the Sierra Nevada in California's Central Valley. Uncommon to fairly common at dams and large lakes locally in E Washington and Oregon. Winters rarely north to N Bering Sea (when ample open water present). In Asia, winters south to Japan. Numbers in southern part of winter range (i.e., California) may vary substantially from year to year. In S California they are typically uncommon, but some years they are fairly common, and other years they can be rare. Farther south, they are mostly uncommon to very rare in NW Mexico, casual in central Mexico, and accidental in El Salvador. Very rare in N Nevada and Salton Sea, SE California, and a casual but increasing visitor (mostly fall and winter) throughout much of the remainder of interior W North America east to the Colorado Front Range and the Canadian Prairies. Fairly regular to Utah, where it is no longer a review list species. Very rare to casual east to Hudson Bay and Great Lakes region, Missouri, New York, Illinois, Ohio, Pennsylvania, Texas, Wisconsin, Alabama, Ontario, Newfoundland, NW Europe, Canary Islands, and Morocco, as well as to Hawaii and S China.

Northbound migration commences in February–March. A few immatures linger well south into May, fewer still into June, and only rarely or very rarely later in summer in California, mostly from central California northward. Accidental in summer east to Nunavut.

IDENTIFICATION

Adults: A large, stocky, and well-proportioned gull, with a relatively short wing projection. Upperparts are pale gray with a flat bluish hue. At rest, a broad secondary skirt is regularly seen with a broad tertial crescent. The extent of pigment on the wingtip pattern often matches the upperparts in color or is moderately darker; others with wingtips slightly paler. Clean white head in alternate plumage. In basic

---

1 Juvenile. Pale, buffy-brown aspect recalls Glaucous or Iceland Gull at times. Large head with relatively small eye often placed higher on the face. Deep breast and relatively plain upperparts. AMAR AYYASH. ANCHOR POINT, ALASKA. AUG.
2 1st cycle. Greater coverts finely patterned on this individual. Less common are dark diamond tips on the primaries, a feature suspected of being influenced by Glaucous Gull, although this appears to be unfounded. AMAR AYYASH. BRITISH COLUMBIA. MARCH.

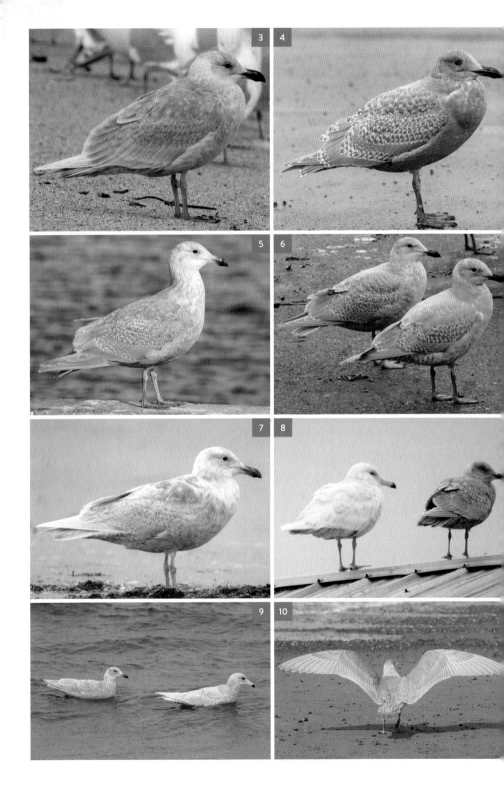

30 GLAUCOUS-WINGED GULL

3   1st cycle. Pale primaries on an otherwise perfect-looking Glaucous-winged. Plump body, classically plain upperparts, and strong bill with noticeable gonys expansion. Rich-purplish legs common. ALVARO JARAMILLO. CALIFORNIA. JAN.

4   1st cycle. A dark individual largely in juvenile plumage, and remarkably fresh for this date (perhaps of northernmost origins). Such strongly patterned birds are often suspected of being hybrids. Here it is presumed variation. AMAR AYYASH. OREGON. JAN.

5   1st cycle. An unknown (small) percentage of 1st cycles may show a paling bill base with a diffuse pattern, especially late in the season. In some cases, this is attributed to outside influence (see Glaucous-winged × Glaucous hybrid, e.g.). LIAM SINGH. BRITISH COLUMBIA. FEB.

6   1st cycles. Postjuvenile molt less commonly includes lower scapulars, which remain largely unpatterned. Extent of molt and coarse barring on wing coverts similar to some hybrids with American Herring, which typically show darker primaries. PAUL SUCHANEK. JUNEAU, ALASKA. DEC.

7   1st cycle. A pale individual. Note darker underparts, vent, and undertail coverts. Evident wear and perhaps bleaching on what may have been an already naturally pale individual. Such birds are fairly common throughout the Aleutians, where Glaucous influence may be a factor. FREDERICK LELIÈVRE. BRITISH COLUMBIA. APRIL.

8   1st cycle with 2nd-cycle Glaucous Gull (left). A dark individual with brown coloration heightened by its contrast to an all-white bird. Bill pattern important. MATT GOFF. SITKA, ALASKA. FEB.

9   1st cycle with similar-aged Glaucous Gull (right). Much overlap in size. At times, plumage quite similar, in which case bill pattern is largely decisive. Glaucous-winged shows a grayish-brown aspect, averages a darker head and neck and darker flight feathers. AMAR AYYASH. BRITISH COLUMBIA. MARCH.

10 Juvenile. Overall low contrast between flight feathers and upperparts. An all-dark tail with heavily marked uppertail coverts is fairly standard. AMAR AYYASH. HOMER, ALASKA. AUG.

11 1st cycle. A slightly darker individual with some barring on the tail base, as found in some Thayer's Gulls. Broad wing, dark smudged head and neck, and noticeably large bill point away from that taxon. AMAR AYYASH. CALIFORNIA. JAN.

12 1st cycle. Largely dark tail with densely patterned uppertail coverts. Broad wing, thick neck, and large, powerful bill. Overall uniform look to plumage, but primaries noticeably pale on this individual. AMAR AYYASH. BRITISH COLUMBIA. MARCH.

13 1st cycles. Variable tones of grayish brown, broad wings, thick neck, large bill, and uniform appearance to dorsal surface. Leftmost individual likely of Western Gull influence. STEVE G. MLODINOW. WASHINGTON. MAY.

30 GLAUCOUS-WINGED GULL

14 1st cycle. Milky gray aspect. Pale underside to primaries; broad wing and thick black bill. On the upperwing, note the darker tips extending onto the inner primaries. AMAR AYYASH. BRITISH COLUMBIA. MARCH.
15 1st cycle with similar-aged Iceland Gull (left; ssp. unassigned). Overall, Glaucous-winged is larger, with a bigger head and bill and a deeper belly. The upperwing also averages more muted patterns. AMAR AYYASH. BRITISH COLUMBIA. MARCH.
16 1st cycle. Noticeably pale flight feathers, with milky-gray aspect, similar to Kumlien's Gull, but that taxon shows narrower wings and more compact proportions. ALEX LAMOREAUX. IDAHO. APRIL.
17 2nd cycle. Plain grayish-brown appearance, with low contrast throughout. Pale-edged primaries fairly typical. Thick neck and long powerful bill begins to slowly pale at this age. AMAR AYYASH. BRITISH COLUMBIA. MARCH.
18 2nd cycle. A paler individual with more extensive gray scapulars (2nd alternate). Overall low contrast from primaries to tertials and wing coverts. Deep breast and thick bill. LIAM SINGH. BRITISH COLUMBIA. FEB.
19 2nd cycle. Primaries and coverts worn and faded. Similar to Kumlien's Gull, but note deeper breast and stout bill with short wing projection. 2nd cycles commonly show darker bill like this. Dark leg color distinctive when present. ALEX RINKERT. CALIFORNIA. APRIL.
20 2nd cycle. Relatively pale flight feathers showing darker outer webs on primaries. Rounded primary tips helpful for aging. Mostly dark tail typical. Broad wings, thick neck, and large bill. AMAR AYYASH. WASHINGTON. JAN.
21 2nd cycle. Even grayish-brown tones throughout, with low contrast. Note venetian-blind pattern on primaries. Gray scapulars and white uppertail coverts presumably 2nd alternate. AMAR AYYASH. OREGON. JAN.
22 3rd cycle and 2nd-cycle type (left). Note broad white tips to secondaries and tertials on 3rd cycle (right). Open wing needed to safely age the individual on the left. AMAR AYYASH. BRITISH COLUMBIA. MARCH.
23 3rd cycle. Large body, big head, and thick bill. Yellowish bill showing some red here. Grayish-brown primaries, often with white on the tips, but highly variable at this age. AMAR AYYASH. BRITISH COLUMBIA. MARCH.
24 3rd-/4th-cycle type. A fairly compact individual (female?). Thick bill with swollen tip and deep breast. Upperparts adult-like, but note gray markings on tail base, small apicals, and diffuse pattern on primaries. AMAR AYYASH. WASHINGTON. JAN.

25 3rd cycle. Similar to 30.21, but note broad white tips to secondaries with adult-like gray centers. Ghost mirror on p10. Sharply bicolored bill more expected in 3rd cycle. AMAR AYYASH. ALASKA. MARCH.
26 3rd cycle. Densely marked head and mostly black bill not very rare at this age. Outermost primaries regularly show darker tips with ghosted string-of-pearls pattern. Compare to Glaucous-winged × Western hybrids, which typically show darker pigments on outer primaries and tail. ANTHONY RODGER. OHIO. JAN.
27 3rd cycle. Strikingly pale outer primaries showing little contrast. Bold p10 mirror and string-of-pearls. Variable gray on tail at this age. Noticeable broad wing, bull neck, and powerful bill. AMAR AYYASH. ALASKA. MARCH.
28 3rd cycle. Fairly advanced. Darker outer primaries darkest at tips. Smudged head pattern, light gray upperparts, and fairly broad white tips to inner primaries. AMAR AYYASH. BRITISH COLUMBIA. MARCH.
29 3rd cycle. Noticeably broad-winged, with a large body. Dirty wing linings and dark, grayish-brown tips to outer primaries. Densely marked head, thick bull neck, and large bill distinctive. LIAM SINGH. BRITISH COLUMBIA. FEB.
30 Adults. Breeding pair at breeding site on Protection Island. Pink orbital, dark eye, and small red gonys spot. Female presumably in front. Primaries regularly darker than upperparts, sometimes due to outside influence. JAMES L. HAYWARD. WASHINGTON. JUNE.
31 Adult. Pale gray upperparts, with little contrast shown between outer primaries (p9–p10 old). Deep-purplish leg color distinctive. AMAR AYYASH. ANCHOR POINT, ALASKA. AUG.
32 Adult. Stout bill with noticeable, bulbous-tipped shape. Gray on primaries to p5. LIAM SINGH. BRITISH COLUMBIA. NOV.
33 Adult. Rounded head (due to posture) and fairly straight bill recall Kumlien's Gull. Glaucous-winged averages darker gray, bulkier body, thick legs, and prominent white secondary skirt merging with bold tertial crescent. JEFF POKLEN. CALIFORNIA. JAN.
34 Adult. Strikingly pale outer primaries largely unpatterned. Whether this individual represents a pale extreme in Glaucous-winged or is influenced by Glaucous Gull is uncertain. All other features typical of Glaucous-winged. LIAM SINGH. BRITISH COLUMBIA. DEC.

plumage, the head can range from minimally streaked to heavily clouded with horizontal undulations. Eye color is generally dark, especially from a distance, but with some underappreciated variation. Up to 10% may be classified as having honey-colored eyes with some paling upon close inspection, and 1%–2% having pale irises approaching those of Herring Gull, at times with some dark flecking (Liam Singh, pers. comm.). Pale-eyed birds should be carefully scrutinized for other irregular field marks, as some of these are undoubtedly hybrids, but others are perfectly acceptable Glaucous-wingeds (i.e., typical wingtip color and pattern, gray upperparts, pinkish orbital, and size / structure). Magenta-pink to purplish orbital ring. Pink legs. A strapping yellow bill, with male types showing a deep gonys, and bill therefore appearing bulbous shaped. Female types may have surprisingly smaller and more compact bill with little gonys expansion. In nonbreeding condition, adults commonly show a rather dull yellow bill with some black markings emerging on or above the red gonys spot. In flight, the wings appear broad, with a long-necked appearance. The outer wingtip will show contrasting pale gray subterminal bands from below, but beware harsh light, which sometimes gives the impression of an all-white wingtip. In neutral lighting, the underside of the gray remiges contrasts with the white wing linings and white trailing edge. On the upperside, the primary pattern can be uniformly colored or show contrasting darker pigment around the tips, which gradually become paler toward the feather base. P10 usually has a medium-sized to large mirror and sometimes an entirely white tip. The mirror on p9 may completely merge with the inner web, resembling the well-known *thayeri* pattern; other times, p9 is without a mirror entirely. A string-of-pearls is not uncommon in adults, mostly from p6 to p8. Subterminal band on p5 variable, at times absent or complete, and often with a diffuse and faded pattern on the outer edges. *Similar Species:* Glaucous Gull has all-white wingtips, paler gray upperparts, and pale yellow eyes. Iceland Gulls may have wingtips similar in color and pattern, specifically Kumlien's Gull, but structure is sufficiently different, and the two are seldom found together. Typically, Kumlien's should show less pigment on inner webs of outer primaries. Iceland Gulls average more petite bills with little expansion at the gonys, have a slimmer body, and appear longer winged. See also Glaucous-winged × Western and Glaucous-winged × Herring hybrids accounts.
1st Cycle: Variable plumage aspect ranges from muddied brown to milky gray, and to near white in birds that are prone to bleaching, sometimes very early in the season. Overall, upperparts are low contrast and plain, but some with thin, wavy barring. The scapulars appear lightly scalloped on fresh juveniles, some with light notching. Once worn, the upperparts become more uniform and muted. Postjuvenile scapulars—typically not extensive in hatch year—have variable brown-gray pattern showing faint contrast with juvenile feathers. By late winter, more likely to show advanced gray aspect to 1st alternate scapulars, at times heavily contrasting with worn juvenile wing coverts. The tertials and primaries show marginal contrast and duly give the impression of a white-winged rather than a black-winged species. Some have primaries slightly darker than the upperparts; others have primaries appearing bleached white and paler than the upperparts. Pale tips are commonly found on the tertials, and the primaries regularly have pale edging, which may wear down by midwinter. Dark eyes. Heavy, strong bill appearance. The bill typically

---

35 Adults. Can appear rather short-legged (in part due to posture). Individual on right banded in Cordova, July 2020, and ruling out American Herring influence uncertain, although note very pale underside to far wing. Image slightly underexposed. NAT DRUMHELLER. GUSTAVUS, ALASKA. NOV.
36 Adult. Likely male, given massive bill, thick bull neck, and large head. Pattern on head and neck distinctive, appearing smudged, but with underlying horizontal streaks. AMAR AYYASH. WASHINGTON. JAN.
37 Adult. Most adults are decidedly dark eyed, but some variation exists. 1%–2% have pale eyes approaching Herring Gull. Whether this is normal variation or due to outside influence is not known. AMAR AYYASH. BRITISH COLUMBIA. MARCH.
38 Adult. Darker wingtips are fairly common in Asian populations and often accepted as "pure." In the Pacific Northwest of North America, such birds are evaluated more cautiously due to widespread hybridization. AKIMICHI ARIGA. JAPAN. APRIL.
39 Adult with similar-aged California Gull (left). Shown here to demonstrate relative sizes. Glaucous-winged is generally paler, but some extremes can overlap with *albertaensis* California Gulls. AMAR AYYASH. BRITISH COLUMBIA. MARCH.
40 Adult with 1st cycle. Small red gonys spot, dark eye, and saturated purple-pink leg color. Eyes are placed relatively high on the face in this species, which, overall, can appear small-headed relative to its large, deep body. AMAR AYYASH. BRITISH COLUMBIA. MARCH.
41 Adult. Broad wing. Subdued pattern on outer primaries showing little contrast with gray upperparts. Some have all-gray p9 with no mirror. Broad wing, large head and long neck, and powerful bill, unlike Kumlien's. AMAR AYYASH. CALIFORNIA. JAN.
42 Adult. Another pale-winged adult with all-white tip to p10, large p9 mirror, and bold string-of-pearls on p6–p8. LIAM SINGH. BRITISH COLUMBIA. FEB.

remains all black through much of 1st cycle, but not uncommon to see some pink paling around the bill base, although not extensive or sharply demarcated. Pink legs can appear almost purplish with a saturated quality, darker than similar-sized gulls found throughout its range. Often shows distinctively dark shin markings in 1st cycle. On the open wing, the outer primaries have a gentle two-toned pattern with paler inner webs and darker outer webs. Inner primaries are generally paler than rest of wing, but only slightly so, and sometimes not contrasting with remaining flight feathers. The tail is typically solidly dark and muted, with plain patterned uppertail coverts. From below, the body and wing linings are warm colored and rather plain. The undertail coverts show variable barring, usually with a paler base color than uppertail coverts. *Similar Species:* Overall, compare to other white-winged taxa. Hybrids with Western, Herring, and Glaucous Gulls; see those hybrids accounts for details. Glaucous Gull is similar in build but overall has a paler juvenile plumage with paler remiges. By October, most 1st-cycle Glaucous Gulls show a clean pink bill base with a sharply demarcated black tip, while many Glaucous-wingeds retain much black in the bill throughout 1st cycle. Thayer's Gull shows more contrast between darker primaries and body, whereas Glaucous-winged is overall more uniform throughout. Kumlien's Gull is less of a problem, since they seldom overlap, although increasingly overlapping in British Columbia, and strays are found throughout the Great Basin and Colorado Front Range. Bleached birds in late winter present a greater challenge, so assessment of size and structure is critical. Glaucous-winged is burly in comparison, with a shorter wing projection, and has a stockier bill with larger gonys expansion. It also averages higher eye placement on a larger face, whereas the Iceland Gulls often have a more centered eye and gentler expression.

Usually not an identification concern with Slaty-backed Gull, but there are darker Glaucous-winged Gulls and paler-winged Slaty-backed Gulls that can look remarkably similar in structure. Overall, Glaucous-winged has a noticeable grayish aspect with plainer scapulars (be aware that a small percentage of Glaucous-wingeds do show thin, dark shaft streaks on their postjuvenile scapulars). More problematic is when the

43 Adult. Darker wingtip on this individual. Undeveloped p9 mirror, complete band on p5, and small mark on p4 (rare). IVAN MUNKRES. CALIFORNIA. JAN.
44 Adult. An interesting combination of all-white tip to p10 and no p9 mirror. Darker wingtip showing some contrast with gray upperparts. Such birds are not rare and are presumed pure Glaucous-wingeds. LIAM SINGH. BRITISH COLUMBIA. MARCH.
45 Adult. Noticeably broad wings and, on this individual, bold string-of-pearls pattern. Stout bill. Honey-colored iris not very rare. AMAR AYYASH. JUNEAU, ALASKA. MARCH.

flight feathers have begun to fade and bleach, especially in late winter, making it difficult to ascertain specific patterns (such as those found in E Bering Sea and W Alaska in summer). In general, Slaty-backed will have darker secondaries that should show more contrast with the rest of the wing. The remiges are more uniformly pale on Glaucous-winged. The greater coverts on Slaty-backed often suffer more wear and fading in comparison to the median and lesser coverts, whereas on Glaucous-winged, the entire wing panel seems to wear equally. Some birds with feathers completely bleached will not be safely identified.

2nd Cycle: Sometimes rather similar in appearance to 1st cycle and may retain much black in the bill. Compared to 1st cycle, note the rounded primary tips, as well as plainer tertial and greater covert pattern. Overall, looks more muddied compared to 1st cycle. The mottled head and neck are variably marked. May become mostly white headed and gray backed, usually by late winter, via 2nd prealternate molt. On the open wing, similar to 1st cycle, but more uniform gray, gray-brown wings. The uppertail coverts show a paler base color and may be white and minimally marked. The tail pattern can resemble that of 1st cycle but is often more defined and may show a broad, contrasting gray tailband. Pink legs, but many with dark tonal quality. Those with palest bill pattern can show pale nail and extensive pinkish bill base, but again, not commonly sharply demarcated, showing obvious dark segment along the cutting edge. *Similar Species:* Glaucous Gull and Glaucous-winged Gull may have similar bill patterns at this age, but the former often with much less black around the tip, with a sharper demarcation and paling nail. Glaucous Gulls in 2nd cycle typically start to show a paling iris with a creamier and paler plumage aspect. Glaucous-winged is a duskier gray brown throughout, with darker 2nd basic and 2nd alternate scapulars. Bill size and structure average smaller and straighter on Glaucous, especially in W and N Alaska. Iceland Gulls usually separated by combination of smaller bill and body size, more delicate head proportions, and longer wings.

3rd Cycle: Body and head largely white with mottling around head, neck, and breast in basic plumage. Bill pattern highly variable, but now generally with much more pinkish-yellow base with black tip, some becoming adult-like, usually in late winter and early spring showing small red gonys spot. Eyes dark with raspberry-pink orbital ring on some individuals. Pink legs. Bluish-gray upperparts with variable brown hue. Secondaries and tertials show noticeable white tips resembling the adult's, but secondary skirt reduced in size, often with some brownish-gray peppering. Outer primaries vary from advanced, adult-like, with sizable white apicals and smaller p10 mirror, to showing little to no white on tips and resembling 2nd cycle on most-delayed birds. The wing linings are patterned with variable dusky brown markings and rarely completely white. Uppertail coverts virtually all white, with variable fragmented dark gray pigment on rectrices, rarely forming a complete tailband. *Similar Species:* Rather distinctive at this age, with putative Glaucous-winged hybrids presenting the most problems. See hybrids accounts H1, H2, and H4.

MOLT Extent of 1st prealternate molt variable, but typically limited to scapulars, head, and body. Some 1st-cycle individuals begin showing an appreciable amount of 2nd-generation scapulars in early fall (relegated to 1st alternate), while others are largely juvenile through late winter to early spring, possibly related to populations (more study needed). A small minority of 1st-cycle birds in late summer through early winter already have extensive white head and neck, with contrasting, typical warm brown body (e.g., as early as late Aug. and early Sept. in Alaska; Dec. and Jan. in Washington and California; Ayyash, pers. obs.). More study needed to determine if some of this is molt related, bleaching, intraspecific variation, or a combination of any of these factors (see plate 30.3). Often, alternate feathers found on the lower neck, and especially upper flanks and sides of breast, are a darker grayish brown, but this is not the case in these white-headed individuals.

HYBRIDS Much gene flow exists between Glaucous-winged and Western Gulls in Washington and Oregon, between Glaucous-winged and Herring Gulls in S Alaska in the Cook Inlet region, and, to a much lesser known extent, between Glaucous-winged and Glaucous Gulls along the E Bering Sea coast of SW Alaska (Swarth 1934; Strang 1977). A few scattered reports of hybridization with Slaty-backed Gull (see that species account for details) and presumably with Vega Gull (in 2018 on St. Matthew Island; Robinson et al. 2020). Fragmented hybridization reports and occurrences throughout the Pacific Northwest need detailed documentation.

31 GLAUCOUS GULL *Larus hyperboreus*

L: 22"–29" (56–74 cm) | W: 56"–63" (142–160 cm) | Four-cycle | SAS | KGS: 2.5–5.0

OVERVIEW The ultimate white-winged larid. Glaucous Gull is the second-largest gull species in the world and the only four-cycle species with a circumpolar breeding distribution in the northern hemisphere. A hardy gull found both below and well above the arctic circle. Noticeably short winged. Adult gray among the palest of the gulls. Breeds in smaller groups or singly, both cliffside and on the tundra. Regularly nests in mixed colonies of cliff-dwelling seabirds, eiders, and geese, which it often depredates. Consider the following report by Bowman et al. (2004), who surveyed Glaucous Gull depredation on goslings in SW Alaska: "We estimated that a minimum of 21,000 Emperor Goose, 34,000 Canada Goose, and 16,000 White-fronted Goose goslings were consumed by 12,600 Glaucous Gulls during the brood-rearing period on the Y-K Delta in 1994." An all-around generalist with a wide diet, ranging from marine invertebrates to fish, eggs, and chicks of cohabitants; small mammals; carrion; and human discards. Birds breeding inland are known to consume more terrestrial prey than those breeding coastally (Schmutz & Hobson 1998). Its diverse diet also includes isopods, berries, and mussels. *Hyperboreus* is in reference to its being a denizen of the far north, where it commonly feeds on polar bear kills, including seals and walruses, as well as beached whales. Slightly less gregarious than other gulls, and interactions between conspecifics are sometimes highly agonistic.

TAXONOMY Banks (1986) described four subspecies: nominate *hyperboreus*, *leuceretes*, *pallidissimus*, and *barrovianus*, based on size differences and upperpart darkness. However, the sample sizes were unclear, and the study used a confusing method of tabulating data. The smallest and darkest, *barrovianus*, breeds on mainland Alaska. The largest and palest, *pallidissimus*, occupies several known Bering Sea islands, including St. Matthew and Walrus Islands (Alderfer & Dunn 2014). Reported increase in bill and tarsus measurements from Alaska east through the Holarctic and around to Siberia, with the two extremes being separated by the Bering and Chukchi Seas and Bering Strait (Olsen & Larsson 2004; Petersen et al. 2015). Nominate *hyperboreus* is intermediate in size and upperpart coloration. Banks (1986) proposed *leuceretes* for birds from Franklin Bay, Mackenzie, the Canadian archipelago, south in Hudson Bay to the Belcher Islands, and on the Atlantic Coast to N Labrador, Greenland, and provisionally Iceland. Banks described *leuceretes* as being slightly paler than nominate and longer billed. Detailed study is lacking in this regard, and other authors simply attribute this population to nominate, and the two are likely inseparable in the field (Olsen 2018). Samples collected away from the breeding grounds have contributed to this ambiguity. Noteworthy, though, are several striking differences between Canadian birds and those wintering in W Europe. Canadian winter populations show substantially fewer head markings in basic plumage as adults, and juveniles are paler overall with flight feathers showing noticeably less distinct dark markings throughout.

The degree of overlap between *hyperboreus* and *pallidissimus* on the N Taimyr Peninsula is not entirely clear, but an intergrade zone is suspected (Stepanyan 2003). In addition, overlap between *pallidissimus* and *barrovianus* in the E Bering Sea is suspected. For instance, a recent summer survey on St. Matthew Island, where the largest and palest *pallidissimus* breed, included a short-winged and darker *barrovianus* type, which was likely breeding here, although it was cited as a "salvaged specimen" (Robinson et al. 2020). Whether small numbers of *pallidissimus* are breeding on mainland Alaska is not clear, and, similarly, how many visit the mainland as nonbreeders is unknown. Birds nest-trapped and fitted with trackers from the North Slope of Alaska between Wainwright and Colville Delta were found to migrate over the Bering Sea, and they wintered on the coasts of Kamchatka and in the Sea of Okhotsk (Declan Troy, pers. comm.). This region is inhabited by *pallidissimus* in the breeding season, but both races overlap here in winter. In years with heavy ice, those *barrovianus* that were tracked appeared to move farther south, to Sakhalin Island, the Kurils, and N Japan. Therefore, wintering Glaucous Gulls in E Asia are attributed to both *pallidissimus* and *barrovianus*. A recent study that tracked adult Glaucous Gulls on Coats Island, N Hudson Bay, found the majority of individuals traversed the Hudson Strait and ended up in the N Atlantic for the winter (Baak et al. 2021). These were presumably *leuceretes*. One individual, however, migrated west, to the Sea of Okhotsk in Siberia, for two consecutive winters (Julia Baak, pers. comm.). Therefore, an east–west movement of arctic gulls may be more common than once thought. A summary of range distributions as provided by Weiser and Gilchrist (2012) is as follows.

L.h. hyperboreus (KGS: 3–4): Nominate race. Breeds from NW Siberia west to N Scandinavia and Spitzbergen. Winters in N Europe.

L.h. leuceretes (KGS: 3–4): Breeds throughout N Canada, east through Greenland and Iceland. Winters in the NW Atlantic, from Newfoundland and Labrador south to NE United States and presumably through parts of the interior, from the Great Plains south to Texas and Florida.

L.h. pallidissimus (KGS: 2.5–4.0): Breeds in N Siberia, east of Taimyr Peninsula, through Chukotka to St. Matthew, St. Lawrence, and Walrus Islands (footprint on mainland W Alaska unclear). Winters from NW China to Japan.

L.h. barrovianus (KGS: 4–5): Breeds along coastal west-central and N Alaska, east through N Yukon and Mackenzie Delta. Winters from the North Slope to S Alaska, and some likely reach south to N Baja California, Chukchi and Bering Seas, west to Kamchatka, Sea of Okhotsk, and south to Japan.

RANGE

Breeding: Holarctic (circumpolar), breeding in the Arctic and wintering south to midlatitudes. In North America, breeds often on cliffs and sea stacks, but also on the ground in the tundra and on islands and islands in lakes. Occasionally in trees—namely, spruce trees up to 15 meters high. Near and along coasts from W and N Alaska east across arctic Canada to central Labrador and to Greenland. Southernmost breeders in N Hudson Bay, few on Belcher Island in S Hudson Bay, and irregularly in smaller numbers on coasts of James Bay. Small population on St. Matthew Island increasing: it outnumbers Glaucous-winged Gull by approximately five to one. Is not known to breed on the Pribilof or Aleutian Islands, although common here in nonbreeding season.

Nonbreeding: Frequents habitats favored by other large *Larus* gulls, including marine waters well offshore, coastlines, harbors, larger lakes and reservoirs, fish-processing plants, large rivers, dams, and landfills. Migrants start moving by August and September, but they are not found well south of the breeding range (from S Canada southward) until November or December. Winters as far north as appreciable open water is present, in good numbers only off W and SW Alaska, in Eastern Canada, and off SW Greenland. Small numbers may also overwinter in open water (polynyas) near the Belcher Islands in Hudson Bay. Some N Alaska breeders winter along the Asian coast south to N Japan. South of the core wintering area in North America, generally very uncommon to very rare, following a north-to-south gradient. Small numbers reliably found on all the Great Lakes. Increasing in the N Great Plains, especially at large lakes along the Missouri River. Largest numbers in United States typically at landfills and nearby loafing areas, but even there the highest one-day counts are usually only in the single digits and occur between December and mid-March. Rare to very rare across the southern states, and casual in Arizona, N Mexico, Bermuda, and to Hawaii. Northbound migrants occur mainly from late February to early April. A very few individuals linger into May. Summering nonbreeders are regular in small numbers in SE Canada but are casual in United States as far south as S California and Florida; reports from here often prove to be of leucistic or very bleached large white-headed gulls.

IDENTIFICATION

Adult: Noticeably pale gray upperparts with all-white wingtip. At rest, appears brawny with a short wing projection. Bill is relatively stout, deepest at its base, with no noticeable expansion to the gonys. Bill base often a dull pink, especially in nonbreeding condition. Pale yellow eye. Yellow to orange orbital, at times appearing reddish (reportedly in *barrovianus*). Legs pinkish and not commonly bright. Many adults in basic

31 GLAUCOUS GULL

1. Juvenile. Sandy tones to plumage, with noticeably pale primaries showing dark diamonds near the tips. Bill still developing, with dark base remaining for only a very short time. Presumed *pallidissimus* by range, but migrant *barrovianus* possible. STEVE HEINL. GAMBELL, ALASKA. SEPT.

2. Juvenile with similar-aged Iceland Gull (right). Thick bill base now largely pink with hint of diffuse black, becomes sharply demarcated typically by late Sept. to early Oct. Beady eye on large head with stocky body. Presumed *leucerctes* by range. JOHN CHARDINE. GREENLAND. AUG.

3. 1st cycles. Plumage aspect ranges from ghostly white to warm buffy-brown. Sharply bicolored bill now with pink bill base. VERNON BUCKLE. LABRADOR. JAN.

4. 1st cycle. Some individuals in Europe have surprisingly dark bodies and heads, and may show well-patterned tails and flight feathers. Dark diamond centers on primary tips common. Presumed nominate *hyperboreus* by range. KRIS WEBB. ENGLAND. JAN.

5. 1st cycle. Large pale body with piggish look to head and bill. Deep chest, thick legs, and short wing projection. AMAR AYYASH. INDIANA. JAN.

6. 1st cycle. An individual with densely patterned coverts and warmer underparts, recalling Glaucous-winged Gull. Pale primaries and bill pattern distinctive. ALEX A. ABELA. CALIFORNIA. JAN.

7. 1st cycle. By mid- to late winter, some birds are heavily bleached and worn, losing all patterning to their upperparts. Short wing projection and wide bill base help rule out leucistic individuals of other species. AMAR AYYASH. FLORIDA. JAN.

8. 1st cycle with American Herring. Appreciable size variation could be subspecific or sex-related. Kumlien's Gull is ruled out by bill pattern, short wing projection, thick neck, and large head with sloping forehead. AMAR AYYASH. ILLINOIS. DEC.

9. 1st cycle with similar-aged Glaucous-winged Gull. Glaucous is distinctively paler and never shows an all-dark bill this late in the season. The bill tip is fairly straight (not bulbous). CAROLINE LAMBERT. CALIFORNIA. NOV.

10. 1st cycle. Long, thick neck, broad wings, and pale flight feathers. The uppertail shows variable degrees of pigment (average here) with dense patterning on the uppertail coverts. CHRISTOPH MONING. NORWAY. APRIL.

11. 1st cycle. It's typical to retain juvenile upperparts into spring, but this individual's are surprisingly fresh for this time of year. Note dark diamonds on primary tips, often faded by now. Iceland Gull is ruled out by the large, long bill and bull neck. JOSH JONES. IRELAND. FEB.

12. 1st cycle. A large and pale individual with broad wings, bull neck, and large head. Bill likely still growing. Presumed *pallidissimus*. AARON LANG. ALASKA. SEPT.

31 GLAUCOUS GULL

13 1st cycle. A pale individual with largely unpatterned upperparts (likely some fading and wear already settled in). The uppertail is sometimes contrastingly darker. ALEX EBERTS. OHIO. NOV.
14 1st cycle. Large body with broad wings, thick neck, and powerful bill. The wing linings are overall pale, with variable light barring often found on the undertail coverts. DANNY BALES. FLORIDA. MARCH.
15 2nd cycle. Aged by molting inner primaries (see 31.21). Eye typically begins to pale at this age. Scrappy appearance, with 2nd-generation upperparts often contrastingly dark. AMAR AYYASH. WISCONSIN. JULY.
16 2nd cycle. Heavyset bird with pale, creamy plumage and little patterning on wing coverts. Aging points include rounded primary tips, paling iris, and pale nail. Presumed *leuceretes* by range, although uncertain. PATRICK HORAN. NEW YORK. DEC.
17 2nd cycle. Presumed nominate by range and overall appearance. A darker individual with extensive vermiculations and patterning on wing coverts and tertials when compared to most North American birds. Some show darker tail and outer primaries, inviting thoughts of Herring influence. MERIJN LOEVE. THE NETHERLANDS. JAN.
18 2nd cycle and 1st cycle (right). The two ages are often strikingly similar in appearance, especially in mid- to late winter, when patterns have faded. But when seen well, eye color, marbled coverts, and rounded primary tips distinctive in 2nd cycle. DAVE BROWN. NEWFOUNDLAND. JAN.
19 2nd cycle. Noticeably pale eye now, with reduced black on bill tip. Some adult-like gray (2nd alternate) scapulars are now visible. Aged with 2nd-generation-like inner primaries (not visible). AMAR AYYASH. ILLINOIS. MARCH.
20 2nd-cycle type. More solid gray, adult-like scapulars, with advanced yellow bill and red on gonys. Pale eye with dark orange-red orbital. Such birds are best aged by an open wing. Presumed *barrovianus* by range. NIGEL VOADEN. UTQIAGVIK, ALASKA. JUNE.
21 2nd cycle. Same individual as 31.15. Long, broad wing and strong bill compared to Iceland. Undergoing 2nd prebasic molt. Worn tail feathers and p8–p10 are juvenile (=1st basic). AMAR AYYASH. WISCONSIN. JULY.
22 2nd cycle. Averages paler and plainer than 1st cycle at this time of year. Some have entirely pale eye, as seen here, but others may still be darkish. Deep belly, bull neck, broad wings, and thick bill help rule out Iceland Gull. AMAR AYYASH. INDIANA. DEC.

23 3rd cycle and 1st-cycle Great Black-backed. Glaucous is only slightly smaller, on average, with noticeable overlap. Brown markings down the neck, breast, and belly, with little gray on the coverts and a few tertials. Commonly with clear iris at this age. AMAR AYYASH. MASSACHUSETTS. APRIL.

24 3rd cycle with adult Thayer's. Presumed *barrovianus* based on smaller size and relative darkness of gray upperparts, although uncertain. The wing panel is seldom all gray at this age. LIAM SINGH. BRITISH COLUMBIA. MARCH.

25 3rd cycle with Thayer's and Glaucous-wingeds in backdrop. An impressively large and pale individual, presumed to be *pallidissimus*. LIAM SINGH. BRITISH COLUMBIA.

26 3rd cycle. A compact individual with short legs, short wing projection, and fairly short bill. A pinkish bill base with dark tip is fairly common, even at this age. Small beady eye on large face. VERNON BUCKLE. LABRADOR. JAN.

27 3rd cycle. Adult-like gray centers to flight feathers with relatively broad white tips to trailing edge eliminates 2nd cycle. A few brown blemishes on tail and heavily marked body. AMAR AYYASH. ILLINOIS. DEC.

28 3rd cycle. Heavy bill, with long projection. Less commonly shows an all-white tail. Outermost primary tips are all white and show no ghosting or mirrors (compared to some Kumlien's and Glaucous-winged Gulls). AMAR AYYASH. WISCONSIN. MARCH.

29 Adult. Large head with bull neck and pale, beady eyes. Short wing projection and pale gray upperparts (no other large white-headed gull beside nominate Iceland Gull matches this gray). Some dusky mottling on the lower neck is fairly common. AMAR AYYASH. ILLINOIS. FEB.

30 Adult. Presumed *barrovianus* with slightly darker upperparts (in part due to lighting). Compact with relatively long wing projection, recalling nominate Iceland Gull, but note rich, dark orange orbital, which sometimes appears red in *barrovianus*. MARTEN MULLER. SOUTH KOREA. FEB.

31 Adult. Heavy head and neck markings recall *hyperboreus* or European Herring × Glaucous hybrids from Iceland (so-called Viking Gull). An open wing revealed extensive white down the feather centers with no contrasting pigments, but the darker gray upperparts and bill pattern may, in fact, be *argentatus* influence. KIRK ZUFELT. NEWFOUNDLAND. JAN.

32 Adult with similar-aged American Herring Gull. Glaucous Gull is noticeably paler in most lighting conditions, shows no pigment on the wingtip, and is generally bulkier. Brighter bill transitioning to breeding condition. KATHY MARCHE. NEWFOUNDLAND. MARCH.

33 Adult with Glaucous-winged Gull (left). Shown here to demonstrate how massive and pale *pallidissimus* may be. Variation in size may also be sex-related. PAUL FRENCH. JAPAN. FEB.

34 Adult. Large squarish head, smaller eye with dark orange orbital (breeding condition), and straight yellow bill. Short wing projection and pale gray upperparts. DAVID TURGEON. QUEBEC. APRIL.

plumage show some light dusky streaking on the head, hindneck, and commonly down to the sides of the upper breast, but typically not heavily marked in North America. In the Old World, some adults show remarkably heavily marked heads, especially in Asia. Others largely white headed in December–February. On the spread wing, the outermost primaries have extensive white tips with no pigment. Broad winged. Flight pattern is labored but stealthy and powerful. In general, there is considerable variation across subspecies, with *barrovianus* and *pallidissimus* forming extremes. *Similar Species:* Iceland Gulls with all-white wings are most similar. Nominate Iceland Gull is closest, especially in matching gray upperparts and pale yellow eyes. Bill size and shape differences critical. Glaucous has a wider bill base, noticeably shorter wing projection, and a deeper chest. Iceland averages shorter legs and a more rounded head, but beware opposite-sex overlap. Some female-type Glaucous and male-type Iceland Gulls may appear similar in size, but Iceland overall is sleeker with noticeably longer wings at rest. In flight, Iceland's wings are narrower, with shorter neck and head projection. Bare-part colors helpful when discernible. Iceland has pinkish-purple orbital and sometimes with a notable green cast to the bill, whereas Glaucous generally shows an orange-yellow orbital with pinkish cast to bill base in nonbreeding condition. Pale-winged Kumlien's Gull should appear slightly darker gray on upperparts and often has variable gray markings on outermost primaries. Kumlien's has similar bare-part colors to those of nominate Iceland, and a good percentage of Kumlien's will show honey-colored to darkish eyes, which helps eliminate adult Glaucous. Also see Glaucous-winged Gull account.

35 Adult (center) with American Herring (right) and 3rd-cycle Thayer's Gull (left). Remarkable size difference may be sex-related. Glaucous commonly shows a wide bill base. CHRISTIAN ARTUSO. NUNAVUT. JULY.

36 Adults and 3rd-/4th-cycle types. Assigned to *leuceretes* by range, although most Glaucous Gulls on the wintering grounds (except for extremes and highly distinctive individuals) are not safely assigned to subspecies in many cases. JOHN CHARDINE. GREENLAND. AUG.

37 Adult. Distinctive. All-white wingtips with zero pigment. Broad wing with pale gray coloration. Large powerful bill and heavy chest. AMAR AYYASH. ILLINOIS. MARCH.

38 Adult. The ultimate white-winger. Nominate Iceland Gull ruled out by broad wings, bull neck, and long powerful bill. RYAN SANDERSON. ILLINOIS. FEB.

39 Adult. Thick neck, beady eye on large face, and fleshy bill-base coloration are helpful in ruling out Iceland Gull. AMAR AYYASH. ILLINOIS. FEB.

40 Adult. All-white throughout underparts, with much bill protruding past head. Large body, broad wings, and proportionally long tail. JIM TAROLLI. NEW YORK. FEB.

1st Cycle: Fresh juveniles show light brown scaly pattern throughout the upperparts on a creamy-colored to grayish-white body. Dark bill with base beginning to pale before young have completely fledged. Majority with bright pink bill base and sharply demarcated black tip by early to mid-October. Relatively small, dark eye against a large, pale face sometimes gives this age group a peculiar pig-eyed feel. Dull pink-colored legs. Tertials can be moderately dark with subterminal barring. Primaries often show little contrast with body, but at times, they may show strong dusky wash throughout with darker diamond-tipped centers surrounded by pale chevrons. These individuals often show strong marbling on uppertail and inner primaries. Others have surprisingly whitish primaries a month or two after fledging. Underwing coverts warm buffy color to pale, some with noticeable barring, especially on axillaries. By late winter, many have faded or bleached entirely, to the extent that little to no dark patterning is detected, often resulting in an entirely milky-white plumage aspect. By early to mid-spring, some individuals show contrasting 2nd-generation scapulars, usually patchy and randomly distributed, extending onto sides of breast and flanks (likely 1st alternate or beginning of 2nd prebasic). A few start to show faint traces of a paling eye and might develop a pale nail by end of 1st cycle. ***Similar Species:*** Nominate Iceland Gull is smaller overall, with more petite bill, longer wings, and overall gentler appearance with a proportionately larger eye on a smaller face. Many nominate Iceland Gulls show bicolored bill pattern, but typically with some dark blemishes along cutting edge and not as sharply demarcated as found in Glaucous. Kumlien's Gull often shows duskier brown pigments on tertials and flight feathers, especially outer primaries. Again, size and structure critical. Other species when completely bleached in early summer may suggest Glaucous. Compare size, structure, and bare parts. Similarly, leucistic individuals of other four-year species—often Herring—may cause confusion, but those individuals fairly often show some darker

creamy regions on uppertail or secondaries, whereas on Glaucous Gull those feathers tend to bleach more evenly. Also see Glaucous-winged Gull and Glaucous Gull hybrids accounts. There are growing reasons to suspect that 1st-cycle Glaucous Gulls—namely, those between Asia (*pallidissimus / barrovianus*) and the western shores of North America—may naturally show some diffuse pigment bleeding onto the cutting edge and basal two-thirds of the bill. Some of these individuals are suspected to be Glaucous × Glaucous-winged hybrids (see that account), but others may prove to be pure Glaucous Gulls (more study needed). Many bleached Glaucous-wingeds in Alaska often misidentified as Glaucous in spring / summer. Bill pattern and shape are useful, as Glaucous Gull typically lacks a bulbous tip.

2nd Cycle: Generally pale bodied, much like 1st cycle, but with plainer pattern to upperparts. Wing coverts marbled with dark freckling in fresh birds; fainter on others, due to wear and bleaching. Head, neck, and flanks more blotchy and coarsely marked than in 1st cycle. Dull pink legs. Eyes are generally pale with dark specks on most individuals, but some may remain dark into 2nd cycle. Note, however, less defined black tip on bill when compared to 1st cycle, as well as other age-related clues, such as rounded primary tips and more muted pattern on primary coverts. Variable light gray on scapulars and mantle by late winter through late spring, and some entirely gray via 2nd prealternate molt. Replaced wing coverts usually limited to a few inner median or lesser coverts. *Similar Species:* Again, compared to Iceland Gull, Glaucous is heavyset and larger in appearance with a thicker bill and shorter wing projection. Kumlien's generally shows more pigment on outer primaries, tertials, and tail at this age. Also see Glaucous-winged Gull and Glaucous Gull hybrids accounts.

3rd Cycle: Bill color variable, from adult-like yellow with black subterminal marks to dull bone color, and some appearing rather 2nd-cycle-like and pink. Eyes decidedly pale. Head and body largely white with minimal to moderate markings on lower neck and breast. Pink legs. Patchy white and adult-like gray scapulars. Gray found on upperwing coverts, although seldom fully gray on all three tracts simultaneously. Birds with extensive gray wing panel usually show noticeable lighter beige hues throughout. On the open wing, the flight feathers have weakly contrasting gray centers with broad white tips (although this may be absent on outermost primaries with retarded 2nd-generation aspect). Underwing appears sullied with light brown markings. Uppertail largely pale with variable freckling and barring. *Similar Species:* As for similar species for adult, but note that bare parts are quite useful here. Many Glaucous Gulls have pinkish bill base, while nominate Iceland Gull has greenish-gray cast to bill base. Size and structure trump plumage in confusing individuals. See also Glaucous-winged Gull and Glaucous Gull hybrids accounts.

MOLT Inserted molt in 1st cycle appears to be a partial prealternate molt. Quite limited at times and may be absent in some individuals that overwinter at southern latitudes (Ayyash, pers. obs.; Howell 2010). More study needed on larger samples in late spring after migration. Deciding which body feathers were replaced under tail end of 1st prealternate or start of 2nd prebasic molt not always clear. Similarly, 2nd prealternate and 3rd prebasic limits not clear at times, in part due to low-contrast plumes in this species.

HYBRIDS Hybridizes with all three species of Herring Gulls: with European Herring in Iceland (Ingolfsson 1970, 1987; Ingolfsson et al. 2008), with American Herring in the Mackenzie Delta of NW Canada (Spear 1987), and presumably with Vega Gull in E Siberia (Kessel 1989). Hybridization with American Herring also strongly suspected in E and NE Canada, but breeding sites largely unknown. Recorded annually in St. John's, Newfoundland, in small numbers, and scattered reports annually throughout Eastern Seaboard and Great Lakes. Origins of these birds unknown, and identification of backcrosses is speculative. A mixed pair identified as Vega / Glaucous in Bluff, Alaska, produced young in 1977 (Kessel 1989). Also reported annually in E Asia and Russia.

Putative hybrids with Great Black-backed Gull recorded annually in Newfoundland (fewer than 10 per winter; Bruce Mactavish, pers. comm.). Rare outside Newfoundland. Unknown origins. Also recorded in small numbers in Europe. Hybridization with Glaucous-winged Gull in SW Alaska reported historically; see that species account for details. Reported hybridizing with a single adult Slaty-backed that returned for several summers near Prudhoe Bay (Delcan Troy, pers. comm.; Karl Bardon, pers. comm.).

32 ICELAND GULL (LARUS GLAUCOIDES) COMPLEX

The Iceland Gulls are among the most alluring larids in the world. They have raspberry-pink orbital rings and are generally more refined in structure when compared to larger congeners. Watching them throughout the winter season is the muse of many gull enthusiasts. They remain the least known of our large gulls, breeding patchily in remote regions throughout the Canadian Arctic and Greenland. They commonly nest on narrow ledges on high sea cliffs, often in colonies of 50–100 pairs. Despite the English name, Iceland Gulls do not breed in Iceland. Interestingly, they don't appear to freely hybridize with other species—an exception to the rule when it comes to four-cycle gulls of the far north. This is not to say they don't or won't hybridize, but records remain unsubstantiated.

When we hear "Iceland Gull," often the inference is the species as a whole. The complex consists of three subspecies, adults being distinguished chiefly by the pigment on their wingtips. The darkest subspecies is Thayer's Gull (*Larus glaucoides thayeri*, KGS: 5–6), the highly variable subspecies is Kumlien's Gull (*L.g. kumlieni*, KGS: 4–5), and the palest (and rarest) subspecies in the ABA Area is nominate Iceland Gull (*L.g. glaucoides*, KGS: 3–4), sensu stricto.

Observers familiar with this complex agree on what typical *thayeri* and typical *glaucoides* look like as adults. The two can never be confused. However, a small percentage of individuals in the Iceland Gull complex are not comfortably identified to subspecies. The highly variable form, *kumlieni*, remains poorly defined, and future studies must expand on its status. This taxon presumably overlaps with *thayeri* and *glaucoides* in every observable field mark, and where one begins and another ends is at the crux of the matter. Some authorities seem to have forged their own arbitrary lines using winter ranges to make such determinations, while others have refused to recognize any lines, staunchly arguing that all members of the complex represent points on a cline. The latter view has been held by some of the foremost authorities on this complex— some who have been to the breeding grounds and have parsed copious museum specimens (Godfrey 1986; Snell 1989; Snell et al. 2020). The clinal-species theory appears to be an oversimplification, however, but some evidence for nonassortative breeding, uncertain breeding range boundaries, and a continuum of phenotypes on display makes it impractical to disassociate Thayer's from the Iceland group given current knowledge.

In this guide, all three taxa are treated as Iceland Gulls, with the admission that the American Ornithological Society lacked absolute data in its most recent recommendation to lump Thayer's, and that this complex is inherently knotted (Chesser et al. 2017). Each taxon is given its own account, with additional details in its "Taxonomy" section. Further commentary on the complex as a whole follows the accounts.

1   From top to bottom: Thayer's (Washington), Kumlien's (Massachusetts), Kumlien's (Massachusetts), Kumlien's (Newfoundland), nominate *glaucoides* (southwestern Iceland). The countless gradations of Kumlien's appear to bridge Thayer's and nominate.

2   Adult Kumlien's Gulls. An excellent example of the wide variation presumed in this taxon. Adults can have wingtips that more or less match Thayer's at the dark end and nominate *glaucoides* at the pale end. DAVE BROWN. NEWFOUNDLAND. FEB.

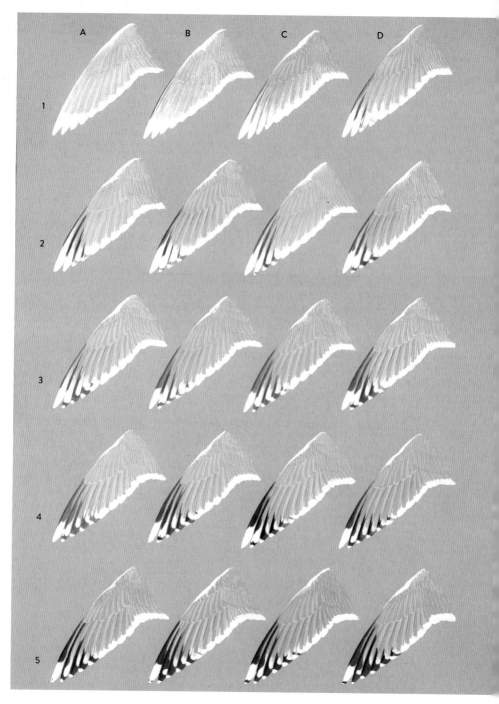

3 A small sample of the myriad patterns found in this complex. The first (1A) and last (5D) represent nominate *glaucoides* and *thayeri*, respectively. The line between Thayer's and Kumlien's is somewhat arbitrary and may prove tenuous. Widely accepted Thayer's patterns resemble row 4 and row 5, with row 3 blurring the lines. Whether some *glaucoides* can match 1B or 1C (or beyond) has not been fully established and requires careful investigation. ARTWORK BY HANS LARSSON.

32A THAYER'S GULL Larus glaucoides thayeri

L: 18"–23" (45.5–58.0 cm) | W: 48"–55" (122–140 cm) | Four-cycle | SAS | KGS: 5–6

OVERVIEW Of the Iceland Gulls, Thayer's is the darkest subspecies, showing the most pigment on the wingtip, with slightly darker upperparts, and is said to average larger proportions than *kumlieni* and nominate *glaucoides*. The wingtip pattern is appreciably variable, with the pattern on p9 being most noteworthy (the so-called *thayeri* pattern). Thayer's often shows reduced pigment on the wingtip when compared to Herring Gull—noticeably so on the underside—and hence qualifies as a white-winged gull.

Thayer's Gull has long been the subject of debate. Its checkered taxonomic history has cast a certain spell over the birding community, alternately delighting and vexing us. The bird has acquired this mystique because we're not sure what it is (Ayyash 2017). It is among our northernmost nesting gulls in the high Arctic. That we know. But its breeding range is highly fragmented, and the bird is spread thinly over a vast number of large islands and remote fjords. Many colonies, especially those at the northernmost latitudes, have not been thoroughly surveyed, and some have only been surveyed incidentally by water and aircraft. Other sites have not been reported on from the 1950s to the time of this writing (Snell et al. 2020).

Inventories of breeding sites are without question lacking, and in all likelihood current population estimates suffer from sources of error, in which, for example, Glaucous and American Herring Gulls were not safely ruled out, and separation from Kumlien's Gull is not clear. For instance, several specimens collected from E Ellesmere Island in the early 20th century were originally labeled *kumlieni* by J.S. Warmbath and then relabeled *thayeri* by W.S. Brooks (Bent 1921). Bruggemann and others called birds in this same region "herring gulls" (Parmelee & MacDonald 1960). Herring Gull, and presumably Kumlien's Gull, is not found at these latitudes, and these individuals were later assigned to Thayer's Gull (Snell & Godfrey 1991). Or, for instance, reported "Thayer's Gulls" breeding on Coats Island in the late 1950s were all but gone by the early 1990s, seemingly replaced by Kumlien's Gulls (Gaston & Elliot 1990; Gaston & Ouellet 1997; Snell et al. 2020).

Specimens collected from Fosheim Peninsula (central Ellesmere Island) had wingtips ranging from nearly white to light gray to nearly black, showing much variation and differing from the type specimen (Parmelee & MacDonald 1960). It is not clear if specimens collected here were of breeding birds, however. Nonetheless, Ellesmere Island is believed to be in the core of the breeding range of "pure" Thayer's, and therefore, these specimens raise several important questions: Might such birds be Kumlien's? Are such birds *kumlieni* / *thayeri* intergrades? Is Thayer's more variable than the description popularized in bird-identification literature?

The enigma surrounding Thayer's stems from ambiguous beginnings: it was never thoroughly described across its breeding range. Consequently, there is no fixed criteria pertaining to its wingtip pattern. This is problematic, since wingtip pattern is generally a key feature used to separate large white-headed gulls (see Ayyash & Singh 2023). A mismatch in current population estimates also underscores the problem. For example, Snell et al. (2020) estimated 6,300 adult pairs of Thayer's Gulls from the breeding grounds, and Olsen and Larsson (2004) provided a similar value of 4,000–6,000 pairs. But observations during the nonbreeding season from Vancouver Island alone seem to suggest significantly greater numbers. The number of adult Thayer's Gulls approximated in the Salish Sea in early March 2001 was 25,000–30,000, and just between Nanoose Bay and Deep Bay the number was believed to be around 20,000 in 2003 (Guy Monty, pers. comm). These figures were somewhat lower in 2023, as Pacific Herring numbers continue to decrease in this region, resulting in less spectacular herring runs. Still, one can conservatively estimate the number of adult Thayer's present here in mid- to late March to be 15,000–20,000 individuals, and this excludes any synchronized effort to fully survey large parts of N Vancouver Island (Ian Cruickshank, pers. comm.). Given that the numbers don't take into account the remaining Thayer's scattered throughout the West Coast from California to Alaska or those found inland to the Great Lakes, the actual population of adult Thayer's Gulls may be closer to 30,000, if not greater.

Richards and Gaston (2018), who have carried out extensive surveys in the Canadian Arctic, provide an estimate of 50,000–100,000 breeding birds in Nunavut but do not specify whether these are all "dark" Thayer's, as we've defined them on the wintering grounds, or an aggregate of *thayeri* and *kumlieni* types. Note that some workers—particularly those who have visited the breeding grounds—maintain there is complete gradation across several axes with these taxa and therefore do not typically assign a subspecies to many Iceland Gulls. And although this is perplexing to readers who might demand black-and-white answers in an age when anything and everything seem possible, surveying Iceland Gulls in the Arctic ranks among the most challenging feats in modern ornithology. The most recent numbers suggested by Gaston and colleagues exclude regions north of the Foxe Basin (Gaston et al. 2007; Richards & Gaston 2018). This critical region is still largely unexplored, and surveys here are lacking. Therefore, the number of breeding adult Thayer's is not known, for two reasons: first, no comprehensive effort has been made to survey this population of arctic gulls across the entire Canadian Arctic; and, second, the phenotypic limits of Thayer's have yet to be clearly defined on the breeding grounds. One might say the first point cannot be resolved because of the second point, but the two are complexly intertwined. More disturbing is what seems to be a moving target, with colonies shifting and shrinking at a relatively rapid rate, so that some colonies studied in one region in the 1970–1980s appeared to be all but gone or replaced within a couple of decades (Knudsen 1976; Snell 1989; Snell et al. 2020).

TAXONOMY Type specimen collected by J.S. Warmbath from Buchanan Bay, east-central Ellesmere Island in 1901. Described and named by W.S. Brooks in 1915, who dedicated the bird to Colonel John E. Thayer. Dwight (1917) assumed Thayer's was a dark-eyed Herring Gull that showed greater white in the wingtips, and the American Ornithologists' Union treated it as such from 1931 to 1973. Interestingly, in his colossal monograph, "The Gulls (Laridae) of the World," Dwight (1925, 195) expressed some apprehension: "Surprising as it may seem, a large series shows that with increase of the white element whatever it is, as indicated by paler and paler primaries and larger areas of white, there is complete intergradation, first between *Larus argentatus thayeri* and the bird now known as *Larus kumlieni*, and then between *Larus kumlieni* and *Larus leucopterus* [i.e., *L. glaucoides glaucoides*]. There is no escaping from this fact, explain it as we may."

It was later demonstrated that Thayer's is not a Herring Gull, as the two did not interbreed (Macpherson 1961). Thayer's and Herring have largely disjoint breeding ranges, although there are reports of the two nesting within the same colonies (Smith 1966; Knudsen 1976). The notion that Thayer's was a dark, high-arctic population of Iceland Gull had been gaining traction for decades, and in fact it was already being treated as such by some workers (Salomonsen 1950; Manning et al. 1956; Macpherson 1961). Instead, Thayer's was given species status in 1973 based on Neal Smith's (1966) now widely rejected study published by the American Ornithologists' Union. Smith's conclusions were almost immediately called into question and ultimately discredited by many authorities (Sutton 1968; Weber 1981; Godfrey 1986; Snell 1989, 1991; Sibley & Monroe 1990; American Ornithological Society 2017). Smith's (1966) work was grievous and led to decades of misdirection on *thayeri*. In 2017, the American Ornithological Society (formerly the American Ornithologists' Union) reversed its 1973 decision, and Thayer's was lumped with the Iceland Gulls as *L.g. thayeri*. (For a detailed historic timeline of Thayer's Gull, see Pittaway 1999.)

Unanswered questions raised by Howell (1999) and Banks and Browning (1999) remain relevant to this taxonomy. Essential questions awaiting clarification include, What makes a Thayer's Gull a Thayer's Gull? Is it defined by geographic location (i.e., the colony it originates from), or is it defined by a set of arbitrary phenotypic traits? Or both? Do some Thayer's have light gray or predominantly white wingtips, and do some lack black markings on p6/p7 entirely, like those specimens collected from Ellesmere Island in the core of "Thayer's" breeding range (Parmelee & MacDonald 1960)? The problem, at root, was perfectly articulated by Howell (1999, 308), who commented on field-identification papers that attempted to describe adult and 1st-cycle Thayer's Gulls on the wintering grounds: "These papers all proceed from the supposition that Thayer's Gull is a species, or at least a taxon, and for reasons outlined above, such a premise may be circular: Thayer's Gulls look like this because this is what Thayer's Gulls look like … demonstrates a fundamental obstacle in seeking to define something that has yet to be defined." Howell (1999) certainly hit the nail on the head but, strangely, later proposed Thayer's should continue to be treated as a species, based on data collected "at close range" on the wintering grounds (Howell & Elliott 2001). A consequence of describing taxa without extensive baseline data from known-origin birds is that a "faith-based" system is constructed which may or may not reflect what is found throughout the breeding colonies. This approach, when employed exclusively, creates a bedrock of shifting sands, and with this come inherent implications of uncertainty.

Gay et al. (2005) sampled three *"thayeri"* specimens (which came from the wintering grounds and were therefore of unknown origin) that suggested, genetically, Thayer's may be closely related to Glaucous-winged Gull (*L. glaucescens*). The results of this study were difficult to interpret, as the relationships of these "arctic species" remain unresolved, showing a high incidence of shared lineages with low divergence. Of interest is that the results of Gay et al.'s study seem to invalidate genetic data reported by Crochet et al. (2002) on a winter *"thayeri"* specimen from Louisiana. The latter results may be attributed to horizontal gene transfer or misidentification of the specimen, which evidently showed a close relation to Glaucous Gull (*L. hyperboreus*). At any rate, sequencing longer segments of the genome of known-origin individuals remains fundamental.

RANGE
Precise mapping of the breeding (and nonbreeding) range of this taxon will prove difficult without comprehensive field studies across the Canadian Arctic and Greenland.
Breeding: Breeds in colonies on coastal cliffs in NE Northwest Territories and Nunavut from Banks and Victoria Islands east to Ellesmere, NE Keewatin, N Southampton, and central Baffin Island. Status throughout the N Queen Elizabeth Islands is not entirely known (Richards & Gaston 2018). Very scarcely found in nominate Iceland Gull colonies in NW Greenland. For example, in 2013, two adults were reportedly found in a colony of nominate Iceland Gulls, along with three Kumlien's Gulls, at Nordenskiold Glacier, around N 76° (Boertmann & Huffeldt 2013). Critical surveys in this region are few, however, obscuring this taxon's standing here.

Nonbreeding: Fall movement commences in central Canada during August, and the first wave of migrants arrives in SE Alaska already by the end of August or early September. In a first of its kind, a study from the breeding grounds tracked four adults with satellite transmitters from St. Helena Island (NW Devon Island; Gutowsky et al. 2020). These birds all departed in early September, moving southwest over the Northwest Territories, migrating over multiple mountain ranges, and eventually reaching the Pacific coastline. From this small sample, it does not appear that Thayer's (regularly?) migrates around the North Slope of Alaska and down the Pacific Coast, but more data is needed, especially from colonies farther north. Of note is one individual named THGU4, tracked for two consecutive years, which took divergent pathways on its southbound journeys. Typical arrival is mid-October (uncommon to fairly common by end of month). The first birds are in California by late October, but not in numbers until well into November. Locally fairly common in winter from SE Alaska south to N California; rare to locally uncommon in S California and extreme NW Mexico. Adults are abundant on Vancouver Island and become less and less common moving south along the Pacific Coast in winter.

Throughout the winter, Thayer's frequents the same coastal habitats as other large *Larus* gulls, such as nearshore ocean waters, beaches, mudflats, harbors, estuaries, lakes, parks, wet farm fields, sewage-treatment ponds, and landfills—including interior valleys west of the Cascades and Sierra Nevada. Also found well offshore out to 60 miles, occasionally 100 miles or more. In the interior, most likely to be found at large lakes and reservoirs, marinas, river mouths, dams, and landfills.

In late fall and winter, somewhat rare but regular in much of interior North America, from S California to Nevada, Colorado, Kentucky, and Great Lakes (most numerous on W Great Lakes, but regular in small numbers to E Great Lakes and the Niagara River). Rare to very rare south to S New Mexico, Texas and Gulf Coast, and central Florida. Casual to NE Mexico, and very rare to casual along East Coast north to Newfoundland, although separation from dark Kumlien's often unresolved. One to three consistently found annually in SW Nova Scotia, where there is high observer awareness, and this raises the question of whether some Thayer's in the Maritime Provinces are overlooked in American Herring flocks. Once casual along the coasts of the Seward Peninsula, the Bering Sea islands, and the North Slope, but reports increasing of late, mostly in summer and fall, perhaps due to more coverage. Rare in East Asia, but found in small numbers annually, primarily in Japan and Korea, although some are likely overlooked as Vega. Multiple records from Ireland, but casual to Britain and other parts of N Europe, where identification is uncertain, due to overlap with dark-end Kumlien's. Exceptional was a returning winter individual to Spain, starting in 2008 as a 1st cycle and continuing until at least March 2020.

Northbound movements begin in February but occur particularly during March and early April. A month or two before staging in SE Alaska on their northbound journey, large concentrations of Thayer's are found fueling up at herring runs on Vancouver Island, where they peak in March. Lingering nonbreeders in late spring and early summer are rare to mostly casual south of the Arctic, but young stragglers can be found summering throughout the continent annually.

IDENTIFICATION

The "winter handbook" on identifying Iceland Gulls is revised every decade or so. Earlier literature that sought to describe this taxon to birders presented some field marks that were either overemphasized or misconstrued. Some of the obvious misconceptions can be rectified as follows: First, eye color in adults is variable. Thayer's often shows dark eyes (especially from a distance), but many are honey colored with dark speckling or completely pale (although uncommonly as pale as Herring Gull's). Also, Thayer's does not always show a round and gentle dove-like head. A good percentage have sloping foreheads, but with heads that are relatively small. The bill may be substantially long and the body equal in stature to that of other large gulls (i.e., male types). In addition, Thayer's does not always show bright pink legs. Although some do, this is only a supporting field mark. As for 1st cycles, they often retain an all-black bill on the wintering grounds, but it is not unusual for a few to start showing a paling bill base in the fall, and certainly by the new year. By late winter, a paler bill base is not terribly rare, especially when observed in good lighting at a reasonable distance. A sharply bicolored bill is not expected in 1st cycles. Finally, it has long been thought that 1st-cycle Thayer's Gulls do not replace juvenile scapulars on the wintering grounds; some do, however, although this is typically limited and not as extensive as in southern-breeding taxa. Also, the newer scapulars don't often show high-contrast patterns with the remaining juvenile plumage, making them more difficult to detect (see this account's "Molt" section).

Adult: Sometimes described as a medium-sized gull, but, in fact, Thayer's is a large four-cycle gull. It is often smaller and more compact than adjacent Herring Gulls but larger than Ring-billed Gull. Structure-wise, Thayer's is well proportioned and similar to California Gull. Pale to medium gray upperparts, pink legs, and a yellow bill with a relatively small orange-red gonys spot. The legs can be shorter than those of surrounding large white-headed gulls, giving the bird a more squat appearance. Eye color is variable, with many showing some pigmentation. In high breeding condition, the orbital ring is raspberry pink, the bill becomes bright yellow, and the legs more vividly pink. In low breeding condition, the orbital ring is a dull

---

1 Juvenile. Medium-brown aspect with neat pale edging throughout, including the outer primaries. Scapulars often have small centers with prominent pale fringes, which differs from the long, notched, holly-leaf pattern found in many American Herring Gulls (see 20.2). PHIL PICKERING. OREGON. OCT.
2 1st cycle. Often with compact head and relatively large eye on smaller face. Commonly shows gradual, three-toned brown appearance, from primaries being darkest, slightly paler tertials, and even paler wing coverts and scapulars (several uppermost scapulars already replaced here). PHIL PICKERING. OREGON. NOV.
3 Juvenile. Compared to American Herring, Thayer's averages thinner bill, frostier pattern to greater coverts, and paler tertials, often with more marginal patterning. Note consistent pale edging on brownish primaries. AMAR AYYASH. ILLINOIS. DEC.
4 1st cycle with similar-aged American Herring (left). Thayer's overall neater with less contrast, often retaining juvenile scapulars and largely black bill. Compare also pale edging on primaries and frostier tertials on Thayer's. AMAR AYYASH. MICHIGAN. DEC.
5 1st cycle. Can recall Slaty-backed, but Thayer's averages more compact appearance with smoother head, neck, and underparts, more patterned tertial tips, and busier pattern on scapulars (some postjuvenile here with thin dark barring). AMAR AYYASH. ILLINOIS. JAN.
6 1st cycle. A small-billed individual with noticeably pale and uniform aspect recalling Kumlien's Gull, which averages paler secondaries and primaries. Some 1st-cycle Thayer's may show a paling bill base on the wintering grounds, although not sharply demarcated. AMAR AYYASH. CALIFORNIA. JAN.
7 1st cycle. Noticeably more faded than 32a.6, with stronger bill. Very little patterning and pigment across upperparts on what may have been an already pale and plain juvenile. Prominent pale edging on otherwise dark primaries. ALEX A. ABELA. CALIFORNIA. JAN.
8 1st cycle. The primaries and tertials can become entirely bleached by late winter into early spring. Most scapulars (and flanks) are 1st alternate. Note vivid leg color, compact head with relatively large eye well-centered on small face. PHIL PICKERING. OREGON. APRIL.

9   1st cycle (same as 32a.3). Frosty grayish-brown aspect. A pale window is expected. The outer primaries show paler inner webs. Dark tail can show variable patterning along base. AMAR AYYASH. ILLINOIS. DEC.
10 1st cycle. An individual with darker tail and secondaries, recalling American Herring, but well within range for Thayer's. Consider the paler outer primaries with two-toned pattern, compact head, relatively large eye, and thin bill. AMAR AYYASH. WISCONSIN. NOV.
11 1st cycle. Dilute, milky appearance, recalling Kumlien's Gull. Note contrasting dark tail, largely dark secondaries, and darker outer primaries. AMAR AYYASH. WISCONSIN. MARCH.
12 1st cycle. Importantly, note pale underside to primaries and paler underside to tail on this individual. On the far wing, noticeable venetian-blind pattern on outer primaries (dark outer webs with paler inner webs). Disproportionally large eye on small face. AMAR AYYASH. OREGON. JAN.
13 1st cycle. Paler aspect with noticeable silvery appearance to underside of flight feathers. Commonly shows contrasting dark tips to outer primaries. Busy pattern on undertail coverts, typically well-marked throughout vent region. AMAR AYYASH. ILLINOIS. FEB.
14 1st cycle and adult. Pale grayish-brown aspect with frosty upperwing and uppertail coverts. Smooth, even-brown underparts and underwing linings. AMAR AYYASH. BRITISH COLUMBIA. MARCH.
15 2nd cycle. Fairly delayed appearance to scapulars. Distinguished from 1st cycle by plain tertials and upperwing coverts, particularly the outer greater coverts. Also note more rounded outer primaries (compare to 32a.7). Eyes are typically dark at this age, with variable bill pattern. AMAR AYYASH. CALIFORNIA. JAN.
16 2nd cycle. Large dark eye, small head, and thin bicolored bill. Fairly slim body with attenuated look to rear. Aged by noticeable gray on some scapulars, extensive stippling on greater coverts and tertials, rounded primary tips, and ghost mirror on underside of p10 (far wing). AMAR AYYASH. ILLINOIS. NOV.
17 2nd cycle. An icy individual with noticeably paler primaries than 32a.16 and plainer wing coverts. Structural features include compact body, straight bill, short legs, and relatively long wing projection. LIAM SINGH. BRITISH COLUMBIA. NOV.
18 2nd cycle. A stockier bird that may recall Cook Inlet, in part due to posture and packed crop (distended esophagus). Consider the relatively large eye to shallow bill-depth ratio. Glaucous-winged types often show smaller eye with deeper bill base. AMAR AYYASH. BRITISH COLUMBIA. MARCH.
19 2nd cycle. A darker individual with heavily marked underparts. Appears intermediate between American Herring and Kumlien's. Consider the short legs, small head, and petite bill (female?). The tertials are mid-brown, and the darker primaries regularly show pale fringes. AMAR AYYASH. BRITISH COLUMBIA. MARCH.
20 2nd cycle. The mid-brown tertials with broad pale tips and the bleached wing panel may recall Vega, which is overall a bulkier bird (see 22.18). Vega has blacker outer primaries and typically lacks such pale fringes. On the open wing, Thayer's often maintains a venetian-blind pattern, with paler underside to the wingtip. WOODY GOSS. WISCONSIN. MAY.

21 2nd cycle. Fairly typical mid-brown tones to upperwing, with venetian-blind pattern on outer primaries. Inner greater coverts often have noticeable stippling. All-dark tail and dark secondaries are expected. AMAR AYYASH. INDIANA. FEB.
22 2nd cycle. A swarthy individual with noticeable marbling on tips of inner primaries and secondaries. A p10 mirror is not rare at this age, although averages smaller and less common than in Kumlien's. AMAR AYYASH. OREGON. JAN.
23 2nd cycle. A paler-winged bird, but showing expected all-dark tail, darker secondaries, and entirely dark outer webs on outer primaries (typically from p5/6 to p10). AMAR AYYASH. BRITISH COLUMBIA. MARCH.
24 2nd cycle. Striking venetian-blind pattern on outer primaries. Note paler underside to outer primaries (far wing) with dark tips. Dark secondaries, muted pattern on greater coverts and commonly found paler median coverts. LIAM SINGH. BRITISH COLUMBIA. MARCH.
25 3rd cycle with adult. Similar to American Herring, but primaries appear washed out and typically not true black. Helpful here is the white on the inner webs of p7–p8 hooking around the adjacent primary. Vivid pink legs variable. Eye color and bill pattern also variable at this age. AMAR AYYASH. BRITISH COLUMBIA. MARCH.
26 3rd-cycle types. Grayish-black primaries with fairly small apicals. A genuine size difference, not unexpected, but partly due to posture. LIAM SINGH. BRITISH COLUMBIA. NOV.
27 Subadult type with adult American Herring (right). Thayer's averages slimmer and more compact. Note extensive white on underside of far wing and noticeable white on inner webs of p6–p10. Aged by bill pattern and diffuse dark pigment on primary coverts. AMAR AYYASH. ILLINOIS. JAN.
28 3rd cycle. Delayed individual with outer primaries and tail pattern virtually identical to 2nd basic. But compare adult-like inner primaries, innermost secondaries, and bill pattern to 32a.23. AMAR AYYASH. BRITISH COLUMBIA. MARCH.
29 3rd cycle. Outer primaries are blackish (rarely true black), with dark outer webs and paler inner webs. Pale underside to far wing differs from American Herring. Sometimes with noticeably broad white tips to the inner primaries. AMAR AYYASH. WISCONSIN. MARCH.
30 3rd cycle. Wingtip more adult-like and blacker than 32a.29. Small- to medium-sized mirror on p9, and complete subterminal band on p5 fairly common at this age. Blemished wing linings, age-related. Pale, silvery underside to outer primaries. AMAR AYYASH. WISCONSIN. MARCH.
31 Adult. Standard individual. Compact head with smudged pattern, straight bill, and honey-colored iris. Considerable gray on base of primaries, grayish band on p5, and much white on underside of far wing. AMAR AYYASH. ILLINOIS. DEC.
32 Adult. Greenish cast to bill and vivid leg color found regularly. Thayer's can appear fairly long-winged at times. Wingtip pattern is rather variable, here showing reduced black, with a small mark on the inner web of p6 and rift in p7 band. AMAR AYYASH. BRITISH COLUMBIA. MARCH.

pink to pinkish gray, and the bill often has a dull greenish cast to it, frequently with a contrasting yellow "saddle spot" near the culmen. Head streaking in basic plumage is a brownish wash and can be very dense, particularly on the lower neck and back of the head. Short, diffuse streaks are common on the crown and face but can appear blotted. Structurally, the bill is straight and relatively thin for a four-year gull. The eye is relatively large and well centered on the face. Dark-eyed individuals appear big eyed and portray an innocent expression. Some possess a frown to the gape-line that may rival that of California Gull. Male types can show a bulging gonys angle (although rarely disproportionate to the bill length and size). The pigment on the wingtip is variable, ranging from jet black to grayish black and to dark gray in the palest birds (more study needed from known-origin birds). P10 has a large white tip or a large mirror that commonly extends from edge to edge. P9 often has the signature *thayeri* pattern but may have a small to medium-sized mirror enclosed in black (seen only when primaries fully spread). On rare occasions, adults are without a mirror on p9. P5 pattern ranges from complete subterminal band to partially marked and to completely white. Not terribly rare for perfectly typical looking adults to have small marks on p4, usually limited to the outer web. A small percentage shows a slightly broken p6 band or a paler p6 band with a faded look. Some extreme individuals with otherwise typical structural features and gray upperparts may show unmarked or nearly unmarked p6 (Ayyash & Singh 2023). White tongue tips on p5–p8 variable in width, but often bold and eye-catching. Some individuals show noticeably broad white tips to the inner primaries eating into the feather shaft, as found in many Slaty-backed Gulls, and in these birds, the white tongue tips to p5–p8 are usually much more prominent. Underside of the wingtip varies from almost entirely white to blackish subterminal bands in the darkest birds. *Similar Species:* In the West, the primary confusion taxa are American Herring Gull and Glaucous-winged × Herring hybrids (see those accounts). Wingtip pattern, body size, and overall structure are used to separate most birds. However, male-type Thayer's can be bigger than female-type Herring Gulls on some occasions. Female-type Thayer's can approach the size of Ring-billed Gull. The rarest lookalike is Vega Gull, which has darker upperparts but is at times remarkably similar in shade (based on museum specimen comparisons). Compared to Vega, Thayer's shows, on average, a smaller size with relatively thinner appearance, a more petite bill, shorter legs, and a paler underside to the wingtip. Orbital color is crimson red in Vega, purple pink in Thayer's.

In the East, compare to American Herring Gulls showing reduced black on the wingtip (see that species account). Compared to Kumlien's, average differences are apparent, with Thayer's being longer billed with slightly darker upperparts (although the two commonly appear to have matching KGS values on the Great Lakes; Ayyash, pers. obs.). The wingtip in Thayer's has more extensive pigment and may show a complete black subterminal band on p10 and p5 (features not thought to be combined on any Kumlien's). However, some Thayer's, as noted above, also have entirely white-tipped p10 and unmarked p5. P9 on Thayer's averages more black on base of outer web, relatively broad subterminal band often joined with continuous black along outer edge, and black on medial band crossing over to the inner web. Both webs of p9 mirror often white from edge to edge on many Kumlien's, although variable, and sometimes found in Thayer's. Some adult Thayer's have similar white from web to web, and these types often show reduced medial band where *thayeri* pattern is subsumed with pigment only found on outer web. Kumlien's types can also show white infringing on the outer web of p8 to the extent that only a thin sliver of pigment is restricted to the outer edge of the primary with white across the shaft. Such birds typically have much-reduced pigment on the wingtip overall and should not present confusion with

---

33 Adult. At least 17th cycle here, Yellow E3 was banded as an adult on St. Helena Island, Nunavut (June 2007). Thayer's from this region migrate overland through Canada's northern interior, and winter primarily along the Pacific Coast. LIAM SINGH. BRITISH COLUMBIA. FEB.
34 Adults. Eye color in Thayer's Gull is quite variable, from completely dark to pale. A bright, all-yellow iris, as seen on the right, is not unexpected, although uncommon. LIAM SINGH. BRITISH COLUMBIA. DEC.
35 Adults. A small percentage (around 15%) have wingtips that are noticeably slate black, as seen here (left). Note also the broken p6 band and incomplete p9 band on that individual. AMAR AYYASH. BRITISH COLUMBIA. MARCH.
36 Adult. Alternate plumage. Rich-purplish orbital ring noticeable upon close inspection. Gonys spot is orange-red. OWEN STRICKLAND. ONTARIO. MAY.
37 Adult with American Herring (left). Typically becomes white-headed by April. Paler iris appears dark from a distance. Importantly, white on inner webs of outer primaries often wraps around preceding primary. AMAR AYYASH. ILLINOIS. MARCH.

typical Thayer's. Separation from darkest Kumlien's Gulls can be guesswork at times, and it is not known how pale the wingtip can be on Thayer's (Ayyash & Singh 2023).

1st Cycle: Juveniles have variable plumage aspect that ranges from light coffee brown to dark brown (as in many Herring Gulls). Commonly medium brown. Overall, the plumage shows a gradual and low-contrast appearance on the perched bird, with primaries being the darkest feather group, followed by slightly paler but dark-centered tertials and, often, paler upperparts. Bill relatively straight and lacks noticeable bulge at gonys. Jet-black bill and dingy pink legs on the earliest arrivals. Scapular pattern variable from short, intricate barring to tricolored pattern of paler bases showing flared centers that become progressively dark tipped. Scapulars regularly show dark diamond- or spade-shaped centers. Many with an upside-down three-point crown pattern on scapular centers. Regularly shows neat pale fringes to the upperparts early in the season, but these are reduced by wear throughout the winter season. At rest, the primary edges generally show noticeable pale crescents, but in a minority of birds, this can be limited to an insignificant pale tip or in a very small percentage all dark. Greater coverts largely dark but with fine internal markings and at times show bold checkering with extensive pale regions. The underparts are a uniformly textured soft brown. On the open wing, the outer primaries have a gradual venetian-blind effect, with paler inner webs and darker outer webs exhibiting a clear two-toned pattern. A pale window is found on the inner primaries, which may show some concentrated splotches or marbling, mostly near the feather tips. The

38 Subadult type undergoing prebasic molt. Aged by residual dark marks on tail. Old faded-brown primary is p9, showing distinct *thayeri* pattern. JOHN CHARDINE. NORTH ARM FJORD, NORTHEASTERN BAFFIN ISLAND. AUG.
39 Adult. Many are still completing the prebasic molt on arrival at the winter grounds. Here at the end of Oct., p10 is still old (frayed white tip,) with a large molt gap in the secondaries. LIAM SINGH. BRITISH COLUMBIA. OCT.
40 Adult. A small percentage of Thayer's show slate-gray wingtips. This wingtip has fairly restricted pigment, with a large white tip to p10, mirror on p9 covering both webs, no pigment on the inner web of p9, broken p6 band, and bold white tongue tips to p6–p8. AMAR AYYASH. BRITISH COLUMBIA. MARCH.
41 Adult. More-typical blackish wingtip coloration. Bold white tongue tips similar to 32a.40, as well as no pigment on inner web of p9. Symmetric band on p6 and weak grayish band on p5. AMAR AYYASH. OREGON. JAN.

secondaries commonly have dark centers similar in shade to the outer primaries. The tail can range from mostly dark with slight barring near the outer tail feather bases to a relatively defined tailband with heavily vermiculated webs on the outer tail feathers. Some have a distinctive paler pattern across the tips of the tail feathers, resulting in a thin contrasting border. The uppertail and undertail coverts are generally densely barred, but the undertail coverts may have dark bars that are widely spaced. *Similar Species:* Compare to American Herring Gull and Glaucous-winged × American Herring hybrids (see those accounts). Also compare to Glaucous-winged Gull, which often occurs where Thayer's is found on the Pacific Coast.

42 Adult. Slate-colored wingtip showing classic *thayeri* pattern on p9. A sharp "spiked" pattern is sometimes found on p6–p8, as seen in some American Herrings. AMAR AYYASH. BRITISH COLUMBIA. MARCH.
43 Adult. Well-marked wingtip with no p9 mirror (uncommon). The dark eye, small, orange-red gonys spot, and head markings recall Vega Gull. That taxon has a reddish orbital and darker gray upperparts, and averages more black on the inner web of p9 and p10 (mostly gray on this individual as seen on underside of far wing). AMAR AYYASH. BRITISH COLUMBIA. MARCH.
44 Adult. Smudged head, dark iris, and subtle greenish cast to bill with orange gonys spot. Relatively extensive black on underside of p10 (far wing), but note gray inner web to p9 (near wing) with *thayeri* pattern. AMAR AYYASH. OHIO. NOV.
45 Adult with presumed Great Lakes American Herring type (front). Pigment on underside of p10 is mostly restricted to outer web (compare to 32a.44). Prominent string-of-pearls on p6–p8 on this individual, with typical black restricted to tips. AMAR AYYASH. ILLINOIS. FEB.
46 Adult with Glaucous Gull (left). Can often be found associating and nesting in similar habitats in the Canadian Arctic, although Thayer's averages a longer migration, favoring warmer and shallower waters farther south along the Pacific Coast. THIERRY GRANDMONT. POND INLET. BAFFIN ISLAND. JUNE.

Plumage patterns on the upperparts are sufficiently different; also compare bills, eye placement, and wing projection. Glaucous-winged is larger in most cases, with bulkier proportions, a shorter wing projection, and a deep, bulbous-tipped bill. Kumlien's Gull is generally paler throughout, with lighter outer primaries, secondaries, and tertials, which show less contrast with the upperparts than in Thayer's. In Kumlien's, the tertials are more patterned with less solid centers, and the uppertail is often intricately patterned with a weakly defined tailband, showing little contrast with the pattern to the upperparts in the lightest birds. Considerable overlap between pale Thayer's and dark Kumlien's types, which can't be easily labeled. Recent field observations relating to p6 show some promise (Noah Arthur, pers. comm.). On Thayer's, the darker pigment on the tip of this feather consistently covers the outer web and wraps around to the inner web. Conversely, in Kumlien's, the shaft is dark, with pigment mostly restricted to the subterminal region along the shaft, followed by a pale area proximally, and then a graduated dark spot or diamond near the very tip. Also, the outer edges to the outer primaries have distinct pale borders on Kumlien's, whereas on average Thayer's, the outer edges to the outer primaries are mostly dark with less paling. These features are only assessable on fresh primaries that have not suffered wear and/or fading.

2nd Cycle: Intermediate in appearance between a neat Herring and a dark Kumlien's. Most are dark eyed, but some show a paling, honey-colored iris. A short postocular line is detected at times, with a tender expression against a high forehead and pinched-in appearance to the bill base. The bill ranges from all black to an all-pale base with black tip, but the majority display an admixture of a paling bill base with dark smudges. The bill base ranges from grayish pink to grayish yellow. Pink legs. The upperwing coverts are largely brown, with a frosty-gray quality to them, often showing some faint marbling or light barring. The scapulars range from advanced adult-like gray to patchy gray and brown to nearly all brown-and-tan barring in the least advanced birds. The tertials are dark centered, typically darker than the wing coverts, with some grayish-brown stippling sometimes found near the feather tips and edges. Less advanced individuals may show light barring against plain tertials. The primaries are decidedly brown, with many individuals showing pale edging, boldest around the feather tips. On the open wing, the primaries maintain the two-toned venetian-blind pattern seen in 1st cycles, mostly on p6–p10, with some dark crossing over the feather shaft and onto the inner webs. The inner primaries maintain a pale window. P10 can show a diffuse to bold mirror, typically small, and generally limited to the inner web. In flight, the secondaries range from paler centers with dark peppering to dark brownish-gray centers in more pigmented birds. The greater coverts are typically uniformly patterned. The tail is largely dark, matching the darkest primaries and secondaries, with many showing slight marbling near the feather bases. The uppertail coverts can be light brown with faint barring. In more advanced birds, the uppertail coverts may be mostly white. The undertail coverts are dark with some barring, but variable. On the underside of the primaries, a silvery sheen is visible, with a contrasting dark band across the tips of the mid- to outer primaries. The wing linings are predominantly light brown and unpatterned. *Similar Species:* Compare to Herring Gull and Glaucous-winged × Herring hybrids (see those accounts). Glaucous-winged is larger and bulgier, averaging a more uniform look between the upperparts and flight feathers. Compared to Kumlien's, Thayer's has more solidly pigmented wing coverts, tertials, and uppertail and has an overall darker plumage aspect. Kumlien's averages less extensive pigment on the outer primaries, with largely pale secondaries. On Kumlien's, p10 is more likely to show a ghost mirror extending across both webs which averages more extensive than that seen in Thayer's (Gibbins & Garner 2013). It should be stressed that the presence of a p10 mirror does not make an individual any more or less likely to be a Thayer's or a Kumlien's. Kumlien's also more likely to show diffuse would-be tongue tips on the inner webs of p6/p7–p9, commonly so on p9, mimicking a small mirror. These differences are tentative and require more study from known-origin birds beyond the wintering grounds.

3rd Cycle: Similar to adult plumage aspect in most cases, but bill pattern often with more extensive black, from light smudging around the gonys to a complete black tip. Yellow to grayish-yellow bill base, and on rare occasions still pinkish, as in 2nd cycle. A sickly orange-yellow coloration develops around the tip in some individuals, at times more boldly red and similar to the adult's. Pink legs. Eye color variable, but generally pigmented. Head markings sometimes extend well below the neck and onto the lower breast, upper belly, and flanks. The wing coverts and tertials can be heavily marked with brown and gray marbling or almost entirely adult-like gray in advanced birds. At rest, the 3rd-generation primaries generally have smaller white apicals than the adult's and on rare occasions are completely absent. Those not showing white apicals at rest often have primaries that suggest 2nd cycle, but when seen, the inner primaries and secondaries should be adult-like (and this is generally true with delayed, 3rd-generation outer primaries; e.g., see plate 30.25). Also at rest, the uppertail may reveal some markings. In flight, more blackish brown on the central tail feathers is often revealed, some black on the primary coverts, and more pigment on the outer primaries than is found in definitive adults. Advanced birds (4th-cycle types?) can have all-white tail and entirely adult-like gray upperparts but with a few brownish-black streaks on the primary coverts or less developed primary pattern. P7–p10 average more pigment on the inner webs and more pigment extending to the primary coverts than the adult's, commonly with a noticeable brown tone. P9 can show the signature *thayeri* pattern or at times a very small mirror or entirely black. P10 generally with smaller mirror than adult's, but variable. The underwing coverts are tainted with variable light brownish-gray wash in many birds. Finally, the underside of the wingtip averages more pigment than the adult's, with more extensive black along the tips of p5–p10. *Similar Species:* Compare to American Herring Gull and Glaucous-winged × Herring hybrids (see those accounts). Most "typical" Kumlien's can be separated from Thayer's at this age by the former's noticeably paler outer primaries. The outer primaries average less pigment on the inner webs, and p9–p10 average larger mirrors on Kumlien's (more data needed from known-origin birds). Some Kumlien's will show marginally paler gray upperparts, but difficult to assess at times, due to brown wash found in 3rd cycles and to individual variation. Side-by-side comparisons needed to appreciate this field mark.

MOLT Majority of adult early arrivals in late September through October are still growing outer primaries (p8 dropped, p9–p10 old) and mid- to inner secondaries. Some still growing p9–p10 in early to mid-November (e.g., approximately 42% of presumed *thayeri* adults in S Lake Michigan region in a sample of 112; Ayyash, pers. obs.). P10 still growing in small percentage at the end of November. The partial inserted molt in 1st cycle is attributed to prealternate. This molt is limited on the wintering grounds, with many retaining juvenile scapulars into late winter. Some 1st-cycle individuals, especially those wintering farther south and associated with coastal areas, appear to replace more scapulars than populations at the northern edge of the wintering range. Personal observations by Ayyash and Singh in 2022 suggest some Thayer's Gulls may molt a variable number of tertials, secondaries, and, in extreme individuals, inner primaries, in what is presumed to be a prealternate molt in late winter on the nonbreeding grounds or possibly a protracted prebasic molt (more study needed).

HYBRIDS None confirmed. Neal Smith's 1966 study on isolating mechanisms featured 55 hybrid Glaucous × Thayer's pairings. Smith reported that he was able to induce hybridization by drugging adult birds and painting their orbital rings, which convinced the opposite species they were in the company of conspecifics. This study is now widely rejected, and the conclusions are believed to be erroneous. No authority has found evidence of these two hybridizing. In fact, no authority has found evidence of any Iceland Gull successfully hybridizing in the wild. Smith apparently found a mixed Thayer's / Herring pair on White Island, which he collected (both specimens now housed at American Museum of Natural History (787401, 787402). It is completely justified to view this account with skepticism.

Forbes et al. (1992) reported an unsuccessful Thayer's / Herring Gull "nest" on Igloolik Island in June 1985. After speaking with the lead author, who was present at this site, I found the report somewhat unclear (G.K. Forbes, pers. comm.). No adults were ever observed together, on or off a nest. The evidence used for hybridization was an adult Thayer's and a Herring Gull being present in the same general area. In any case, a single egg was found dislodged from the nest in nearby water, with no young hatched.

Thayer's and Kumlien's presumably share (or shared) a narrow contact zone on E Southampton Island (Gaston & Decker 1985) and on E Baffin Island, at Home Bay (Snell 1989). Definitions used to describe these taxa are discussed vaguely, with few details (see the introductory text to the "Iceland Gull (*Larus glaucoides*) Complex" section). Those who have visited these sites reported nonassortative breeding between dark-winged and pale-winged birds. This data is now several decades old, which opens many new questions, especially given the apparent increase in Kumlien's Gull and / or intermediate phenotypes across the wintering grounds.

32B KUMLIEN'S GULL *Larus glaucoides kumlieni*

L: 18"–23" (45.5–58.0 cm) | W: 48"–55" (122–140 cm) | Four-cycle | SAS | KGS: 4–5 (tentative), (3–6?)

OVERVIEW A highly variable four-cycle gull which breeds in the lower SE Canadian Arctic. Adult wingtip pattern unmatched in variability. First described as a species by Brewster (1883), who named it Lesser Glaucous-winged Gull (*Larus kumlieni*), and subsequently known as Kumlien's Gull. Pronounced "Koom-leens." Like Thayer's, Kumlien's Gull is a poorly defined taxon, retaining the most ambiguous status of all North American gulls. Brewster himself did not have a complete grasp of the taxon, nor did Ludwig Kumlien, who described the young as being darker than *argentatus*. Later workers attempted to employ eye color and/or wingtip characteristics to distinguish Kumlien's from Thayer's, features which have time and again proven inconsistent and unreliable. Furthermore, using eye color to separate white-winged Kumlien's from nominate Iceland may prove to be just as undependable (more data needed from known-origin birds).

Kumlien's Gull seems to be the key taxon to untangling the Iceland Gull complex. Nonassortative breeding between *thayeri* and *kumlieni* has been reported on both Southampton and Baffin Islands, but precise methods and definitions relating to how these two were distinguished are not entirely clear (Macpherson 1961; Knudsen 1976; Gaston & Decker 1985; Snell 1989). For example, some workers adopted broad definitions based on Smith's (1966) work to assign *thayeri* and *kumlieni* on Southampton Island (Gaston & Decker 1985). Snell (1989, 14) relegated all taxa to *glaucoides* and simply referred to individuals on E Baffin Island as "melanistic or nonmelanistic *glaucoides*." He prefaces his reports with, "I do not distinguish between the subspecific variants of *L. glaucoides*; these taxa are not phenotypically discrete." In this same study, we're told, "Despite these problems, field identification at these Home Bay colonies of *glaucoides* individuals with distinctly melanistic primaries, as well as distinctly melanistic × nonmelanistic *glaucoides* pairs (i.e., *kumlieni* × *thayeri* hybridization), is straightforward." Birds with wingtip melanin of *thayeri* and others with wings "just as those which breed in Greenland" (i.e., nominate types) were reportedly found here by Snell. Although identification of these individuals may have been straightforward in the eyes of Snell, it is difficult for readers to appreciate the scope of introgression based on his writings.

Some later workers argued that there is actually little evidence for extensive interbreeding between Thayer's and Kumlien's (Adriaens 2012; Browning 2022). And although this seems to be the case when reading these works, it is statistically unlikely that the two sites where close ground studies have been conducted (i.e., Bell Peninsula and Home Bay) are the only two sites in the Arctic where *thayeri* and *kumlieni* types seem to be in contact. Therefore, one would have to give serious consideration to revising the phenotypic limits of *thayeri*; see, for instance, the light-winged adults described by Bray (1943) and by Parmelee and MacDonald (1960). Consequently, whether Kumlien's is an intergrade population, a hybrid between Thayer's and nominate, or worthy of species rank remains unknown. Others have suggested there could be two types of "Kumlien's": those that may be a product of very recent and/or ongoing hybridization between *thayeri* and nominate and those that are now a stable— albeit highly variable—population of Kumlien's Gull (Howell & Elliott 2001; Howell & Mactavish 2003; Adriaens 2012).

Virtually all that has been written about Kumlien's Gull on the wintering grounds refers to birds of unknown origin (Zimmer 1991; Weir et al. 1995; McGowan & Kitchener 2001; Howell & Mactavish 2003; Adriaens 2012). Perhaps more reassuring are studies providing details of specimens of known origins (Gosselin & David 1975; Snell & Godfrey 1991; Snell et al. 2020), but whether these are confirmed breeders is unknown. Admittedly, these descriptions lack rigorous analysis methods, reinforcing the convictions of the latter workers that *thayeri* and nominate are bridged by Kumlien's Gull.

TAXONOMY Type specimen collected by Ludwig Kumlien on a polar expedition in 1878 in Cumberland Sound. Originally, he believed it to be an eastern form of Glaucous-winged Gull. This idea was erased by Brewster (1883), who later described it more formally as a species, using the common name Lesser Glaucous-winged Gull and the scientific name *Larus kumlieni*. Dwight (1925) was perplexed by *kumlieni* and suspected it was a hybrid between *thayeri* and nominate *glaucoides*. Rand (1942) was the first to treat *kumlieni* as a subspecies of Iceland Gull, but it is clear when reading through his analysis that he was uncertain of what, if any, relation it had to *thayeri*. Banks and Browning (1999) raised the possibility of Kumlien's and Thayer's forming a species distinct from nominate, but this was a hypothetical point, rallying for the need to go back to the type specimens as a basis for definitions. Howell and Mactavish (2003) were uncertain whether Kumlien's deserved full species rank or should continue to be treated as a subspecies of Iceland. One thing that becomes apparent when reading Howell and Elliott's (2001) and Howell and Mactavish's (2003) works is that very little consideration has been given to the slew of puzzling birds spread throughout the Great Lakes region, where an estimated 25%–35% of individuals are intermediate between Pacific-like Thayer's and St. John's–like Kumlien's (P. Svingen & K. Bardon, pers. comm.; S.C.G. Haas, pers. comm.; K. Brock, pers. comm.; J. Iron & R. Pittaway, pers. comm.; J. Brumfield, pers. comm.; Ayyash, pers. obs.). For instance, in Howell and Elliot's (2001) work, photo 8, which is left unidentified, is a rather common Great Lakes Iceland Gull. Such birds have been largely neglected in the literature and would be a critical component in better understanding this complex. Although looking for overall patterns in the densest populations is vital, intermediate types situated in the center of the winter range are just as relevant to the complex.

RANGE
Precise mapping of the breeding (and nonbreeding) range of this taxon will prove difficult without comprehensive field studies across the Canadian Arctic and Greenland.
Breeding: Reportedly breeds in colonies on coastal cliffs in SE Nunavut on E and S Baffin Island, E Southampton Island, south to Digges Sound and Coats Islands (Gaston & Elliot 1990), and on the Ungava Peninsula in N Quebec, southwest to Broughton Island. Extent of breeding range requires further study, particularly on N Baffin Island, and from White Island to Lyon Inlet, and north along the east coast of the Melville Peninsula. Perhaps a rare breeder in W to SW Greenland, as far north as approximately N 76°, where three individuals were reported in a colony of nominate *glaucoides* and Glaucous Gull, as well as two adult Thayer's, in 2013 (Boertmann & Huffeldt 2013; Boertmann, pers. comm.). See also work by Weir et al. (2000) for historic accounts of suspected breeding. Of interest is an intriguing report of two Iceland Gull chicks captured in SW Greenland which were raised in captivity in Austria, only to reveal adult wingtip patterns of Kumlien's Gull and olive-brown irises (Goethe 1986). The odds of this event occurring at random must be so long that one wonders what might be discovered if a more thorough and full-scale on-the-ground survey were conducted in these remote colonies. Keep in mind that most breeding surveys conducted on the Iceland Gulls have been made from a distance, from either airplane or watercraft, making it nearly impossible to detect traces of lighter gray pigment in the outer primaries on what are already worn, faded, and molting flight feathers. Critical surveys in far NW Greenland lacking.
Nonbreeding: Frequents the same habitats as other large *Larus* gulls, both coastal and inland. In southern part of winter range, found at landfills, dams, marinas, and fish-processing plants. Fall migrants begin arriving in Atlantic Provinces, New England, and Great Lakes by late October, but not regular in many areas until November, and not fairly common until December. Largest winter numbers in Atlantic Canada, where most common in St. John's, Newfoundland, and locally common on Prince Edward Island, Nova Scotia, and throughout Gulf of St. Lawrence. Found well offshore in winter, particularly in the N Atlantic off Labrador and Newfoundland. Rare to uncommon south to mid-Atlantic (e.g., North Carolina) and west to W Great Lakes, where most consistently found at dams and, especially, landfills. Higher counts at favored sites west to E Great Lakes around Toronto and Niagara River, south to Delaware River in Pennsylvania/New Jersey and Massachusetts, particularly from Cape Cod south to Nantucket. Rare but regular in singles south to east-central Florida and east to Bermuda, where almost all associated with landfills and nearby loafing spots. Rare to very rare west across Prairie Provinces and Great Plains states to Nebraska, Colorado Front Range, south to interior Southeast and

Texas. Since 2012, more frequently recorded in late fall to early spring in California and points north to S British Columbia, where separation from pale-end Thayer's remains problematic in some cases, but pale individuals recorded regularly in small numbers. For years, Kumlien's had a low acceptance rate in California, due to taxonomic uncertainties as well as poorly understood identification criteria in the Iceland Gull complex (Hampton 2013). Therefore, its true status in the West is obscured by field-identification criteria at odds with historic range maps. A small percentage of birds wintering between SE Alaska and British Columbia have wingtip patterns that, also, are inconsistent with traditional identification criteria for *thayeri*, and they may be *kumlieni*. Further study is required, but see Ayyash and Singh's (2023) work. Also winters east to Iceland, primarily the western shores of the island, where numbers seesaw, but regular in small numbers (approximately 10%–15% of wintering birds). Similarly, to the Faroes, where it is scarcer, and British Isles, where it is rare but regular. Apparently unrecorded in British Isles until around 1900. Rare until 1915, but becoming quite regular here after 1950. Select samples of Iceland Gulls here ranged from 33% to 80% *kumlieni* (Weir et al. 1995). Its separation from nominate *glaucoides* in these regions is problematic at times (particularly in younger individuals) and requires further study. Very rare in other parts of Europe. Spring movement commences in February and peaks in March and early April. Nonbreeders are regular in late spring and summer south to S Hudson Bay but are very rare south of here, with occasional young stragglers overwintering.

Also associated with sea ice in migration and likely offshore S to SW Greenland. Found regularly in August–May in SW Greenland in small numbers and appears to be increasing (Boertmann 2001; David Boertmann, pers. comm.). It is very likely that the number of *kumlieni* wintering with the masses of nominates in S Greenland is greater than historically believed. To support this, we can use Iceland (the country) for analysis. Reports from Iceland—a minor extension of nominate's winter range—suggest an increasing presence of *kumlieni*, which make up approximately 10% of Iceland Gull flocks, and up to 15% in some years (Adriaens 2012; Yann Kolbeinsson, pers. comm.). Incidentally, Iceland as a whole receives much more coverage, with fairly regular winter gull reports from all around the island. As we move west along the N Atlantic, the likelihood of encountering more *kumlieni* should increase, not decrease, especially if Iceland's *kumlieni* supposedly originate strictly from Canada. Note that no Kumlien's have been recorded breeding or wintering in E Greenland (David Boertmann, pers. comm.). This disjunct winter range of *kumlieni* is curious, to say the least. How can the small flocks of nominate *glaucoides* wintering in Iceland and the Faroe Islands combine approximately 95% of N Atlantic and European *kumlieni* reports? This, in my opinion, can only be explained by an observer bias throughout the Reykjanes Basin and west to S Greenland. Although this is highly presumptuous, the increasing numbers of *kumlieni* in Iceland is verifiable, although the numbers fluctuate from year to year (Yann Kobielson, pers. comm.; Ayyash, pers. obs.). Whether there's an even higher percentage of *kumlieni* veiled as nominates wintering here is not known, due to the lack of clarity on phenotypic limits. A similar argument can be made with respect to pale-winged "Thayer's Gulls" in the Pacific Northwest: Might some of these be *kumlieni* veiled as *thayeri*, due to the lack of clarity on phenotypic limits? See the "Commentary" at the end of the Iceland Gull complex section.

IDENTIFICATION
The identification data outlined here is largely accepted from patterns observed on wintering populations (of birds with unknown breeding origins).
Adult: Pale gray upperparts, pink legs, and yellow bill with a small red gonys spot. Bill is relatively short and straight, often appearing blunt tipped, with little expansion to the gonys. Bill can show a greenish tone in nonbreeding condition. Eye color is variable, from all pale to all dark, although paler eyes appear to be slightly more common than all-dark eyes (Ayyash, pers. obs.). Raspberry-pink orbital ring in high breeding condition. Structurally, many individuals have a steep forehead and rounded head, although male types can have flat heads and stronger bills. Wingtip coloration ranges from all white to medium gray and to charcoal black. Wingtip pattern highly variable, with some individuals identical to nominate *glaucoides* and others matching *thayeri*. Average Kumlien's, as diagnosed on the wintering grounds, have light to medium gray wingtip markings generally restricted to the outer webs of the outermost primaries. P10 commonly

1. 1st cycle. Pale upperparts often show V-shaped barring on scapulars and patterned tertials. Note contrasting grayish-brown primaries with subterminal diamonds on primary tips. AMAR AYYASH. NEWFOUNDLAND. JAN.
2. 1st cycle. More boldly patterned scapulars, resembling those of Thayer's, but with broader white regions. Darker and plainer tertial centers than 32b.1. AMAR AYYASH. NEWFOUNDLAND. JAN.
3. 1st cycle. A paler individual with whiter head and pale bill base, which is not very rare, especially later in the winter season. AMAR AYYASH. NEWFOUNDLAND. JAN.
4. 1st cycle. Exceedingly pale primaries with pale aspect. Nominate averages paler bill base and more delicate appearance, although with much overlap and not separable without open wing (see 32c.3). AMAR AYYASH. NEWFOUNDLAND. JAN.
5. 1st cycles. Extremes in plumage patterns and coloration. The individual on the left approaches a small Thayer's, but with paler primaries, outer tail feathers, and tertial centers. AMAR AYYASH. NEWFOUNDLAND. JAN.

6 1st cycle. Fairly dark primaries with broad pale edges. Tertials with noticeable patterning. Small bill already paling at base. AMAR AYYASH. NEWFOUNDLAND. JAN.

7 1st cycle. A worn and faded individual showing a number of postjuvenile upper scapulars. Note petite bill, small, rounded head with relatively large eye, and long wing projection. AMAR AYYASH. FLORIDA. JAN.

8 1st cycles. Plumage aspect is highly variable, in both color and pattern. AMAR AYYASH. NEWFOUNDLAND. JAN.

9 1st cycle with similar-aged Glaucous Gull (right). Glaucous shows a small eye on a large face. Compare overall body girth, eye-to-head ratios, bill size, and bill pattern. WAYNE FIDLER. NEW YORK. JAN.

10 1st cycle. Largely bleached with 1st alternate upper scapulars and flanks. 2nd-generation scapulars typically have low-contrast pattern with thin brown barring. Slight bulge to bill tip with thinner base. BEN LUCKING. ONTARIO. MAY.

11 1st cycle. Fairly uniform, pale upperwing, but often with darker outer webs to the outer primaries. AMAR AYYASH. NEWFOUNDLAND. JAN.

12 1st cycle. Dark wash across flight feathers, with some marbling on inner primaries that have distinct diamonds on tips. Compact head and bill. AMAR AYYASH. NEWFOUNDLAND. JAN.

13 1st cycle. A darker individual showing relatively dark secondaries. Finely patterned inner-primary tips, as in some nominates, but note darker outer webs to outer primaries. AMAR AYYASH. NEWFOUNDLAND. JAN.

14 1st cycle. Slightly darker tail feathers, and less commonly seen, contrasting white uppertail coverts. Small bill and narrow wing differ from Glaucous-winged. AMAR AYYASH. INDIANA. JAN.

15 1st cycle. Pale underside to flight feathers, with slightly darker underparts. Undertail coverts show variable barring, but they are sparsely marked on this individual. JIM TAROLLI. NEW YORK. FEB.

16 1st cycle. Noticeably pale flight feathers and plain uppertail. Recalls nominate, which is dismissed more by range than any particular feature. Fading presumed on outermost primaries, but curious. AMAR AYYASH. FLORIDA. JAN.

17 1st cycle. Recalls nominate, but such a pale appearance is quite common in presumed Kumlien's, especially at this date. Note also solidly dark pigment on tail. AMAR AYYASH. MASSACHUSETTS. APRIL.

18 2nd cycle. Retarded appearance recalls 1st cycle, but aged by rounded primary tips, as well as fine marbling on greater coverts and tertials. AMAR AYYASH. NEWFOUNDLAND. JAN.

19 2nd cycle. Plain, icy-gray scapulars with ghostly-white wing panel and primaries. Diffuse bill pattern fairly common at this age. Paler iris. AMAR AYYASH. NEWFOUNDLAND. JAN.
20 2nd cycle. Darker grayish-brown outer primaries, marbled tertials, and greater coverts, pale gray upper scapulars (2nd alternate), and thin, bicolored bill. AMAR AYYASH. NEWFOUNDLAND. JAN.
21 2nd-cycle duo with 3rd cycle (far right). The individual on the right has relatively dark primaries that show high contrast with the tertials. AMAR AYYASH. NEWFOUNDLAND. JAN.
22 2nd cycle (center), 2nd-cycle American Herring (left) and 1st-cycle Ring-billed (right). Intermediate size here suggests a female type. AMAR AYYASH. MICHIGAN. NOV.
23 2nd cycle. A remarkably dark individual with solidly filled upperparts. Note petite bill and finely marbled tertial tips. BRUCE MACTAVISH. NEWFOUNDLAND. JAN.
24 2nd cycle. Darker primaries, as found in some Thayer's, which typically shows darker tertials. Note long wing projection, thin bill, and small head. AMAR AYYASH. NEWFOUNDLAND. JAN.
25 2nd cycle. May easily be mis-aged as 1st cycle, but note finely patterned greater coverts with freckling rather than bold barring, and rounder primary tips. Darker outer primaries. AMAR AYYASH. FLORIDA. JAN.
26 2nd cycle (same as 32b.20). Fairly narrow and long wings, small head and bill, and overall uniform upperparts. AMAR AYYASH. NEWFOUNDLAND. JAN.
27 2nd cycle. A pale individual with muted pattern. Outer primaries and tail feathers show dark contrast. Pale bill base and paling iris (seen at closer distance). AMAR AYYASH. NEWFOUNDLAND. JAN.
28 2nd cycle (same as 32b.24). Fairly advanced with large ghost mirror on p10 and mirror on p9, but with typical 2nd-generation-like inner primaries. Perhaps not safely separated from pale Thayer's, which typically shows darker remiges, and often with dark on outer webs to p5–p7 reaching primary coverts. AMAR AYYASH. NEWFOUNDLAND. JAN.
29 2nd cycle. A fairly typical individual. Outer primaries often darkest from p7/p8 outward. Ghost mirror on p10. Pale underside to primaries. CHRIS VAN RIJSWIJK. NEWFOUNDLAND. JAN.
30 3rd cycle. Darker outer primaries with prominent pale edging. 3rd-generation secondaries visible here (below tertials and greater coverts). Honey-colored iris. AMAR AYYASH. NEWFOUNDLAND. JAN.

white tipped or with partial subterminal band. Complete subterminal band on p10 is rare but does occur in a small percentage of individuals (more study needed from known-origin birds). P9 often with large mirror across both webs from edge to edge, pale inner web, and limited pigment reaching feather base on inner web, but highly variable. P5 typically unmarked. Complete band on p5 very rare, and when present, appears faded and not black. More expected, albeit rare, is for p5 to show a dark marking on the outer web or a broken band (Bruce Mactavish, pers. comm.). In basic plumage, head markings appear cinnamon brown overall, with some showing dense blotting on lower neck and upper breast, and others with light diffuse streaking limited to the hindneck. *Similar Species:* Rather distinctive throughout its range, but separation from Thayer's sometimes problematic. Kumlien's averages shorter bill, more symmetrically rounded head (at least in female types), and less pigment on wingtip, although variable. Adult Kumlien's should not, in theory, combine the following features: complete black subterminal band on p10 with largely dark inner web, black to base of outer web of p9 with continuous black subterminal band, and complete black band on p5 (these features are pending further study from known-origin individuals). Eye color cannot be used to separate *kumlieni* and *thayeri*. Kumlien's averages paler gray upperparts, but this is difficult to ascertain in the field, especially on the Great Lakes, where presumed Thayer's and Kumlien's can be found side by side. An important variable at work when comparing wingtip patterns is age-related characteristics: subadult Kumlien's will show more pigment on the wingtip. It is expected that an advanced 3rd-cycle-type or delayed 4th-cycle-type Kumlien's may have a wingtip pattern quite similar to that of Thayer's but appearing more dilute or slate gray, often with brownish wash appearance. Such birds should have their bill pattern and plumage scrutinized closely for age-related clues. Many intermediate types show darker wingtips at rest, due to the pigment on their wingtip overlapping in layers (and therefore no light being reflected off each feather). Once an open wing is secured, the pigment may look noticeably paler and reduced, which complicates initial impressions of perched birds. Therefore, an open wing is critical, and even then, subspecific labels are not advisable for every individual encountered. Some *kumlieni* appear shorter winged or with blunter tipped wings in flight, although this may be sex related, a function of flight behavior, or simply individual variation (more study needed).

Compared to nominate *glaucoides*, majority of Kumlien's show some pigment on wingtip. While convention in the United States and Canada is to reject any potential nominate if even the slightest pigment is detected on the wingtip, some apparent nominates wintering in Iceland fairly regularly show a contrasting, darker gray demarcation along the outer web of p9 and p10—often not visible at rest and from various angles in flight (Peter Adriaens, pers. comm.; Ayyash, pers. obs.). Nominate averages paler gray upperparts but overlaps with some Kumlien's. It is also generally accepted that nominate only has clear, cream-yellow eyes as an adult, but upon close inspection, a fair number show obvious dark flecks on the iris (Ayyash, pers. obs.). Others show warmer, less-than-yellow eyes. An all-pale iris is not considered a reliable feature in adult types found in Greenland and must be further investigated on the breeding

---

31 3rd cycle. White-winged individual resembling nominate. Compared to 32b.19, more adult-like bill, with paler eye and more extensive gray on the upperparts. See 32b.35 for open wing. AMAR AYYASH. NEWFOUNDLAND. JAN.
32 3rd cycle. White-headed now, so presumably 3rd alternate. Adult-like bill pattern but with extensive dark ring near tip. Pale iris and apparent white tail. JEAN IRON. ONTARIO. APRIL.
33 3rd cycle (Yellow H8, banded as 1st cycle in St. John's, Newfoundland). Extensive pigment on outer primaries, with mostly unmarked tail. Olive-gray iris. LANCY CHENG. NEWFOUNDLAND. JAN.
34 3rd cycle. Retarded appearance, aged by adult-like innermost primaries and innermost secondaries. Compared to Glaucous-winged (30.27), Kumlien's averages a smaller bill and head, is paler gray, and often has fine marbling and stippling on greater coverts and tail. Ghost mirror on p10 helps exclude nominate Iceland. AMAR AYYASH. NEWFOUNDLAND. JAN.
35 3rd cycle (same as 32b.31). Adult-like inner primaries and inner secondaries used for aging. Remarkably white outer primaries; overall pale aspect makes separation from nominate difficult, if not impossible. AMAR AYYASH. NEWFOUNDLAND. JAN.
36 3rd cycle. Robust bill, heavily marked neck, and darker wingtip resemble Thayer's, which typically shows continuous black on outer web of p9 (white here), dark medial band on inner web of p9, and more black on base of outer web of p7–p8. AMAR AYYASH. NEWFOUNDLAND. JAN.
37 3rd cycle. Average appearance, with medium gray on wingtip and large ghost mirror on p9 and p10. Underside of wingtip shows string-of-pearls effect. Grayish-brown pigment on wing linings, variable. AMAR AYYASH. ONTARIO. DEC.
38 Adult. Color and pattern on outer primaries highly variable. Fragmented markings of slate-gray are seen on p6–p8 here. Petite bill, honey-colored iris, and at times bright pink legs. AMAR AYYASH. NEWFOUNDLAND. JAN.

grounds (Lars Witting, pers. comm.). Nominates may have pigmented eyes for several reasons: age-related, introgression with Kumlien's, or individual variation that has gone underappreciated.

Kumlien's is said to average slightly larger head and bill proportions, but extensive sex-to-sex comparisons needed from known-origin birds. Compare also to Glaucous and Glaucous-winged Gulls (see those species accounts). Hybrid Glaucous × Herring can have a superficial resemblance to many Kumlien's, especially given the variability of their wingtip patterns, but this hybrid is larger overall with noticeably bulkier bill and head proportions and orange-yellow orbital ring (when seen well). Nonetheless, identification is not always straightforward, especially if presented with a lone, small, female hybrid, and especially with 1st cycles.

1st Cycle: Plumage often grayish brown but highly variable, ranging from ghost white to fawn-colored tones. Finely patterned upperparts appear frosted in the freshest juveniles. Black bill and pink legs. Many start to show a lightly paling bill base by midwinter, and by mid- to late winter, the bill base may become noticeably pale, although not sharply bicolored. Others retain all-black bill until spring. Upperparts often with dark wavy barring or thin V-shaped bars. Overall, tertials lack solidly dark centers, although the lower tertials and tertial bases can have dusky plain centers, generally with darker barring, notching, and/or peppering around the edges and tips. Juvenile scapulars often persist on the wintering grounds, although not uncommon to show some 2nd-generation scapulars beginning in midwinter. At rest, the primaries are typically concolorous with the upperparts or gently darker brown, with contrasting pale chevrons on the outer edges. May show outer primaries paler than the upperparts, some due to bleaching, others naturally so early in the season before bleaching has set in. The underparts are uniformly textured. On the open wing, the outer primaries are generally darker than the inners, although variable, with some appearing plain and uniformly colored. Others have light speckling or marbling near the tips of the inner primaries, with contrasting brown diamond tips, while showing dark outer webs to the outer primaries. Secondaries variable: some uniformly pale as the coverts, others slightly darker than the wing coverts and primaries. Similarly, the tail feathers can be uniformly pale, as found across the upperwing, while others are contrastingly dark and often among the darkest feathers seen in flight. The uppertail and undertail coverts show light to moderate barring. First cycles are highly susceptible to bleaching, particularly those wintering at the southern end of the range. The palest birds and those that become heavily bleached range from a creamy beige to simply white. *Similar Species:* Compare to Glaucous and Thayer's Gulls and Glaucous × Herring hybrids (see those accounts). Compared to nominate *glaucoides*, Kumlien's averages a plainer and darker uppertail, as well as darker outer primaries. Primaries on nominate often go from dark to pale distally, with more intricately patterned primary tips, secondaries, and uppertail, but variable. Some not safely separated, particularly those showing all-plain flight feathers. Further, once bleaching and wear have set in, juvenile plumage patterns are impossible to discern, and it's safest to leave problematic birds unidentified or default to the expected taxon. Nominate averages a noticeably paler bill base earlier in the fall/winter season, at times approaching the sharply bicolored pattern of Glaucous,

---

39 Adult. Overall resembles nominate, with no pigment on outer wingtip. Identified as Kumlien's by dark eye and darker gray upperparts. Note slight greenish cast to bill, which can be found in all subspecies of Iceland Gull. DAVE BROWN. NEWFOUNDLAND. JAN.
40 Adult type. A larger individual. Diffuse grayish-brown pigment on primary coverts (exposed below secondary tips) and brownish tones on outer edge of primaries suggest subadult. CHRIS VAN RIJSWIJK. NEWFOUNDLAND. JAN.
41 Adults with 2nd cycle (front). Immense variation in extent and color of pigment on outer primaries. These two represent extremes, and assigning subspecies out of range would prove difficult. AMAR AYYASH. MASSACHUSETTS. NOV.
42 Adult with American Herring (right). A pale-eyed male type that can easily rival Herring Gull in size. Note thinner bill and slightly paler upperparts (which may overlap at times). AMAR AYYASH. NEWFOUNDLAND. JAN.
43 Adult with similar-aged Great Black-backed. Included here for scale, showing a typical size comparison. Petite bill, small, rounded head, and vivid leg color are common features. RONNIE D'ENTREMONT. NOVA SCOTIA. FEB.
44 Adult. All-white tip to p10 fairly common, as well as white from edge to edge on p9 mirror, although highly variable. Some show very broad white tips to inner primaries. AMAR AYYASH. NEWFOUNDLAND. JAN.
45 Adult. Face and bill strongly recall nominate with lemon-yellow iris and petite bill. Extensive pigment on p9–p10 and bill color may be age-related. AMAR AYYASH. NEWFOUNDLAND. JAN.
46 Adult. All-white tip to p9 and unmarked p6. Kumlien's Gulls wintering in Iceland are not rare, making up approximately 10% of population, but varies annually. HANS LARSSON. ICELAND. MARCH.

32B KUMLIEN'S GULL

47 Adult. Slate-colored wingtip matching some Thayer's (which typically has more pigment on base of outer web of p9). Very faint and incomplete markings on p5 band. AMAR AYYASH. NEWFOUNDLAND. JAN.
48 Adult. Pale extreme with lighter gray upperparts. Darker gray on outer web of p8–p9 presumably eliminates nominate (more study needed). AMAR AYYASH. NEWFOUNDLAND. JAN.
49 Adult. Dark but restricted pigment on wingtip. AMAR AYYASH. NEW YORK. DEC.
50 Adult. A strong-billed individual with sloping forehead and bull neck (recalling Glaucous × Herring hybrid). Greenish bill base and pinkish orbital seen at closer distance. AMAR AYYASH. WISCONSIN. JAN.

although variable, as some nominates remain dark billed. Structurally, many nominates appear smaller headed with a pinched-in appearance to the bill base (in part due to a more paling bill base).

2nd Cycle: Quite distinctive in its range. Often, the challenge with 2nd-cycle Kumlien's is not separating it from other species, but separating it from 1st- and 3rd-cycle Kumlien's. This age group often has darkish eyes, but a minority have eyes appearing clear and honey colored when viewed up close. Pink legs. Bill pattern ranges from sharply bicolored with pink to grayish-yellow base and to almost entirely black. Scapulars range from plain light brown with moderate barring, as in some 1st cycles, to largely plain gray, approaching some 3rd cycles. Plumage aspect is grayish brown with finely patterned wing coverts and tertials, consisting of thin barring and prominent marbling on the upperparts. Some upperwing coverts and one to two upper tertials can be adult-like gray and are presumably 2nd alternate. At rest, the primary tips commonly show pale V-shaped edges similar to those of 1st cycles, but with wider pale border and dark centers appearing dilute. Underparts are typically uniform with the head and neck, with some light speckling. Head markings appear somewhat dense at times, but fine streaking is found in some individuals, particularly on the head and face. On the open wing, a muted pattern is often found on p6/p7–p10, with dark outer webs and paler inner webs, but variable. Also, a p10 mirror is not uncommon, and p6/p7–p9 can have diffuse pale tongue tips on the inner webs, showing a pseudo-string-of-pearls pattern (a ghosted adult pattern). Secondaries commonly marbled light brown with some barring and peppering. The uppertail is often similar in tone to the outermost primaries or slightly darker with a brownish wash and patterned with fine vermiculations. Uppertail coverts range from well marked to largely white. *Similar Species:* Compare to Glaucous and Glaucous-winged Gulls and Glaucous × Herring hybrids (see those accounts). Compared to Thayer's, Kumlien's averages a shorter bill, paler wingtips, and paler tertials. On the open wing, Thayer's has more bold, dark secondary centers and uppertail, as well as darker primary coverts and, at times, darker greater coverts. Overlap exists in a small percentage of individuals believed to be intergrades (more study of known-origin birds needed). Compared to nominate *glaucoides*, Kumlien's averages dark outermost primaries and may show faint mirror on p10 with dark subterminal band. On nominate, the outer primaries are normally light, often boldly whitish, contrasting with darker and more patterned inner to mid-primaries.

3rd Cycle: Pale gray upperparts appear similar to adult's, but variable brown wash with some dark regions to wing coverts and tertials. Less advanced birds can show considerable marbling on wing coverts and tertials. Scapulars may match adult's, although often with intermittent white edging. Adult-like scapular and tertial crescent commonly seen. Bill pattern ranges from dull pink to yellow, often with dark subterminal markings or even complete black band. Eye color variable. Pink legs. Lower neck and upper breast average more markings than adult in basic plumage. At rest, brownish-gray primaries seen on the closed wing in many individuals, with reduced apicals not typically matching adult's in size. Less advanced birds have 2nd-generation-like outer primaries at rest and may be impossible to age without an open wing (note, however, adult-like scapular and tertial crescent and / or drooping white secondary tips). In flight, a brownish tone may be found throughout the upperparts, particularly the primary coverts and alula. The wingtip is variably pigmented, with p7/p8–p10 showing more pigment approaching up the base compared to adult's. Wingtip also includes broader subterminal bands, especially on p6–p7, when compared to definitive adult's. White rump with variably marked tail shows dusky gray barring and peppering. *Similar Species:* Kumlien's with medium to dark gray wingtips have similar plumage aspect to Glaucous-winged Gull and Glaucous-winged hybrids. Size and structure, especially with respect to bill, are usually sufficient in settling such questions, but side-by-side comparisons needed. Lone individuals can be surprisingly tricky to confidently nail down, and even more so out of range. Pure Glaucous-winged has darker gray upperparts, but this difference is obscured at times, due to brown wash found in the upperparts of many 3rd-cycle Kumlien's.

Compared to Thayer's, Kumlien's generally has paler and more limited pigment on the inner webs of the wingtip pattern, as well as paler, dusky uppertail markings. The wingtip and tail markings on 3rd-cycle Thayer's are brownish black and average darker, but a small percentage of individuals show some overlap. Third-cycle nominate *glaucoides* can be similar, but many show noticeably paler wingtip at this age. Kumlien's often shows large white mirror on p9/p10, which isn't expected in nominate. Also, nominate often takes on a staring pale yellow eye at this age, but not diagnostic by any means.

MOLT As in Thayer's and nominate, the partial inserted molt in 1st cycle is attributed to prealternate. This molt is limited on the wintering grounds, with many retaining juvenile scapulars into late winter. Some 1st-cycle individuals, especially those wintering farther south and associated with coastal areas, replace more scapulars than those at the northern edge of the wintering range (Ayyash, pers. obs.). Adult birds arriving in St. John's in late October through November appear to be farther along in flight feather molt than Thayer's found to the west. First arrivals typically show p9 being 90%–100% grown and p10 being 70%–95% grown. By the first week of December, virtually all adults from a sample of 230 in St. John's were with full p10 (Bruce Mactavish, pers. comm.). The earliest adult arrivals on the W Great Lakes in October have flight feather molt pattern virtually identical to Thayer's (see that account).

HYBRIDS None documented. Interbreeding with darkest subspecies (*thayeri*) reported on Southampton Island (Gaston & Decker 1985) and Baffin Island (Knudsen 1976; Snell 1989), although ambiguous definitions used to diagnose various subspecies. Isolated reports of interbreeding with nominate *glaucoides* in very small numbers in Greenland (Weir et al. 2000; Boertmann & Huffeldt 2013; Boertmann & Rosing-Asvid 2014). An adult individual from St. John's, Newfoundland, reported in the winter of 2018 had characteristics of what was suspected of being a Lesser Black-backed × Kumlien's hybrid (Alvan Buckley, pers. comm.). It is worth noting that nominate Iceland and Lesser Black-backed Gulls have hybridized in captivity, producing two clutches of three eggs in which the young were reared to adulthood (Lonnberg 1919).

32C ICELAND GULL *Larus glaucoides glaucoides*

L: 18"–23" (45.5–58.0 cm) | W: 48"–55" (122–140 cm) | Four-cycle | SAS | KGS: 3–4

OVERVIEW The nominate subspecies. Averages smallest and palest of the Iceland Gulls and is believed to breed strictly in Greenland. Evidently the most stable race, with adults having immaculate white wingtips and pale eyes. Described as a miniature Glaucous Gull in the earliest records, and its specific scientific name, *glaucoides*, is a reference to that species. This elegant gull nests on coastal cliffs throughout low-arctic Greenland and winters primarily off S Greenland. Nests singly, in small colonies, or in larger colonies with over 1,000 birds. At times, mixes in Glaucous Gull and/or Black-legged Kittiwake colonies, although in segregated tiers, with Glaucous favoring the highest cliffs, nominate Iceland the mid-elevations, and Black-legged Kittiwake the lower elevations. This taxon is primarily an intraregional migrant. When adults reach the breeding grounds as early as March, many of the fjords below are still covered in ice. Surrounding waters continue to harbor large ice floes in some regions well into the egg-laying and incubation period. Total breeding population of the nominate race is estimated at 50,000–100,000 pairs (Snell et al. 2020).

TAXONOMY Type specimen collected in Iceland, but no breeding records here. Earlier literature confused Glaucous and nominate *glaucoides* and incorrectly described the latter's breeding range east to Iceland (the country). Generally the least problematic of the Iceland Gulls in terms of identification, but phenotypic limits disputed and need further study. Convention is to reject any potential nominate adult candidate if the slightest pigment is detected on the wingtip, but Snell et al. (2020) insist some individuals can show variable gray or lightly patterned wingtips (Hørring & Salomonsen 1941; Ingolfsson 1967; Goethe 1986). Therefore, individuals showing ever so light contrast on the outer webs of p9/p10 with otherwise typical looking nominate *glaucoides* attributes (i.e., pale eye, strikingly pale upperparts, and diminutive structure) may be *glaucoides* but are problematic out of range. Additionally, some observers in Greenland, Iceland, and the Faroes regularly record white-winged nominate types with pigmented eyes and do not consider eye color a reliable feature for separating nominate *glaucoides* from Kumlien's (Lars Witting & Silas Olofson, pers. comm.). Whether these are *kumlieni*, intergrades between *kumlieni* and nominate, or proper nominates remains to be answered.

Bear in mind that very little has been written on the sum characteristics of wintering nominate *glaucoides* populations off S Greenland, and furthermore, extensive in-hand studies of known-origin birds are largely lacking. Suffice to say that less than a handful of observers in all of Greenland actively report on gulls, and those few who do are concentrated around the capital city of Nuuk (David Boertmann, pers. comm.). Points outside Nuuk are not reported on annually, and comprehensive studies here are much needed. Unusual is the account of two Iceland Gull chicks captured in SW Greenland and raised in captivity in Austria, only to reveal adult wingtip patterns of Kumlien's Gull and olive-brown irises (Goethe 1986). With the thousands of nominate chicks that could have been retrieved here, the odds of randomly selecting *kumlieni* types—in Greenland—should give anyone with a genuine interest in this complex serious pause. This well-documented account suggests that the nominate race may be more variable than popularized in bird-identification literature, or, more likely, that (pale) Kumlien's Gulls breeding in Greenland may be overlooked.

RANGE
Precise mapping of the breeding (and nonbreeding) range of this taxon will prove difficult without comprehensive field studies across the Canadian Arctic and Greenland.
Breeding: Breeds on coastal cliffs and stacks in W Greenland north to Balgoni Øer (around N 76°), abundantly along the southwestern and southern rim of Greenland, and sparsely in E Greenland north to around N 69° (Boertmann & Huffeldt 2013; Boertmann & Rosing-Asvid 2017). Critical surveys lacking.
Nonbreeding: Largely maritime. Primarily found in N Atlantic waters off SW and S Greenland. Generally avoids heavy ice cover, feeding on aquatic prey around open leads and polynyas. Often follows trawlers; also found in fishing ports, harbors, and even

landfills, where it regularly associates with larger gulls. Moderate numbers winter east to Iceland (up to 10% of population), smaller numbers to the Faroes, and more scarcely to N and W Europe. Wintering numbers outside Greenland routinely and unpredictably fluctuate from year to year.

In Canada and the United States, very rare to casual, but regular to St. John's, Newfoundland, where keen observers report 1–3 almost annually. Of 2,512 banded individuals (presumed nominate) from Greenland, only 6 recoveries were recorded outside the country, 1 from Newfoundland and Labrador, which is statistically significant (Snell et al. 2020). White-winged adults recorded annually in the Maritime Provinces, some very likely nominate *glaucoides*, whereas others cannot be safely distinguished from pale Kumlien's, or late-season Kumlien's that have bleached. Regarding these pale birds, the late Ian McLaren—renowned birder and biologist from Nova Scotia—remarked, "I am struck by the number of purely-white-winged Iceland Gulls lingering at the mouth of the Halifax Harbour" (Maybank 2009). Several records from Ontario and Quebec, where nominate occurs very rarely. Casual to accidental in fall and winter in other regions of North America, including west to MN, YT, AK, BC, OR, WA, and several records from CA. A possible 1st cycle found 5 miles offshore east-central Florida is the most convincing southern occurrence I am aware of (Ayyash, pers. obs.). Committees in the ABA Area take a conservative approach when it comes to separating white-winged Kumlien's from nominate, and black-winged Kumlien's from Thayer's. Therefore, the true status of nominate is obscured by difficulty separating this subspecies from some pale "Kumlien's Gulls", but by no means is it common anywhere in Canada and the United States. Older accounts from the Atlantic Coast seldom provided details regarding gray tone of upperparts, which is crucial (Howell & Dunn 2007). Many reports of 1st-cycle birds are doubtful, particularly bleached individuals in late winter and early spring, in which the bill pattern is also more likely to show a pale base. Plumage patterns of younger birds is most discernible in earliest arrivals, typically between October and late November. Subspecific identification once these patterns fade or show wear is not advisable. Also recorded in Bermuda. Several accounts from Japan, but separation from pale Kumlien's problematic (Ujihara & Ujihara 2019).

IDENTIFICATION
Adult: Noticeably pale gray upperparts with all-white wingtips. Most adults have distinctly pale, creamy-yellow eyes. A few with amber-like quality to iris, with light speckling. Raspberry-pink orbital, at times reddish in high breeding condition. Pink legs. Fairly petite yellow bill with small red gonys spot, may show greenish tone in low breeding condition. Generally compact in structure with a diminutive impression, especially when compared to most other large white-headed gulls. Male types bulkier, with stronger bill and flatter head profile. In basic plumage, head markings appear cinnamon gray overall, with some showing dense blotting, at times with horizontal texturing. Some birds show a densely marked hood effect; others have light diffuse streaking. *Similar Species:* Primary confusion species is Glaucous, which typically has matching pale gray upperparts and pale yellow eyes. Bill size and shape differences critical, although some quite similar. Glaucous has a wider bill base and noticeably shorter wing projection with a deeper chest. Iceland averages shorter legs with a more rounded head, but beware opposite-sex overlap. Some female-type Glaucous and male-type Iceland Gulls may appear similar in size, but Iceland is overall sleeker with noticeably longer wings at rest. Glaucous averages longer legs. In Iceland, the wing projection is typically greater than the length of the bill, while in Glaucous, the wing projection appears about the same length (or shorter) than the bill's length. In flight, Iceland's wings are obviously narrower, with shorter neck and head projection. Orbital-ring and bill colors helpful when discernible: Iceland has pinkish-purple orbital and sometimes a notable green cast to the bill, whereas Glaucous generally shows an orange-yellow orbital with pinkish cast to bill base in nonbreeding condition. Despite this, some individuals—particularly lone birds with no direct comparison—can be surprisingly tricky to nail down. Also compare to white-winged Kumlien's Gulls, from which it may be impossible to distinguish at times (see that account). Nominate *glaucoides* averages paler gray and more diminutive overall head and bill, but considerable overlap exists. Eye color in nominate adults should be investigated more thoroughly on the breeding grounds, as it is not considered a reliable feature for separating some birds from Kumlien's in Greenland (Lars Witting, pers. comm.). It is not clear whether this is due to introgression with Kumlien's

32C ICELAND GULL

1. Juvenile with presumed *leuceretes* Glaucous Gull (by range). Overall pale aspect, showing long pale wings. Dark bill expected on young juveniles. JOHN CHARDINE. GREENLAND. AUG.
2. 1st cycle. Thin barring across upperparts, with much white on feather centers. Commonly shows noticeably pale outer primaries and pale bill base. CHRIS GIBBINS. SCOTLAND. NOV.
3. 1st cycle. A darker individual. Sooty breast, flanks, and several upper scapulars are 1st alternate. Less contrast with outer primaries invites thoughts of Kumlien's, but open wing often decisive. STEVE RAY. ENGLAND. DEC.
4. 1st cycle. White wingtips on an overall pale individual. Neatly organized barring on undertail coverts. Diffuse black on bill base. AMAR AYYASH. ICELAND. JAN.
5. 1st cycle. Kumlien's Gull not safely ruled out on perched bird. Identified using open-wing photos, which showed slightly darker pattern on inner primaries and finely vermiculated tail pattern. KENT MILLER. OHIO. NOV.
6. 1st cycle with similar-aged Glaucous (right). Iceland typically shows a rounder head, proportionally larger eye on smaller face, petite bill, and longer wing projection. DAVID DILLON. IRELAND. FEB.
7. 1st-cycle type. Thin bill, petite head, and long white wings. Postjuvenile scapulars (presumably 1st alternate) have dark, thin bars. Open wing needed to determine if 2nd prebasic molt has commenced. JOHN CHARDINE. ICELAND. JUNE.
8. 1st cycle. Pale aspect with low-contrast plumage, showing fine pattern throughout. Importantly, the inner primaries are often darker and more patterned than pale, plain outer primaries. AMAR AYYASH. ICELAND. JAN.
9. 1st cycle. Highly patterned plumage with thin barring throughout. Note tips to tail feathers and primaries—a pattern not typically found in Kumlien's. Compare to 32b.13, which shows plain outer primaries with darker outer webs. PETER SOER. THE NETHERLANDS. JAN.
10. 1st cycle. A paler individual with dark bill and plain flight feathers. Kumlien's Gull not safely separated, and such birds are typically assigned by range or left without a trinomial. AMAR AYYASH. ICELAND. JAN.
11. 1st cycle. Same as 32c.2. A darker and plainer tail pattern is less common, but not rare. Faint but visible dark diamond tips on innermost primaries. CHRIS GIBBINS. SCOTLAND. NOV.
12. 1st cycle. A warmer-brown individual showing a fairly classic pattern on the primaries. The inner primaries show dark wash with fine patterning, becoming paler and plainer at the outer primaries. HANS LARSSON. ICELAND. MARCH.
13. 1st cycle. A mid-color individual. Note pattern on inner to outer primaries, which may become less noteworthy later in the season if the outer primaries show signs of wear or bleaching. Fine vermiculations on tail feathers. JOSH JONES. IRELAND. FEB.

32C ICELAND GULL

14 2nd cycle. Eye color variable at this age, but generally shows noticeable paling. Pale bill base now with sharper demarcation. Greater coverts show fine stippling, and some pale gray on scapulars. CHRIS GIBBINS. SCOTLAND. JAN.
15 2nd cycle with 1st cycle (left). Note plain and more muted pattern to upperparts, with some pale gray on upper scapulars. Paling iris upon close inspection. PAUL SLADE. IRELAND. MARCH.
16 2nd cycle. Noticeable gray on upper scapulars and finely patterned greater coverts and tertials. Whitish wingtips common. Grayish-green bill base. Kumlien's Gull cannot be safely eliminated. AMAR AYYASH. ICELAND. JAN.
17 2nd cycle. Suggests 1st cycle, but distinguished by rounded outer primaries with ghosted pattern, plain greater coverts, and muted pattern to upper scapulars. Some have noticeably dark tail. AMAR AYYASH. ICELAND. JAN.
18 2nd cycle. Finely patterned upperwing coverts with noticeable vermiculations throughout. Icy-gray scapulars and strikingly white primaries. JOSH JONES. ENGLAND. JAN.
19 2nd cycle. Glaucous ruled out by diminutive head, small bill, and long slender wings. Kumlien's more commonly shows paler inner primaries and darker outer webs to outer primaries. AMAR AYYASH. ICELAND. JAN.
20 3rd cycle. Most pale eyed at this age. Compared to adult, duller bill, and noticeable vermiculations commonly found on greater coverts and tertials. Adult-like p6–p7 with dusky, brownish-gray outer primaries. CARL BAGGOTT. SCOTLAND. FEB.
21 3rd cycle with adult-type Kumlien's Gull (background). Strikingly pale gray upperparts, but note patchy white appearance to upperwing coverts (solidly gray in adults). Adult also ruled out by dusky gray-brown wash on outer primaries and primary coverts (found on open wing). AMAR AYYASH. NEWFOUNDLAND. JAN.
22 3rd cycle with presumed adult-type Kumlien's Gull (right). Slight dark wash on outer primaries expected, but typically less extensive than Kumlien's. The individual on the right identified as Kumlien's by darker gray upperparts and entirely dark outer edge to p9 (perhaps age-related?). HANS LARSSON. ICELAND. MARCH.
23 3rd cycle with adult Iceland Gull. Some have extensive, solid dark neck (and head) pattern. Note how the hand can appear short and blunt-tipped at times, but the wing is fairly narrow with a short neck, small head, and thin bill. AMAR AYYASH. ICELAND. JAN.
24 Adult. All-white wingtips, noticeably pale gray upperparts, and pinkish orbital. This individual shows dark spotting on iris. Grayish-brown wash can extend to lower neck. CHRIS GIBBINS. SCOTLAND. JAN.

or in part due to age or underappreciated variation (similar misconceptions existed about eye color in *thayeri* when that taxon was less known, decades ago). It is now accepted that apparent nominate adults can show a darker gray edge to the outer web of p10. In need of further study is whether nominate can also show a similar pattern on both p10 and p9, which is generally accepted by some observers in Iceland and the Faroes. A smaller percentage appear to show only a small dark outer edge on p9 with unmarked outer edge to p10 (which is unrecorded in Kumlien's). Answers to such questions cannot be addressed with fidelity based solely on wintering adults of unknown origins.

1st Cycle: Plumage consists of various shades of uniform light brown. Fresh juveniles have a frosted, dusky gray-brown aspect to their plumage, with a pale base color to the upperparts. Upperparts often with dark wavy barring or thin V-shaped bars. Overall, tertials lack solid dark centers, although occasionally some tertials can have dusky plain centers, generally with darker barring, notching, and/or peppering around the edges and tips. Dark eyes and pink legs. A relatively small and compact appearance to the bill, which commonly takes on a pale base early in the fall/winter. At times, the bill is predominantly pale, approaching the bicolored pattern of Glaucous. Others retain a mostly black bill into late winter. On the open wing, the inner primaries often appear darker and more marked than the outer primaries. The primaries become progressively pale distally, to the extent that the outer primaries can be all plain and unpatterned, some due to bleaching, others naturally before bleaching has set in, with few dark distal markings. On the inner to mid-primaries, light brown speckling and marbling often seen, a unique pattern with distal diamonds near the feather tips. On heavily marked birds, this pattern sometimes carries over to the outer primaries and secondaries. In these individuals, the uppertail shows a similar pattern with much barring and vermiculations. Uppertail typically speckled and with intricate pattern of barring against a pale base color. The undertail coverts often show bold, organized barring. *Similar Species:* Glaucous Gull has larger bill with larger proportions overall, appearing pig-eyed against a large face. Iceland is more compact in appearance with a proportionately large eye on a smaller face. Many nominate *glaucoides* show bicolored bill pattern, but typically with some dark and diffuse blemishes along cutting edge, and not sharply demarcated as found in Glaucous. Also compare to Kumlien's Gulls at this age, which are often distinguished by their darker outer primaries. Some pale individuals, however, may not be readily identified to subspecies.

2nd Cycle: Similar to 1st cycle but with overall more muted and uneven patterns on the upperparts. Plumage aspect is grayish brown, with finely patterned wing coverts and tertials, consisting of thin barring and prominent marbling on the upperparts. Eyes range from all dark to noticeably pale, although many dusky yellow. Bill also similar to that of 1st cycle, but with base appearing more grayish yellow, and more individuals showing sharply demarcated tip. Scapulars range from plain light brown with light vermiculations, similar to those of 1st cycles, to largely plain whitish gray, approaching those of some 3rd cycles. Plumage variable, with some being ghostly white, others surprisingly dark brown and appearing weathered. A few random upperwing coverts can be adult-like gray, but typically limited. The underparts, breast, and neck are mottled with variable buffy-gray markings; some advanced individuals have nearly

---

25 Adult with similar-aged European Herring Gull. Diminutive (female?), with short bill and compact appearance. Blocky head due to posture. Pale gray upperparts and white wingtips key. ARIE OUWERKERK. THE NETHERLANDS. MARCH.
26 Adult type. White-headed now with brighter yellow bill, a rather typical look late in the winter season. Long wing projection and petite bill, unlike most Glaucous Gulls. DAVID DILON. IRELAND. MARCH.
27 Adult. Sleek, long-winged appearance. Often with small head and shallow forehead. AMAR AYYASH. ICELAND. JAN.
28 Adult. All-white wingtips from above and below. Diminutive head and fine bill. CHRIS GIBBINS. SCOTLAND. FEB.
29 Adult (left) with similar-aged Kumlien's Gulls. Assigned to nominate by all-white wingtips coupled with noticeably pale gray upperparts and pale iris. Small greenish bill helps rule out Glaucous Gull. LANCY CHENG. NEWFOUNDLAND. JAN.
30 Adult. Typical individual with long, narrow wings, small head, and thin bill. Glaucous has a broader wing, blockier head, and smaller eye on bigger face. AMAR AYYASH. ICELAND. JAN.
31 Adult. Brilliantly pale gray upperparts with narrow wings and shallow forehead. JOSH JONES. IRELAND. FEB.
32 Adult. Some show a densely marked head resulting in a hooded appearance. Compact head and short neck. Hand may appear blunt-tipped. Outer edge to p9 (and p10) can show slight contrast with rest of primaries. AMAR AYYASH. ICELAND. JAN.
33 Adult with similar-aged Kumlien's Gull (right). Identified to subspecies by noticeably pale gray upperparts, white wingtips, and pale iris. LIAM SINGH. NEWFOUNDLAND. JAN.

34 Adult with similar-aged Kumlien's Gull (left). Noticeable dark gray on the outer webs of p9–p10 on the Kumlien's, which also shows slightly darker gray upperparts. Note that the gray upperparts on these two can overlap. HANS LARSSON. ICELAND. MARCH.

white head and neck. The open wing reinforces these various plumage aspects, with pale birds having nearly ghost-white outer primaries and upperwing. Dark birds have dark barring on their secondaries and diffuse brown on the centers of the mid-primaries, often appearing unrefined and dingy. The palest flight feathers on these types are the outermost primaries, which show contrastingly dark primary coverts. However, some individuals have dark markings at random, even on the outer primaries, appearing patchily patterned and untidy. The uppertail can range from largely dark with much barring and fine patterning to plain and lightly patterned with a dusky gray peppering. The uppertail coverts are variable, ranging from prominent brown spotting to very faint barring. The undertail coverts show variable barring, and the wing linings generally show an even brown wash throughout. *Similar Species:* More readily distinguished from Glaucous at this age, not only by size and structure, but also with tangible plumage differences. Glaucous in basic plumage is less likely to show much gray on the scapulars and averages less of a contrasted appearance on the wing coverts. However, birds undergoing complete molt may be difficult to interpret, especially in Greenland, where the two are found together. Compared to Kumlien's, nominate *glaucoides* averages paler outer primaries and should not show ghosted p10 mirror as seen in some *kumlieni*. The outermost primaries in many Kumlien's are commonly two toned and darker than the mid-inner primaries, although overlap exists, and some birds defy identification. In Kumlien's, the tailband may be more solidly dark and sometimes shows a more noticeable contrast against the uppertail coverts, while in nominate, the tail is on average fainter and more finely patterned, but highly variable.

3rd Cycle: Much like adult plumage aspect, but commonly shows brown tones on the upperparts, particularly on the wing coverts. Some marbling and fine barring still present in some, especially on the greater coverts and tertials, resembling pattern on 2nd cycles. Bill predominantly yellow, although less vibrant than adult's, often showing dusky markings around the tip; some rather advanced individuals have orange-red gonys spot and adult-like yellow bill. Iris is noticeably pale on most individuals, with light speckling or light amber coloration. More extensive breast and belly markings than adult. The flight feathers may show a grayish-brown wash with variable gray peppering on the uppertail. The undertail coverts, and underparts in general, regularly show a dingy brown wash. *Similar Species:* Again, the primary confusion taxa are similar-age Glaucous and white-winged Kumlien's. The former is shorter winged at rest and bulkier and averages a more pinkish bill base. Nominate *glaucoides* is typically more compact, with a slender bill. Pale-end Kumlien's Gull generally has darker gray upperparts and more noticeable dark wash on outer primaries, but overlap exists. Kumlien's more likely to show dark iris at this age, but highly variable. More study needed from known-origin birds.

MOLT The partial inserted molt in 1st cycle is attributed to prealternate. This molt is limited on the wintering grounds, with many retaining juvenile scapulars into late winter. More study needed on adult molt patterns.

HYBRIDS None known in the wild. However, successfully hybridized with Lesser Black-backed Gull in captivity: two clutches of three eggs were reared to adulthood (Lonnberg 1919). Adult hybrids said to have pink legs of female (Iceland), with intermediate to dark upperparts similar to gray tone of male, with variable white apicals and mirrors (McCarthy 2006; Snell et al. 2020).

32D COMMENTARY

Despite all that has been said and written about this complex, at the time of this work, no comprehensive effort to closely survey the Iceland Gulls has been made. The few studies that have been conducted on the breeding grounds do not lend themselves to meaningful interpretation. For instance, a study which found dark-winged birds and white-winged birds interbreeding on Baffin Island lacked operational definitions, insisting the Iceland Gulls here were not morphologically discrete (Snell 1989). Other reports from Home Bay employed *thayeri* and *glaucoides* designations for birds found interbreeding but also failed to provide explicit data on distinguishing these forms (Knudsen 1976). Reports of interbreeding between *kumlieni* and presumed *thayeri* on Southampton Island similarly lacked rigorous methods, employing ambiguous definitions that were used by previous workers (Gaston & Decker 1985). Finally, the quantitative data from all of these reports is meager, to say the least, and the reports certainly don't paint a compelling picture of vast and wide-ranging introgression of a polytypic species (Adriaens 2012; Browning 2022).

All of these reports do, however, have one theme in common: they suggest some evidence for potential nonassortative breeding in the Arctic, as little as it may be. Browning (2022) commented that this evidence has been overinterpreted, and Howell and Dunn (2007) noted that measuring the extent of interbreeding would be impossible based on the employed definitions, or lack thereof. Godfrey (1986, 262), who subscribed to the clinal-species theory, noted, "Additional reasons for treating *thayeri*

Exposed p9 pattern suggests Thayer's Gull, as does location. An underappreciated difficulty in studying Thayer's and Kumlien's Gulls on the breeding grounds is the condition of their flight feathers from June to Aug.
KATELYN LUFF. CAMBRIDGE BAY, NUNAVUT. AUG.

Although Thayer's Gull is expected here, confidently excluding a 4th-cycle-type "Kumlien's Gull" is not possible due to the condition of the outer primaries in the boreal summer. Previous workers who visited the breeding grounds in the late 20th century did not clarify how they reconciled such variables.
KATELYN LUFF. CAMBRIDGE BAY, NUNAVUT. AUG.

as a subspecies of *L. glaucoides* include abundant specimen evidence from widely separated localities that colour and pattern differences between *thayeri* and *kumlieni* are completely bridged by individual variation." The question of why Kumlien's was appointed a subspecies of "Iceland Gull" to begin with has never been clearly answered (Banks & Browning 1999). Consider the fact that Kumlien's and nominate *glaucoides* are separated by the Davis Strait, with no notable contact zone known (although see Boertmann & Huffeldt 2013; Weir et al. 2000).

The deeper one digs, the more apparent it becomes that our knowledge of how the Iceland Gulls behave on the breeding grounds is very much in its infancy and that this taxonomy is truly encumbered by a dearth of data. Unanswered questions abound. How much, if any, contact exists between *thayeri* and *kumlieni* on E Baffin Island? Is there any contact between *thayeri* and nominate *glaucoides* throughout NW Greenland? Although little to no overlap is the standing assumption, the extent of their breeding ranges has not been adequately outlined or recently updated (Boertmann, pers. comm.). What do Thayer's Gulls *really* look like near the type locality on Ellesmere Island? Are some gray winged with limited pigment on the outer primaries? Can some nominates have dark eyes, and is it outside the realm of possibility that some may occasionally show contrasting gray streaks on the outermost primaries? How far north and east does Kumlien's extend as a breeder? Where do Kumlien's Gulls wintering in Iceland come from, and are these the same Kumlien's Gulls found in Atlantic Canada? Finally, where does the highly intermediate population of Great Lakes birds originate, and how are we to interpret their position on the wintering grounds?

There have been very few case studies of known-origin birds mapped to specific colonies in the Canadian Arctic. The same is true for Greenland. Practically no modern in-hand studies of live birds have been published, and few biometrics exist—certainly none that can serve as discriminating measurements from one population to the next (although see Snell et al. 2020). There is some indication that Thayer's has sufficiently different long and flight calls from those of Kumlien's Gull (Pieplow 2017). However, published spectrograms come from very small samples (three birds) retrieved away from the breeding

3 Presumed Kumlien's (left) nesting in nominate *glaucoides* colony. Can its mate on the right be classified as a white-winged Kumlien's? Note the lack of difference in gray upperparts. Or can the bird on the left be classified as nominate simply by geography? Answers to these questions are in need of genuine research on the breeding grounds. DAVID BOERTMANN. QAMAVIK, GREENLAND. JULY.
4 Presumed adult Thayer's (upper left) and Kumlien's Gull. Although it is popularized that Thayer's and Kumlien's are largely absent from Greenland, the few informal surveys that have been conducted here in the last two decades have turned up both taxa in very small numbers. It is not known to what extent these two are found in Greenland due to a lack of comprehensive field work. DAVID BOERTMANN. QAASUITSUP MUNICIPALITY, GREENLAND (~78.5N -59.2W). AUG.
5 Kumlien's Gulls. The individual on the right has a pale iris, paler gray upperparts, and white wingtips. Nominate *glaucoides* immediately comes to mind, but all of its field marks, including the paler-gray upperparts, are within range for Kumlien's. ALVAN BUCKLEY. NEWFOUNDLAND. FEB.
6 Nominate *glaucoides* (left) and Kumlien's Gull (right). The Kumlien's shows contrasting dark gray on the outer edge of p9–p10; this is currently touted as the most reliable field mark for separating the two when eye color and gray upperparts overlap. LEFT: HANS LARSSON. ICELAND. MARCH. RIGHT: AMAR AYYASH. ILLINOIS. DEC.
7 A dark-eyed individual with pale gray upperparts and all-white wingtip. Kumlien's or nominate? Some observers from Greenland, Iceland, and the Faroes do not consider eye color a reliable feature for separating white-winged Kumlien's from nominate. This is not unlike the problem North American observers dealt with several decades ago before accepting variable eye color in Thayer's and Kumlien's. HANS LARSSON. ICELAND. MARCH.
8 Presumed Kumlien's Gulls. The individual in the front has noticeably pale gray upperparts, with obvious gray subterminal bands on p7–p8. Is this expected variation found in Kumlien's, or is nominate *glaucoides* involved? LANCY CHENG. NEWFOUNDLAND. MARCH.
9 Presumed Thayer's Gull (left) and Kumlien's Gull. Birds wintering on the Great Lakes present a number of subspecific ID challenges. Much overlap exists with wingtip patterns and gray upperparts completely bridged by individual variation. JEAN IRON. TORONTO, ONTARIO. JAN.
10 Iceland Gull (*sensu lato*) with adult Ring-billed Gull (right). The head and bill on this individual are very *thayeri*-like, as is the extensive pigment on the inner web of p9 and complete band on p5. However, the lighter-gray upperparts and grayish wingtip suggest Kumlien's. Such birds do not neatly fit any label. AMAR AYYASH. ILLINOIS. NOV.

grounds (Nathan Pieplow, pers. comm.). Winter ecology differences have also been noted. Overall, Thayer's is highly coastal in the nonbreeding season, while the other two subspecies are said to be more pelagic (Howell & Mactavish 2003). But Kumlien's Gulls wintering on the Great Lakes and in the E United States and Canada are just as coastal and occupy a niche very similar to Thayer's (i.e., urban settings that include warm-water outflows, fish-processing plants, landfills, open fields, large lakes, and riverways). The desperate need for genetic data is clear. Currently, there are no substantive genetic findings to speak of—a dilemma common to most of the large white-headed gulls of the far north. The few sequences that have been randomly published show substantial allele sharing. Most of these underdeveloped write-ups leave us with more questions than answers.

Despite this uncertainty, a number of theories have been formulated on the wintering grounds over the last century that have attempted to classify and distinguish these arctic denizens. These theories, no matter how well intentioned, equate to lab reports that have been written before the lab work is completed. It is for this reason that the field-identification criteria used to diagnose Iceland Gull winter populations have liberally been revised and modified multiple times over the last half-century (Gosselin & David 1975; Lehman 1980; Godfrey 1986; Zimmer 1991; Howell & Elliott 2001; Howell & Mactavish 2003; Adriaens 2012; Snell et al. 2020; Ayyash & Singh 2023).

It should be understood that labels currently employed are agreeable constructs that are supported by noticeable patterns found on the winter populations that we are witness to. This is "good enough" for the average field observer, but for the eager ornithologist, these taxa raise many interesting questions. What has driven the species process in this complex? Are the Iceland Gulls undergoing a period of secondary contact after some separation? Are there conclusive isolating mechanisms? Although there are striking and unmistakable features unique to those birds at opposite ends of the winter range, the Iceland Gull complex cannot be demystified strictly through observations at opposite ends. The notion that they segregate as three winter populations was once used to support "two" good species, yet this is no more logical than proposing "three" species. Making such determinations on winter populations without any redirection to the breeding grounds is unsound and leaves much to be desired.

Additionally, the phenotypic end points of nominate *glaucoides* and, particularly, *thayeri* do not appear to be positively established (Hørring & Salomonsen 1941; Ingolfsson 1967; Godfrey 1986; Snell 1989; Snell & Godfrey 1991; Howell 1999; Weir et al. 2000; Snell et al. 2020; Ayyash & Singh 2023). Other unanswered questions remain: of greatest importance, perhaps, is why *kumlieni* types appear to be increasing (Weir et al. 2000; Adriaens 2012; Harrison et al. 2021). Observer awareness and increased interest are playing a part in this, no doubt, but if we were to designate every *thayeri* and

---

11 Presumed Thayer's Gulls. The darker two individuals (left) were collected in British Columbia (March), while the paler three are from Cambridge Bay, Nunavut (June). Whether the differences in gray here represent seasonal effects (i.e., fading) or genuine variation is unknown. Such data on breeding populations in Arctic Canada is sorely lacking. AMAR AYYASH. BURKE MUSEUM. WASHINGTON.
12 Adult Thayer's Gulls with a noticeably paler individual. How we interpret this variation is in need of study. Thayer's has KGS of 5–6, with many matching the gray upperparts of American Herring, while others can match Vega Gulls. LIAM SINGH. BRITISH COLUMBIA. MARCH.
13 Adult American Herring (back) with unknown Iceland Gull. The pigment on the primaries is extensive, but far from black. The gray upperparts are noticeably pale, making this individual difficult to place, especially without viewing p5 and p9. AMAR AYYASH. ILLINOIS. MARCH.
14 Unknown Iceland Gull. Slate-gray wingtip with nearly unmarked p6 and pigment on p9 limited to outer web, not reaching the primary coverts. Such patterns were reported near the type locality of Thayer's Gull, but historic reports are unclear, if not nebulous. LIAM SINGH. BRITISH COLUMBIA. MARCH.
15 Thayer's Gull (right) with unknown Iceland Gull (left). Grayish pigment on p7–p8 with no visible markings on p6 and all-white tip to p9/p10. A trinomial is not safely assigned here, but could such birds originate from Thayer's Gull colonies? AMAR AYYASH. BRITISH COLUMBIA. MARCH.
16 Adult Iceland Gull, not safely assigned. Overall the grayish wingtip and paler gray upperparts are in line with Kumlien's, but the pattern on p9 and near-complete subterminal band on p5 are curious. Such birds are fairly common on the Great Lakes and labeling them is arbitrary. AMAR AYYASH. WISCONSIN. DEC.
17 Nominate *glaucoides* (right) and presumed Thayer's based on jet-black wingtip, *thayeri* pattern on p9, and extensive black on p5. The small head and bill cannot be used to make any determination, but how a dark Kumlien's Gull is safely eliminated is unknown. YANN KOLBEINSSON. ICELAND. APRIL.

nominate showing a single one-off feature as *kumlieni*, then Kumlien's Gull would be a substantially larger population than currently accepted, and its winter range would have to be extended to include the Pacific Coast of North America. If we label such birds intergrades (or hybrids), we further reinforce the unknowns entrenched in this complex.

Intermediate types are all but neglected in current identification literature, and for good reason: they are an inconvenient mystery. Although they arguably only form a small segment of the population, to not address them is to not address the Iceland Gull complex. An intuitive interpretation of *kumlieni* is to attribute it to a hybrid swarm. Hybrid populations generally show extensive phenotypic variation, as is evident throughout several well-known hybrid zones. We can draw comparisons to the Glaucous-winged × Western Gull phenomenon in the Pacific Northwest or the Glaucous × Herring hybrids in Iceland, but this brushes over several important differences. These swarms are tangible, so to speak, and bounded by parentals that are well defined (Hoffman et al. 1978; Ingolfsson 1987; Bell 1996, 1997). Attempts to define and distinguish Thayer's and Kumlien's Gulls have been rooted largely in assumptions and impregnated with circular logic (Howell 1999; Howell & Mactavish 2003; Howell & Dunn 2007; Ayyash & Singh 2023).

Weir et al. (2000), who champion the hybrid theory, begin with mostly the same arbitrary definitions used by earlier workers. However, Weir et al. (1995) and McGowan & Kitchener (2001) do catalogue Thayer's with reduced pigment on the outermost primaries and entertain records of Kumlien's being spread thin as a breeder in Greenland. In their analysis, these workers all expound on expanding and contracting breeding ranges, which, admittedly, are established on uncertain baselines. Predictably, they reach the conclusion that Kumlien's is a hybrid, based on a sample of museum specimens, and go one step further, to suggest it is an unstable hybrid. Whether these patterns are truly mirrored on the breeding grounds has yet to be confirmed. If a hybrid swarm is at work, it very much does not appear to be linear. An unavoidable theory—but even so, does a gradient in phenotypes necessarily equate to a gradient in ancestry? A growing opinion is that Kumlien's may be an incipient species, and this is as valid as any other viewpoint (Howell & Mactavish 2003; Adriaens 2012; Harrison et al. 2021; Adriaens et al. 2022), although it is not a new viewpoint, having been first proposed by Taverner (1933) close to a century ago.

Not long ago, Thayer's was thought to be a Herring Gull, and before this, Kumlien's was suspected of being a hybrid between Glaucous-winged and *glaucoides*. Although great strides have been made since the time of Brewster (1883) and Dwight (1925), at present, it behooves us to acknowledge that there is no definitive answer that resolves *kumlieni*. Resolving *kumlieni* is a requisite to untangling the knot. But let's suppose extensive work is done on the breeding grounds and genetic data eventually clarifies these forms. We will likely never have black-and-white answers for every individual encountered on the wintering grounds: such is the nature of large-gull identification.

18 Iceland Gull (*sensu lato*). A rather in-the-middle wingtip with slate-colored pigment. p9 shows complete subterminal band with continuous dark on the outer edge, no pigment on the inner web, with pigment not reaching primary coverts on the outer web. Faded and broken p6 band may exist in some Thayer's. Small dark spot on outer web of p5. AMAR AYYASH. ILLINOIS. MARCH.
19 Adult with grayish wingtip showing all-white tip to p10, much pigment on the inner web of p9, rift in p7 band, and unmarked p6. A remarkably odd pattern, although assuming Thayer's is unavoidable. LIAM SINGH. BRITISH COLUMBIA. MARCH.
20 Adult Iceland Gull (*sensu lato*) with American Herring. A rather typical appearance on the Great Lakes. With folded wings, this bird would be classified as Thayer's from a Western perspective and dark Kumlien's in the East. AMAR AYYASH. ILLINOIS. MARCH.
21 Adult Iceland Gull (*sensu lato*) photographed moments apart from 32d.20. Unmarked p5 and small rift in p6 band. AMAR AYYASH. ILLINOIS. MARCH.
22 An otherwise typical 1st-cycle Thayer's Gull, although when found on the Atlantic, such birds give observers pause. Can Kumlien's ever be this dark? DAVE BROWN. NEWFOUNDLAND. FEB.
23 A brilliantly pale individual with noticeably pale bill base, identified as Kumlien's based on location. In Iceland this would be identified as nominate *glaucoides*, however. An open wing shows rather plain primaries similar to 32b.16. CHRIS VAN RIJSWIJK. NEWFOUNDLAND. JAN.
24 A 1st-cycle Kumlien's type based on the paler primaries, but the dark solid tertial bases and scapular pattern invite thoughts of Thayer's, especially given the location. Can such birds be pale extremes of Thayer's? LIAM SINGH. BRITISH COLUMBIA. NOV.
25 Typical Thayer's Gull (front). The individual in the back is identified as a pale-winged Thayer's, although in the East it might be identified as a "beefy" Kumlien's. PHIL PICKERING. OREGON. FEB.

For some, lumping Thayer's with the Iceland Gulls has upgraded the problem until convincing fieldwork can shed more light on their relation. For others, it seems a major step backward. The lump has provided needed respite from pretending to know where lines are cemented and how these taxa behave in the Arctic. One day, we may learn that these "three" are not each other's closest relatives or that they very much are.

Since the lump, observers have not struggled any more or any less to find and identify "Thayer's Gulls." Lumping Thayer's did not erase its presence, and it remains an identifiable form. As does Kumlien's. As does nominate *glaucoides*. Also real are the many paradoxical individuals regularly identified with two different labels, depending on who is doing the labeling and where the birds are found. My principal observation is that the evidence put forward regarding the affinities of these three taxa is fragmented to the extent that a consistent position on their treatment is currently unworkable. Anyone who asserts otherwise will resort to trying to justify their own observer biases.

26 Intermediate 1st cycle that raises questions in the center of the winter range. In the West it would be an acceptable, pale Thayer's, and in the East a somewhat darkish Kumlien's. AMAR AYYASH. ILLINOIS. JAN.
27 1st-cycle Kumlien's type (right), pictured here with two nominate Iceland Gulls. It has a dark aspect that is Thayer's-like with outer primaries matching Kumlien's. Can nominate *glaucoides* ever be this uniformly plain and dark, and what would such a bird be labeled in North America? AMAR AYYASH, ICELAND, JAN.
28 This 2nd cycle shows intermediate features that make it difficult to comfortably assign to subspecies. In the West it would homogenize in a Thayer's flock, and Kumlien's in the East. AMAR AYYASH. WISCONSIN. MARCH.
29 Assigning subspecies to 3rd-cycle types can be difficult due to undeveloped wingtip patterns. The increased amount of pigment on the primaries is in part age-related. AMAR AYYASH, MICHIGAN, NOV.

SECTION 3

HYBRIDS

Hybridization in gulls is more prevalent than in most other bird groups, specifically within the northern large white-headed gulls. Some hybrids are genetic dead-ends, but in other instances they may be catalysts of evolution. The high incidence of hybridization in gulls is often viewed as a labyrinth of darkness in the avian world; however, we accept that organisms don't exist in stasis, and birds are perpetually evolving.

A hybrid of two distinct parental species is known as an F1 hybrid. The offspring of F1 hybrids are known as F2 hybrids, followed by F3 hybrids, and so on. And although a perfectly intermediate hybrid may be first generation, we can't be sure, unless we have DNA evidence or other life-history details.

There are various degrees of hybridization. Some hybrids are highly pervasive, such as the Western × Glaucous-winged Gull of the Pacific Northwest (Olympic Gull). These two make up a dense *hybrid zone* which is on full display for everyone to witness. The offspring are quite viable (not sterile), with remarkable adaptations, thriving just as well as any other species. Generations upon generations of backcrosses have resulted in a medley of phenotypes that range from having one slightly atypical feature to being perfectly intermediate birds. With Western × Glaucous-winged hybrids, it is important not to assume that one parent is a pure Glaucous-winged and the other a pure Western. These populations exhibit increased levels of gene flow and form a *hybrid swarm*. In general, breeding ranges of active hybrid swarms are contained within restricted boundaries where the parental ranges are adjacent or slightly overlapping. Birds that essentially resemble one of the parents but are suspected of having outside influence are sometimes referred to as "types." For example, if an individual overall tends toward Glaucous-winged Gull, we may call it a "Glaucous-winged type."

Less dense but regularly encountered hybrid pairings are reported annually in small numbers, such as American Herring × Glaucous Gull (Nelson's Gull). This hybrid is relatively rare, and estimating the size of its populations is difficult. They originate from various geographic areas that are somewhat remote, and those in the East may look appreciably different from those in the West. Other hybrids are extremely rare and seldom seen. For example, Kelp × American Herring hybrids, which are restricted to the Chandeleur Islands, off the coast of Louisiana, are a recent, very contained phenomenon and have only been studied by a small number of observers. Fortunately, specimens of various ages have been collected and preserved for future research. If it weren't for field ornithologists and biologists documenting these rare hybrids, most people wouldn't know they existed. Finally, there are hypothetical hybrids, which have been suspected but never confirmed at a nest site. Some of these are very convincing and point directly to two obvious parentals. Nonetheless, labels for such birds are presumptions and educated guesses.

Much of the available information about gull hybrids is anecdotal, and you would be pressed to find known hybrid individuals in North America that have been followed from the nest to adult plumage. With the exception of studied hybrid swarms (e.g., those of the Olympic Peninsula and Cook Inlet region), sample sizes are often too small for intelligent understanding of the amount of variation that can be expected in backcrosses. Given the amount of variability found in large white-headed gulls at the species level, it should be no different with hybrids. Our understanding of what hybrids look like often derives not from a prodigious database of known-origin birds but instead from inferences and processes of elimination. We have a set of accepted characteristics believed to be found in "typical" individuals that have been accepted, mostly by their plausibility, and it bears repeating that hybrid gulls in North America are greatly underresearched.

Most hybrid identifications involve eliminating a known species or deciding which hybrid it is. Some hybrids do a very good job of mimicking known species. For instance, Glaucous-winged × American Herring Gull can resemble Thayer's Gull (*L.g. thayeri*). The former would be exceptional away from the West Coast, and the expectation is to eliminate Thayer's Gull first. Sometimes it's the opposite, and eliminating a hybrid is the expectation. For instance, Lesser Black-backed × American Herring Gull can resemble Yellow-legged Gull (*L. michahellis*). The latter is unexpected just about anywhere on the continent, and the expectation is to eliminate the hybrid first. Other hybrids, such as Glaucous-winged × Western Gull and Glaucous-winged × American Herring Gull, can sometimes show remarkably similar structural and plumage features to each other, rendering identification highly subjective and dependent on where and when a bird is found and on the observer's experience. Such determinations are far from

HYBRIDS

1. The individual on the left looks like an otherwise typical Glaucous-winged, except for its darker primaries. At the center of the hybrid zone, we suspect this can only be explained by introgression, but refer to such individuals as a Glaucous-winged "type." STEVE HAMPTON. WASHINGTON. APRIL.
2. Adult Kelp × American Herring. This very rare hybrid is seldom encountered and has likely decreased in numbers since Hurricane Katrina. Long, grayish-yellow legs, large, bulbous-tipped bill, and slate-black upperparts contrast with its black primaries. Not seen here is a thick, white trailing edge, as in Kelp, as well as a small, square-shaped p10 mirror. AMAR AYYASH. INDIANA. OCT.
3. Adult California × American Herring (left) and California Gull. Another rare hybrid that is very sporadically seen. The *albertaensis* subspecies of California Gull may see larger and paler individuals, but the combination of paling eyes and pinkish-yellow legs points to outside influence. STEVE MLODINOW. COLORADO. DEC.
4. Adult Laughing × Ring-billed Gull. Returning for over 20 years, this hybrid—dubbed The Colonel—was noticed several years after an adult Laughing Gull was found breeding in a nearby Ring-billed Gull colony. AMAR AYYASH. ILLINOIS. MARCH.
5. Lesser Black-backed × Iceland Gull (*kumlieni* ssp?). Highly conjectural, but very much appears to be a refined, white-winged Lesser Black-backed Gull. This pairing has never been confirmed in the wild, although it has successfully hybridized in captivity. ALVAN BUCKLEY. NEWFOUNDLAND. FEB.

6 Lesser Black-backed × Ring-billed Gull with adult Black-headed Gull. Another hybrid pairing that remains unconfirmed at a nest site. One or two are reported in western Europe annually. DELFIN GONZALEZ. SPAIN. FEB.

7 Glaucous × American Herring Gull. 1st cycle. Classic bicolored bill pattern, paler upperparts with low contrast, and noticeable pale edging on brownish primaries. Individuals showing intermediate features are regularly identified with no dispute. BRUCE MACTAVISH. ST. JOHN'S, NEWFOUNDLAND. JAN.

8 Juvenile Glaucous-winged × American Herring (AK; Aug.) and Thayer's Gull (IL; Dec.). A "confusion pair" that sometimes presents a challenge. Thayer's (right) averages smaller and more refined. Month for month, Thayer's also appears more dapper, with limited postjuvenile molt. AMAR AYYASH.

9 Adult Great Black-backed × American Herring (center) with Great Black-backed (right of center) and American Herrings. A rare, but regularly recorded hybrid throughout the Great Lakes and less so on the Atlantic. Younger age groups are poorly known. AMAR AYYASH. WISCONSIN. MARCH.

---

scientific. Naturally, observers tend to emphasize and weigh field marks differently. Although this can be a source of frustration, the keen observer uses encounters with potential hybrids to help crystallize what they know and don't know about the suspected parentals. For this reason, you will find highly experienced gull-watchers who prefer to leave a number of gulls unidentified rather than settle on a presumptuous hybrid label. Indeed, being an expert does not give one the ability to identify the unidentifiable. Although many hybrids are routinely identified, labels must be assigned with some pliability. The photographic examples in this guide's hybrid accounts provide a basic introduction to some of the more widespread and popular hybrids. All identifications are declared assumptions and should be understood as putative.

H1 GLAUCOUS-WINGED × WESTERN (OLYMPIC) GULL

The most well known gull hybrid in North America. Ranges primarily as a breeder along the Pacific from far SW British Columbia to central Oregon. Relatively high concentrations of this hybrid are also found slightly inland, centered around Puget Sound and south to the Columbia River near Portland. An estimated 70% of breeders in Washington and Oregon are attributed to Olympic Gull (Bell 1996, 1997). As a nonbreeder, moves slightly to the north as far as SE Alaska and south to S California, but overall is a short-distance migrant. Inland numbers in the nonbreeding season are likely underestimated, but can be found east to the Central Valley, E Oregon, and up the Columbia River in SE Washington. In the interior West, sporadic reports of singletons from NV, ID, UT, and CO, and as far east as TX and IL (Ayyash 2014).

Hybrids run the gamut in appearance, from *almost* Western to *almost* Glaucous-winged. Therefore, identification main points often revolve around separation from parentals. Separation from Western Gull is fairly straightforward, and

1 Juvenile. Overall plumage patterns similar to Western, but primaries and tail feathers decidedly grayish-brown with pale edging. Remarkably fresh plumage for the end of Dec. Compare to juvenile Slaty-backed. AMAR AYYASH. OREGON. DEC.
2 1st cycle. Paler-gray aspect closer to Glaucous-winged, but dark primaries and strongly patterned wing coverts typical of this hybrid. Note bulbous-tipped bill, high eye placement on large face, and deep breast. AMAR AYYASH. OREGON. JAN.
3 1st cycle. Recalls Slaty-backed Gull, which sometimes has dark bases to the greater coverts. Slaty averages a shallower bill tip, neater pale bases to the median coverts, and brighter legs, but appreciably variable. AMAR AYYASH. OREGON. JAN.
4 1st cycle. Brown flight feathers with pale fringes and pale underside to p10. The pale greater coverts with dark-tipped median coverts recall Slaty-backed Gull, but note the Western-like blocky head, beady eye, and robust bill. AMAR AYYASH. CALIFORNIA. JAN.

440 H1 GLAUCOUS-WINGED × WESTERN (OLYMPIC) GULL

5 1st cycle. Identification difficult. Glaucous-winged × American Herring averages paler face with streakier pattern, thinner bill tip, and barred postjuvenile scapulars (although variable). The dark shaft streaks recall Slaty-backed, but that taxon typically lacks strong spotting on the greater coverts, especially this late in the season. CONOR SCOTLAND. OREGON. FEB.
6 1st cycle. Stature is that of a typical Glaucous-winged Gull with the face of a Western Gull. Strong-patterned coverts now distorted by wear. Primaries too pale for Western and too dark for Glaucous-winged. STEVE ROTTENBORN. CALIFORNIA. JAN.
7 1st cycle. The extent of fading suggests the primaries are too dark for Glaucous-winged. The paler postjuvenile scapulars and longer wings make separation from Glaucous-winged × American Herring difficult. Overall short, thick bill and blocky head with densely marked face and neck favor this hybrid, but speculative. STEVEN MLODINOW. WASHINGTON. APRIL.
8 1st cycle with similar-aged Thayer's Gull (left). Compare eye placement as well as bill size and shape. This hybrid averages bulgy, with messier pattern to coverts, and often with contrasting postjuvenile scapulars. AMAR AYYASH. WASHINGTON. JAN.
9 Juvenile. Suggests Western Gull, but with pale contrasting window and suspiciously pale outer greater coverts. AMAR AYYASH. WASHINGTON. AUG.
10 Juvenile. Overall appearance is that of a washed-out Western Gull, with pale window and brown (not black) flight feathers. AMAR AYYASH. WASHINGTON. AUG.
11 1st cycle. Tending toward Western Gull, but with brown outer primaries showing noticeable venetian-blind pattern. Noticeable notching on outer edge of outer tail feather. AMAR AYYASH. BRITISH COLUMBIA. MARCH.
12 Juvenile. Remarkably intermediate in every respect. Broad wing and overall appearance similar to Glaucous-winged × American Herring, which often shows more patterned base to outer tail, with less contrast to uppertail coverts. AMAR AYYASH. WASHINGTON. AUG.
13 Juvenile. Strongly barred uppertail coverts, as found in Western, but obviously paler. Juvenile scapulars can appear scalloped. AMAR AYYASH. OREGON. SEPT.
14 1st cycle. Darker gray postjuvenile scapulars, densely patterned head, and large bulbous-tipped bill suggest this hybrid, but separation from Glaucous-winged × American Herring or a strongly marked Glaucous-winged is difficult. AMAR AYYASH. OREGON. JAN.
15 2nd cycle. Subdued grayish-brown aspect, with intermediate color to incoming primaries (completing 2nd prebasic). All-black bill is presumably a Glaucous-winged trait, as Western typically shows a paling bill base by now and paler underparts. AMAR AYYASH. WASHINGTON. AUG.

H1 GLAUCOUS-WINGED × WESTERN (OLYMPIC) GULL 443

16 2nd cycle. More commonly confused with Glaucous-winged × American Herring, which averages streakier face and paler gray scapulars, but variable. AMAR AYYASH. WASHINGTON. JAN.
17 2nd cycle. Suggests sooty Glaucous-winged but combination of darker gray scapular and darker flight feathers difficult to write off as variation in the hybrid zone. AMAR AYYASH. OREGON. JAN.
18 2nd cycle. Flight feathers with intermediate color. Glaucous-winged × American Herring averages slighter head and neck with streakier pattern and less bulbous-tipped bill, but variable. STEVE HAMPTON. WASHINGTON. DEC.
19 3rd cycle. Deep breast and blocky head with smudged pattern. Gray upperparts intermediate. Delayed aspect to outer primaries, but note white, adult-like tips to secondaries. AMAR AYYASH. OREGON. JAN.
20 3rd-/4th-cycle type. Overall adult-like, but tricolored bill and dark marks on exposed primary coverts suggest otherwise. Darker gray upperparts and darkness of p6–p7 point away from (pure) Glaucous-winged. AMAR AYYASH. CALIFORNIA. SEPT.
21 3rd cycle. Intermediate gray upperparts. Western has black markings on the tail (not grayish), and paler in Glaucous-winged. AMAR AYYASH. WASHINGTON. JAN.
22 3rd cycle. Approaching Glaucous-winged, but combination of bill and head pattern, as well as darker wingtip, typical of this hybrid. CHRISTOPHER LINDSEY. OREGON. DEC.
23 Adult. Overall close to Glaucous-winged, but with dark wingtip and slightly darker gray upperparts. Hybrids too can show surprisingly large, bulbous-tipped bills, often with black marks near the tip. AMAR AYYASH. OREGON. JAN.
24 Adult. Suggests a Western Gull, but with densely marked head and slaty tones to the wingtip. Hybrids often show an admixture of pinkish-yellow orbital, as seen here (yellow in pure Western). BLAKE MATHESON. CALIFORNIA. JAN.
25 Adult. Upperpart coloration, broad tertial crescent, and secondary skirt recall Slaty-backed Gull, but with robust "Western Gull" bill and smudged head markings (streakier on Slaty-backed). STEVEN MLODINOW. WASHINGTON. JAN.
26 Adult. A smaller individual, with upperparts and wingtip moderately darker than Glaucous-winged. Olympic hybrids often show this dingy head pattern with bright, banana-yellow "Western" bill. AMAR AYYASH. WASHINGTON. JAN.
27 Adult. Suggests nominate *occidentalis* with suspicious head markings. However, known-origin, adult-type Westerns are occasionally recorded with similarly smudged head and neck (see 26.35). AMAR AYYASH. WASHINGTON. JAN.

many are routinely distinguished from Glaucous-winged. However, the lack of agreement on what a "pure" Glaucous-winged looks like makes this a subjective, if not a likely impossible, task at times.

Other identification issues pertain to Glaucous-winged × American Herring hybrids (Cook Inlet Gull). The differences between adults and 1st cycles are explained here. Olympic Gull averages shorter wing projection with a stockier body, blockier head, and brighter yellow bill, which is often thicker with a more pronounced bulbous tip. Head and face markings on Olympic can be noticeably dense and smudged, while Cook Inlet averages streakier (presumed Herring influence). Both hybrids have variable eye color. Orbital-ring color may be an admixture of pinkish yellow in either hybrid, but more commonly found blended in Olympic, and the predominant color in Cook Inlet is dull pink (Ayyash, pers. obs.). Olympic averages darker upperparts (due to Western influence), and on the open wing, less contrast is seen between the dark wingtip and gray upperparts (although pale individuals regularly overlap with Cook Inlet). Finally, Olympic is less likely to show a p9 mirror, while in Cook Inlet, a variable-sized p9 mirror (often small) is found fairly regularly.

For 1st cycles, employ structure and shape comparisons noted for the adults, especially the heftier bill, blockier head, and shorter wings of Olympic. Average differences in plumage include more intricate and finer barring on wing coverts of Cook Inlet, as well as more streaked face. Olympic averages darker ground color overall, often with more extensive postjuvenile molt of scapulars by late fall, and more wear by early winter. Oddly, a few 1st-cycle "Olympics" remain immaculately fresh into late winter; these individuals may prove to be well-patterned Glaucous-winged Gulls (more study needed). When found with a paling bill base in winter, Cook Inlet averages stronger, paler regions. The tail base on Olympic averages darker and plainer, while on Cook Inlet, there is a tendency for the tail base to be paler with some light patterning, but much overlap exists. Finally, Olympic may show more well-defined and bolder barring on the uppertail coverts, mimicking a Western Gull pattern (see plates H1.10 and H1.13). See also Slaty-backed Gull account.

28 Adult with similar-aged Western Gull (back). Paler upperparts are okay for Western, but the wingtip is too pale and the orbital is pink. PHIL PICKERING. OREGON. FEB.
29 Adult with similar-aged Western Gull (front). The paler gray upperparts may be found in palest *occidentalis*, but note the obviously pale underside to p10 (should be black in pure Western). ALVARO JARAMILLO. CALIFORNIA. DEC.
30 Adult with 3rd-cycle Thayer's Gull (right). Thick, bulbous-tipped bill and paler upperparts suggest a pale *occidentalis*. However, the orbital is purplish, and the wingtip shows diffuse, grayish black pattern. LIAM SINGH. BRITISH COLUMBIA. NOV.
31 Adult. Wingtip too dark for Glaucous-winged and too pale for Western. The pale underside to the wingtip, thin white tongue tips on p6–p8, and limited pigment on the base of p8 are presumably Glaucous-winged influence. AMAR AYYASH. WASHINGTON. JAN.
32 Adult. Darker upperparts, as in some pale Westerns, but with limited, grayish-black pigment on the wingtip. AMAR AYYASH. OREGON. JAN.
33 Adult. Overall suggests Glaucous-winged but with curiously dark wingtip and polluted bill pattern typical of many hybrids. AMAR AYYASH. OREGON. JAN.
34 Adult with similar-aged Glaucous-winged (above). Bright yellow "Western Gull" bill, slightly darker gray upperparts, and grayish-black wingtip. AMAR AYYASH. OREGON. JAN.
35 Adult. A fairly straightforward hybrid with grayish-black wingtip and dark gray upperparts, which don't align with either parental. CHRISTOPHER LINDSEY. OREGON. DEC.
36 Adult. Close to a pale Western Gull, but with grayish-black wingtip and heavily marked head. AMAR AYYASH. OREGON. JAN.
37 Adult. Paler gray upperparts, with grayish-black wingtip and white tongue tips to p8. The polluted, robust, yellow bill is typical of this hybrid, but separation from Glaucous-winged × American Herring difficult. That hybrid averages thinner and longer bill and streakier head, but variable. AMAR AYYASH. OREGON. JAN.

H2 GLAUCOUS-WINGED × AMERICAN HERRING (COOK INLET) GULL

The second most well known and populous of the Glaucous-winged hybrids. Cook Inlet Gulls breed in far south-central Alaska in the Anchorage region (i.e., Cook Inlet) and slightly southeast into the Glacier Bay basin (Patten 1980). The nucleus of the hybrid zone is the Kenai Peninsula, where an estimated 30,000 pairs breed around the Kenai River mouth (Laura Burke, pers. comm.). The hybrid swarm here, like the Olympic Gull swarm farther south, puts on a spectacular display of phenotypes, and it is doubtful that

any breeders here are genetically "pure." Farther inland to the north and east, phenotypes approaching American Herring Gull are found, and farther south and east along the coast of Alaska, Glaucous-winged types become the predominate large gull (Weiser & Pohlen 2022).

In the nonbreeding season, hybrids begin to disperse at the same time as parentals, migrating south along the West Coast as early as late September and early October (Ahlstrom et al. 2021). Winters from SE Alaska through British Columbia and south along the Pacific Coast of the United States. In the Puget Sound region, Cook Inlet

---

1 Juvenile. Suggests a burly Thayer's, with duller wing covert pattern appearing less frosted. Paler-brown primaries, especially bases, inconsistent with American Herring. AMAR AYYASH. KENAI PENINSULA, ALASKA. AUG.
2 Juvenile. Extensive pale edging across upperparts, but note disorganized pattern to lesser and median coverts. Pale greater coverts with dotted pattern recall Vega, which shows blackish primaries. AMAR AYYASH. KENAI PENINSULA, ALASKA. AUG.
3 Juvenile with Glaucous-winged (left). The hybrid approaches American Herring, but with densely patterned coverts with much pale fringing and brown primaries showing pale tips. Both individuals show notched holly-leaf pattern on lower scapulars. AMAR AYYASH. KENAI PENINSULA, ALASKA. AUG.
4 Juveniles. Fairly plain upperparts as in Glaucous-winged, with long wing projection and Herring-like body shape. Although both individuals are likely products of the Cook Inlet swarm, only the dark-winged bird is labeled a hybrid. AMAR AYYASH. KENAI PENINSULA, ALASKA. AUG.

H2 GLAUCOUS-WINGED × AMERICAN HERRING (COOK INLET) GULL

Gulls are well outnumbered by Olympic Gulls, although some reports are conflicting and likely in error, with estimates showing half of the hybrids found here as Cook Inlets (more study needed). Becomes more common along the coast, south to San Francisco Bay, where it regularly outnumbers Thayer's and Herring Gulls and at times locally outnumbers Olympic Gulls (Alvaro Jaramillo, pers. comm.). Cook Inlet numbers drop off significantly once inland and farther south to S California, while Herring Gull numbers increase locally inland. On the Oregon coast around Lincoln City, Thayer's is typically more common and more expected than Cook Inlet, suggesting a real latitudinal difference in distribution or perhaps unexplained observer biases (Phil Pickering, pers. comm.).

The range of this hybrid, especially inland, seems to have been overlooked until the late 1970s and in all likelihood is still not fully mapped. For instance, the first reports from E Oregon and E Washington only began trickling in around 2004 (Mlodinow et al. 2005). Found fairly regularly in the Columbia Basin in small numbers, and farther north and east to the Okanogan Valley. Whether this is due to a real expansion of this hybrid or increased observer awareness is unknown. Farther east on the Colorado Front Range, a sharp increase in reports began in the winter of 2011–2012, primarily in areas with dense coverage by experienced birders, with reports as far east as Illinois and Toronto, suggesting it may regularly wander farther in the interior than known (Mlodinow 2012). Still, Cook Inlet Gull is very rare to casual in most parts of the interior. Confusion occurs with eliminating a large Thayer's or establishing whether a hybrid or pure Glaucous-winged Gull is at hand.

Identification criteria not as well worked out as that for Olympic Gull, in part because of this hybrid's more remote breeding range. Also, there is an increased variation component, due to American Herring Gull. In general, American Herring is not very well understood in the West, being clouded by perpetual incertitude over Glaucous-winged influence. Separation of some individuals from parentals often proves more difficult for Cook Inlet Gull than for Olympic Gull. Those with darkest wingtips blur the lines with American Herring, and paler-winged birds can be equally convincing as Glaucous-winged (for which

5   1st cycle with 2nd-cycle Olympic Gull (right). Plumage patterns as well as size and shape are used simultaneously to approximate these hybrids. Olympic averages a more rotund body, shorter wing projection, and more bulbous-tipped bill, but much overlap exists. STEVE HAMPTON. WASHINGTON. JAN.

6   1st cycle with Thayer's (left). Size differences can be absorbed by intraspecies variation, but the less-organized wing covert pattern and more-advanced postjuvenile molt (upper scapulars) support a hybrid. The petite bill and docile face suggest Thayer's, however, and such birds may defy ID. LIAM SINGH. BRITISH COLUMBIA. DEC.

H2 GLAUCOUS-WINGED × AMERICAN HERRING (COOK INLET) GULL

7   1st cycle. Glaucous-winged type with curiously dark primaries and tertials. Some Slaty-backeds approach this appearance, but average warmer brown tones, with greater coverts often the palest and plainest covert tract. AMAR AYYASH. BRITISH COLUMBIA. MARCH.
8   1st cycle. Solidly dark with contrasting pale head. Some 1st-cycle Glaucous-wingeds approach this look, but with milky, brown aspect. The increased melanin here is attributed to American Herring. Olympic Gull averages stockier hind parts, fuller bill base, and more smudged head pattern. STEVE ROTTENBORN. CALIFORNIA. DEC.
9   1st cycle. Approaching American Herring, but with paler, washed-out tail and primaries, and matte gray postjuvenile scapulars. Some regularly show a contrasting pale forehead with paling bill base. Patterned white tips to greater coverts and tertials fairly common. AMAR AYYASH. OREGON. JAN.
10 1st-cycle Cook Inlet with unidentified 1st cycle (right). The left bird is a paler version of H2.9 (see H2.13 for open wing). The right bird approaches Thayer's, but the ill-defined pattern on the upperparts, long bill, and obscure body shape can fit a smaller hybrid. AMAR AYYASH. OREGON. JAN.
11 Juvenile. Tail and outer primaries too dark for Glaucous-winged, but decidedly paler than American Herring. Olympic Gull averages bolder and more defined barring on uppertail coverts and paler tips to rectrices. AMAR AYYASH. KENAI PENINSULA, ALASKA. AUG.
12 Juvenile. Notched scapulars as in American Herring and indistinct pattern on wing coverts attributed to Glaucous-winged. Slightly paler tail base, with white outer edge to outer tail feather. AMAR AYYASH. KENAI PENINSULA, ALASKA. AUG.
13 1st cycle. Same as H2.10. Impression of head shape and bill size changes greatly with behavior. Venetian-blind pattern approaches Glaucous-winged, but with contrasting dark tips. Silvery underside to primaries similar to Thayer's, with dark bar across tips. AMAR AYYASH. OREGON. JAN.
14 1st cycle. Suggests a washed-out American Herring or midwinter Thayer's. Paling bill base, contrasting pale head, postjuvenile upper scapulars, and broader wing support Cook Inlet, but not straightforward. Slaty-backed can approach this look, but fine patterning on greater coverts is typical of Cook Inlet. LIAM SINGH. BRITISH COLUMBIA. NOV.
15 2nd cycle. Shape suggests Glaucous-winged but with progressively darker tertials and outer primaries (growing). Bill more commonly has pink base compared to Olympic. AMAR AYYASH. KENAI PENINSULA, ALASKA. AUG.
16 2nd cycle. Muted pattern to upperparts common at this age. Thayer's appears smaller-headed, with shallower chest and body. Olympic Gull averages darker gray scapulars (see lowest scapular), with smudgier head and thicker bill. AMAR AYYASH. WASHINGTON. JAN.
17 2nd cycle. Similar to American Herring, but note unpatterned wing panel and slightly darker, matte-gray scapulars. Primaries and tail average browner than Herring. Face and crown commonly show underlying streaky pattern. AMAR AYYASH. OREGON. JAN.
18 2nd cycle. Similar to some Thayer's Gulls at this age, but Cook Inlet averages a broader wing in direct comparison. Thayer's more commonly shows fine vermiculations on greater coverts and secondary tips, and paler tail base, but variable. AMAR AYYASH. OREGON. JAN.

H2 GLAUCOUS-WINGED × AMERICAN HERRING (COOK INLET) GULL

19 2nd cycle. Same as H2.17. Again, can approach a dark, muddied Thayer's or even American Herring. The outer primaries are fairly uniform with the upperparts, but average darker on American Herring. AMAR AYYASH. OREGON. JAN.
20 2nd cycle. Same as H2.18. Two-toned venetian-blind pattern carries over to many 2nd-cycle Cook Inlet Gulls. Importantly, the underside of the primaries is often distinctly pale. AMAR AYYASH. OREGON. JAN.
21 3rd cycle. Approaching American Herring, with suspiciously small eye placed higher on a large face. Averages a darker eye at this age, but variable. NOAH ARTHUR. CALIFORNIA. DEC.
22 3rd cycle. Smudged head pattern and vivid legs attributed to Glaucous-winged. Brownish tertial center common at this age, and upperparts may appear darker than adult. STEVE HEINL. KETCHIKAN, ALASKA. FEB.
23 3rd cycle. Powerful bill with pinkish base and brownish outer primaries. Highly variable at this age, and some with darker outer primaries likely overlooked as Herring or brawny Thayer's. BLAKE MATHESON. CALIFORNIA. DEC.
24 Adult. Many "identifiable" adults suggest American Herring, but often with darkish eyes, fleshy orbital, and slate-black wingtips. Concentrated markings on the lower hindneck often produce a noticeable shawl in adults. AMAR AYYASH. BRITISH COLUMBIA. MARCH.
25 Adult. A delicate individual with wingtip similar to Thayer's. Note the smaller eye against a large face. The bill shows a subtle bulge at the tip that, more often than Thayer's, has black markings. AMAR AYYASH. OREGON. JAN.
26 Adult. An ungainly individual with blocky head showing smudged pattern (as in Glaucous-winged). Slate-gray wingtips and noticeably pale underside to p10. Olympic Gull averages thicker bill tip and shorter wing projection, but not always safely separated. STEVE HEINL. KETCHIKAN, ALASKA. FEB.
27 Adult with Glaucous-winged (right). Can recall Vega Gull (which very much recalls Thayer's at times). Vega averages darker gray upperparts with reddish orbital. The underside to p9–p10 should be blackish on Vega, not gray. LIAM SINGH. BRITISH COLUMBIA. NOV.

there is no unanimous description). Such identifications carry regional biases and inevitable circular reasoning, especially with out-of-range birds. Observers should learn average field marks of parentals and deliberately compare these to potential hybrids. Despite the inherent guesswork involved with labeling some individuals, Cook Inlet Gull is by and large a recognizable taxon and one of our most widely encountered hybrids.

Apart from Glaucous-winged × Western hybrids (see that account), the main confusion taxon is Thayer's Gull. The differences between adults and 1st cycles are explained here. In adult Thayer's and Cook Inlet, the gray upperparts are quite similar and are of little use for identification. Separation of these two begins with size and structure. At rest, note the overall larger build to Cook Inlet, with longer and stronger bill, bigger and sometimes flatter head, and shorter wing projection. In flight, Cook Inlet is more barrel-chested and broader winged. With that said, some male-type Thayer's can appear equally muscular and large, in which case, other average differences should be considered. Eye color is variable in both taxa and of little use. However, Cook Inlet averages a duller, washed-out pink orbital (with yellow at times), while Thayer's shows a more vivid and saturated orbital, appearing purplish in breeding condition. The bill on Thayer's often shows a greenish cast, while Cook Inlet averages a duller bill and regularly has dark subterminal marks around the gonys. On the wingtip, Thayer's averages more white, which is usually distinctive enough to remove any uncertainties. However, some Thayer's have more extensive pigment on

28 Adult. Overall very much approaches Glaucous-winged, although the combination of paling iris and contrasting, murky gray wingtips suggests a hybrid, likely with Herring, based on the smaller head and bill, but conjectural. COLIN TALCROFT. CALIFORNIA. DEC.
29 Adult with Thayer's Gull (front). Eye color highly variable in both taxa. Underside to p10 pale in both birds. Body and bill size often the first indication of this hybrid. Thayer's commonly has a greenish cast to the bill. ALVARO JARAMILLO. CALIFORNIA. DEC.
30 Adult Cook Inlet (left), Glaucous-winged (center), and presumed hybrid (far back) with exposed underside to p10 pale. MATT GOFF. SITKA, ALASKA. MARCH.
31 Adult. Completing prebasic molt with old p9–p10. Emerging p5–p6 show slate-black color. Pale iris. A pinkish orbital is the dominant orbital color in the center of the hybrid zone. AMAR AYYASH. KENAI PENINSULA, ALASKA. AUG.
32 Adult completing wing molt, with old p9 showing *thayeri* pattern. Vivid yellow bill typically becomes dull in nonbreeding season (unlike many Olympic hybrids). Dark eye and pink orbital eliminate American Herring. AMAR AYYASH. ANCHORAGE, ALASKA. AUG.
33 Adult. Wingtip may approach eastern American Herrings, but unexpected in western population to show pale grayish underside to p10 with broken subterminal band and long p9–p10 tongues (see underside of wingtip), as well as incomplete p5 band. Flesh-colored orbital observed in the field. LIAM SINGH. BRITISH COLUMBIA. MARCH.
34 Adult. Similar to some Olympic adults, the wingtip can be slate-colored, regularly with no p9 mirror. Complete band or mark on p5 fairly common. Thick bill with black mark and blurred streaky pattern on face. LIAM SINGH. BRITISH COLUMBIA. MARCH.
35 Adult. Slate-black wingtip with reduced string-of-pearls on p6–p9. Recalls Thayer's but with broader wing. The black on the bases of p8–p9 often appears more diffuse and messier compared to Thayer's (here bleeding onto the inner webs). AMAR AYYASH. WASHINGTON. JAN.
36 Adult with fairly black wingtip, approaching American Herring. Grayish underside to wingtip and dark eye attributed to Glaucous-winged. Thayer's averages cleaner and more demarcated pattern on base of p8–p9, and seldom shows such black on the bill. AMAR AYYASH. OREGON. JAN.

the outer primaries (perhaps age related). On these well-marked Thayer's, the bases to p8–p9 average more neatly demarcated black and a more prominent p9 mirror, if present. Cook Inlet is more likely to be without a p9 mirror, to show extensive pigment on the inner web of p9, and/or to show a small mark on p4. The underside to p9–p10 averages paler, and sometimes largely white, on Thayer's.

In 1st cycle, use size and structure comparisons noted for the adults. In the trickiest individuals, comparing the respective ratios of eye diameter to bill depth and overall length can be useful: Thayer's averages a relatively large eye to shallow bill base, whereas Cook Inlet averages a proportionately small eye (placed slightly higher on the head) to deeper bill base. The ratio of eye diameter to bill base used with the impression of bill length seems to be fairly reliable, even with larger Thayer's and smaller Cook Inlets. Month for month, Cook Inlet shows messier scapulars and tends to replace more mantle and scapular feathers well before spring migration. Some maintain fairly crisp juvenile plumage, however, which increases the potential for confusion. The wing coverts do not appear as nicely frosted as in Thayer's, with Cook Inlet averaging darker and plainer patterns. The tertial tips also average plainer on Cook Inlet. On the open wing, Thayer's generally has a cleaner, unblemished venetian-blind pattern on the outer primaries and often shows more patterning on the tail base. See also 1st-cycle description in Slaty-backed Gull account.

H3 GLAUCOUS × AMERICAN HERRING (NELSON'S) GULL

In North America, interbreeding between Glaucous and American Herring Gulls has been documented around the Mackenzie Delta and likely occurs elsewhere. In contrast to the hybrid zones of Olympic Gull and Cook Inlet Gull, around the Mackenzie Delta, hybrids occur at low frequency, making up 3%–10% of local flocks (Spear 1987). An average count of hybrids at the Inuvik landfill in the breeding season is 50–100, and these hybrids are presumed to be *L.h. barrovianus* × *L. smithsonianus*. Hybridization is also strongly suspected farther east in the lower Arctic and NE Canada, but breeding sites are unknown and remain unconfirmed. Nonetheless, hybrids in the East are attributed to *L.h. leucerotes* × *L. smithsonianus* (more study needed).

In the nonbreeding season, recorded annually in small numbers from Newfoundland and Labrador, and usually as singletons from the Maritime Provinces, Great Lakes, and Eastern Seaboard, south as far as Florida. Scattered reports throughout the interior are rare to very rare, but regular. Found less commonly on the West Coast than on the Atlantic, in part due to Glaucous having a limited coastal range on the Pacific. It's likely hybrids originating from N Alaska and the Northwest Territories migrate west over the Bering Sea, and some may winter to E Asia (like pure Glaucous Gulls from this region).

Adult Nelson's Gulls show orange-yellow orbitals with pale eyes, and pinkish legs. Upperpart coloration shows slight variation, but often tending to Glaucous (Spear 1987). Wingtip pattern is highly variable, with some adults showing limited dark streaks on the outer primaries and others with outer primaries approaching those of American Herring but decidedly slate colored. Body size is variable, with a noticeable increase in size eastward (presumably due to both Glaucous and American Herring being larger in the East). Compare to Thayer's, Kumlien's, Glaucous-winged × Glaucous, and Glaucous-winged × American Herring Gulls.

Identification of 1st cycles dependent on bill pattern, coloration of flight feathers, and plumage condition. The bill is generally bicolored, with plumage aspect recalling that of other white wingers, showing crisp patterning and variable shades of light brown. The outer primaries can recall Kumlien's or Thayer's, showing low contrast and venetian-blind pattern. Others do not necessarily show the sharply demarcated bill pattern of Glaucous (see plate H3.6) but generally have dilute appearance to the upperparts and paler remiges. A fresh juvenile in late September and a 1st alternate bird in early April might give two entirely different impressions, however. A common identification pitfall beginning in late winter is for bleached American Herrings to be reported as this hybrid. Also consider the possibility that American Herring × Glaucous hybrids in E North America (presumably of *leucerotes* parentage) may average different characteristics from those found in the West (birds in the East being larger and more patterned). In some circumstances, this may have implications for how Glaucous-winged × Glaucous and American Herring × Glaucous Gulls are identified.

With respect to differences across the breeding range, this hybrid presents a nomenclature dilemma. Glaucous Gull comprises several races with a circumpolar breeding range, and the English name Nelson's

---

1. 1st cycle. Primaries and tertials too pale for American Herring and too dark for Glaucous. Short wing projection and high eye placement. Finley patterned greater coverts closer to Glaucous. PHIL PICKERING. OREGON. NOV.
2. 1st cycle. Crisply patterned plumage and thinner bill suggest a pale Thayer's, but rather bulky body with proportionally short wings. Eye placement also suggests Glaucous. Exceptionally fresh plumage for this date. LIAM SINGH. BRITISH COLUMBIA. MARCH.
3. 1st cycle. Most commonly recorded bill pattern at this age is the classic bicolored "Glaucous Gull" bill. Intermediate color and pattern to primaries and tertials. BRUCE MACTAVISH. NEWFOUNDLAND. MARCH.
4. 1st cycle. Many have darker brown primaries with noticeable pale edging. Some bleached, late-season American Herrings can show whitish upperparts, with similar bill pattern, in which case the tertial pattern and underwing should be assessed. CHRIS HILL. SOUTH CAROLINA. FEB.
5. 1st cycle. Overall aspect, structure, and pale tertials suggest Glaucous, but with genuinely dark outer primaries. Contrasting scapulars are postjuvenile (1st alternate). LIAM SINGH. BRITISH COLUMBIA. MARCH.
6. 1st cycle (left) with similar-aged American Herrings. Plumage similar to H3.3, but with diffuse black pattern to the bill base, recorded less frequently, especially later in the winter season. AMAR AYYASH. FLORIDA. JAN.
7. 1st cycle. Venetian-blind pattern on brown (not black) outer primaries, inconsistent with American Herring. Pale window expected. The rectrices are often the darkest feathers seen with variably barred uppertail coverts. BRUCE MACTAVISH. NEWFOUNDLAND. JAN.
8. 1st cycle (right) with American Herrings. A large bird towering over surrounding Herrings when perched. Importantly, the pale flight feathers are intact and relatively fresh, and not due to bleaching. AMAR AYYASH. FLORIDA. JAN.

H3 GLAUCOUS × AMERICAN HERRING (NELSON'S) GULL

9 1st cycle. Underwing evenly pale, sometimes with faint barring on wing linings. Dark tips on underside of primaries may form a contrasting dark bar (not found in pure Glaucous). ALIX D'ENTREMONT. NOVA SCOTIA. MARCH.

10 1st cycle. Darker tailband with pale, patterned base may recall some Old World taxa. Venetian-blind pattern on brownish primaries and frosted upperparts expected. BRUCE MACTAVISH. NEWFOUNDLAND. FEB.

11 2nd cycle. Aged by rounded primary tips and pale iris (typically pale eyed at this age). Fairly "classic" eastern bird (2nd cycle and older less commonly recorded in the West). BRUCE MACTAVISH. NEWFOUNDLAND. MARCH.

12 2nd cycle with 1st-cycle American Herring (back). In addition to the large head and short wing projection, note pale tertials, tail, and underside to far wing. AMAR AYYASH. WISCONSIN. DEC.

13 2nd cycle. Some American Herrings show finely patterned tertials and greater coverts at this age, but the extensive white edging on brownish primaries and short wing projection suggest Glaucous influence. Proportions and structure wrong for Thayer's. IAN DAVIES. NEW YORK. MARCH.

14 2nd cycle. Suggests a washed-out American Herring with finely patterned upperparts. Compare to Thayer's and Kumlien's, which average shorter neck, with smaller head and thinner bill. ED COREY. NORTH CAROLINA. FEB.

15 2nd cycle. Somewhat advanced wingtip pattern with relatively large p10 mirror, although secondaries and inner primaries suggest 2nd basic. Regularly found with smaller, indistinct p10 mirror. BRUCE MACTAVISH. NEWFOUNDLAND. FEB.

16 3rd cycle. Suggests a Glaucous Gull with dark primaries and dark tail. Note subdued, adult-like gray across upperparts, but aged with open wing (see H3.18). BRUCE MACTAVISH. NEWFOUNDLAND. NOV.

17 3rd cycle. Shows short wing projection, robust body, and often-seen wide bill base of Glaucous Gull. Adult-like gray averages paler than typical American Herring, with decidedly pale primaries. BOB CUNNINGHAM. NEW JERSEY. JAN.

18 3rd cycle (same as H3.16). Brownish outer primaries maintain venetian-blind pattern. Similar to 2nd cycle, but inner secondaries and p1 show adult-like gray centers with broad white tip. BRUCE MACTAVISH. NEWFOUNDLAND. NOV.

19 3rd cycle. Adult-like inner primaries and secondaries with fairly large p10 mirror. Not well documented at this age. Likely overlooked as Kumlien's, which averages paler tail with finer markings and smaller bill. KATE SUTHERLAND. NORTH CAROLINA. FEB.

Gull needs a stricter definition. The name has been used to denote a species (Henshaw 1884) now invalid; hybrids of Vega Gull and Glaucous Gull (Dwight 1925); and, its common early-21st-century meaning, hybrids of Glaucous × American Herring (Jehl 1987). Common names of hybrids are usually of no consequence, but they must facilitate precise communication. Take, for example, a mixed Vega × Glaucous pair that nested in Bluff, Alaska, and successfully produced young (Kessel 1989). Should the hybrids be labeled Nelson's Gull? Vega × Glaucous hybrids, presumably of the *pallidissimus* subspecies, occur regularly in E Asia and Russia, and a distinction, if only by name, should be maintained. This needed distinction is not dissimilar to using the name Viking Gull for the well-known European Herring × Glaucous hybrid found in Iceland. This hybrid, especially in 1st and 2nd cycles, looks sufficiently different from birds found in North America (see "Other Hybrids" account). One alternative is to spell out presumed crosses (e.g., *barrovianus* × *smithsonianus*, *pallidissimus* × *vegae*), although, admittedly, this may prove to be equally convoluted. As is common practice now in North America, this guide uses the name Nelson's Gull to imply Glaucous Gull crossed with American Herring.

20 Adult. Pigment on outermost primaries reduced to thin, blackish streaks. Glaucous-like head and facial expression, with strong pink cast to bill (as found in many Glaucous). BRUCE MACTAVISH. NEWFOUNDLAND. NOV.
21 Adult. A smaller individual with darker pigment on primary bases. Kumlien's is ruled out by bill size and shape and overall body proportions (and orbital color, not obvious here). KEITH LEONARD. NEW JERSEY. FEB.
22 Adult. Intermediate in size and structure with shorter legs, as found in *barrovianus*. Bare-part colors very helpful with this identification. Duller legs, golden-yellow iris, and orange-yellow orbital rule out Thayer's. Cook Inlet often has pinkish orbital. CHRISTOPHER ESCOTT. INUVIK, NORTHWEST TERRITORIES. JUNE.
23 Adult. Remarkably small proportions recalling Thayer's/Kumlien's. Pictured here at a known stronghold for this hybrid, where smaller *barrovianus* breeds, thus identified primarily by location. Orange orbital noted. CAMERON ECKERT. NORTHWEST TERRITORIES. JUNE.
24 Adult. A husky bird with massive bill and large blocky head (presumably *leuceretes/hyperboreus* influence). Paler gray upperparts and reduced black on outer primaries expected. JOHN CASSADY. INDIANA. MARCH.
25 Adult with American Herring (left). On the Great Lakes, almost always strikingly paler and noticeably larger than smaller *smithsonianus* population. AMAR AYYASH. ILLINOIS. FEB.
26 Adult between American Herrings. Large head with disproportionally small eye. Primaries slate-black, with much white and noticeable gray bases. KENT MILLER. OHIO. JAN.
27 Adult (same as H3.24). Commonly shows sharp and neatly demarcated blackish markings on the outer primaries, with limited pigment on the inner webs. JOHN CASSADY. INDIANA. MARCH.
28 Adult (same as H3.22). Slate-gray wingtip pattern can suggest Thayer's or Kumlien's. A small percentage show even paler, diluted gray pigment on wingtip. Faint but nearly complete band on p6, with old p5–p10. CHRISTOPHER ESCOTT. INUVIK, NORTHWEST TERRITORIES. JUNE.
29 Adult. Largely white-headed in winter, with markings often restricted to lower hindneck. Pigment to p6 here (p5 rarely marked, which may be an observer bias). VERNON BUCKLE. LABRADOR. FEB.
30 Adult (same as H3.20). Essentially, a Glaucous Gull with short, black segments on the outer edges of p8–p10, a pattern more common in European Herring × Glaucous hybrids (e.g., Viking Gull). BRUCE MACTAVISH. NEWFOUNDLAND. NOV.

H4 GLAUCOUS × GLAUCOUS-WINGED (SEWARD) GULL

Not as well known as North America's other two Glaucous-winged hybrids. Reports in the literature are brief, with an almost-complete lack of study of known-origin birds. Historically found interbreeding in small numbers along the Bering Sea coast of SW Alaska (Swarth 1934; Strang 1977) and subsequently on Aniktun Island, off Cape Romanzof (McCaffery et al. 1997). An isolated account of an adult Glaucous-winged and adult Glaucous paired up and nesting in Deadhorse in 2001 is curious (Declan Troy, pers. comm.).

1   Juvenile. Plumage closer to Glaucous-winged, but with strikingly pale remiges (see H4.6 for open wing). Intermediate structure. The pale ghosting on the bill base is attributed to Glaucous. TOM JOHNSON. ST. PAUL ISLAND. OCT.
2   1st cycle. Suggests Glaucous, but diffuse black pattern on pink bill base and several replaced scapulars uncharacteristic of that species in Nov. Diffuse black on the bill base is an accepted feature in Asian Glaucous Gulls, although in North America, a sharply demarcated tip is expected by mid-Oct. See H4.7 for open wing. LIAM SINGH. BRITISH COLUMBIA. NOV.
3   1st cycle. An intermediate bird with darker lesser and median covert markings, and darker gray postjuvenile scapulars attributed to Glaucous-winged. The pale primaries with diamond tips and paling bill are attributed to Glaucous. MALIA DEFELICE. CALIFORNIA. JAN.
4   1st cycle. A curious bird with a disproportionately thin bill and small head. Body structure and gray postjuvenile scapulars attributed to Glaucous-winged. Barring on upperparts Glaucous-like, but with contrasting copper-brown coloration. JEFF POKLEN. CALIFORNIA. DEC.
5   1st cycle with 2nd-cycle Glaucous-winged (rear). Many 1st-cycle Glaucous-wingeds from the Aleutians and the Pribilofs have this bleached aspect in late spring. However, the noticeably pale 1st alternate scapulars and paler bill base may be Glaucous influence, but conjectural. SIMON COLENUTT. ST. PAUL ISLAND. MAY.
6   Juvenile. Same as H4.1. The combination of pale flight feathers (which is not unheard of in Glaucous-winged) and paling bill base at this time of year suggest Glaucous influence. TOM JOHNSON. ST. PAUL ISLAND. OCT.
7   1st cycle. Same as H4.2. Plumage overall acceptable for Glaucous. Whether the diffuse bill pattern is natural variation or due to outside influence is unknown. 1st-cycle Glaucous Gulls in eastern North American and Europe do not show this bill pattern, however. LIAM SINGH. BRITISH COLUMBIA. NOV.
8   1st cycle. Similar to H4.7. Assigned to Glaucous-winged × Glaucous, but the darker brown tail makes it difficult to rule out residual American Herring influence. CONOR SCOTLAND. OREGON. JAN.
9   2nd cycle. Overall approaches Glaucous-winged, but paling iris, icy-gray scapulars, and white primaries and tail presumably expressions of Glaucous. LIAM SINGH. BRITISH COLUMBIA. MARCH.

As a nonbreeder, ranges patchily throughout the Bering Sea, but rare on most visited islands, including St. Lawrence, and east to Nome (Paul Lehman, pers. comm.). Some likely evade identification due to seasonal effects on plumage and overall low observer coverage. Increasing reports from Bethel, Alaska, and more regularly in recent years from Vancouver Island, where it is annual in very small numbers from late fall through early spring. Along the Pacific Coast of the United States, a handful of 1st-cycle and adult individuals are reported annually, mostly from Oregon and California. Second- and 3rd-cycle age groups require careful separation from parentals and need clarification.

Identification challenges are twofold: first, separating presumed hybrids from parentals; and, second, separating them from American Herring × Glaucous hybrids tending to Glaucous. The extent of variation in adult and 1st-cycle Glaucous-winged Gulls is not fully comprehended or agreed upon, meaning some individuals should remain unidentified. The ideal adult Seward Gull candidate shows gray upperparts paler than those of Glaucous-winged, paling eyes with pinkish orbital, and faint gray markings on the outer primaries. Identification of 1st cycles is heavily dependent on bill pattern, extent of patterning on the upperparts, and paleness of flight feathers, all of which must be interpreted in context, such as geography and time of year, as is found in the plates, which should stimulate questions that help build a better search image for this lofty hybrid.

10 2nd cycle. Same as H4.9. The secondaries and primaries show contrasting white fringes with dusky subterminal pigment, suggesting these are genuinely pale remiges and not bleached. This may be attributed to Glaucous, especially considering a paling iris at this age. LIAM SINGH. BRITISH COLUMBIA. MARCH.
11 3rd cycle. Largely unknown at this age. Very pale eye and straighter bill, as expected in Glaucous, but with solid brown wash to wing panel and tertials, and less-than-white tail and primaries. STEVE ROTTENBORN. CALIFORNIA. FEB.
12 Adult with similar-aged Glaucous-winged Gulls. Smudged head and bulbous-tipped bill of Glaucous-winged, but pale eye, paler gray upperparts, and washed-out primary pattern attributed to Glaucous. MALIA DEFELICE. CALIFORNIA. DEC.

13 Adult type. Unlikely a definitive adult, given the noticeable light gray edging on the upperparts. Pale iris, pink bill base, as well as shape and structure closer to Glaucous. But the faint gray primary pattern and purplish legs attributed to Glaucous-winged. JEREMY GATTEN. BRITISH COLUMBIA. DEC.
14 Adult with Western Gull (left) and California Gull (right). Pale eye and reduced gray on primaries presumably Glaucous influence. Smudged head pattern and gray upperpart color typical of Glaucous-winged. ALEX RINKERT. CALIFORNIA. DEC.
15 Unidentified adult with Glaucous Gull type (right). The left bird is likely a hybrid Glaucous × American Herring based on its Herring-like build, location (Inuvik), and time of year. Out of context, Glaucous-winged × Glaucous may be suspected, reinforcing the need for caution when labeling these two hybrids. CAMERON ECKERT. NORTHWEST TERRITORIES. JUNE.
16 Adult. Dark eye, purplish orbital, and gray upperparts expected on an otherwise small Glaucous-winged with proportionally small bill. The largely white primary pattern is problematic, however. Whether this is influenced by Glaucous Gull or an extreme in Glaucous-winged is unknown (see 30.34). LIAM SINGH. BRITISH COLUMBIA. MARCH.
17 Adult (same as H4.16) with Glaucous-wingeds. Identification problematic. Suggests Glaucous-winged, except for its wingtip, which has faint gray subterminal bands and prominent white pearls. It's unknown if this is Glaucous influence or an extreme in variation. LIAM SINGH. BRITISH COLUMBIA. MARCH.
18 4th-cycle type. About as good as it gets for this rare Bering Sea hybrid. Ghostly pale gray upperparts and pale eye typical of Glaucous, but with gray subterminal bands on the outer primaries (likely 3rd basic). Purplish orbital noted. CAT RAYNER. KAMCHATKA. JUNE.

H5 GLAUCOUS × GREAT BLACK-BACKED GULL

Breeding sites unknown and likely occur scarcely rather than on a large scale. Found most commonly as a wintering bird in Newfoundland, where a handful are reported annually in St. John's. Becomes rarer down the St. Lawrence River and throughout the Great Lakes, where one to two are reported annually. Relatively common around Nuuk, Greenland, where it outnumbers hybrids of European Herring × Glaucous (Viking Gull). Very rare in Iceland, despite overlap in breeding ranges here. Seen patchily in Atlantic Europe, with 1st cycles showing a peculiar liking for Portugal and Spain.

Without question our "easiest" large white-headed hybrid gull to identify, with little to no overlap with other pairings. This is also our only large hybrid which has not been "tarnished" with a colloquial name. The main identification question is whether some adult types may be aberrant Great Black-backeds suffering from a reduction or complete absence of melanin on the outer primaries (see plate H5.17).

---

1   1st cycle. Darker bill and wingtip than H5.2. Extensive pale edging on primaries and extensive white on upperparts. BRUCE MACTAVISH. NEWFOUNDLAND. JAN.
2   1st cycle. Suggests a very frosty Great Black-backed at this age, with paler primaries and highly patterned tertials. Paling bill here presumably Glaucous influence. Fairly dark 1st alternate scapulars, similar to Great Black-backed. DANIEL LOPEZ VELASCO. SPAIN. DEC.
3   1st cycle with similar-aged Great Black-backed (right). Size is variable, with some smaller and others larger than adjacent Great Black-backeds. BLAIR DUDECK. NEWFOUNDLAND. FEB.
4   1st cycle. Flight feathers a diluted brown with relatively broad white tips. Much white on uppertail. BRUCE MACTAVISH. NEWFOUNDLAND. JAN.
5   1st cycle. Suggests a washed-out Great Black-backed with brown (not black) outer primaries. Wing linings weakly barred. BRUCE MACTAVISH. NEWFOUNDLAND. NOV.
6   2nd cycle. Very much tending to Great Black-backed, with several blackish 2nd alternate scapulars. Washed-out brown outer primaries with extensive pale edging attributed to Glaucous Gull. BRUCE MACTAVISH. NEWFOUNDLAND. DEC.
7   2nd cycle. A strikingly solid dark wing, as found in some Great Black-backeds at this age, but with uniformly pale outer primaries. A p10 mirror is regularly found on hybrids at this age. CHUCK SLUSARCZYK JR. OHIO. NOV.
8   3rd cycle. Similar to Great Black-backed, but with whitish outer primaries. Tail feathers average more pigment than 1st cycle. See H5.9 for open wing. BRUCE MACTAVISH. NEWFOUNDLAND. MAY.
9   3rd cycle. Same as H5.8. Inner primaries have shed (3rd prebasic molt), but all visible remiges (as well as solid brown tail feathers) are 2nd generation (2nd basic). Fairly pale underwing, presumably due to Glaucous influence. BRUCE MACTAVISH. NEWFOUNDLAND. MAY.
10 3rd-cycle type. Great Black-backed-like, overall. Aging is difficult without an open wing (see 28.22 and 28.28). Whitish primaries are presumably Glaucous influence and not bleached. BRUCE MACTAVISH. NEWFOUNDLAND. NOV.
11 3rd-/4th-cycle type. Aged by extensive brown wash on primary coverts and bill pattern. The upperparts appear dilute, with progressively paler pigment on the primaries. Great Black-backed wing in frame (left). KIRK ZUFELT. NEWFOUNDLAND. JAN.

H5 GLAUCOUS × GREAT BLACK-BACKED GULL

12 4th-cycle type with similar-aged Great Black-backed (left). Less commonly seen with this hybrid are individuals with intermediate gray upperparts, which is curious. JIM PAWLICKI. NEW YORK. OCT.
13 Adult type. Suggests Great Black-backed with extensive white on outer primaries. BRUCE MACTAVISH. NEWFOUNDLAND. JAN.
14 Adult. Tending toward Great Black-backed, but with greatly reduced pattern on outer primaries and paler upperparts, presumably due to Glaucous influence. JEREMIAH TRIMBLE. MASSACHUSETTS. JAN.
15 Adult with similar-aged Kumlien's Gull (left). Intermediate gray upperparts darker than Kumlien's here. Largely pale underside to wingtip. PETER ADRIAENS. NEWFOUNDLAND. JAN.
16 Adult. Same as H5.14. Limited pigment on outer primaries and slightly paler upperparts than found in Great Black-backed. JEREMIAH TRIMBLE. MASSACHUSETTS. JAN.
17 Adult. An otherwise typical Great Black-backed with unpatterned outer primaries. Whether these types represent extreme variation in Great Black-backed, color aberrants, or indeed hybrids with Glaucous Gull requires study. BRUCE MACTAVISH. NEWFOUNDLAND. DEC.

H6 AMERICAN HERRING × GREAT BLACK-BACKED (GREAT LAKES) GULL

A low-frequency hybrid noticed first in the late 1950s, and increasingly by the late 1970s, mostly between New York and Ontario (Jehl 1960; Andrle 1972; Foxall 1979). Found most commonly as a nonbreeder on the Great Lakes, with around 10 adults reported annually. Several breeding records from Lake Huron and Michigan's upper peninsula, and suspected from Door County, Wisconsin, but unconfirmed. Nesting also suspected on the Four Brothers Islands, in the Lake Champlain Basin, with several sightings beginning in the breeding season of 2019 (Derek Rogers, pers. comm.). Drops off farther east: 1–3 are reported annually in the nonbreeding season from the mid-Atlantic and points north to Newfoundland and Labrador, but no documented breeding in the Northeast. Interestingly, this hybrid is fairly rare in St. John's, where both parentals winter together in large numbers and other large hybrids are reliably found.

On the Atlantic, closer to American Herring in body and bill size. On the Great Lakes, often towers over American Gulls, although some presumed backcrosses are smaller and tend to Herring in size (see plate H6.12). Adult most likely to be confused with Slaty-backed Gull, especially on the Great Lakes, but open wing and orbital color sufficiently separate the two. Orbital color on the hybrid is closer to American Herring and typically not reddish. Compare also to Western Gull and American Herring × Lesser Black-backed hybrids. Identification of 1st cycles poorly understood.

1st cycle. Not well known at this age, but a combination of features suggests this hybrid, including paling bill base, well-marked belly, and warmer brown centers to the upperparts. PETER ADRIAENS. NEWFOUNDLAND. JAN.

1st cycle (same as H6.1–H6.4). Larger size and frosted appearance recall Great Black-backed × Glaucous, but note solid dark tertial centers with fairly prominent white tips and dark primaries. PETER ADRIAENS. NEWFOUNDLAND. JAN.

3   1st cycle (same as H6.1–H6.4). Limited pigment on the undertail coverts with white base color, and contrasting tailband appearance, influence of Great Black-backed. PETER ADRIAENS. NEWFOUNDLAND. JAN.
4   1st cycle (same as H6.1–H6.3). Subdued window, as in Great Black-backed, as well as barred greater coverts and axillaries. The upper tail is entirely dark and attributed to American Herring. PETER ADRIAENS. NEWFOUNDLAND. JAN.
5   2nd cycle with similar-aged American Herring (back). Great Black-backed influence includes slate-gray scapulars, blocky head, and smaller eye placed high on the face. The smooth belly patch, solid brown tertial bases, and long wing projection are presumably Herring influence. KIRK ZUFELT. ONTARIO. NOV.
6   3rd cycle. Yellow-orange orbital and pink legs expected in this hybrid. Intermediate gray upperparts suggest a pale Lesser Black-backed. Powerful bill, deep breast, and overall large size inconsistent with American Herring × Lesser Black-backed hybrid. JEN BRUMFIELD. OHIO. MARCH.
7   3rd cycle with similar-aged American Herring (left). Smaller-bodied individual, but note relatively small eye on blocky head. Slaty-backed averages broader white scapular crescent and shorter wing projection, and often has noticeable eye patch in basic plumage. KIRK ZUFELT. ONTARIO. NOV.
8   3rd cycle with Great Black-backeds. Recalls Lesser Black-backed, but note broad wing, massive bill, and largely unmarked p5. Slaty-backed should have noticeably broader white trailing edge. CHUCK SLUSARCZYK JR. OHIO. FEB.
9   Adult with Great Black-backed (right). Obvious contrast between slate-black upperparts and primaries. A large individual with thick bill and blocky head. Speckled iris, although eye color is variable in adults. Dark orange orbital in breeding condition. STEVE ARENA. MASSACHUSETTS. MARCH.
10 Adult with similar-aged Great Black-backed (right). In basic plumage, shows more head and neck streaking than Great Black-backed. Upperpart coloration and short wing projection similar to Slaty-backed, which averages a broader tertial crescent and secondary skirt, thinner bill, and brighter legs. JARED CLARKE. NEWFOUNDLAND. JAN.
11 Adult with American Herring. Suggests Great Black-backed, but with paler upperparts and more extensive head markings. Orbital is a dull orange-yellow here. JOEL TRICK. WISCONSIN. FEB.
12 Adult with 3rd-cycle-type American Herring (left). A paler and smaller individual tending to American Herring, but note stout bill, blocky head, and slate-gray upperparts attributed to Great Black-backed. Likely a backcross. AMAR AYYASH. ILLINOIS. JAN.

13 Adult in alternate plumage. Suspected local breeder from the Four Brothers Islands on Lake Champlain. Thick bill, orange orbital, and pale eye. Dull legs. Upperparts similar to pale Lesser Black-backed. DEREK H. ROGERS. NEW YORK. JULY.
14 Adult (same as H6.11). Overall wingtip here mimics Great Black-backed, with white tip to p10, moderate p9 mirror, and unmarked p5. Broad wing, blocky head, and massive bill. AMAR AYYASH. WISCONSIN. MARCH.
15 Adult (same as H6.14). Wingtip inconsistent with Lesser Black-backed (e.g., unmarked p5 and broad white tongue tips to p6–p7). The trailing edge is narrow, and the white tips to the inner primaries are far too thin for Slaty-backed Gull. AMAR AYYASH. WISCONSIN. MARCH.
16 Adult (same as H6.9). Bluish-black upperparts, thick neck, and large head, with bright yellow bill and swollen tip, recall Western Gull or Yellow-footed Gull, but wingtip pattern positively wrong. STEVE ARENA. MASSACHUSETTS. MARCH.
17 Adult type with 3rd-cycle American Herring (front left) and Great Black-backeds in background. Intermediate in upperpart coloration as well as bill and body size. NATHAN DUBROW. MAINE. SEPT.
18 Adult type (same as H6.17) with Great Black-backed. Suggests an athletic Lesser Black-backed, but compare wing width and neck protrusion to Great Black-backed. Broken p5 band, white tongue tip to incoming p6, and gray bases to p8–p9 also point away from Lesser. NATHAN DUBROW. MAINE. SEPT.

H7 AMERICAN HERRING × LESSER BLACK-BACKED (APPLEDORE) GULL

The most frequently encountered hybrid gull on the Atlantic, outpacing Glaucous × American Herring since 2020. It is unknown where American Herring and Lesser Black-backed Gulls hybridize in North America, and to what extent. The only known record of these two successfully producing young is from a single male Lesser Black-backed (Green F05) with American Herrings on Appledore Island (Ellis et al. 2014). Hybridization was recorded here for several summers, with adults and young banded and thoroughly documented. Three of the hybrid offspring were occasionally sighted along the Atlantic Coast and were followed for a few years before disappearing (none were observed as definitive adults). See "Hybrids" section in Lesser Black-backed Gull account.

Outside the breeding season, adults are found regularly in small numbers in Newfoundland, particularly in St. John's, which is almost certainly due to increased coverage. In the United States, found with the highest frequency in Bucks County, Pennsylvania; Cape Cod and Nantucket, Massachusetts; Outer Banks, North Carolina; and throughout Atlantic Florida. A consistent increase in reports from the Maritime Provinces and Midwest correlates with the sharp rise in Lesser Black-backed numbers. Currently, 1st and 2nd cycles remain elusive. The recent surge in adult types suggests younger ages are simply being overlooked, which can be explained by the immense variation shown by both parentals. First-cycle hybrids may be just as highly variable and not intermediate in appearance, which presents quite the identification conundrum. More study needed from known hybrids.

For adults, identification confusion revolves around separation from pure parentals and from much less expected Yellow-legged and Azores Gulls. Adults are relatively straightforward to separate from parentals: they are too dark for American Herring but too pale for Lesser Black-backed. Separation from Yellow-legged Gull depends on the subspecies (see Yellow-legged and Azores Gulls accounts for details). The three most useful features are head streaking from late November through February, obscure leg color, and wingtip pattern. Wingtip pattern is by far the most permanent and reliable feature, while the

1   Juvenile F07 with Lesser Black-backed father, F05. Long wing projection, thin hind parts, and paler undertones to breast and belly suggest Lesser Black-backed influence. The notched tertial edges and inner greater coverts more common in *smithsonianus*. LAUREN KRAS. APPLEDORE ISLAND. AUG. 2011.

other two are strongly supportive when present. The average wingtip in adult hybrids approaches a well-marked American Herring from the Atlantic population with moderate black on the outer webs of p8–p9, fairly large p10 mirror, smaller p9 mirror, thin white tongue tips on p6–p7/p8, variably marked p5 (although often with thin band), and unmarked p4. Orbital color tends to be dark orange, orange red, or reddish, and seldom yellow. Also consider darker European Herring Gulls (namely, northern *argentatus*).

2   1st-cycle Lesser Black-backed Gull shown here for comparison with H7.3. AMAR AYYASH. ILLINOIS. NOV.
3   1st cycle. Putative. Largely unknown at this age. Thin straight bill and grayish-brown postjuvenile scapulars recall Lesser Black-backed. Notching on lowest scapulars and inner coverts more common in American Herring (although see 24.2). Lesser also lacks this very smooth texture to the face, lower hindneck, and sides. AMAR AYYASH. ILLINOIS. DEC.
4   1st cycle. Same as H7.3. Dark outer greater coverts, as in Lesser, with largely white tail base and uppertail coverts and contrasting tailband. Window with darker outer webs and dark tips to inner primaries, subdued compared to American Herring. AMAR AYYASH. ILLINOIS. DEC.
5   1st cycle. Same as H7.3 and H7.4. Plain, dark, centered upperwing coverts with extensive pale edging, as in Lesser Black-backed. Subtle barring on wing linings, intermediate in pattern. AMAR AYYASH. ILLINOIS. DEC.
6   1st cycle with similar-aged American Herrings. Dark face mask, head and bill shape, and elongated hind parts recall Lesser Black-backed. Patterned tertial tips and solid brown belly attributed to American Herring. MICHAEL BROTHERS. FLORIDA. DEC.
7   1st cycle. Same as H7.6. Upperwing strongly recalls Lesser Black-backed, with dark outer greater coverts and darker inner primaries. MICHAEL BROTHERS. FLORIDA. DEC.
8   1st cycle. Same as H7.6 and H7.7. Darker body with smooth brown texture and largely dark tail with densely marked uppertail coverts suggest American Herring. MICHAEL BROTHERS. FLORIDA. DEC.
9   2nd cycle. Known-origin hybrid F02 from Appledore Island (offspring of Lesser Black-backed F05). Sighted several times in migration in 2009–2010, from Massachusetts to Florida. Would be identified as a Lesser without life history. BLAIR NIKULA. MASSACHUSETTS. DEC. 2009.
10 2nd cycle. Same as H7.9. Long wing projection and darker gray (2nd alternate) upperparts typical of Lesser Black-backed (but see *lusitanius* Yellow-legged Gull 23a.23). MITCHELL HARRIS. FLORIDA. FEB. 2010.
11 3rd cycle. The pallid gray upperparts are suspiciously pale for *graellsii*. Although some Lessers can have pinkish legs at this age, here leg color may be supportive of a hybrid. MICHAEL BROTHERS. FLORIDA. JAN.
12 3rd cycle. Same as H7.11. Obvious gray on the inner webs of p7–p9 (underwing). 3rd-cycle Lesser Black-backed is expected to have more black at this age. A faint white tongue tip is also visible on the inner web of p7. MICHAEL BROTHERS. FLORIDA. JAN.
13 4th cycle. Green F07. Same as H7.1. Upperparts too dark for American Herring and marginally paler than *graellsii*. Smallish head, reddish orbital, and grayish-yellow legs with pinkish feet. LAUREN KRAS. APPLEDORE ISLAND. JULY 2014.

H7 AMERICAN HERRING × LESSER BLACK-BACKED (APPLEDORE) GULL

14 4th cycle. Same as H7.12. 4th prebasic molt underway. 3rd-generation p10 shows medium-sized mirror, and p7–p10 show entirely black outer webs. New p4 with black on outer edges and complete band on growing p5. LAUREN KRAS. APPLEDORE ISLAND. JULY 2014.
15 Adult. Intermediate gray upperparts, too dark for American Herring and too pale for Lesser Black-backed. Variable head streaking common throughout the winter. AMAR AYYASH. FLORIDA. JAN.
16 Adult. Head markings and bill pattern strongly recall Lesser, but with paler upperparts, and leg color blended with pink and yellow. AMAR AYYASH. FLORIDA. JAN.
17 Adult. Grayish-blue upperparts and all-white tip to p10 recall *argentatus* with yellowish legs, but note obvious pinkish feet and pink streaks on tarsus. *Argentatus* averages coarser head and neck markings at this date and a stronger bill with duller base, but variable. BRUCE MACTAVISH. NEWFOUNDLAND. DEC.
18 Adult with Ring-billed Gull. Suggests a Lesser Black-backed, but with salmon-colored legs, not reasonably explained by variation and thus labeled a hybrid. AMAR AYYASH. FLORIDA. JAN.
19 Adult with a rather small and dark subadult Lesser Black-backed Gull (right). Note similar bill structure, pear-shaped head, and attenuated look to hind parts. ROY NETHERTON. FLORIDA. JAN.
20 Adult with American Herring Gull (left). A more compact individual, with subtle but noticeably darker gray upperparts, which is often the first indication of this hybrid. AMAR AYYASH. PENNSYLVANIA. DEC.
21 Adult (right) with adult American Herring (left) and 1st-cycle American Herring (center) in flock of Lesser Black-backed Gulls. TOM JOHNSON. PENNSYLVANIA. MARCH.

H7 AMERICAN HERRING × LESSER BLACK-BACKED (APPLEDORE) GULL

22 Adult (center) with American Herring (left) and Laughing Gulls (right). Adults paler than Laughing Gull are comfortably labeled as hybrids. AMAR AYYASH. FLORIDA. JAN.

23 Adult. May recall Great Black-backed × American Herring, but Appledore Gull usually has paler slate-gray upperparts, smaller head, slimmer bill, and longer wing projection (outer primaries growing here). Note leg color. BRUCE MACTAVISH. NEWFOUNDLAND. NOV.

24 Adult with American Herring (left). Commonly has streaking down hindneck and upper breast from fall through midwinter. Old p10. KEITH LEONARD. NEW JERSEY. NOV.

25 Adult leg color is often intermediate, with a variable blend of pink and yellow tones. However, some can be all yellow or all pink, particularly in breeding condition, beginning in late winter and early spring. WILL CHATFIELD-TAYLOR. FLORIDA. DEC.

26 Adult. Alternate plumage. Upperparts approaching pale *graellsii*, but dull salmon-colored legs evident. VERNON BUCKLE. LABRADOR. APRIL.

27 Adult. Alternate plumage. Reddish orbital. May show vivid yellow legs in high breeding condition. Identified as a Yellow-legged Gull at the time, this individual returned to the same beach the following Dec. with moderate head and neck markings and pinkish-yellow legs, which ruled out a much rarer Yellow-legged Gull. JEFFREY OFFERMANN. MASSACHUSETTS. APRIL.

H7 AMERICAN HERRING × LESSER BLACK-BACKED (APPLEDORE) GULL

28 Adult. Same as H7.27. Wingtip matches some nominate *michahellis*, but this race is largely unexpected in North America. Interestingly, there is a small mark on p4, something seldom seen on Appledore hybrid adults. The pattern is not diagnostic for Azores Gull, however, mostly due to the p9 mirror. The large p10 mirror and reduced black on the inner web of p8 (underside) are undesired in a vagrant Azores Gull. JEFFREY OFFERMANN. MASSACHUSETTS. APRIL.

29 Adult. Same as H7.15. A small percentage of hybrids lack a mirror on p9. Black pattern on p8–p9 is far too limited for Lesser Black-backed. AMAR AYYASH. FLORIDA. JAN.

30 Adult. Streaked hood pattern is found in some hybrids in winter, superficially recalling Azores Gull, but note white forehead and lores, and, above all, time of year. Overall the upper wingtip pattern is *smithsonianus*-like. BRUCE MACTAVISH. NEWFOUNDLAND. FEB.

31 Adult with American Herring. Darker gray upperparts and densely streaked head recall Lesser Black-backed, but reduced black on p8–p9 and noticeable white tongue tips on p6–p8 point directly to this hybrid. ALVAN BUCKLEY. NEWFOUNDLAND. JAN.

32 Adult. Same as H7.17. The combination of a long, all-white tip to p10, coupled with a large p9 mirror, is not typically found in this hybrid (p9 mirror averages small). Northern *argentatus* comes to mind, but in North America, that taxon should ideally have unmarked p5 and more gray on the inner web of p9 (see 21.36). BRUCE MACTAVISH. NEWFOUNDLAND. DEC.

33 Adult. Small p9 mirror expected in this hybrid. Reduced black on base of p8 and noticeable white tongue tips on p6–p7 rule out Azores and Lesser Black-backed Gull. Some hybrids have broad white tips to secondaries and tertials (tertial crescent at rest), which is not typically seen in Yellow-legged taxa. BLAIR DUDECK. NEWFOUNDLAND. FEB.

H8 OTHER HYBRIDS

The following plates feature a number of hybrids relevant to the ABA Area, although they are much rarer and less known than those discussed in the individual hybrid accounts. As with any hybrid, the identifications should be read as putative.

1 Adult Kelp × American Herring with Lesser Black-backed Gull (left). Regular features include long, grayish-yellow legs, blocky head, and long bill with swollen tip. RYAN SANDERSON. INDIANA. OCT.
2 Adult-type Kelp × American Herring with Black Skimmers. Upperparts typically paler than outer primaries (worn apicals). Orange orbital with large gonys spot and long, grayish-yellow legs. Great Black-backed shows considerably larger p10 mirror at this age. BRYAN WHITE. MISSISSIPPI. JUNE.
3 Adult Kelp × American Herring. Extensive black on p8–p10, with small, squarish p10 mirror. Few black streaks on primary coverts, and markings to p1 do not appear to be age-related, as this returning adult is at least in its 14th cycle here. AMAR AYYASH. INDIANA. MARCH.
4 Adult Kelp × American Herring (same as H8.1 and H8.3). Thin white tongue tips on p5–p7, presumably American Herring influence. Light head streaking, more densely concentrated on hindneck. Old p9–p10 and p8 growing at the end of Nov. AMAR AYYASH. INDIANA. NOV.
5 Subadult. Recalls Western Gull or Yellow-footed, but averages a smaller p10 mirror set farther back from the feather tip. Leg color is crucial for this ID. OSCAR JOHNSON. CHANDELEUR ISLANDS. MAY.
6 Adult Glaucous-winged × Slaty-backed. Suggests Slaty-backed, but with smudged head pattern and paler slate-gray upperparts. PAUL FRENCH. JAPAN. FEB.

7 Adult Glaucous-winged × Slaty-backed with Slaty-backed Gull (right). Outer primaries are slate-colored and paler than H8.6. Upperparts typically paler (or at least grayer) than pale-end Slaty-backed. JOHN MARTIN. JAPAN. MARCH.
8 Adult Glaucous-winged × Slaty-backed (bottom) with Slaty-backed Gulls. Eye color is variable in this hybrid. Note paler upperparts and largely white underside to p10. PAUL FRENCH. JAPAN. FEB.
9 Adult-type Glaucous-winged × Slaty-backed Gull (center) with "Slaty-backed Gulls." Most are undergoing their 2nd prebasic molt (1 year of age) and are identified largely by range. The desire to find and label such birds in North America in the summer months is an acquired taste. MALTE SEEHAUSEN. KAMCHATKA. JUNE.

10 Adult. Dark-winged individuals such as this with slightly darker upperparts are considered pure Glaucous-wingeds by some in eastern Asia. However, the paler eye and overall appearance look foreign from a North American perspective, and thus such birds are best labeled as hybrids or left unidentified. JOSH JONES. JAPAN. JAN.

11 Adult Glaucous-winged × Slaty-backed Gull. White trailing edge averages narrower than Slaty-backed, particularly on the inner primaries. Noticeably pale underside to p10. PAUL FRENCH. JAPAN. FEB.

12 Adult Slaty-backed × Vega Gull (right), 3rd-cycle Slaty-backed (left), and Black-tailed Gulls (foreground). Shows intermediate features between Vega and Slaty-backed Gull (including orbital color). Such birds have been recorded from Alaska to California. Average wingtip differences should be used, but some will defy ID. AMAR AYYASH. JAPAN. JAN.

13 Adult Slaty-backed × Glaucous Gull. The reduced pigment on the primaries and suitably dark upperparts are attributed to this pairing (for which there is precedent; see Glaucous account). DANIEL NELSON. ALASKA. JUNE.

14 Same as H8.13 with subadult Glaucous Gull (left). Slightly smaller and more refined than Glaucous, with all-pale underside to p10. DANIEL NELSON. ALASKA. JUNE.

15 Adult Glaucous × Vega Gull. Resembles Glaucous × American Herring Gull, but averages darker gray (based on museum specimens). Currently, there is no reliable way to distinguish these two hybrids in the field. Some "Asian" birds may have their origins in northern Alaska. YANN MUZIKA. JAPAN. FEB.

16 1st-cycle Glaucous × Vega Gull. Averages more diffuse bill pattern, with paler vent and paler undertail coverts compared to Glaucous × American Herring, but highly variable and not safely distinguished in the field, given current knowledge. ROB FRAY. JAPAN. FEB.

17 Adult Glaucous × European Herring (so-called Viking Gull). Wingtip pattern highly variable, but seen somewhat regularly are distinctive black segments limited to the outer webs of p8/p9–p10, with largely white-tipped outer primaries. See H3.30. DAN OWEN. IRELAND. FEB.

18 1st-cycle Glaucous × European Herring. Fairly distinctive, resembling some dark Glaucous Gulls (ssp. *hyperboreus*) from Europe, but with noticeably darker bases to the primaries, swarthy wing coverts, barred scapulars, and, importantly, a contrasting white eye-ring. AMAR AYYASH. ICELAND. JAN.

19 1st-cycle Glaucous × European Herring with similar-aged Iceland Gull (right). Appears glaucous-like in coloration, with Herring undertones in structure, face streaking, and patterning on the upperparts. Replaced upper scapulars show thin *argenteus*-like barring. AMAR AYYASH. ICELAND. JAN.

20 Adult Lesser Black-backed × Ring-billed Gull (Lesser Black-backed in background). A novel hybrid showing convincing intermediate features. The gray upperparts are too dark for Ring-billed (which never shows red on the bill). DELFIN GONZALEZ. SPAIN. JAN.

21 Same as H8.20. Fairly extensive black on p8–p10, which is attributed to Lesser Black-backed. Unmarked p5 and thin white tongue tips on p6–p7 presumably Ring-billed influence, but obscure. DELFIN GONZALEZ. SPAIN. JAN.
22 Adult Common × Ring-billed Gull. A long-staying individual banded in a Common Gull colony in Ireland (2004). Shows intermediate bill size and narrower black ring than Ring-billed. Intermediate dusky iris. Compact head, broader tertial crescent, and slightly longer wings attributed to Common Gull. JEFF B. HIGGOTT. IRELAND. DEC.
23 Same as H8.22. Wings more compact and slightly darker than Ring-billed. Fairly large p9–p10 mirrors and broad trailing edge to secondaries, with very thin tips to inner primaries, presumably Common influence. JEFF B. HIGGOTT. IRELAND. DEC.
24 Adult Laughing × Ring-billed with Ring-billed Gulls. Similar in size but with darker gray upperparts, slate-colored hood, and orange-red legs and bill. AMAR AYYASH. ILLINOIS. MARCH.
25 Same as H8.24. p10 mirror and black ring on the bill tip attributed to Ring-billed Gull. The hood is not solid dark, often showing a white-powdered appearance around the lores and face. AMAR AYYASH. ILLINOIS. APRIL.
26 Same as H8.24 and H8.25. Undergoing prebasic molt with remnants of hood pattern. The legs and bill remain dark orange, but the bill averages more dark along the cutting edge. AMAR AYYASH. ILLINOIS. AUG.
27 1st-cycle Laughing × Ring-billed with two similar-aged Ring-billed Gulls (back). Note the orange-brown bill base, quasi hood pattern, darker legs, and marginally darker upperparts (difficult to appreciate here). JANICE FARRAL. OHIO. MAY.
28 Adult Black-headed × Ring-billed Gull. Recalls Laughing × Ring-billed, but smaller, with paler gray upperparts and long white edge on the primary bases. KEITH MUELLER. CONNECTICUT. MARCH.
29 Adult Black-headed × Ring-billed with Ring-billed Gulls. The back of the hood is often held vertically, as commonly found in Black-headed Gull. Note extensive white on the underside of p10 and paler gray upperparts. JEANNETTE LOVITCH. NEW YORK. MARCH.
30 Adult Black-headed × Ring-billed Gull. Recalls hybrid with Laughing, but note small, tapered bill, pale gray upperparts, and prominent mirror on p9. KEVIN MCGOWAN. NEW YORK. FEB.
31 Adult Black-headed × Ring-billed. Absence of a white trailing edge points to *Chroicocephalus* species. The wingtip pattern very closely matches other presumed Black-headed × Ring-billed hybrids observed in alternate plumage. LOUIS BEVIER. MAINE. FEB.

APPENDIX

Given the steady increase of Lesser Black-backed Gulls in W North America, the question of whether some may be of Russian origin has become relevant. So far, there is no real indication of a transpacific expansion. The spread of Lesser Black-backed Gulls in the interior over the last few decades can be mapped from east to west. The earliest itinerants found wintering in places like the Texas Gulf Coast, the Colorado Front Range, and the Salton Sea do not show any collective features that point away from *L.f. graellsii*. Furthermore, any noteworthy movement from Asia would likely be detected in Alaska through British Columbia and south. This has not been the case, and this region remains the last to see an uptick in Lesser Black-backed numbers. Although we can't rule out the occasional waif crossing the Pacific, 1st-cycle individuals recorded in the West overall look like standard *graellsii* types, reinforcing that this is the subspecies currently found in the West. Observers should keep close watch on the growing numbers of Lesser Black-backed Gulls here, however, and be on the lookout for Heuglin's and Taimyr Gulls. The latter is the more likely of the two, as it migrates and winters farther east in E Asia, where it regularly reaches the Pacific Coast. Heuglin's Gull winters mainly to the Middle East, with small numbers noted in E Asia. Noteworthy reports of these two in North America, along with brief descriptions, are provided below. Also included is a short summary of Pallas's Gull, which was added to the ABA Checklist in 2020 (Pyle 2020), and Gray Gull, which was documented in 2023 in the ABA Area (Pranty, forthcoming).

1. Juvenile. Many remain largely juvenile into Jan. Long black wings, straight bill, and largely plain greater coverts, although some can show barring here. Eliminating Lesser Black-backed in North America unlikely. HANS LARSSON. OMAN. NOV.
2. 1st cycle. Similar to Lesser Black-backed, but note purely white underparts, whitish head, marked hindneck, and bicolored bill. All scapulars and a number of inner-wing coverts and tertials replaced. JAMES KENNERLEY. OMAN. FEB.
3. 1st cycle. Advanced individual with all upperparts replaced in postjuvenile molt on nonbreeding grounds (1st prealternate molt). Slate-gray scapulars and broad white fringes throughout. Visible primaries are juvenile. Yellowish legs and bill base. See A1.6 for open wing. BENGT BENGTSSON. SWEDEN. APRIL.
4. 1st cycle. Dark inner primaries. Some can show paler aspect and more Herring-like window. Pale body, often with spotted pattern on sides, and barring on wing linings. Whitish head and paling bill base average more advanced than Lesser Black-backed at this age. PETER KENNERLEY. UNITED ARAB EMIRATES. FEB.

A1 HEUGLIN'S GULL *Larus fuscus heuglini*

RANGE Breeds in N Russia in the open tundra between the Kola and Yamal Peninsulas. Winters primarily from the Red Sea east to the W Indian Ocean. Small numbers winter east to Japan, where it is far less common than Taimyr Gull. Found in a variety of habitats in winter, including coastlines, beaches, fish-processing plants, and landfills. Separation of paler *heuglini* and darker *taimyrensis* problematic; therefore, its true status in E Asia may be somewhat obscured (Ujihara & Ujihara 2019).

Reports in North America: A specimen collected at Icy Cape, in N Alaska, in 1921 was assigned to *heuglini* by Bailey (1948) but viewed as problematic by Gibson and Kessel (1997). An adult type observed at sea around 80 miles west-northwest of Gardner Pinnacles, in Northwestern Hawaiian Islands, on 9 October 2010 is attributed to *L.f. heuglini* (Pyle & Pyle 2017). An adult female collected on Shemya Island, in the Aleutians, on 15 September 2015 was attributed to Heuglin's Gull (Gibson & Byrd 2007). Photos of this specimen are convincing, with its white head in September, limited inner-primary molt (p1–p3 new, p5–p10 old), broad white trailing edge to the secondaries, black reaching primary coverts on outer web of p8, and thin white tongue tips on p6–p8 (Ayyash, pers. obs.). A small mirror on the inner web of p9 is also present. Bare-part color descriptions are also consistent with Heuglin's.

5   1st cycle. Postjuvenile molt limited in many, here restricted to scapulars. A fair number of Heuglin's remain surprisingly crisp well into March, pointing to their northern origins. Eliminating *graellsii* type unlikely in North America, but note white head with contrasting neck boa. HANNU KOSKINEN. FINLAND. APRIL.
6   1st cycle. Same as A1.3. Some replace a variable number of flight feathers in 1st prealternate molt. Tail and secondaries replaced, with active inner-primary molt. Such a sequence would be extraordinary in North America (but consider *intermedius/fuscus*, and compare 25.15 and 25.16). BENGT BENGTSSON. SWEDEN. APRIL.
7   2nd cycle. Very much resembles 3rd cycle, but visible (brownish) primaries are 2nd generation. 2nd alternate upperparts show brown cast. Such advanced birds can often show a small p10 mirror. 3rd-cycle Lesser Black-backed averages paler iris and blackish, adult-like outer primaries. HANNU KOSKINEN. FINLAND. MAY.
8   2nd cycle. Advanced bill and overall aspect recall 3rd-cycle large gull, and these types may lend themselves to closer identification. Fine markings often limited to hindneck, although some show substantially more marked head, similar to *graellsii*. See A1.10 for open wing. ARI KUUSELA. FINLAND. APRIL.

A1 HEUGLIN'S GULL

TAXONOMY Heuglin's Gull (or Siberian Gull) is treated as a subspecies of Lesser Black-backed Gull by a number of authorities and taxonomies (Adriaens et al. 2022; Gill et al. 2024), although several workers have assigned it species status (Olsen 2018; Harrison et al. 2021). Much more complex is the treatment of Taimyr Gull and its relationship to Heuglin's and Vega Gulls (see Taimyr Gull account). The American Ornithological Society doesn't have an official position on these forms, in part because Lesser Black-backed Gull isn't a confirmed breeder in mainland North America. Heuglin's has a longer evolutionary history and appears to be genetically more variable than the western Lesser Black-backeds, although it is said to show intrinsic gene flow restriction with nominate *fuscus*, with only a slight step between the two (Collinson et al. 2008b). In this book, it is treated as a form of Lesser Black-backed Gull with the caveat that a future split is likely, and Lesser Black-backed Gull here generally refers to *graellsii* (the expected subspecies in North America), while Heuglin's Gull refers to *L.f. heuglini*.

IDENTIFICATION Heuglin's Gull is the Russian equivalent of Lesser Black-backed Gull, with KGS 8–13. In North America, it's most likely to be overlooked as *graellsii* Lesser Black-backed Gull. Identifying single out-of-range individuals may not be possible, as there is much overlap with *graellsii* / *intermedius*.

9 2nd cycle. Noticeable brown greater coverts and 2nd-generation-like primaries. Separation from *graellsii* type would prove more difficult in North America, but note largely white body and head with thin, sharp streaking on lower hindneck. MICHIAKI UJIHARA. UNITED ARAB EMIRATES. FEB.

10 Presumed 2nd cycle. A likely sequence is p9–p10 1st alternate (replaced previous winter/spring on nonbreeding grounds), p1–p3 2nd basic (presumably replaced on breeding grounds and then suspended), and adult-like p4–p8 via active 2nd prealternate molt. Small mirror not rare on 2nd-generation p10, and also note white crescents on p5–p6. ARI KUUSELA. FINLAND. APRIL.

11 2nd cycle with nominate *fuscus* type (below) and European Herring wingtips (above). Although *intermedius* may be considered, the combination of p10 mirror, randomly replaced p8, and slate-gray upperparts point to Heuglin's. ARI KUUSELA. FINLAND. MAY.

12 Adult. Upperpart coloration most similar to *graellsii* or Dutch intergrade. Bill often appears longer and more slender. Dusky iris, white head, and fine streaking on lower hindneck. Note late primary molt (p6–p7 new, p10 old). MICHIAKI UJIHARA. UNITED ARAB EMIRATES. FEB.

However, at the population level, there are average differences that, when combined, may suggest a *heuglini* type. Complete molt in adults is later than *graellsii*, with primary molt regularly suspended before southbound migration to the wintering grounds (Gibbins 2004). Growing p9–p10 or having old p10 in January is not unusual (complete molt is slightly earlier in subadults, however). Heuglin's is as variable in body size as many other large gulls, but smaller individuals average a sleeker look with a smaller head and weaker gonys angle, averaging a thinner bill, which appears longish with a slight taper. In nonbreeding condition, the gonys spot may be smaller and restricted to the lower mandible, averaging more black around the tip, with a dull yellow bill showing a dull pink color to the base. The iris is more likely to show some dark peppering, or simply dark, in Heuglin's. Orbital ring is red. The mirror on p10 averages small and often appears set farther back from the feather tip, and therefore the black tip to p10 appears larger.

13 Adult. Similar to A1.12, but shorter and stockier bill, high head profile, and dark iris. Combination of thin streaks on lower hindneck and active primary molt should garner attention at this date if seen in North America. CHRIS GIBBINS. UNITED ARAB EMIRATES. JAN.
14 Adult. Primaries appear fully grown (see A1.18 for open wing). Dull yellowish bare parts and noticeable black on bill tip not uncommon in nonbreeding condition. Separation from *graellsii* type would prove problematic, especially without an open wing. OSCAR CAMPBELL. UNITED ARAB EMIRATES. FEB.
15 Adult with nominate *fuscus* (left). Heuglin's more robust with paler upperparts. These two undergo the longest migrations observed by large gulls and have similar late molts (as seen here). Still, a leg band is needed for positive identification of out-of-range birds. AMIR BEN DOV. ISRAEL. NOV.
16 Adult type. Inner secondaries and p9–p10 are still growing in late Jan., which is expected. Commonly shows black on p4 and, at times, on p3. White tongue tips on p5–p7 point away from other Lesser Black-backed taxa. CHRIS GIBBINS. UNITED ARAB EMIRATES. JAN.

A p9 mirror is reportedly only found in less than one-third of adults and is generally small (Olsen 2018). Heuglin's more commonly shows markings to p3/p4, although all of these features can also be found in *graellsii*. However, narrow white tongue tips are routinely found on p6–p7/p8, a feature not regularly expected in *graellsii*, and when present is very subtle (Adriaens et al. 2022). Head streaking appears to be finer in Heuglin's, with dark streaking often confined to the lower hindneck, with predominantly whitish head, while basic *graellsii* very often shows much streaking around the eyes, face, and nape. There is no single diagnostic field mark to separate *graellsii* from *heuglini*.

Some 1st- and 2nd-cycle out-of-range birds may be detected by molt pattern. Heuglin's can undergo an extensive postjuvenile molt from midwinter through early summer (1st prealternate). Advanced birds can have entirely replaced upperparts, secondaries, tail feathers, and a variable number of primaries. This molt is highly variable, however, and consideration should be given to an *intermedius* or *fuscus* type. Paler gray coloration, fine streaking on the lower hindneck, and the presence of a p10 mirror would be combined in an ideal candidate showing advanced molt. (See Adriaens et al. 2022 for various Heuglin's "types" classified by molt schemes.) Keep in mind that this "northern" gull will be relatively fresh and largely juvenile throughout much of its hatch year and into early January, and many 1st cycles retain very pristine upperparts and flight feathers, only replacing scapulars, much like *graellsii*, in which case positive identification will prove difficult, if not impossible.

17 Adult type. Like other members in the Lesser Black-backed group, seemingly adult individuals sometimes show black marks on the outer-primary coverts, often sharply demarcated near the tips, and not long or extensive diffuse streaks found in typical subadults. Relatively small p10 mirror and black to p4. MICHIAKI UJIHARA. UNITED ARAB EMIRATES. FEB.

18 Adult. Same as A1.14. Less common, a small mirror is sometimes found on p9. The relatively large white tongue tip on p7 and white head with finely spotted necklace point away from *graellsii* Lesser Black-backed Gull. OSCAR CAMPBELL. UNITED ARAB EMIRATES. FEB.

19 Adult. Red orbital, orange-red gonys spot. An expected wingtip with black extending to p3, entirely black outer webs to p8–p10, extensive black on the underside of p10 (which is still growing), and broad black subterminal band on p10. EMMANUEL NAUDOT. OMAN. FEB.

A2 TAIMYR GULL *taimyrensis*

RANGE Breeds on the Taimyr Peninsula in arctic Russia, ranging from the Lena Delta west to the Yamal Peninsula. Winters in E Asia from E China, where presumably it's the most abundant large gull in Hong Kong, through the coasts of the Yellow Sea, Korea, and east to the Pacific Coast of Japan. Wintering birds generally begin to arrive in Korea and Japan earlier than Vega, often with a noticeable presence by October. Favors tidal flats, riverways, beaches, and estuaries in winter, with less of a pelagic tendency compared to other similar large gulls.

Reports in North America: At least two 1st cycles from Midway Atoll have been attributed to *taimyrensis*, believed to be associated with the N Pacific winter of 2016–2017 (Peter Pyle, pers. comm.). An adult first found in Sonoma County, California, in December 2016 and continuing until December 2020 showed very promising features for *taimyrensis*, including late wing molt, yellowish leg color, finer head streaking, paler upperpart coloration, and striking structure to the culmen. A 1st cycle from Monterey County, California, discovered on 25 January 2017 at the Marina Landfill and continuing until at least 24 March 2017 was first found largely in juvenile plumage and later observed with postjuvenile scapulars and body feathers that strongly pointed to *taimyrensis* (although ruling out Heuglin's is difficult). An adult type found in Glasgow, Nova Scotia, on 4 January 2021 checked all the boxes for *taimyrensis* and remains, arguably, the only convincing example of this taxon in E North America (D'Entremont & MacDonald 2022).

TAXONOMY The taxonomic status of *taimyrensis* is unresolved and fundamentally perplexing. Some regard it a hybrid between *heuglini* and *vegae* (Ujihara & Ujihara 2019). Collinson et al. (2008b) proposed that *taimyrensis* be viewed as a member of the Lesser Black-backed Gull, in which case it's synonymized with *heuglini*, but also suggested that it may be a transient hybrid of *heuglini* and *vegae*. Olsen (2018) treated it as a subspecies of Heuglin's Gull (*L.h. taimyrensis*), in which case Heuglin's is not a member of the Lesser Black-backed group. Some consider *taimyrensis* to be the same thing as *birulai*: a western clinal form of Vega Gull (*L. vegae birulai*; Harrison et al. 2021). Finally, other workers maintain there is an apparent self-sustaining population of breeding birds at several colonies on Taimyr, which they've elevated to *L. taimyrensis* (van Dijk et al. 2011). Breeders in these colonies were said to show some variation in gray upperpart coloration and leg color, but with no evidence of assortative breeding. No firm position on the taxonomic status of *taimyrensis* is taken in this book, and it's important to stress that the plates that follow are of birds of unknown origin, photographed away from the breeding grounds.

IDENTIFICATION Taimyr Gull is a large white-headed gull with KGS 6–9 (Olsen 2018; more study needed). Adults are darker than American Herring Gull but sometimes only marginally so. At the dark

---

1   1st cycle. Overall appears frostier than Vega and Lesser Black-backed Gull, with scaly upperparts and pale fringes throughout. Shape often gives a certain "Lesser Black-backed" impression, but variable. Note pale base color to body with blotched and spotted pattern. ALVARO JARAMILLO. JAPAN. FEB.
2   1st cycle. Brownish upperparts recall Vega Gull, with similar notching on scapulars and barred greater coverts. The underparts average paler, however, and here the bird shows longer wing projection, imparting a more elegant and svelte appearance. Paling, pinkish bill base regularly seen. MICHIAKI UJIHARA. JAPAN. FEB.

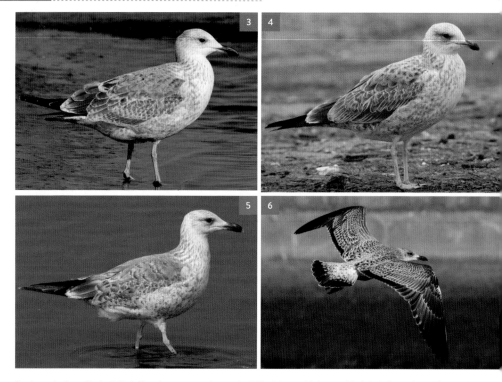

3 1st cycle. A stockier individual. First alternate scapulars can be fairly plain, grayish-brown with dark shaft streaks and faint anchors. Commonly has dark greater covert bases. MICHIAKI UJIHARA. JAPAN. FEB.
4 1st cycle. Relatively large, robust individual found in late Jan. and continuing until late March. Checks many boxes for *taimyrensis*, with paler body showing mealy texture to the sides, paling bill base, pale-tipped tertials, and scaly pattern to wing coverts. Separation from Heuglin's difficult. BRIAN SULLIVAN. CALIFORNIA. MARCH.
5 1st cycle. Often has plain tertials with noticeable white tips, seen well here, even at this date. Plain, medium gray bases to 1st alternate scapulars, tipped with thin brown anchors. Fairly lanky body and small head. MICHIAKI UJIHARA. JAPAN. APRIL.
6 1st cycle. Frosted appearance, appearing more scaly than Vega and Lesser Black-backed Gull. Pattern on greater coverts varies from plain brown (outers), becoming highly barred and notched (inners). Narrow black tailband with largely white base. MICHIAKI UJIHARA. JAPAN. FEB.
7 1st cycle. Suggests something intermediate between Vega and Lesser Black-backed Gull. Some show more zigzag barring on base of outer tail feathers. Much pale fringing on upperwing, and note regularly found pale subterminal capsules on the inner primaries (p1–p4). MICHIAKI UJIHARA. JAPAN. MARCH.
8 1st cycle. More heavily barred greater coverts on this individual, but becoming solid brown on the outer tract. Wings average longer and narrower compared to Vega, often with a subtle inner-primary window. MICHIAKI UJIHARA. JAPAN. MARCH.
9 1st cycle. Some have noticeable two-toned, venetian-blind pattern on inner primaries, suggesting a window, although typically subdued. Largely white uppertail coverts, whiter body, and commonly found blotchy spots on sides. The underwing averages paler than Vega. MICHIAKI UJIHARA. JAPAN. JAN.
10 2nd cycle. Note ambiguous leg color, which may be the first tip-off. Upperparts dark gray, between Heuglin's and Vega. Fairly compact head and bill, with long wings (in part due to posture here). MICHIAKI UJIHARA. JAPAN. APRIL.
11 2nd cycle. Out-of-context identification difficult. Greater coverts can range from mostly white to almost solidly brown, overlapping with Vega. Inner-primary window averages darker than Vega and paler than Heuglin's, but highly variable. MICHIAKI UJIHARA. JAPAN. APRIL.
12 3rd-/4th-cycle type. Old p9–p10 (small mirror) may be 3rd generation, and p5–p7/p8 new, likely 4th generation. Nonetheless, note late primary molt, intermediate gray upperparts, and leg color. MICHIAKI UJIHARA. JAPAN. JAN.
13 3rd cycle. Tail largely unmarked at this age. Pinkish-yellow legs helpful. Compact head and bill and long wings impart an elegant Lesser Black-backed impression. Others (male types) are bulky and render a Herring feel. MICHIAKI UJIHARA. JAPAN. JAN.
14 Adult. Vivid yellow legs and bill in breeding condition. Note blunt tip to bill. Body shape can be intermediate between Heuglin's and Vega Gull. Orange-red orbital and dusky iris. MICHIAKI UJIHARA. JAPAN. APRIL.
15 Adult. Upperparts darker than American Herring but paler than Lesser Black-backed. Long, yellow-pink legs, late primary molt (p9–p10 old), dusky iris, blunt bill tip, and fine head markings restricted to hindneck support this Asian taxon. JASON DAIN. NOVA SCOTIA. JAN.

extreme, it may resemble a paler Lesser Black-backed (i.e., *graellsii*) or Heuglin's Gull. Size and shape can be similar to Vega Gull, averaging darker, but some are more elegant and sleeker, displaying a Lesser Black-backed impression. On the wintering grounds, many Taimyr Gulls can be picked out by the orange-yellow quality to their legs, at times with hints of pink, and these are the most straightforward individuals in terms of identification. Birds sampled on the Taimyr Peninsula were said to show variation in leg color, with many being dull yellow, some grayish pink, and others pink, with no evidence of assortative breeding (van Dijk et al. 2011). Eye color is variable. Easternmost winter populations believed to average darker eyes, but more study needed. The orbital ring is red or orange red. The bill is blunt tipped or noticeably tapered at times, lacking a strong gonydeal expansion, and regularly has its orange-red gonys spot bleed over to the upper mandible on the wintering grounds (gonys spot averages smaller on Vega). Head streaking is variable, averaging coarser and more extensive than that of *heuglini* and at other times with a well-marked face and neck, but generally without the heavily hooded look seen on well-marked Vegas. Fine streaking on the lower hindneck, as found in some Heuglin's, is common. The wingtip shows a relatively large p10 mirror, at times with p9 mirror, and black subterminal markings average one to two primaries more than Vega, commonly to outer web of p4. It averages thinner white tongue tips than Vega on p5–p7, and sometimes completely absent. Primary molt is late but variable, approximately one month later than Vega's, with subadults completing molt earlier (Ujihara & Ujihara 2019). Molt should be used judiciously for identification, especially with vagrants.

16 Adult with Vega Gull (left). Marked head averages finer pattern than Vega. Unmarked head with streaking limited to hindneck more common in late winter and early spring. Thin bill with tapered tip recalls Heuglin's here. Note late primary molt at this date (p6 dropped, p7–p10 old). MICHIAKI UJIHARA. JAPAN. NOV.
17 Adult with presumed Heuglin's Gull (right). Although highly variable, Taimyr averages paler upperparts with more Herring build and a blunter bill tip, with ambiguous leg color. Both taxa observe late wing molt. NEIL DAVIDSON. JAPAN. NOV.
18 Obscure subject bird at right with Vega Gulls. May represent a dark Taimyr Gull or Heuglin's Gull. Open wing revealed growing p9 with mirror and fairly noticeable white tongue tips on p6–p7, which is more common in *taimyrensis*, but does not exclude Heuglin's. Primary molt often lags behind Vega Gull (p8 longest visible primary here). MASAMI YOSHIMURA. JAPAN. JAN.
19 Adult. Intermediate wingtip, somewhere between Heuglin's and Vega. A p9 mirror is common, and black on outer web of p8 regularly falls short of primary coverts. Unmarked p4 is less common. MICHIAKI UJIHARA. JAPAN. APRIL.

Identification of 1st cycles will prove more difficult than Heuglin's, due to Taimyr Gull having a more traditional and limited 1st prealternate molt. There are average differences, however, that can tip off North American observers to something not quite *graellsii* (see plates A2.1 and A2.2). Overall, 1st cycles appear frostier in juvenile plumage, possessing a scaly appearance to the upperparts, and can show a fairly prominent inner-primary window. The tailband can appear blacker than in Vega Gull, with whiter uppertail coverts, and the wing linings often have white undertones, generally paler than Vega's. The impression is a bird with dark upperparts and contrasting paler underparts. Size and shape are variable but recall a peculiar Lesser Black-backed with Herring tendencies (or vice versa).

The more formidable challenge when dealing with out-of-range 1st-cycle and adult birds is confidently eliminating Heuglin's. The poorly defined status and shifting treatments of Taimyr's (i.e., a good species, a subspecies of Heuglin's, a form of Vega, or a hybrid population) have produced varying field-identification criteria. Upperpart coloration in adults reportedly varies across parts of the breeding grounds, and this data requires analysis. For instance, van Dijk et al. (2011) describe upperpart coloration of breeding birds at Mys Vostochny, Taimyr Peninsula, as being "similar" to that of Yellow-legged Gulls from the Mediterranean (*L.m. michahellis*, KGS: 5–7), with reports of birds farther west from Medusa Bay being darker. Taimyr Gulls found wintering in Japan are often darker than Vega Gulls, which average darker than *michahellis*. Further, it appears adults wintering in South Korea and Japan display some observable differences with respect to average size and shape, eye color, and upperpart coloration, but more study is needed from known-origin birds. Ideally, size and shape, wingtip pattern, molt, leg color, upperpart coloration, head streaking, bill shape, and extent of red on the gonys would have to be combined in a vagrant for proper identification. Safely separating an out-of-range "dark-backed" Taimyr Gull with yellow legs from Heuglin's Gull is not likely (see plate A2.18). Finally, a "pink-legged" adult Taimyr Gull in North America would prove impossible to separate from a dark Vega Gull given current knowledge. These statements speak to the intermediate nature of this taxon but also reinforce the suspicion that Taimyr Gull may not be a valid taxon.

20 Adult. Relatively small p10 mirror. Noticeable white tongue tips on p5–p7 average broader than Heuglin's, but thinner than Vega. Often with markings to p4 and regularly to p3. MICHIAKI UJIHARA. JAPAN. MARCH.

21 Adults with Slaty-backed Gull (center). Note leg color, whitish head with streaking on hindneck and blunt bill tip. This individual shows fairly extensive black on the underside of p10 and moderate black on p9. MICHIAKI UJIHARA. JAPAN. APRIL.

22 Adults with overall similar wingtip patterns, showing less black on the underside of p9–p10. The individual on the right has narrow white tongue tips on p5–p7. Without leg color, such a bird may be suspected to be a dark Vega, although that taxon averages denser and blotchier neck markings. MICHIAKAI UJIHARA. JAPAN. APRIL.

A3 PALLAS'S (GREAT BLACK-HEADED) GULL *Ichthyaetus ichthyaetus*

RANGE Breeds patchily in E Europe, SW Russia, Kazakhstan, NW Mongolia, and west to NW China. Largest breeding numbers in central Asia but in steep decline. Winters mainly from Red Sea, Caspian Sea region, east to central Asia, and south to India. Recorded recently in Alaska, where a lone alternate-plumaged adult was found on Shemya Island, Aleutian Islands, 2–4 May 2019. Found dead 10 days later. The specimen was salvaged and is now housed at the University of Alaska Museum (UAM43000; Gibson et al. 2023). Only confirmed record in ABA Area.

TAXONOMY Monotypic with no known geographic variation. Four-cycle gull. Largest *Ichthyaetus* species.

1. 1st cycle. Largely in juvenile plumage, which is fairly plain and neat. Steep sloping forehead, long bill, and grayish greater coverts. ZBIGNIEW KAJZER. POLAND. AUG.
2. 1st cycle. First alternate scapulars regularly show large brown centers. White body and head, with contrasting dark eye mask that highlights white eye crescents. Variable bill color at this age. SUDIP GHOSH. INDIA. OCT.
3. 1st cycle. Can be fairly advanced in appearance, especially by late March. Yellowish bare parts and many replaced wing coverts expected. MICHALIS DRETAKIS. GREECE. MARCH.
4. Juvenile. Neat grayish-brown upperparts, white uppertail coverts, and narrow black tailband. Superficially recalls a jumbo juvenile Common Gull. AREND WASSINK. KAZAKHSTAN. SEPT.
5. 1st cycle. Narrow tailband evident, with largely white underparts. Wing linings often dotted with dark spots. PAVEL SIMEONOV. BULGARIA. MARCH.
6. 1st cycle. Striking long wings, large bill with black tip, dark eye mask, and blotchy hindneck. The resemblance to other *Ichthyaetus*, such as Mediterranean, is well displayed here. PAVEL SIMEONOV. ROMANIA. FEB.
7. 2nd-cycle type. Many are fairly advanced at this age (similar to 3rd cycle of four-cycle species). Much adult-like gray on the upperparts. Black wingtip, often with mirror, on p9 and p10. EMMANUEL NAUDOT. OMAN. FEB.
8. 3rd-cycle type. Age uncertain, but cleaner gray upperwing coverts with more white on outer hand compared to A3.7. Bill pattern typically has yellow base with black ring and variable red. PAUL SCHRIJVERSHOF. INDIA. FEB.
9. Adult type. Striking. Unmistakable due to size, combined with black hood, white eye crescents, and yellow legs. Upperparts pale gray. SUDIP SIMHA. INDIA. MARCH.

A3 PALLAS'S (GREAT BLACK-HEADED) GULL

IDENTIFICATION Unmistakable large, hooded gull, approaching size of Great Black-backed, but slightly smaller. Pale gray back (KGS: 4–5). Steeply sloping forehead, white eye crescents, and red orbital. Strong yellow bill with black ring and proximal red. Long, yellow legs. Much white on outer primaries with reduced black on p5–p10. Crown, lores, and chin are largely white in basic plumage with variable black around eye and nape.

10 Adult with advanced 1st cycle (right) and 2nd-/3rd-cycle type (left). Already fully black-headed. Note yellow legs. FAREED MOHMED. INDIA. FEB.
11 Adult type with Black-headed Gulls for scale. Noticeable eye crescents against mottled head. Difficult to confuse due to massive size. CSABA BARKOCZI. HUNGARY. NOV.
12 Adult. Wingtip shows reduced black with large p9–p10 mirrors, and large white wedge on outer hand. EMMANUEL NAUDOT. OMAN. FEB.
13 Adult type. 1st North American record. Initially found alive, then dead 10 days later. JACK WITHROW. SHEMYA ISLAND, ALEUTIANS; UNIVERSITY OF ALASKA MUSEUM. MAY.

A4 **GRAY GULL** *Leucophaeus modestus*

RANGE A unique colonial ground breeder, nesting mostly in the desolate Atacama Desert region of N Chile. One adult commutes to feed, typically to the Pacific Coast, returning at night to relieve its nest-guarding partner (Medrano et al. 2022). Preference for mole crabs, Peruvian anchovy, sardines, and nereid worms. Nesting sites are volatile, with regular shifts and abandonment documented (Aguilar-Pulido et al. 2021). Further, it's suspected that a number of colonies are unknown, given the discrepancy between adult numbers found on the coast and those surveyed at nest sites (Medrano et al. 2022). Breeding failure and extralimital reports are both correlated with strong El Niño years (Guerra, Aguilar, et al. 1988; Guerra, Fitzpatrick, et al. 1988; Howell & Dunn 2007; Aguilar et al. 2016).

Mostly coastal in the nonbreeding season, ranging from Ecuador to S Chile. Regularly found at sea exploiting the rich Humboldt Current. Vagrant records have increased recently from Argentina (2015), Brazil (2019), Panama (approximately five), Costa Rica (Slud 1967), Mexico (2019, 2021, and 2023), Guatemala (2022), Galapagos (2023), and a first confirmed record in the ABA Area from Florida (2023).

1 Largely juvenile with a few upper scapular and mantle feathers replaced (dark gray). Fairly plain, grayish-brown aspect with pale edging on upperparts. Warm, smooth texture to head, neck, and breast. GIANNIRA ALVAREZ ALFARO. CHILE. APRIL.
2 1st cycle. Moderate wear to upperparts, but largely juvenile. Long, thin bill remains black in all ages. White freckling on forehead helpful for ID. Uniformly brown underparts, unlike Laughing Gull, which is contrastingly pale here. ETIENNE ARTIGAU. CHILE. OCT.
3 1st cycle. Late 1st-cycle/early 2nd-cycle individuals commonly have replaced scapulars and wing coverts and, here, some replaced tertials. Recalls Heermann's, which has a thicker bill, with a pale base. PETER KENNERLEY. CHILE. OCT.

An unaccepted record from Louisiana (1987) was rejected due to concerns of a melanistic Laughing Gull not being eliminated (no open-wing photos; Shawneen Finnegan, pers. comm.). The first Galapagos record (adult, July) and the first US record (2nd- / 3rd-cycle type, June / Sept.), as well as Mexico's third record (Sept.), all occurred during the intense El Niño event of 2023. The Florida bird was discovered in early September 2023 (Walton County) but apparently was present at nearby Camp Helen State Park in early June 2023 (Marvin Bojo Reil, pers. comm.). Photos from June showed what appeared to be a 2nd alternate individual, which presumably, by September, had lost its pale hood and by then had completely worn tips to the secondaries and was undergoing its 3rd prebasic molt. Whether the June 2023 and the September 2023 sightings represent the same individual cannot be positively confirmed, due to a lack of needed detail on both the secondaries and the tail feathers, but the odds of being two different individuals are very low. The September 2023 bird appears suspiciously immature in appearance and overall resembles a Gray Gull undergoing its 2nd prebasic molt, but its leftmost tail feathers show a contrasting tailband pattern, and its outermost secondaries on the right wing show fairly prominent white edges, despite being extensively worn. This individual drifted west on two occasions, making it as far as Baldwin County, Alabama, in late November 2023 and back to Walton County, Florida; then back to Alabama through April 2024.

TAXONOMY Monotypic. No known geographic variation. Three-cycle "hooded" gull, believed to be most closely related to Dolphin Gull (*Leucophaeus scoresbii*; Pons et al. 2005).

IDENTIFICATION Adult upperparts are dark slate gray (KGS: 8.0–9.5) with slightly paler gray body. Brownish head in basic plumage, with contrasting white eye-ring, and often with white freckling on the forehead and around the base of the bill. Fairly long, thin black bill. Dark eyes and blackish legs. In alternate plumage, the head is largely white, fading to light gray on the nape and hindneck. Black primaries. At rest, note the broad white tips to the secondaries, which are the flashiest feathers in flight. The uppertail coverts are gray with a poorly defined black tailband and contrasting pale tips.

Second cycles (and presumably some 3rd-cycle types) have noticeably narrower white tips to the secondaries, with a more obvious brown cast to the wing coverts. Aging criteria need closer study, and caution should be taken when aging adults, as some may show old, faded coverts that are brownish (particularly the greater coverts), and some adults appear to regularly show blackish-brown primary coverts. Adults with worn secondary tips may resemble 2nd-cycle types, and aging may prove difficult if not impossible at times (as in the case of the first ABA record, Florida 2023). Second-cycle averages more black on the rectrices and shows a less demarcated tailband, although variable. Some apparent adults show extensive black approaching bases of tail feathers. First cycles should be compared to Heermann's Gull, (melanistic) Laughing Gull, and the much rarer Lava Gull (*Leucophaeus fuliginosus*).

4   1st cycle. Solid, dark brown tail. Grayish uppertail coverts and scapulars (1st alternate). The thin, all-black bill is distinctive and an important field mark. ARNOLD WIJKER. PERU. NOV.
5   Juvenile. Velvety grayish-brown underparts and solid brown wing linings. Blackish outer primaries and dark tail. DAVID F. BELMONT. PERU. MAY.

6   Start of 2nd prebasic molt (p1 dropped). Paler underwing and sides now, but note dark tail base, thin black bill, sooty-gray flanks, and head. GARY THOBURNS. PERU. DEC.
7   Start of 2nd prebasic molt (p1–p2 dropped), but all visible flight feathers, tail, and upperwing coverts are juvenile. Noticeable, but thin, white tips to secondaries. LINDA J. NUTTALL. CHILE. NOV.
8   2nd cycle. Retained secondaries (brown) and outermost primaries are juvenile, but otherwise a typical individual undergoing 2nd prebasic molt. Note whitish forehead and pale eye-ring. RICHARD BONSER. CHILE. APRIL.
9   2nd cycle. Grayish-white head in alternate plumage, fades to medium gray neck and breast. Aged by thin white tips to secondaries and blackish tertial centers. TRISTAN JOBIN. CHILE. NOV.
10 2nd cycle. Brownish flight feathers, unlike black on adult. Noticeably thin white tips to secondaries helpful in aging. PETER KENNERLEY. CHILE. OCT.

11 3rd-cycle type, undergoing 3rd prebasic molt (p1–p3 new, with narrow white tips). Retained outer primaries, worn secondaries and tail, presumably 2nd basic. RICHARD BONSER. CHILE. APRIL.
12 3rd-cycle type with 1st-cycle Laughing Gull (right). Gray Gull averages slightly bulkier body but with noticeably thinner bill. Blotchy grayish-black neck pattern expected in subadults. AMAR AYYASH. FLORIDA. SEPT.
13 3rd-cycle type (undergoing 3rd prebasic). Same as A4.12. A tricky individual to age due to extreme wear on secondaries, but photos of presumably the same individual from June 2023 reveal a 2nd-cycle type. Note contrasting tailband (all dark in juvenile). AMAR AYYASH. FLORIDA. SEPT.
14 2nd- or 3rd-cycle type. Aging difficult, but a definitive adult is ruled out by fairly narrow white tips to outer secondaries and extensive brown wash across upperparts. Noticeably thin black tailband expected more in older individuals, suggesting 3rd cycle, but study needed. ROGER AHLMAN. PERU. JUNE.
15 2nd- or 3rd-cycle type. Alternate plumage. Dark tail suggests 2nd cycle. Adult ruled out by narrow white tips to outer secondaries, and brownish greater coverts and primaries (although possibly due to fading?). GARY THOBURNS. PERU. DEC.
16 Adult type (basic plumage). Black primaries, broad white tips to secondaries, and demarcated brown "hood" pattern. Some residual dark blotching on back of head expected, even at this age. Note pale forehead and eye-ring. ETIENNE ARTIGAU. CHILE. JULY.

17 Adult. Alternate plumage. Distinctive. Grayish-white head fades to solid gray neck. Solid gray upperparts. Black secondaries with broad white secondary skirt. Black legs and black bill in all ages. DUSAN M. BRINKHUIZEN. PERU. SEPT.

18 3rd-cycle type or adult. Aging difficult, but adults commonly show blackish-brown primary coverts. White tips to outer secondaries are fairly broad and consistent with adult type. White eye-ring sometimes absent in basic plumage. BEN LOEHNEN. PERU. AUG.

19 Adult. Again, blackish-brown primary coverts are expected in adults. The combination of defined tailband and broad white tips to the secondaries is typical of an adult. MARK MADDOCK. PERU. SEPT.

20 Adult. Alternate plumage. Pale, grayish-white head with dark body. Solid gray underwing coverts with blackish primaries and broad, white trailing edge. PETER KENNERLEY. CHILE. OCT.

GLOSSARY

adult plumage The consummate plumage that a gull acquires. Once an adult, a gull will undergo no marked changes in appearance outside the expected alternate and basic plumages.

advanced Characterized by a molt, plumage, or other feature further along in development than expected (as opposed to retarded or delayed).

alternate plumage The plumage acquired by a second molt in a cycle, known as the "prealternate molt." This plumage alternates with basic plumage.

apicals White terminal spots on primary feathers. Typically used to refer to white tips to the outer primaries of adult white-headed gulls when perched. An all-white tip to p10 is the result of the mirror and apical completing merging.

aspect The general appearance of a gull, such as "breeding aspect" or "nonbreeding aspect," ascertained using a combination of bare-part condition and plumage.

axillaries Loosely known as the "armpits," these feathers sometimes have distinctive patterning that can assist in identification.

bare parts Any unfeathered part of a bird: feet, legs, bill, gape, and eyes.

basal Belonging to or near the base, such as the basal portion of the bill.

basic plumage The plumage acquired by the prebasic molt, believed to be homologous in all bird species.

bleached Describes feathers that are faded and colorless due to prolonged exposure to the sun's ultraviolet light.

body feathers Smaller contour feathers covering the body (as opposed to flight feathers, or remiges).

breeding aspect A breeding appearance, such as vivid bare-part coloration, coupled with alternate plumage.

Complex Alternate Strategy (CAS) Having two molts per cycle, along with an additional inserted molt in first cycle (preformative). Many of our small gulls adhere to this strategy.

Complex Basic Strategy (CBS) Having one molt per cycle (prebasic), along with an inserted molt in first cycle (preformative). No gull is known to exhibit this strategy.

conspecific A member of the same species.

coverts The layer of non-flight feathers that covers the bases of the flight feathers.

diagnostic An unequivocal trait or feature used for identification.

distal Situated or moving away from the body.

dorsal Relating to the upperside.

earspot A dark spot on the face, also known as an "auricular spot."

eye crescent Thin white feathers bordering the eye. Unlike eye-rings, eye crescents are incomplete arcs.

F1 hybrid First-generation hybrid produced by two distinct parentals.

flight feathers The longest feathers on the wing and tail, made up of primaries, secondaries, and rectrices. These stiff feathers are essential for flight.

formative plumage The plumage acquired by the preformative molt. Only found in the first cycle of some species.

gape The unfeathered corner of the mouth, where the upper and lower mandible meet.

ghost mirror A fairly indistinct and diffuse mirror, typically on p10, and often applied with second-cycle large gulls—namely, Kumlien's and Thayer's Gulls (see plate 32b.28).

gonydeal expansion The pointed angle near the tip of the lower mandible. Useful in assessing bill structure in larger gulls when viewed from the side. Also known as "gonydeal angle."

gonys spot The orange/red spot found near the tip of the lower mandible of some adult-type large gulls. Also known as a "gonydeal spot."

ground color The overall underlying color of a bird beneath its visible patterns. First-cycle gulls typically have either a dark or a pale ground color.

Humphrey-Parkes-Howell (H-P-H) A modified version of the Humphrey-Parkes system in which molt terminology reflects perceived homologies.

hybrid The offspring of a mixed-species pair or subsequent offspring of interbreeding hybrids.

inner web The proximal half of a feather. With gulls, used mostly to describe the inner part of a primary feather, separated from the outer part by the feather shaft. With respect to the primaries, the inner web is typically broader than the outer web. Also known as "inner vane."

intergrade Used here to denote the offspring of two subspecies (but note that intergrades can also be the result of gene flow between any two populations).

intraspecific Existing or occurring within a single species (i.e., intraspecific variation).

GLOSSARY

iris The outer part of the eye surrounding the pupil.

juvenile plumage The first plumage acquired (equivalent to first basic plumage). A "juvenile" is a bird strictly in juvenile plumage.

known-age bird A bird that has been banded or tracked from the nest and therefore whose hatch time is known.

known-origin bird A bird that has been banded or tracked from the nest and therefore whose natal origin is known.

Kodak Gray Scale (KGS) A range of swatches numbered from 0 to 19 used to describe relative darkness of a taxon's upperparts.

large white-headed gull Collectively used to refer to large four-cycle *Larus* gulls but also includes some three-cycle species.

Larus The largest (and most problematic) genus of gulls, which encompasses the large white-headed gulls.

mantle The upper back feathers. Sometimes used for the entire back and scapulars (referred to as "saddle" by some).

mirror A white subterminal spot typically found on p9/p10 in large gulls. At times, the entire tip to a primary may be white, in which case the mirror and apical become one. See plate 12.

molt The routine growth of new feathers, resulting in a new plumage. Molts are named using the prefix "pre" (i.e., prebasic, prealternate).

molt cycle The period between one prebasic molt and the next. Undergoing a complete molt is typically an annual process (sometimes referred to as "plumage cycle").

molt strategy The systematized molts that a taxon adheres to. Believed to be paralogous at the species level. See table 5.

monotypic Describes a species having no distinct subspecies, and sometimes used for a genus with a single member (i.e., monospecific).

morphology The size, shape, and structure of a bird, including its outward appearance.

M-pattern A dark "M" found across the upperwing, formed by dark outer primaries and ulnar bar. Found in several first-cycle small gulls.

nape The back of the head.

Nearctic Biogeographical region that includes much of North America.

nominate The subspecies first described when two or more subspecies exist. In the nominate subspecies, the second and third parts of the scientific name are the same (e.g., *Larus californicus californicus*).

nonbreeding aspect The overall appearance associated with nonbreeding birds, such as dull bare parts, coupled with basic plumage.

oiled Feathers that have come into contact with chemicals or oils in the water, usually appearing dark, dingy, and matted. Oiled feathers lose their waterproofing potential and impede flight.

orbital ring A circle of bare skin around the eye, brightest in high breeding condition.

outer web The distal half of a feather. With gulls, used mostly to describe the outer part of a primary feather, separated from the inner web by the feather shaft. With respect to the primaries, the outer web is typically narrower than the inner web. Also known as the "outer vane."

phenotype The observable characteristics of an individual, determined by its genotype but influenced by environmental factors.

polytypic Describes a species with two or more subspecies.

postjuvenile molt The first inserted molt, producing postjuvenile feathers. This term is used widely in Europe but is not used in the HPH system. Often used to reference replaced juvenile scapulars and wing coverts.

postocular line Thin dark line of feathers extending behind the eye.

prealternate molt The molt that produces alternate plumage. In gulls, this inserted molt is typically incomplete.

prebasic molt The molt that produces basic plumage. This molt is typically complete.

preformative molt A one-time inserted molt in first cycle, producing formative plumage. In gulls, this molt is typically incomplete (more study needed).

primaries The outer flight feathers on the wing, which are attached to the hand bone. Gulls have 10 primaries, numbered p1–p10 distally.

primary pattern The wingtip pattern to the outer primaries.

primary projection On the perched bird, this is the length the primaries extend beyond the tertials.

proximal Situated or advancing toward the body.

rectrices (*sing.* rectrix) Tail feathers. Gulls have 12 rectrices, numbered distally from the center as left r1–r6 and right r1–r6.

remiges (sing. remex)   Long, stiff flight feathers on the wings (i.e., primaries and secondaries).

retarded   Characterized by a molt, plumage, or other feature delayed in maturation or appearance (as opposed to advanced).

scapular crescent   Contrasting white tips found on the lowest scapulars of some adult-type gulls.

scapulars   Shoulder feathers that cover the top of the wing at rest and the back and inner wing in flight. These feathers streamline the gap between the wing and body and are often the first visible body feathers to be replaced in first cycle.

secondaries   The inner flight feathers attached to the arm bone. Gulls have 16–23 secondaries, depending on the species. These feathers are numbered proximally.

secondary skirt   Contrasting white tips to the secondaries that droop below the wing panel at rest.

Simple Alternate Strategy (SAS)   Having two molts per cycle, with only one inserted molt in first cycle. Most large gulls appear to adhere to this strategy (more study needed).

Simple Basic Strategy (SBS)   Having a single prebasic molt in each cycle, with no inserted molts in any cycle. Ivory Gull is the only gull believed to adhere to this strategy.

stress bar   A malformed, contrasting bar that runs across a series of feathers. Found primarily on flight feathers in juvenile gulls and can lead to premature feather breakage. Believed to be a response to stress during feather growth (also known as a "fault bar").

string-of-pearls   A series of two to five white tongue tips adjacent to the black subterminal bands on the outer primaries.

subadult   Used here to indicate an individual that is mostly adult-like in appearance but with some obvious imperfections (i.e., dark streaks on the primary coverts, brownish tones to the upperparts, dark blemishes on the tail, and/or a delayed bill pattern).

subspecies   A geographically and/or morphologically distinct population of a species (also known as "race").

subterminal band   A black band proximally adjacent to the tips of the outer primaries (typically from p5 to p10).

taxon (pl. taxa)   A taxonomic unit of any rank, whether a hybrid, subspecies, species, family, etc.

tertial crescent   Contrasting white tips found on the tertials of some adult-type gulls.

thayeri pattern   A distinctive pattern on a primary feather in which the dark medial band is incomplete on the inner web, with white tongue tip and mirror merged. Typically on p9, especially so in Thayer's Gull, but also found in a number of other taxa.

tongue   Used here to denote the pale inner web of a primary, which may have a white tip or terminate at a black subterminal band. See plate 12.

tongue tip   The distal edge of a tongue, adjacent to a black subterminal band. Tongue tips are typically white in adult-type gulls. See plate 12.

type   A distinctive form found in a species (e.g., a cinnamon-type California Gull).

ulnar bar   A dark bar that extends diagonally across the upper-wing coverts from the wrist to the innermost greater coverts (also known as a "carpal bar").

venetian-blind pattern   A contrasting two-toned color pattern of light and dark browns on the primaries of some younger gull taxa, particularly the outer primaries.

W-band   Often used to describe the pattern of the subterminal band on p5 in large white-headed gulls—namely, American Herring Gull.

wear   When feathers break down due to abrasion and other environmental factors.

window   A contrasting region of pale inner primaries. Generally used to describe the inner primary pattern of some first-cycle large gulls.

wing panel   The collection of upperwing coverts, specifically when viewing a gull at rest.

wing projection   On a perched bird, the length the primaries extend beyond the tail tip.

REFERENCES

Adriaens, P. 2012. Iceland and Kumlien's Gulls Photo Guide. *Birdwatch* 238:41–47.

Adriaens, P. 2013. A New Feature for Identifying Adult American Herring Gull. *Birding Frontiers* (blog). April. https://birdingfrontiers.files.wordpress.com/2013/04/a-new-feature-for-identifying-adult-american-herring-gull.pdf.

Adriaens, P., Alfrey, P., Gibbins, C., & Lopez-Velasco, D. 2020. Identification of Azores Gull. *Dutch Birding* 42 (5): 303–334.

Adriaens, P., Ayyash, A., & Muusse, M. 2023. Extensive Prealternate Molts in Peruvian Kelp Gulls. *Western Birds* 54 (3). http://doi.org/10.21199/WB54.3.6.

Adriaens, P., Dubois, J.P., Jiguet, F., & Muusse, M. 2022. *Gulls of Europe, North Africa, and the Middle East: An Identification Guide*. Princeton, NJ: Princeton University Press.

Adriaens, P., & Gibbins, C. 2016a. Field Identification of Russian Common Gull and Its Occurrence in Scotland. *Scottish Birds* 36 (1): 66–85.

Adriaens, P., & Gibbins, C. 2016b. Identification of the *Larus canus* Complex. *Dutch Birding* 38 (4): 1–64.

Adriaens, P., & Mactavish, B. 2004. Identification of Adult American Herring Gull. *Dutch Birding* 26:151–179.

Adriaens, P., Vercruijsse, H.J.P., & Stienen, E.W.M. 2012. Hybrid Gulls in Belgium: An Update. *British Birds* 105, no. 9 (September): 530–542.

Aguilar, R., Simeone, A., Rottmann, J., & Perucci, M. 2016. Unusual Coastal Breeding in the Desert-nesting Gray Gull (*Leucophaeus modestus*) in Northern Chile. *Waterbirds* 39 (1): 69–73.

Aguilar-Pulido, R., Catoni, C., Luna-Jorquera, G., Perucci, M., Dell'Omo, G., Zavalaga, C., & Simeonoe, A. 2021. Distribución, características y situación actual de las colonias reproductivas de la Gaviota garuma (*Leucophaeus modestus*) en el desierto de Atacama, Norte de Chile. *Revista Chilena de Ornitologia* 27 (1): 21–36.

Ahlstrom, C.A., van Toor, M.L., Woksepp, H., Chandler, J.C., Reed, J.A., Reeves, A.B., Waldenström, J., et al. 2021. Evidence for Continental-scale Dispersal of Antimicrobial Resistant Bacteria by Landfill-foraging Gulls. *Science of the Total Environment* 764:144551. https://doi.org/10.1016/j.scitotenv.2020.144551.

Alderfer, J., & Dunn, J.L. 2014. *National Geographic Complete Birds of North America*. 2nd ed. Washington, DC: National Geographic Society.

Alfrey, P., & Ahmad, M. 2007. Short-billed Gull on Terceira, Azores, in February–March 2003 and Identification of the "Mew Gull Complex." *Dutch Birding* 29:201–212.

Allard, K.A., Breton, A.R., Gilchrist, H.G., & Diamond, A.W. 2006. Adult Survival of Herring Gulls Breeding in the Canadian Arctic. *Waterbirds* 29 (2): 163–168.

Allard, K.A., Gilchrist, G.H., Breton, A.R., Gilbert, C.D., & Mallory, M.L. 2010. Apparent Survival of Adult Thayer's and Glaucous Gulls Nesting Sympatrically in the Canadian High Arctic. *Ardea* 98 (1): 43–50.

Altenburg, R.M., Meulmeester, I., Muusse, M., Muusse, T., & Wolf, P.A. 2011. Field Identification Criteria for Second Calendar-year Baltic Gull. *Dutch Birding* 33:304–311.

American Ornithological Society. 2017. *Proposal Set 2017-C(07)*. Washington, DC: American Ornithological Society.

American Ornithologists' Union. 1931. *Checklist of North American Birds*. 4th ed. Lancaster, PA: American Ornithologists' Union.

American Ornithologists' Union. 2003. Forty-fourth Supplement to the American Ornithologists' Union Check-list of North American Birds. *Auk* 120:925.

Anderson, C.M., Gilchrist, H.G., Ronconi, R.A., Shleper, K.R., Clark, D.E., Fifield, D.A., Robertson, G.J., & Mallory, M.L. 2020. Both Short and Long Distance Migrants Use Energy-minimizing Migration Strategies in North American Herring Gulls. *Movement Ecology* 8:26.

Anderson, C.M., Gilchrist, H.G., Ronconi, R.A., Shleper, K.R., Clark, D.E.C., Weseloh, C., Robertson, G.J., & Mallory, M.L. 2019. Winter Home Range and Habitat Selection Differs among Breeding Populations of Herring Gulls in Eastern North America. *Movement Ecology* 7 (8): 2–11.

Anderson, D.W. 1983. The Seabirds. In *Island Biogeography in the Sea of Cortez*, edited by T.J. Case & M.L. Cody, 246–264. Berkeley, CA: University of California Press.

Andrle, R.F. 1972. Another Probable Hybrid of *Larus marinus* and *L. argentatus*. *Auk* 89 (3): 669–671.

Arias, S. 2023. Observation of Oceanic Birds in the Costa Rican Pacific Ocean and the Cocos Marine Conservation Area. *Zeledonia* 27 (1): 202.

Arizaga, J. 2018. The Yellow-legged Gull in the Basque Region: Current Studies, Future Perspectives. Talk presented at the International Gull Meeting, Ruse, Bulgaria, 8–11 February.

Arizaga, J., Herrero, A., Galarza, A., & Hidalgo, J. 2011. First-year Movements of Yellow-legged Gulls (*Larus michahellis lusitanius*) from the Southeastern Bay of Biscay. *Waterbirds* 33 (4): 444–450.

Artukhin, Y.B. 2022. American Herring Gull *Larus smithsonianus* Coues, 1862 Is a New Species for the Avifauna of Russia. *Amurian Zoological Journal* 14 (2): 335–344.

Aubry, Y. 1984. First Nests of the Common Black-headed Gull in North America. *American Birds* 38:366–367.

Ayyash, A. 2013. Rethinking the Lesser Black-backed Gull in North America. *Birding* 45 (1): 34–41.

Ayyash, A. 2014. Chicago's Only Western Gull—Not. *Birding* 46 (1): 40–45.

Ayyash, A. 2015. In Search of Banded Gulls: A New North American Herring Gull Longevity Record. *North American Bird Bander* 40 (3): 92–93.

Ayyash, A. 2016. Bleached, Tattered, Bedraggled: Gulls in Summer. *Birding* 48 (3): 48–53.

Ayyash, A. 2017. Thayer's Gull: A Checkered History Near Its End? *Birding* 49 (3): 36–40.

Ayyash, A., Mactavish, B., Brown, D., & Muusse, M., coordinators. 2015. American Herring Gull (*smithsonianus*). Gull Research Organisation. Last updated 30 October 2015. https://gull-research.org/smithsonianus/ahgrings.html.

Ayyash, A., & Singh, L. 2023. Reconsidering Variation in Winter Adult Thayer's Gulls. *North American Birds* 74 (1): 30–47.

Baak, J., Patterson, A., Gilchrist, H., & Elliott, K. 2021. First Evidence of Diverging Migration and Overwintering Strategies in Glaucous Gulls (*Larus hyperboreus*) from the Canadian Arctic. *Animal Migration* 8:98–109.

Baggott, C. 2022. Algarve, Portugal: A Yellow-legged Gull Melting Pot. *Dutch Birding* 44 (4): 247–261.

Bailey, A.M. 1948. *Birds of Alaska*. Denver, CO: Colorado Museum of Natural History.

Bailey, S.D. 2008. Field Notes: The 2007 Breeding Season. *Meadowlark* 17 (1): 14.

Baird, P.H. 1994. Black-legged Kittiwake (*Rissa tridactyla*). In *The Birds of North America*, no. 92, edited by A. Poole & F. Gill, pp. 1–28. Philadelphia, PA: Academy of Natural Sciences; Washington, DC: American Ornithologists' Union.

Banks, R.C. 1986. Subspecies of the Glaucous Gull, *Larus hyperboreus* (Aves: Charadriiformes). *Proceedings of the Biological Society of Washington* 99:149–159.

Banks, R.C., & Browning, R.M. 1999. Questions about Thayer's Gull. *Ontario Birds* 17 (3): 124–130.

Barton, C. 2017. Irish Rare Bird Report 2016. *Irish Birds* 10: 545–578.

Beck, D.E. 1943. California Gull: A Comparative Plumage Study. *Great Basin Naturalist* 4:57–61.

Behle, W.H., & Selander, R.K. 1953. The Plumage Cycle of the California Gull (*Larus californicus*) with Notes on Color Changes of Soft Parts. *Auk* 70 (3): 239–260.

Bell, D.A. 1996. Genetic Differentiation, Geographic Variation and Hybridization in Gulls of the *Larus glaucescens-occidentalis* Complex. *Condor* 98:527–546.

Bell, D.A. 1997. Hybridization and Reproductive Performance in Gulls of the *Larus glaucescens-occidentalis* Complex. *Condor* 99:585–594.

Bent, A.C. 1921. Life Histories of North American Gulls and Terns. *Bulletin of the United States National Museum* 113:77.

Billerman, S.M. 2023. Treat *Larus smithsonianus* and *L. vegae* as Separate Species from Herring Gull *L. argentatus*. Proposal 2024-A-11. In *Proposal Set 2024-A*, edited by AOS Classification Committee—North and Middle America, pp. 84–91. American Ornithological Society. https://americanornithology.org/wp-content/uploads/2024/01/2024-A.pdf.

BirdLife International. n.d.-a. Swallow-tailed Gull *Creagrus furcatus*. BirdLife International. Accessed 13 February 2024. https://datazone.birdlife.org/species/factsheet/swallow-tailed-gull-creagrus-furcatus.

BirdLife International. n.d.-b. Heermann's Gull *Larus heermanni*. BirdLife International. Accessed 13 February 2024. https://datazone.birdlife.org/species/factsheet/heermanns-gull-larus-heermanni.

BirdLife International. n.d.-c. Red-legged Kittiwake *Rissa brevirostris*. BirdLife International. Accessed 13 February 2024. https://datazone.birdlife.org/species/factsheet/red-legged-kittiwake-rissa-brevirostris.

Blokpoel, H. 1987. *Atlas of the Breeding Birds of Ontario*. Waterloo, ON: University of Waterloo Press.

Blount, J.D., Surai, P.F., Nager, R.G., Houston, D.C., Moller, A.P., Trewby, M.L., & Kennedy, M.W. 2002. Carotenoids and Egg Quality in the Lesser Black-backed Gull *Larus fuscus*: A Supplemental Feeding Study of Maternal Effects. *Proceedings of the Royal Society B* 269:26–36. https://doi.org/10.1098/rspb.2001.1840.

Boertmann, D. 1994. An Annotated Checklist to the Birds of Greenland. *Meddelelser om Grønland, Bioscience* 38.

Boertmann, D. 2001. The Iceland Gull Complex in Greenland. *British Birds* 94:546–548.

Boertmann, D. 2008. The Lesser Black-backed Gull (*Larus fuscus*) in Greenland. *Arctic* 61 (2): 129–133.

Boertmann, D., & Frederiksen, M. 2016. Status of Greenland Populations of Great Black-backed Gull (*Larus marinus*), Lesser Black-backed Gull (*Larus fuscus*) and Herring Gull (*Larus argentatus*). *Waterbirds* 39 (1): 29–35.

Boertmann, D., & Huffeldt, N.P. 2013. *Seabird Colonies in the Melville Bay, Northwest Greenland*. Aarhus University, DCE—Danish Centre for Environment and Energy Scientific Report, no. 45. Aarhus, Denmark: Aarhus University.

Boertmann, D., Petersen, I.K., Nielsen, H.H., & Haase, E. 2019. *Ivory Gull Survey in Greenland*. Aarhus University, DCE—Danish Centre for Environment and Energy Scientific Report, no. 343. DCE. https://dce2.au.dk/pub/SR343.pdf.

Boertmann, D., & Rosing-Asvid, A. 2014. *Seabirds and Seals in Southwest Greenland*. Aarhus University, DCE—Danish Centre for Environment and Energy Scientific Report, no. 117. DCE. https://dce2.au.dk/pub/sr117.pdf.

Boertmann, D., & Rosing-Asvid, A. 2017. *Seabirds and Marine Mammals in Southeast Greenland II*. Aarhus University, DCE—Danish Centre for Environment and Energy Scientific Report, no. 215. Aarhus, Denmark: Aarhus University.

Bonomo, N. 2017. Mew Gulls in Connecticut: Past, Present and Future. *Connecticut Warbler* 37 (2): 32–47.

Bonser, R.H.C. 1995. Melanin and the Abrasion Resistance of Feathers. *Condor* 97 (2): 590–591.

Borrow, N., & Demey, R. 2001. *A Guide to the Birds of Western Africa*. Princeton, NJ: Princeton University Press.

Bowman, T.D., Stehn, R.A., & Scriber, K.T. 2004. Glaucous Gull Predation of Goslings on the Yukon-Kuskokwim Delta, Alaska. *Condor* 106 (2): 288–298.

Bray, R. 1943. Notes on the Birds of Southampton Island, Baffin Island, and Melville Peninsula. *Auk* 60:504–536.

Brazil, M. 2009. *Birds of East Asia: Eastern China, Taiwan, Korea, Japan, and Eastern Russia*. Help Field Guides. London: Christopher Helm.

Brewster, W. 1883. On an Apparently New Gull from Eastern North America. *Bulletin Nuttall Ornithological Club* 8:214–219.

Brinkley, E.S., & Patteson, B.J. 2001. Yellow-legged Gull (*Larus cachinnans* cf. *michahellis*) at Back Bay National Wildlife Refuge, Virginia Beach. *Raven* 72 (1): 66–75.

Brooke, R.K., Allan, D.G., Cooper, J., Cyrus, D.P., Dean, W.R.J., Dyer, B.M., Martin, A.P., & Taylor, R.H. 1999. Breeding Distribution, Population Size and Conservation of the Grey-headed Gull *Larus cirrocephalus* in Southern Africa. *Ostrich* 70:157–163.

Brooks, W.S. 1915. Notes on the Birds of East Siberia and Arctic Alaska. *Bulletin of the Museum of Comparative Zoology* 59:261–413.

Browning, M.R. 2022. Reassessment of Taxonomic Status of Thayer's Gull. *Dutch Birding* 44 (2): 137–144.

Burger, J., & Gochfeld, M. 1984. Seasonal Variation in Size and Function of the Nasal Salt Gland of the Franklin's Gull (*Larus pipixcan*). *Comparative Biochemistry and Physiology Part A: Physiology* 77 (1): 103–110.

Burger, J., & Gochfeld, M. 1996. Family Laridae (Gulls). In *Hoatzin to Auks*, edited by J. del Hoyo, A. Elliott, & J. Sargatal, pp. 686–699. Vol. 3 of *Handbook of the Birds of the World*. Barcelona: Lynx Edicions.

Burger, J., & Gochfeld, M. 2002. Bonaparte's Gull (*Chroicocephalus philadelphia*). In The Birds of North America Online, edited by A. Poole. Cornell Lab of Ornithology. https://doi.org/10.2173/bow.bongul.01.

Burger, J., & Gochfeld, M. 2020. Franklin's Gull (*Leucophaeus pipixcan*), version 1.0. In Birds of the World, edited by A.F. Poole. Cornell Lab of Ornithology. https://doi.org/10.2173/bow.fragul.01.

Byrd, G.V., & Williams, J.C. 1993. Red-legged Kittiwake (*Rissa brevirostris*). In The Birds of North America Online, edited by A. Poole. Cornell Lab of Ornithology. https://doi.org/10.2173/bow.relkit.01.

Campbell, W.R. 2007. New Longevity Record of a Glaucous-winged Gull from British Columbia. *Wildlife Afield* 4 (1): 78–80.

Chardine, J.W. 2002. Geographic Variation in the Wingtip Patterns of Black-legged Kittiwakes. *Condor* 104:687–693.

Charles, D. 2008. Ring-billed Gull Breeding with Common Gull on Copeland Islands Co. Down: The First Confirmed Breeding Record for Ring-billed Gull in the Western Palearctic. *Northern Ireland Bird Report* 28:122.

Chase, C.A. 1984. Gull Hybridization: California × Herring. *Colorado Field Ornithologists' Journal* 18 (2): 62.

Chen, J.Z., Yauk, C.L., Hebert, C., & Hebert, P.D.N. 2001. Genetic Variation in Mitochondrial DNA of North American Herring Gulls, *Larus argentatus*. *Journal of Great Lakes Research* 27 (2): 199–209.

Chesser, R.T., Billerman, S.M., Burns, K.J., Cicero, C., Dunn, J.L., Hernandez-Banos, B.E., Kratter, A.W., et al. 2021. Sixty-second Supplement to the American Ornithological Society's Check-list of North American Birds. *Ornithology* 138, no. 3 (July): 544–560.

Chesser, R.T., Burns, K., Cicero, C., Dunn, J., Kratter, A., Lovette, I., Rasmussen, P., et al. 2017. Fifty-eighth Supplement to the American Ornithological Society's Checklist of North American Birds. *Auk* 134:751–773.

Chin, J. 2020. The Seaside Heermann's Gulls. *Birding* 52 (2): 30–43.

Chisholm, G., & Neel, L.A. 2002. *Birds of the Lahontan Valley: A Guide to Nevada's Wetland Resources.* Reno, NV: University of Nevada Press.

Chu, P.C. 1998. A Phylogeny of the Gulls Inferred from Osteological and Integumentary Characters. *Cladistics* 14:1–43.

Collinson, J.M., Parkin, D.T., Knox, A.G., Sangster, G., & Svensson, L. 2008a. Genetic Relationships among the Different Races of Herring Gull, Yellow-legged Gull and Lesser Black-backed Gull. *British Birds* 94:523–528.

Collinson, J.M., Parkin, D.T., Knox, A.G., Sangster, G., & Svensson, L. 2008b. Species Boundaries in the Herring and Lesser Black-backed Gull Complex. *British Birds* 101 (7): 340–363.

Cotter, C., Rail, J.F., Boyne, A.W., Robertson, G.J., Weseloh, D.V.C., & Chaulk, K.G. 2012. *Population Status, Distribution, and Trends of Gulls and Kittiwakes Breeding in Eastern Canada 1998–2007.* Canadian Wildlife Services Occasional Papers, no. 120. Government of Canada. https://publications.gc.ca/collections/collection_2013/ec/CW69-1-120-eng.pdf.

Coues, E. 1862. Revision of the Gulls of North America: Based upon Specimens in the Museum of the Smithsonian Institution. *Proceedings of the Academy of Natural Sciences of Philadelphia* 1862:291–312.

Coulson, J. 2011. *The Kittiwake.* Poyser Monographs. London: T. & A.D. Poyser.

Cramp, S., & Simmons, K.E.L., eds. 1983. *Handbook of the Birds of Europe, the Middle East and North Africa*, vol. 3. New York: Oxford University Press.

Crochet, P.A., Bonhomme, F., & Leberton, J.D. 2000. Molecular Phylogeny and Plumage Evolution in Gulls (Larini). *Journal of Evolutionary Biology* 13:47–57.

Crochet, P.A., Lebreton, J.D., & Bonhomme, F. 2002. Systematics of Large White-headed Gulls: Patterns of Mitochondrial DNA Variation in Western European Taxa. *Auk* 119 (3): 603–620.

Cutright, N.J., Harriman, B.R., & Howe, R.W. 2006. *Atlas of the Breeding Birds of Wisconsin.* Waukesha, WI: Wisconsin Society for Ornithology.

Czaplak, D. 1990. Checklist S8866530. eBird. https://ebird.org/checklist/S8665520.

Davis, S.E., Maftei, M., & Mallory, M.L. 2016. Migratory Connectivity at High Latitudes: Sabine's Gulls (*Xema sabini*) from a Colony in the Canadian High Arctic Migrate to Different Oceans. *PLoS ONE* 11 (12): e0166043. https://doi.org/10.1371/journal.pone.0166043.

Day, R.H., Stenhouse, I.J., & Gilchrist, H.G. 2001. Sabine's Gull (*Xema sabini*). In The Birds of North America Online, edited by A. Poole. Cornell Lab of Ornithology. https://doi.org/10.2173/bow.sabgul.01.

De Almeida, B.J., Rodrigues, R., Mizrahi, D., & Lees, A.C. 2013. A Lesser Black-backed Gull *Larus fuscus* in Maranhão: The Second Brazilian Record. *Brazilian Journal of Ornithology* 21 (4): 213–216.

Dee, T. 2018. *Landfill: Notes on Gull Watching and Trash Picking in the Anthropocene.* London: Chelsea Green Publishing.

De Knijff, P., Helbi, A., & Liebers, D. 2005. The Beringian Connection: Speciation in the Herring Gull Assemblage of North America. *Birding* 37 (4): 402–411.

Del Hoyo, J., Elliott, A., & Sargatal, J., eds. 1996. *Hoatzin to Auks.* Vol. 3 of *Handbook of the Birds of the World*, edited by J. Del Hoyo, A. Elliott, J. Sargatal & D.A. Christie. Barcelona: Lynx Edicions.

Denlinger, L.M. 2006. *Alaska Seabird Information Series.* Anchorage, AK: US Fish and Wildlife Service.

D'Entremont, A., & MacDonald, A. 2022. Nova Scotia's First Taimyr Gull. *Nova Scotia Birds* 63 (2): 26–29.

Dickinson, E.C., & Remsen, J.V., Jr. 2013. *The Howard and Moore Complete Checklist of the Birds of the World*, 4th ed. Vol. 1. Eastbourne: Aves Press.

Dittman, D.L., & Cardiff, S.W. 2005. Origins and Identification of Kelp × Herring Gull Hybrids. *Birding* 37:266–276.

Dorogoi, I.V., & Zelenskaya, L.A. 2016. Glaucous-winged Gull *Larus glaucescens*: New Nesting Occurrence in the Magadan Region. *Russian Ornithological Zhurn* 25 (1309): 2528–2531.

Drummond, B.A., Orben, R.A., Christ, A.M., Fleishman, A.B., Renner, H.M., Rojek, N.A., & Romano, M.D. 2021. Comparing Non-breeding Distribution and Behavior of Red-legged Kittiwakes from Two Geographically Distant Colonies. *PLoS ONE* 16 (7): e0254686. https://doi.org/10.1371/journal.pone.0254686.

Dubois, P.J. 2001. Atlantic Islands Yellow-legged Gulls: An Identification Gallery. *Birding World* 14:293–304.

Dwight, J. 1917. The Status of *Larus thayeri*, Thayer's Gull. *Auk* 34:413–414.

Dwight, J. 1922. Description of a New Race of the Lesser Black-backed Gull, from the Azores. *American Museum Novitates* 44:1–2.

Dwight, J. 1925. The Gulls (Laridae) of the World: Their Plumages, Moults, Variations, Relationships, and Distribution. *Bulletin of the American Museum of Natural History* 52:63–402.

Edwards, J.L. 1935. The Lesser Black-backed Gull in New Jersey. *Auk* 52:85.

Elias-Valdez, A., Velarde, E., Medina-Quej, A., Castro-Perez, J.M., Navarro, J., & Rosas-Luis, R.

2023. Feeding Ecology of Coexisting Heermann's Gull (*Larus heermanni*) and Elegant Tern (*Thalasseus elegans*) Chicks, Based on Stable Isotope Measurements. *Marine Ecology Progress Series* 712:101–111.
Ellery, T., & Salgado, J.F. 2018. First Confirmed Record of Belcher's Gull *Larus belcheri* for Colombia with Notes on the Status of Other Gull Species. *Conservacion Colombiana*, no. 25, 51–55.
Ellis, J.C., Bogdanowicz, S.M., Stoddard, M.C., & Clark, W.L. 2014. Hybridization of a Lesser Black-backed Gull and Herring Gulls in Eastern North America. *Wilson Journal of Ornithology* 126 (2): 338–345.
Erard, C., Guillou, J.J., & Mayaud, N. 1984. Sur l'identité spécifique de certains laridés nicheurs au Sénégal. *Alauda* 52:184–188.
Ewins, P.J., Blokpoel, H., & Ludwig, J.P. 1992. Recent Extensions of the Breeding Range of Great Black-backed Gulls (*Larus marinus*) in the Great Lakes of North America. *Ontario Birds* 10 (2): 64–71.
Ewins, P.J., & Weseloh, D.V. 2020. Little Gull (*Hydrocoloeus minutus*). In The Birds of the World, edited by S.M. Billerman. Cornell Lab of Ornithology https://doi.org/10.2173/bow.litgul.01.
Fareh, M., Maire, B., Laidi, K., & Ennoury, A. 2020. Les oiseaux rares au Maroc: rapport de la Commission d'homologation marocaine. *Go-South Bulletin* 17:104–120.
Finch, D.W. 1978. Black-headed Gulls Nesting in Newfoundland. *American Birds* 32: 312.
Flint, V.E., Boehme, R.L., Kostin, Y.V., & Kuznetsov, A.A. 1984. *A Field Guide to Birds of the USSR*. Princeton, NJ: Princeton University Press.
Flores-Martínez, J.J., Herrera, L.G., Arroyo-Cabrales, J., Alarcón, I., & Ruiz, E.A. 2015. Seasonal Dietary Differences of the Yellow-footed Gull in Isla Partida Norte, Gulf of California, Mexico. *Revista Mexicana de Biodiversidad* 86: 412–418.
Forbes, G.K., Robertson, K., Ogilvie, C., & Seddon, L. 1992. Breeding Densities, Biogeography, and Nest Depredation of Birds on Igloolik Island., N.W.T. *Arctic* 45 (3): 295–303.
Foxall, R.A. 1979. Presumed Hybrids of the Herring Gull and the Great Black-backed Gull: A New Problem of Identification. *American Birds* 33 (6): 838.
Garcia-Barcelona, S., Senfeld, T., Shannon, T.J., Collinson, J.M., Edelaar, P., Aldalur, A., Martín, G., García-Mudarra, J.L., & Juste, J. 2021. Morphological and Genetic Analyses Reveal the First Mediterranean Occurrence of American Herring Gull *Larus smithsonianus*. *Revista Catalana d'Ornitologia* 37:1–9.
Garner, M., & Quinn, D. 1997. Identification of Yellow-legged Gulls in Britain, Part 1. *British Birds* 90:25–62.

Garvey, M.P., & Iliff, M.J. 2012. Sixteenth Report of the Massachusetts Avian Records Committee. Massachusetts Avian Records Committee. https://maavianrecords.com/annual-reports/16th/.
Gaston, A.J., & Decker, R. 1985. Interbreeding of Thayer's Gull, *Larus thayeri*, and Kumlien's Gull, *Larus glaucoides kumlieni*, on Southampton Island, Northwest Territories. *Canadian Field-Naturalist* 99:257–259.
Gaston, A.J., & Elliot, R.D. 1990. Kumlien's Gull, *Larus glaucoides kumlieni*, on Coats Island, Northwest Territories. *Canadian Field-Naturalist* 104:477–479.
Gaston, A.J., & Ouellet, H. 1997. Birds and Mammals of Coats Island, N.W.T. *Arctic* 50 (2): 101–118.
Gaston, A.J., Smith, S.A., Saunders, R., & Storm, G.I. 2007. Birds and Marine Mammals in Southwestern Fox Basin, Canada. *Polar Record* 43:33–47.
Gay, L., Bell, D.A., & Crochet, P.A. 2005. Additional Data on Mitochondrial DNA of North American Large Gull Taxa. *Auk* 122 (2): 684–688.
George, J. 2005. Salute to a Fine Feathered Friend: Beloved Ahab, the One legged Gull, Returned to San Rafael Marina Year after Year. SFGate. https://www.sfgate.com/homeandgarden/article/salute-to-a-fine-feathered-friend-beloved-ahab-2558666.php.
Gibbins, C.N. 2004. Is It Possible to Identify Baltic and Heuglin's Gulls? *Birding Scotland* 7 (4): 153–186.
Gibbins, C.N., & Baxter, P. n.d. Baltic and Heuglin's Gull: Photo Essay. SurfBirds. Accessed 13 February 2024. https://www.surfbirds.com/mb/Features/gulls/baltic-heuglins-gulls.html#.
Gibbins, C.N., & Garner, M. 2013. The In-between Age of the In-between Gull: Identification of Second-winter Thayer's Gull. Birding Frontiers. http://www.gull-research.org/papers/papers5/cg_mg_identification-of-second-winter-thayers.pdf.
Gibson, D.D., & Byrd, G.V. 2007. *Birds of the Aleutian Islands, Alaska*. Cambridge, MA: Nuttall Ornithological Club; Washington, DC: American Ornithologists' Union.
Gibson, D.D., Heinl, S.C., Tobish, T.G., Lang, A.J., Withrow, J.J., Decicco, L.H., Hajdukovich, N.R., & Scher, R.L. 2023. Fifth Report of the Alaska Checklist Committee, 2018–2022. *Western Birds* 54 (2): 98–116.
Gibson, D.D., & Kessel, B. 1997. Inventory of the Species and Subspecies of Alaska Birds. *Western Birds* 28 (2): 45–95.
Gilg, O., Istomina, L., Heygster, G., Strøm, H., Gavrilo, M.V., Mallory, M.L., Gilchrist, G., Aebischer, A., Sabard, B., Huntemann, M., et al. 2016. Living on the Edge of a Shrinking Habitat: The Ivory Gull, *Pagophila eburnea*, an Endangered Sea-ice Specialist. *Biology Letters* 12:11. https://doi.org/10.1098/rsbl.2016.0277.

Gilg, O., Størm, H., Aebischer, A., & Gavrilo, M. 2010. Post-breeding Movements of Northeast Atlantic Ivory Gull *Pagophila eburnea* Populations. *Journal of Avian Biology* 41 (5): 532–542.

Gilg, O., van Bemmelen, R.S.A., Lee, H., Park, J.-Y., Kim, H.-J., Kim, D.-W., Lee, W.Y., Sokolovskis, K., & Solovyeva, D.V. 2023. Flyways and Migratory Behaviour of the Vega Gull (*Larus vegae*), a Little-known Arctic Endemic. *PLoS ONE* 18 (2): e0281827. https://doi.org/10.1371/journal.pone.0281827.

Gill, F.D., Donsker, D., & Rasmussen, P. 2024. IOC World Bird List, version 14.1. https://doi.org/10.14344/IOC.ML.14.1.

Godfrey, W.E. 1986. *The Birds of Canada*, rev. ed. Ottawa: National Museum of Canada.

Goethe, F. 1986. Zur Biologie, inbesondere Ethographie der Polarmöwe (*Larus glaucoides* Meyer, 1822). *Annalen des Naturhistorischen Museums in Wien. Serie B für Botanik und Zoologie* 88 / 89:112–146.

Goncalves, A. 2010. Mixed Species Breeding by Great Black-backed Gull *Larus marinus*: First Records for Portugal. *Anuario Ornitologico* 7:126.

Goodwin, T. 1984. *The Biochemistry of the Carotenoids*, vol. 2, *Animals*. New York: Chapman and Hall.

Gosselin, M., & David, N. 1975. Field Identification of Thayer's Gull (*Larus thayeri*) in Eastern North America. *American Birds* 29 (6): 1059–1162.

Gosselin, M., David, N., & Laporte, P. 1986. Hybrid Yellow-legged Gull from the Madeleine Islands. *American Birds* 40 (1): 58–60.

Grant, P.J. 1986. *Gulls: A Guide to Identification*, 2nd ed. San Diego: Academic Press.

Guerra, C.G., Aguilar, R.E., & Fitzpatrick, L.C. 1988. Water Vapor Conductance in Gray Gulls (*Larus modestus*) Eggs: Adaptation to Desert Nesting. *Colonial Waterbirds* 11 (1): 107–109.

Guerra, C.G., Fitzpatrick, L.C., Aguilar, R., & Venables, B.J. 1988. Reproductive Consequences of El Niño–Southern Oscillation in Gray Gulls (*Larus modestus*). *Colonial Waterbirds* 11 (2): 170–175.

Gull Research Organisation. 2014. Lesser Black-backed Gull (*graellsii / intermedius*). Gull Research Organization. Last updated 12 October 2014. www.gull-research.org/lbbg1cy/0start.html.

Gull Research Organisation. 2017. Lesser Black-backed Gull (*graellsii* & *intermedius*). Gull Research Organization. Last updated 14 November 2017. https://www.gull-research.org/lbbg/2cyjan.html.

Gull Research Organisation. n.d. Gull Research Organisation (website). Accessed 31 January 2024. https://gull-research.org.

Gustafon, M.E., & Peterjohn, B.G. 1994. Adult Slaty-backed Gulls: Variability in Mantle Colour and Comments on Identification. *Birding* 26:243–249.

Gutowsky, S.E., Davis, S.E., Maftei, M., & Mallory, M.L. 2021. Flexibility in Migratory Strategy Contrasts with Reliance on Restricted Staging and Overwintering Grounds for Sabine's Gulls from the Canadian High Arctic. *Animal Migration* 8 (1): 84–97.

Gutowsky, S.E., Hipfner, J.M., Maftei, M., Boyd, S., Auger-Méthé, M., & Mallory, M.L. 2020. First Insights into Thayer's Gull *Larus glaucoides thayeri* Migratory and Overwinter Patterns along the Northeast Pacific Coast. *Marine Ornithology* 48:9–16.

Hailman, J.P. 1964a. Breeding Synchrony in the Equatorial Swallow-tailed Gull. *American Naturalist* 98:79–83.

Hailman, J.P. 1964b. The Galapagos Swallow-tailed Gull Is Nocturnal. *Wilson Bulletin* 76 (4): 347–354.

Hallgrimsson, G.T., van Swelm, N.D., Gunnarsson, H.V., Johnson, T.B., & Rutt, C.L. 2011. First Two Records of European-banded Lesser Black-backed Gull *Larus fuscus* in America. *Marine Ornithology* 39:137–139.

Hampton, S. 2013. Status of the Iceland Gull at the Yolo County Central Landfill. *CVBC Bulletin* 16 (1): 1–12.

Harris, M.P. 1970. Breeding Ecology of the Swallow-tailed Gull, *Creagrus furcatus*. *Auk* 87 (2): 215–243.

Harrison, P., Perrow, M.R., & Larsson, H. 2021. *Seabirds: The New Identification Guide*. Barcelona: Lynx Edicions.

Hatch, S.A., Byrd, G.V., Irons, D.B., & Hunt, G.L. 1993. *Status and Ecology of Kittiwakes (Rissa tridactyla and R. brevirostris) in the North Pacific*. Ottawa: Canadian Wildlife Service.

Helberg, M., Systad, G.H., Birkeland, I., Lorentzen, N.H., & Bustnes, J.O. 2009. Migration Patterns of Adult and Juvenile Lesser Black-backed Gulls *Larus fuscus* from Northern Norway. *Ardea* 97:281–286.

Henshaw, B. 1992. Ontario Round-up. *Birders Journal* 1:167–175.

Henshaw, H.W. 1884. On a New Gull from Alaska. *Auk* 1 (3): 250–252.

Higgins, P.J., & Davies, S.J. 1996. *Handbook of Australian, New Zealand, and Antarctic Birds*. United Kingdom: Oxford University Press.

Hoffman, W., Wien, J.A., & Scott, J.M. 1978. Hybridization between Gulls (*Larus glaucescens* and *L. occidentalis*) in the Pacific Northwest. *Auk* 95:441–458.

Holt, D.W., Lortie, J.P., Nikula, B.J., & Humphrey, R.C. 1986. First Record of Common Black-headed Gulls Breeding in the United States. *American Birds* 40 (2): 204–206.

Hørring, R., & Salomonsen, F. 1941. Further Records of Rare or New Greenland Birds. *Meddelelser om Grønland* 131, no. 5.

Howell, S.N.G. 1999. Shades of Gray: The Catch 22 of Thayer's Gull. Birders Journal 7:305–309.

Howell, S.N.G. 2000. Letters: Moult and Age of First Year "White-winged" Gulls. British Birds 93 (2): 99.

Howell, S.N.G. 2001a. Molt of the Ivory Gull. Waterbirds 24 (3): 438–442.

Howell, S.N.G. 2001b. Feather Bleaching in Gulls. Birders Journal 10:198–208.

Howell, S.N.G. 2003. Shades of Gray: A Point of Reference for Gull Identification. Birding 35:32–37.

Howell, S.N.G. 2010. Molt in North American Birds. Peterson Reference Guide Series. Boston, MA: Houghton Mifflin Harcourt.

Howell, S.N.G. Forthcoming. First Mexican Records of Swallow-tailed Gull Creagrus furcatus. Cotinga.

Howell, S.N.G., & Corben, C. 2000a. Molt Cycles and Sequences in the Western Gull. Western Birds 31:38–49.

Howell, S.N.G., & Corben, C. 2000b. Retarded Wing Molt in Black-legged Kittiwakes. Western Birds 31:123–125.

Howell, S.N.G., Correa, S.J., & Garcia, J. 1993. First Records of the Kelp Gull in Mexico. Euphonia 2:71–80.

Howell, S.N.G., & Dunn, J. 2007. A Reference Guide to Gulls of the Americas. Peterson Reference Guide Series. Boston, MA: Houghton Mifflin.

Howell, S.N.G., & Elliott, M.T. 2001. Identification and Variation of Winter Adult Thayer's Gulls, with Comments on Taxonomy. Alula 7:130–134.

Howell, S.N.G., King, J.R., & Corben, C. 1999. First Prebasic Molt in Herring, Thayer's, and Glaucous-winged Gulls. Journal of Field Ornithology 70 (4): 543–554.

Howell, S.N.G., Lewington, I., & Russell, W. 2014. Rare Birds of North America. Princeton, NJ: Princeton University Press.

Howell, S.N.G., & Mactavish, B. 2003. Identification and Variation of Winter Adult Kumlien's Gull. Alula 9:2–15.

Howell, S.N.G., & Pyle, P. 2015. Use of "Definitive" and Other Terms in Molt Nomenclature: A Response to Wolfe et al. (2014). Auk 132:365–369.

Howell, S.N.G., & Wood, C. 2004. First-cycle Primary Moult in Heermann's Gulls. Birders Journal 75:40–43.

Humphrey, P.S., & Parkes, K.C. 1959. An Approach to the Study of Molts and Plumages. Auk 76:1–31.

Hunt, G.L., & Hunt, M.W. 1977. Female–Female Pairing in Western Gull (Larus occidentalis) in Southern California. Science 196:1466–1467.

Ingolfsson, A. 1967. The Feeding Ecology of Five Species of Large Gulls (Larus) in Iceland. PhD thesis. University of Michigan.

Ingolfsson, A. 1970. Hybridization of Glaucous Gulls Larus hyperboreus and Herring Gulls L. argentatus in Iceland. Ibis 112 (3): 340–362.

Ingolfsson, A. 1987. Hybridization of Glaucous and Herring Gulls in Iceland. Studies in Avian Biology, no. 10, 131–140.

Ingolfsson, A., Snaebjorn, P., & Vigfusdottir, F. 2008. Hybridization of Glaucous Gull (Larus hyperboreus) and Herring Gull (Larus argentatus) in Iceland: Mitochondrial and Microsatellite Data. Philosophical Transactions of the Royal Society B 363:2851–2860.

Iron, J., & Pittaway, R. 2001. Molts and Plumages of Ontario's Heermann's Gull. Ontario Birds 19:65–78.

Islam, K., & Velarde, E. 2020. Heermann's Gull (Larus heermanni). In Birds of the World, edited by P.G. Rodewald & B.K. Keeney. Cornell Lab of Ornithology. https://doi.org/10.2173/bow.heegul.02.

Iverson, E.N.K., & Karubian, K. 2017. The Role of Bare Parts in Avian Signaling. Auk 134 (4): 587–611.

Jaramillo, A. 2020. First Cycle Slaty-backed Gulls in Japan. North American Birds 71 (2): 30–43.

Jehl, J.R. 1960. A Probable Hybrid of Larus argentatus and L. marinus. Auk 77 (3): 343–345.

Jehl, J.R. 1987. A Review of "Nelson's Gull" Larus nelsoni. Bulletin of the British Ornithologists' Club 107:86–91.

Jehl, J.R., Jr. 1987. Geographic Variation and Evolution in the California Gull (Larus californicus). Auk 140 (3): 421–428.

Jetz, W., Thomas, G.H., Joy, J.B., Redding, D.W., Hartmann, K., & Mooers, A.O. 2014. Global Distribution and Conservation of Evolutionary Distinctness in Birds. Current Biology 24:919–930.

Jiguet, F. 2002. Taxonomy of the Kelp Gull Larus dominicanus Lichtenstein, Inferred from Biometrics and Wing Pattern, Including Two Undescribed Subspecies. Bulletin of the British Ornithologists' Club 122:50–71.

Jiguet, F., Capainolo, P., & Tennyson, A.J.D. 2012. Taxonomy of the Kelp Gull Larus dominicanus Lichtenstein Revisited with Sex-separated Analyses of Biometrics and Wing Tip Pattern. Zoological Studies 51 (6): 881–892.

Jiguet, F., Jaramillo, A., & Sinclair, I. 2001. Identification of Kelp Gull. Birding World 14 (3): 112–125.

Jobling, J.A. 2010. The Helm Dictionary of Scientific Bird Names. London: Christopher Helm.

Joiris, C. 1978. Le Goéland argenté portugais (Larus argentatus lusitanius), nouvelle forme de Goéland argenté à pattes jaunes. Aves 15:17–18.

Jones, R.E. 1980. Interspecific Copulation between Slender-billed and Black-headed Gull. British Birds 73:474–475.

Jonsson, L. 1998. Baltic Lesser Black-backed Gull *Larus fuscus fuscus*: Moult, Ageing and Identification. *Birding World* 11 (8): 295–317.

Jonsson, O. 2011. Great Black-backed Gulls Breeding at Khniffis Lagoon, Morocco and the Status of Cape Gull in the Western Palearctic. *Birding World* 24:69–76.

Joris, C. 1978. Le goéland argenté portugais (*Larus argentatus lusitanius*), nouvelle forme du goéland argenté à pattes jaunes. *Aves* 15:17–18.

Kazama, K., Hirata, K., & Sato, M. 2011. Observation Records of Slaty-backed Gull × Glaucous-winged Gull Hybrid Breeding Pairs on Rishiri Island. *Japanese Journal of Ornithology* 60 (2): 241–245.

Kennard, J.H. 1975. Longevity Records of North American Birds. *Bird-Banding* 46 (1): 55–73.

Kessel, B. 1989. *Birds of the Seward Peninsula, Alaska: Their Biogeography, Seasonality, and Natural History*. Fairbanks, AK: University of Alaska Press.

King, J.R. 2000. Field Identification of Adult *californicus* and *albertaensis* California Gulls. *Birders Journal* 9:245–260.

Kishchinsky, A.A. 1980. *Birds of the Koryak Highlands*. Moscow: Nauka.

Knudsen, B. 1976. Colony Turnover and Hybridization in Some Canadian Arctic Gulls. Abstract in *Pacific Seabird Group Bulletin* 3:27.

Kokubun, N., Yamamoto, T., Kikuchi, D.M., Kitaysky, A., & Takahashi, A. 2015. Nocturnal Foraging by Red-legged Kittiwakes, a Surface Feeding Seabird That Relies on Deep Water Prey during Reproduction. *PLoS ONE* 10 (10): e0138850. https://doi.org/10.1371/journal.pone.0138850.

Kondratyev, A.Y., Litvinenko, N.M., & Kaiser, G.W. 2000. *Seabirds of the Russian Far East*. Ottawa: Canadian Wildlife Service.

Kristiansen, K.O., Bustnes, J.O., Folstad, I., & Helberg, M. 2008. Carotenoing Coloration in Great Black-backed Gull *Larus marinus* Reflects Individual Quality. *Journal of Avian Biology* 37 (1): 6–12.

Kuroda, N.H. 1941. A Hybrid between the Silver and Black-tailed Gull. *Tori* 11:146–150.

Lehman, P. 1980. The Identification of Thayer's Gull in the Field. *Birding* 12:198–210.

Lehman, P. 2019. *The Birds of Gambell and St. Lawrence Island, Alaska*. Studies of Western Birds, no. 4. Camarillo, CA: Western Field Ornithologists.

Lewis, R.H. 1996. First North Carolina Record of Yellow-legged Gull. *Chat* 60:153–156.

Liebers, D., de Knijff, P., & Helbig, A.J. 2004. The Herring Gull Complex Is Not a Ring Species. *Proceedings of the Royal Society of London B* 271:893–901.

Liebers, D., & Helbig, A.J. 2002. Phylogeography and Colonization History of Lesser Black-backed Gulls (*Larus fuscus*) as Revealed by mtDNA Sequences. *Journal of Evolutionary Biology* 15:1021–1033.

Litwiniak, K., Przymencki, M., & de Jong, A. 2021. Breeding-range Expansion of the Caspian Gull in Europe. *British Birds* 114 (6): 331–340.

Lonergan, P., & Mullarney, K. 2004. Identification of American Herring Gull in a Western European Context. *Dutch Birding* 26:1–35.

Lönnberg, E. 1919. Hybrid Gulls. *Arkiv for Zoologia* 12 (7): 1–22.

Lowe, T. 2012. Apparent Hybrid Ring-billed × Lesser Black-backed Gulls in Shropshire and Spain. *Birding World* 25 (4): 159–163.

Macpherson, A.H. 1961. *Observations on Canadian Arctic Larus gulls, and on the Taxonomy of L. thayeri Brooks*. Arctic Institute of North America Technical Paper, no. 7. Calgary: Arctic Institute of North America.

Maftei, M., Davis, S., Jones, I., & Mallory, M. 2012. Breeding Habitats and New Breeding Locations for Ross's Gull (*Rhodostethia rosea*) in the Canadian High Arctic. *Arctic* 65 (3): 283–298.

Maftei, M., Davis, S., & Mallory, M. 2015. Confirmation of a Wintering Ground of Ross's Gull *Rhodostethia rosea* in the Northern Labrador Sea. *Ibis* 157:642–647.

Mallory, M., Allard, K., Braune, B., Gilchrist, H., & Thomas, V. 2012. New Longevity Record for Ivory Gull (*Pagophila eburnea*) and Evidence of Natal Philopatry. *Arctic* 65 (1): 98–101.

Manning, T.H., Hohn, E.O., & Macpherson, A.H. 1956. *The Birds of Banks Island*. Bulletin (National Museum of Canada), no. 143. Ottawa: National Museum of Canada.

Martin, G., Gutierrez, A., Caratão, R., & Muusse, M., coordinators. 2013. *Lusitanius* Yellow-legged Gull. Gull Research Organization. Last updated 12 October 2013. https://gull-research.org/lusitanius/02cynov.html.

Maybank, B. 2009. Atlantic Provinces & St. Pierre et Miquelon Regional Report. *North American Birds* 63 (3): 386.

McCaffery, B.J., Hardwood, C.M., & Morgart, J.R. 1997. First Breeding Records of Slaty-backed Gull (*Larus schistisagus*) for North America. *Pacific Seabirds* 24 (2): 70.

McCarthy, E. 2006. *Handbook of Avian Hybrids of the World*. United Kingdom: Oxford University Press.

McGowan, R.Y., & Kitchener, A.C. 2001. Historical and Taxonomic Review of the Iceland Gull *Larus glaucoides* Complex. *British Birds* 94:191–194.

McKearnan, J. 1999. Possible Hybrid Gull in Delta Co. *Michigan Birds and Natural History* 6:23–24.

McKee, T., Moores, N., & Pyle, P. 2014. Vagrancy and Identification of First-cycle Slaty-backed Gulls. *Birding* 46 (6): 38–51.

McWilliams, G.M., & Brauning, D.W. 2018. *The Birds of Pennsylvania.* Ithaca, NY: Cornell University Press.

Medrano, F.I., Gutierrez, E., & Silva, R. 2022. Gray Gull (*Leucophaeus modestus*), version 2.0. In Birds of the World, edited by S.M. Billerman. Cornell Lab of Ornithology. https://doi.org/10.2173/bow.grygul.02.

Mlodinow, S. 2012. Stealth Gull: Herring × Glaucous-winged Gull Hybrids in the North American Interior. *Colorado Birds* 46 (3): 198–206.

Mlodinow, S., Irons, D., & Tweit, B. 2005. Winter Season: Oregon and Washington Region. *North American Birds* 59 (2): 313–317.

Mlodinow, S., Irons, D., & Tweit, B. 2008. Oregon & Washington: The Regional Reports. *North American Birds* 62 (1): 138–143.

Montevecchi, W.A., Cairns, D.K., Burger, A.E., Elliot, R.D., & Wells, J. 1987. The Status of the Common Black-headed Gull in Newfoundland and Labrador. *American Birds* 41 (2): 197–203.

Moore, C.C. 1996. Ship-attending Movements of Atlantic Yellow-legged Gull in Portuguese Waters. *Dutch Birding* 18 (1): 18–22.

Moores, N. 2011. Taimyr Gull *Larus (heuglini) taimyrensis*: An Update. Birds Korea. http://www.birdskorea.org/Birds/Identification/ID_Notes/BK-ID-Taimyr-Gull.shtml.

Moores, N., Kim, A., & Kim, R. 2014. Status of Birds, 2014. Birds Korea Report on Bird Population Trends and Conservation Status in the Republic of Korea. Birds Korea. http://www.birdskorea.org/Habitats/Yellow-Sea/YSBR/Downloads/Birds-Korea-Status-of-Birds-2014.pdf.

Moynihan, M. 1959. A Revision of the Family Laridae (Aves). *American Museum Novitates*, no. 1928.

Muusse, M., Muusse, T., Buijs, R.J., Altenburg, R., Gibbins, C., & Luijendijk, B.J. 2011. Phenotypic Characteristics and Moult Commencement in Breeding Dutch Herring Gulls *Larus argentatus* & Lesser Black-backed Gulls *L. fuscus*. *Seabird* 24:42–59.

Muusse, T., Muusse, M., Luijendijk, B.J., & Altenburg, R.G.M. 2005. Identification Update: Moult Variability in 3rd Calendar-year Lesser Black-backed Gulls. *Birding World* 18:338–348.

New York State Avian Records Committee. 2009. Report of the New York State Avian Records Committee for 2006. *Kingbird* 59 (1): 18–42.

Noble, G.K. 1916. The Resident Birds of Guadeloupe. *Bulletin of the Museum of Comparative Zoology* 60:359–396.

Olsen, K.M. 2018. *Gulls of the World: A Photographic Guide.* London: Christopher Helm.

Olsen, K.M., & Larsson, H. 2004. *Gulls of Europe, Asia and North America.* London: Christopher Helm.

Olson, C.S. 1976. Band-tailed Gull Photographed in Florida. *Auk* 93:177.

Olson, S.L., & Banks, R.C. 2007. Lectotypification of *Larus smithsonianus* Coues, 1862 (Aves: Laridae). *Proceedings of the Biological Society of Washington* 120 (4): 382–386.

Parmelee, D.F., & MacDonald, S.D. 1960. *The Birds of West-central Ellesmere Island and Adjacent Areas.* Bulletin (National Museum of Canada), no. 169. Ottawa: National Museum of Canada.

Patten, M.A. 2020. Yellow-footed Gull (*Larus livens*), version 1.0. In Birds of the World, edited by A.F. Poole & F.B. Gill. Cornell Lab of Ornithology. https://doi.org/10.2173/bow.yefgul.01.

Patten, S.M. 1980. Interbreeding and Evolution in the *Larus glaucescens–Larus argentatus* Complex on the South Coast of Alaska. PhD thesis. Johns Hopkins University.

Pennsylvania Game Commission. 2018. This Spring, Game Commission Biologists Placed Satellite Transmitters on Pennsylvania Lesser Black-backed Gulls. Facebook. 21 May 2018. https://www.facebook.com/PennsylvaniaGameCommission/photos/a.268877686477464/1871939099504640.

Peters, J.L. 1934. *Checklist of the Birds of the World*, vol. 2. Cambridge, MA: Harvard University Press.

Petersen, A., Irons, D.B., & Gilchrist, H.G. 2015. The Status of Glaucous Gull *Larus hyperboreus* in the Circumpolar Arctic. *Arctic* 68 (1): 107–120.

Pieplow, N. 2017. *Peterson Field Guide to Bird Sounds of Eastern North America.* New York: Houghton Mifflin Harcourt.

Pittaway, R. 1999. Taxonomic History of Thayer's Gull. *Ontario Birds* 17:2–13.

Pons, J.M., Crochet, P.A., Thery, M., & Bermejo, A. 2004. Geographical Variation in the Yellow-legged Gull: Introgression or Convergence from the Herring Gull? *Journal of Zoological Systematics and Evolutionary Research* 42:245–256.

Pons, J.M., Hassanin, A., & Crochet, P.A. 2005. Phylogenetic Relationships within the Laridae (Charadriiformes: Aves) Inferred from Mitochondrial Markers. *Molecular Phylogenetics and Evolution* 37:686–699.

Portenko, L. 1939. On Some New Forms of Arctic Gulls. *Ibis* 14 (3): 264–269.

Portenko, L.A. 1963. The Taxonomic Value and Systematic Status of the Slaty-backed Gull (*Larus argentatus schistisagus* Stejn.). In Proceedings of the

Kamchatka Complex Expedition: The Terrestrial Fauna of the Kamchatka Region, 61–64. Moscow: USSR Academy of Sciences.

Post, P.W., & Lewis, R.H. 1995a. The Lesser Black-backed Gull in the Americas: Occurrence and Subspecific Identity, Part 1: Taxonomy, Distribution, and Migration. *Birding* 27:283–290.

Post, P.W., & Lewis, R.H. 1995b. The Lesser Black-backed Gull in the Americas: Occurrence and Subspecific Identity, Part 2: Field Identification. *Birding* 27:371–381.

Pranty, B. Forthcoming. A Gray Gull (*Leucophaeus modestus*) in Florida and Alabama: The First Record for North America North of Mexico. *Florida Field Naturalist*.

Pyle, P. 2008. *Identification Guide to North American Birds, Part II: Anatidae to Alcidae*. Bolinas, CA: Slate Creek Press.

Pyle, P. 2020. 31st Report of the ABA Checklist Committee. *North American Birds* 71 (2): 8–13.

Pyle, P., Ayyash, A., & Bartosik, M. 2018. Replacement of Primaries during Prealternate Molts in North American *Larus* Gulls. *Western Birds* 49:293–306.

Pyle, R.L., & Pyle, P. 2017. The Birds of the Hawaiian Islands: Occurrence, History, Distribution, and Status, version 2 (1 January). B.P. Bishop Museum. http://hbs.bishopmuseum.org/birds/rlp-monograph.

Rand, A.L. 1942. *Larus kumlieni* and Its Allies. *Canadian Field-Naturalist* 56:123–126.

Remirez, X., del Campo, F., del Campo, J., & Arizaga, J. 2023. Movement Patterns of Immature Yellow-legged Gulls *Larus michahellis* from Gran Canaria, Canary Islands. *Seabirds* 35:1–17.

Richards, J.M., & Gaston, A.J. 2018. *Birds of Nunavut*, vol. 1: *Nonpasserines*. Vancouver: UBC Press.

Ridgway, R. 1912. *Color Standards and Color Nomenclature*. Washington, DC: The author.

Riley, D.J., Charlton, B.N., Burrell, M.V.A., Burrell, K.G.D., Guercio, A.C., Martin, R.D., Read, M.D., & Timpf, A.P. 2021. Ontario Bird Records Committee Report for 2020. *Ontario Birds* 39 (2): 66–95.

Robinson, B.W., Withrow, J.J., Richardson, R.M., Matsuoka, S.M., Gill, R.E., Johnson, J.A., Degange, A.R., & Romano, M.D. 2020. Further Information on the Avifauna of St. Matthew and Hall Islands, Bering Sea, Alaska. *Western Birds* 51 (2): 78–91.

Rollins, N. 1953. A Note on Abnormally Marked Song Thrushes and Blackbirds. *Transactions of the Natural History Society of Northumberland, Durham, and Newcastle-upon-Tyne* 10:183–184.

Ryder, J.P. 1980. The Influence of Age of the Breeding Biology of Colonial Nesting Seabirds. In *Behavior of Marine Animals*, edited by J. Burger, B.L. Olla, & H.E. Winn. Boston, MA: Springer. https://doi.org/10.1007/978-1-4684-2988-6_5.

Salomonsen, F. 1950. *Grønlands fugle / The Birds of Greenland*. Copenhagen: Enjar Munksgaard.

Sangster, G., Collinson, J.M., Helbig, A.J., Knox, A.G., & Parkin, D.T. 2005. Taxonomic Recommendations for British Birds: Third Report. *Ibis* 1 (47): 821–826.

Sangster, G., Collinson, J.M., Knox, A.G., Parkin, D.T., & Svensson, L. 2007. Taxonomic Recommendations for British Birds: Fourth Report. *Ibis* 149:853–857. https://doi.org/10.1111/j.1474-919X.2007.00758.x.

Sauvage, J.-M., & Muusse, M., coordinators. 2013. Herring Gull (*argentatus* & *argenteus*). Gull Research Organisation. Last updated 11 February 2013. www.gull-research.org/hg/hg5cy/5277846apr.htm.

Schmutz, J.A., & Hobson, K.A. 1998. Geographic, Temporal, and Age-specific Variation in Diets of Glaucous Gulls in Western Alaska. *Condor* 100 (1): 119–130.

Semo, L. 2007. The 43rd Report of the Colorado Bird Records Committee—Decision Summary: Acceptance of Kelp Gull to List of Colorado Birds. *Colorado Birds* 41:33–49.

Sepp, T., Rattiste, K., Saks, L., Meitern, R., Urvik, J., Kaasik, A., & Hõrak, P. 2017. A Small Badge of Longevity: Opposing Survival Selection on the Size of White and Black Wing Markings. *Journal of Avian Biology* 48:570–580.

Shutt, J.L., Andrews, D.W., Weseloh, D.V.C., Moore, D.J., Herbert, C.E., Campbell, G.D., & Williams, K. 2014. The Importance of Island Surveys in Documenting Diseases-related Mortality and Botulism E in Great Lakes Colonial Waterbirds. *Journal of Great Lakes Research* 40:58–63.

Sibley, C.G., & Monroe, B.L. 1990. *Distribution and Taxonomy of Birds of the World*. New Haven, CT: Yale University Press.

Sinclair, J.C. 1977. Interbreeding of Grey-headed and Hartlaub's Gulls. *Bokmakierie* 29:70–71.

Slud, P. 1967. The Birds of Cocos Island (Costa Rica). *Bulletin of the American Museum of Natural History* 134 (4): 262–295.

Smith, N.G. 1966. *Evolution of Some Arctic Gulls (Larus): An Experimental Study of Isolating Mechanisms*. Ornithological Monographs, no. 4. Washington, DC: American Ornithologists' Union.

Snell, R.R. 1989. Status of *Larus* Gulls at Home Bay, Baffin Island. *Colonial Waterbirds* 112 (1): 12–23.

Snell, R.R. 1991. Conflation of the Observed and Hypothesized: Smith's 1961 Research in Home Bay, Baffin Island. *Colonial Waterbirds* 14 (2): 196–202.

Snell, R.R., & Godfrey, W.E. 1991. Geographic Variation in the Iceland Gull (*Larus glaucoides*). Abstract, no.

156, for presentation to American Ornithologists' Union Meeting, Montreal.

Snell, R.R., Pyle, P., & Patten, M.A. 2020. Iceland Gull (*Larus glaucoides*), version 1.0. In Birds of the World, edited by P.G. Rodewald & B.K. Keeney. Cornell Lab of Ornithology. https://doi.org/10.2173/bow.y00478.01.

Snow, D.W., & Snow, B.K. 1967. The Breeding Cycle of the Swallow-tailed Gull (*Creagrus furcatus*). Ibis 109 (1): 14-24.

Sonsthagen, S., Wilson, R., Chesser, R., Pons, J.M., Crochet, P.E., Driskell, A., & Dove, C. 2016. Recurrent Hybridization and Recent Origin Obscure Phylogenetic Relationships within the "White-headed" Gull (*Larus* sp.) Complex. *Molecular Phylogenetics and Evolution* 103:41-54.

Spear, L.B. 1987. Hybridization of Glaucous and Herring Gulls at the Mackenzie Delta, Canada. *Auk* 104:123-125.

Spear, L.B., & Anderson, D.W. 1989. Nest-site Selection by Yellow-footed Gulls. *Condor* 91 (1): 91-99.

Stepanyan, L.S. 2003. *Conspectus of the Ornithological Fauna of Russia and Adjacent Territories (within the Borders of the USSR as a Historic Region)*. Moscow: PTC "Akademkniga."

Sternkopf, V. 2011. Molekulargenetische Untersuchung in der Gruppe der Möwen (Laridae) zur Erforschung der Verwandtschaftsbeziehungen und phylogeographischer Differenzierung. PhD dissertation. University of Greifswald.

Sternkopf, V., Liebers-Helbig, D., & Ritz, M.S. 2010. Introgressive Hybridization and the Evolutionary History of the Herring Gull Complex Revealed by Mitochondrial and Nuclear DNA. *BMC Evolutionary Biology* 10:348. https://doi.org/10.1186/1471-2148-10-348.

Stevenson, H.M. 1980. An Early Record of the Band-tailed Gull in Florida. *Florida Field Naturalist* 8 (1): 21-23.

Stoddart, A., & McInerny, C. 2017. The Azorean Yellow-legged Gull in Britain. *British Birds* 110:666-674.

Stotz, D.F. 2008. Fourteenth Report of the Illinois Ornithological Records Committee. *Meadowlark* 17 (2): 48-54.

Strang, C.A. 1977. Variation and Distribution of Glaucous Gulls in Western Alaska. *Condor* 79:170-175.

Sutton, G.M. 1968. Review of "Evolution of Some Arctic Gulls: An Experimental Study of Isolating Mechanisms," by N.G. Smith. *Auk* 85:142-145.

Swainson, W., & Richardson, J. 1831. *Fauna Boreali-Americana; or the Zoology of the Northern Parts of British America*, vol. 2, The Birds. London: J.

Murray. Digitized at New York Public Library Digital Collections. Accessed August 17, 2022. https://digitalcollections.nypl.org/items/510d47d9-6f70-a3d9-e040-e00a18064a99.

Swarth, H.S. 1934. *Birds of Nunivak Island, Alaska*. Pacific Coast Avifauna, no. 22. Los Angeles: The Club.

Taverner, J.H. 1970. A Presumed Hybrid Mediterranean × Black-headed Gull in Hampshire. *British Birds* 63:380-382.

Taverner, P.A. 1933. A Study of Kumlien's Gull (*Larus kumlieni* Brewster). *Canadian Field-Naturalist* 47:88-90.

Tuck, L.M. 1968. Recent Newfoundland Bird Records. *Auk* 85:304-322.

Ujihara, O., & Ujihara, M. 2019. *An Identification Guide to the Gulls of Japan*. Japan: Seibundoshinkosha.

Van Dijk, K., Kharitonov, S., Vonk, H., & Ebbinge, B. 2011. Taimyr Gulls: Evidence for Pacific Winter Range, with Notes on Morphology and Breeding. *Dutch Birding* 33:9-21.

Van Vliet, G., Marshall, B., Craig, D., & Egolf, J. 1993. First Record of Nesting Activity by a Lesser Black-backed Gull (*Larus fuscus*) in North America. *Bulletin (Pacific Seabird Group)* 20 (2): 21.

Vaurie, C. 1965. *Birds of the Palearctic Fauna: A Systematic Reference; Non-passeriformes*. London: Witherby.

Veitch, B.G., Robertson, G.J., Jones, I.L., & Bond, A.L. 2016. Great Black-backed Gull (*Larus marinus*) Predation on Seabird Populations at Two Colonies in Eastern Canada. Special issue, *Waterbirds* 39 (April): 235-245.

Voous, K.H. 1975. On the Inequality of Genera in Birds. *Ardeola* 21:977-985.

Wang, Y., Shiyi, G., Luo, C., Liu, S., Ma, J., Luo, W., Lin, C., Shu, D., & Qu, H. 2023. Variation in BCO2 Coding Sequence Causing a Difference in Carotenoid Concentration in the Skin of Chinese Indigenous Chicken. *Genes* 14 (3): 671. https://doi.org/10.3390/genes14030671.

Watanuki, Y. 1989. Sex and Individual Variations in the Diet of Slaty-backed Gulls Breeding on Teuri Island, Hokkaido. *Japanese Journal of Ornithology* 38:1-13.

Watanuki, Y. 1992. Individual Diet Difference, Parental Care, and Reproductive Success in Slaty-backed Gulls. *Condor* 94:159-171.

Weber, J.W. 1981. The *Larus* Gulls of the Pacific Northwest Interior, with Taxonomic Comments on Several Forms, Part 1. *Continental Birdlife* 2 (1): 1-10.

Weir, D.N., Kitchener, A.C., & McGowan, R.Y. 2000. Hybridization and Changes in the Distribution of Iceland Gulls (*Larus glaucoides / kumlieni / thayeri*). *Journal of the Zoological Society of London* 252:517-530.

Weir, D.N., McGowan, R.Y., Kitchener, A.C., McOrist, S., Zontrillo, B., & Heubeck, M. 1995. Iceland Gulls from the "Braer Disaster," Shetland 1993. *British Birds* 88:15–25.

Weiser, E., & Gilchrist, H.G. 2012. Glaucous Gull (*Larus hyperboreus*). In The Birds of North America Online, edited by A. Poole. Cornell Lab of Ornithology. https://doi.org/10.2173/bow.glagul.01.

Weiser, E., & Pohlen, Z. 2022. Cook Inlet Gull (Herring × Glaucous-winged Gull Hybrid) Identification in Alaska. Alaska eBird. https://ebird.org/ak/news/cook-inlet-gull-herring-x-glaucous-winged-gull-hybrid-identification-in-alaska.

Weseloh, D.V. 1981. A Probable Franklin's × Ring-billed Gull Nesting in Alberta. *Canadian Field-Naturalist* 95:474–476.

Weseloh, D.V. 1994. A History of the Little Gull in Ontario, 1930–1991. In *Ornithology in Canada*, edited by M.K. McNicholl & J.L. Cranmer-Bying, 240–259. Witherby, ON: Hawk Owl Publishing.

Weseloh, D.V., & Clark, D. 2011. Massachusetts-banded Ring-billed Gulls Breeding in Ontario and the Great Lakes. *Ontario Birds* 29:13–24.

Weseloh, D.V., Herbert, C.E., Mallory, M.L., Poole, A.F., Ellis, J.C., Pyle, P., & Pattern, M.A. 2020. Herring Gull (*Larus argentatus*), version 1.0. In Birds of the World, edited by S.M. Billerman. Cornell Lab of Ornithology. https://doi.org/10.2173/bow.hergul.01.

Weseloh, D.V., & Mineau, P. 1986. Apparent Hybrid Common Black-headed Gull Nesting in Lake Ontario. *American Birds* 40:18–20.

Weseloh, D.V., Moore, D., & Morris, R. 2007. First Nest Records of the Great Black-backed Gull on Lake Erie. *Ontario Birds* 25 (3): 124–133.

Wilbur, S.R. 1987. *The Birds of Baja California*. Berkeley, CA: University of California Press.

Wilds, C., & Czaplak, D. 1994. Yellow-legged Gulls (*Larus cachinnans*) in North America. *Wilson Bulletin* 106:344–356.

Winkler, D.W. 2020. California Gull (*Larus californicus*). In Birds of the World, edited by A.F. Poole & F.B. Gill. Cornell Lab of Ornithology. https://doi.org/10.2173/bow.calgul.01.

Wolfe, J.D., Ryder, T.B., & Pyle, P. 2010. Using Molt Cycles to Categorize the Age of Tropical Birds: An Integrative New System. *Journal of Field Ornithology* 81 (2): 186–194.

Wormington, A. 2013. A Migration of Juvenile Bonaparte's Gulls at Wheatley Harbour, Ontario. *Ontario Birds* 31:24–27.

Wormington, A. 2015. Historical Overview, Seasonal Timing and Abundance of Little Gull at Point Pelee, Ontario. *Ontario Birds* 33:3–21.

Wukasch, P. 2014. Dry Land Nesting of Ring-billed Gull in Simcoe County. *Ontario Birds* 32:49–52.

Yésou, P. 2001. Phenotypic Variation and Systematics of Mongolian Gull. *Dutch Birding* 23:65–82.

Yésou, P. 2002. Systematics of *Larus argentatus-cachinnans-fuscus* Complex Revisited. *Dutch Birding* 24 (5): 271–298.

Zawadzki, L.C., Hallgrimsson, G.T., Veit, R.R., Rasmussen, L.M., Boertmann, D., Gillies, N., & Guilford, T. 2021. Predicting Source Populations of Vagrants Using Breeding Population Data: A Case Study of the Lesser Black-backed Gull (*Larus fuscus*). *Frontiers in Ecology and Evolution* 9:637452. https://doi.org/10.3389/fevo.2021.637452.

Zelenskaya, L., & Solovyeva, D.V. 2016. Changes in the Species Composition and Number of Gulls in Tundra Colonies in the Western Chukotka over the Last 40 Years. *Biology Bulletin* 43 (8): 844–850.

Zimmer, K.J. 1991. Plumage Variation in "Kumlien's" Iceland Gull. *Birding* 23:254–269.

Zufelt, K. 2012. "Vega" Herring Gull in Algoma District: A New Taxon for Ontario. *Ontario Birds* 30:13–25.

INDEX

aberration, 54–56
adult plumage, 17–18
age-related, 31
aging, 17
alternate plumage, 19
American Birding Association (ABA), 6
American Herring × Glaucous, 454–459
American Herring × Glaucous-winged, 446–453
American Herring × Great Black-back, 467–470
American Herring × Kelp, 235, 316, 478
American Herring × Lesser Black-back, 471–477
American Ornithological Society (AOS), 4, 214
apical, 8
Appledore Gull, 471–477

backcross, 436
Baltic Gull, 289, 294, 303
bare parts, 12-14
black-backeds, 5
bleaching, 37, plate 16.7, 55
body feathers, 7
breeding aspect, 23, plate 38
breeding condition, 13, plate 20

California Gull × American Herring, 437
Cape Gull, 305
Chandeleur Gull, 316, 478
Chroicocephalus
  cirrocephalus, 121–126
  philadelphia, 106–113
  ridibundus, 114–120
cinnamon type, plate 19.4
complete molt, 20
Complex Alternate Strategy (CAS), 20–22
Complex Basic Strategy (CBS), 20
Cook Inlet Gull, 446–453

Cregraus furcatus, 68–73
cycle, 19

diagnostic, 31

fading, 37, plate 16.7, 55
female type, 40
first alternate plumage, 19
first basic plumage, 18
flight feathers, 7
formative plumage, 19, plate 12.3

gape, 12, 13
Glaucous-winged × Slaty-backed, 478–480
Glaucous × European Herring, 481
Glaucous × Great Black-backed, 464–466
Glaucous × Slaty-backed, 480–481
Glaucous × Vega, 481
gonys spot, 12
Great Lakes Gull, 467–470
gull
  American Herring, 215–235
  Azores, 276–288
  Belcher's, 152–157
  Black-headed, 114–120
  Black-tailed, 158–164
  Bonaparte's, 106–113
  California, 201–212
  Common, 174–182
  European Herring, 236–246
  Franklin's, 135–142
  Glaucous, 376–386
  Glaucous-winged, 364–375
  Gray, 497–501
  Gray-hooded, 121–126
  Great Black-backed, 338–348
  Heermann's, 144–151
  Heuglin's, 484–488
  Iceland, 418–426
  Ivory, 87–91
  Kamchatka, 183–191
  Kelp, 305–316
  Kumlien's, 404–417

  Laughing, 127–134
  Lesser Black-backed, 289–304
  Little, 98–105
  Pallas's, 494–496
  Ring-billed, 192–200
  Ross's, 92–97
  Sabine's, 60–67
  Short-billed, 165–173
  Slaty-backed, 349–363
  Swallow-tailed, 68–73
  Taimyr, 489–493
  Thayer's, 389–403
  Vega, 247–260
  Western, 317–327
  Yellow-footed, 328–337
  Yellow-legged, 261–275

Herring Gull Complex, 213–214
hooded gulls, 5
Humphrey-Parkes-Howell (HPH), 18, 24
hybrid, 436
hybrid F1, 431
Hydrocoloeus minutus, 98–105

Iceland Gull Complex, 387–388, 427–434
Ichthyaetus ichthyaetus, 494–496
immature, 18
incomplete molt, 138, plate 12.16
inner web, 8
inserted molt, 20

juvenile plumage, 18, plate 30

kittiwake
  Black-legged, 74–81
  Red-legged, 82–86
Kodak Gray Scale, 14–16

large white-headed gulls, 5
Larus
  c.albertaensis, 201–212
  a.argentatus, 226–246
  a.argenteus, 236–246

INDEX

m.atlantis, 276–288
d.austrinus, 305–306, 312–314
h.barrovianus, 376–386
belcheri, 152–157
brachyrhynchus, 165–173
c.californicus, 201–212
c.canus, 174–182
crassirostris, 158–164
delawarensis, 192–200
d.dominicanus, 305–316
f.fuscus, 289–304
glaucescens, 364–375
g.glaucoides, 418–426
f.graellsii, 289–304
heermani, 144–151
c.heinei, 174, 181
f.heuglini, 484–488
h.hyperboreus, 376–386
f.intermedius, 289–304
c.kamtschatschensis, 183–191
g.kumlieni, 404–417
h.leuceretes, 376–386
livens, 328–337
m.lusitanius, 262, 264, 268, 271
marinus, 338–348
michahellis, 261–275
v.mongolicus, 247, 250
o.occidentailis, 317–327
h.pallidissimus, 376–386
schistisagus, 349–363
smithsonianus, 215–235
taimyrensis, 489–493
g.thayeri, 389–403
v.vegae, 247–260
d.vetula, 305–306, 313
o.wymani, 317–327

Leucophaeus
  atricilla, 127–134
  modestus, 497–501
  pipixcan, 135–142

male type, 40
masked gulls, 5
medial band, 9
mirror, 8
molt, 18, 20
molt cycle, 19
molt limit, 23, plate 39
molt strategy, 20

Nelson's Gull, 454–459
nonbreeding aspect, 23

Olympic Gull, 439–445
orbital ring, 12, 14
outer web, 9

Pagophila eburnean, 87–91
partial molt, 20
pearl, 8
phylogeny, 5
postjuvenile, 33, plate 49
prealternate molt, 20
prebasic molt, 20
preformative molt, 19

race, 6
Rhodostethia rosea, 92–97
Ring-billed × Black-headed, 483
Ring-billed × Laughing, 437, 482
Ring-billed × Lesser Black-backed, 481–482

Rissa
  brevirostris, 82–86
  tridactyla, 74–81

scapulars, 7–11
seasonal, 31
second cycle, 20
secondary skirt, 7, plate 29.37
Seward Gull, 460–463
shadow bar, plates 19.39–40
Simple Alternate Strategy (SAS), 20
Simple Basic Strategy (SBS), 20
size and structure, 30
string-of-pearls, 360–361, plate 16.25, plate 29.46
subadult, 17–18, plate 27
subspecies, 6
subterminal band, 8

taxonomy, 4
tertials (crescent), 7
tongue (tip), 8, 9
type, 5, 40

ulnar bar, 11, plate 17
variation, 38

wear, 37–38, plate 55
white-winged gulls, 5
window, 221, plate 20.27
wing projection, 40, plate 58
Wolfe-Ryder-Pyle (WRP), 49

Xema sabini, 60–67